C0-ALY-695

Cellular Radio Performance Engineering

The Artech House Mobile Communications Series

John Walker, *Series Editor*

Mobile Communications in the U.S. and Europe: Regulation, Technology, and Markets, Michael Paetsch

Land-Mobile Radio System Engineering, Garry C. Hess

Mobile Antenna Systems Handbook, K. Fujimoto, and J. R. James, eds.

Mobile Information Systems, John Walker, editor

Narrowband Land-Mobile Radio Networks, Jean-Paul Linnartz

Cellular Radio: Analog and Digital Systems, Asha Mehrotra

Cellular Radio Performance Engineering, Asha Mehrotra

The Artech House Intelligent Vehicle Highway Systems Series

John Walker, *Series Editor*

Advanced Technology for Road Transport: IVHS and ATT, Ian Catling, ed.

For a complete listing of *The Artech House Telecommunications Library,*
turn to the back of this book

Cellular Radio Performance Engineering

Asha Mehrotra

Artech House
Boston • London

Library of Congress Cataloging-in-Publication Data
Cellular radio performance engineering / Asha Mehrotra.
Includes bibliographical references and index.
ISBN 0-89006-748-1
1. Cellular radio—Equipment and supplies—Design and construction.
2. Cellular radio—Reliability. I. Title
TK6570.M6M45 1994 94-7673
621.3845'6–dc20 CIP

A catalogue record for this book is available from the British Library

685 Canton Street
Norwood, MA 02062

International Standard Book Number: 0-89006-748-1
Library of Congress Catalog Card Number: 94-7673

10 9 8 7 6 5 4 3 2

In memory of my brother, Narendra K. Mehrotra

Contents

Preface

Cellular radio is a fast-changing field of mobile communications. Over the last decade, dramatic changes have taken place in the operations of mobile communications that have had a great impact on communication and society in general. The early 1980s saw the advent of analog systems that had an uncertain future. Due to exceptional subscriber growth, however, low-capacity analog systems are rapidly being replaced by digital systems throughout the industrialized world. Both Europe and the U.S. have come out with their own systems based on TDMA technology. It is expected that within the next few years the analog systems in most countries will be replaced by their digital counterparts. It is anticipated that digital TDMA systems will not satisfy the growing capacity requirements in the long run and that system operators will be forced to change over to even-higher capacity systems, such as CDMA. By the turn of the century, personal communication systems (PCS), based on microcellular technology, may be a candidate to satisfy an individual's speech and data requirements. The recent allocation of 220 MHz of spectrum in the 1.85 to 2.2 GHz band in the U.S. sets the course in this direction.

No single book can cover all aspects of cellular radio. Therefore, this book will focus on a concise set of performance-related topics: cellular channel characterization, propagation, antenna systems, interference, modulation, diversity, and signal processing aspects. Present and future aspects of cellular radio are also covered through the discussion of microcellular technology, including cordless telephones, PCN, and PCS services.

A brief introduction of probability theory, which is used in cellular communications, is presented by using examples such as the contention problem in the reverse signaling channel, peak traffic estimates, fast and slow fading, and potential mobile interference with television channels. Channel characterization in terms of fast and slow fading, spatial distance between two independent fades, receiver threshold determination based on level crossing rate, and the fading duration are presented mathematically in Chapter 3. Chapter 4 undertakes the analytical and experimental models used for median loss computation in mobile surroundings at

900 MHz. Models by Carey and Okumura are discussed along with the propagation behavior inside buildings, tunnel, garages, and the like.

The study of antenna systems includes logical reasons of the choice of cell site and mobile antennas along with different ancillary devices, such as transmitter combiner, receiver multicoupler, duplexer, transmission lines, and so on. The effects of dead spots, antenna mounting, and intermodulation are treated extensively in Chapter 5, as are problems of cochannel, adjacent channel, and near-and-far-end effects are covered. It has been shown that cochannel problem is of the utmost importance in cellular radio. Chapter 6 proposes some techniques for cochannel interference mitigation. Both analog and digital modulations are discussed. Chapter 7 emphasizes mobile digital modulation requirements and how they are satisfied by MSK, GMSK, and $\pi/4$ DQPSK. Different techniques of diversity and combining (and the preferred approaches) for cell site and mobile are discussed. Chapter 9 discusses the signal-processing and data-formatting aspects of cellular radio. Baseband signal characterization, filtering, the use of nonlinear channel equalizers, and Barker code for frame synchronization are covered. Useful block codes and techniques of majority voting and interleaving to combat the fading effects are discussed. Newer speech coding using RELPC and CELP algorithms are briefly described. Lastly, the requirements of microcellular systems are discussed in Chapter 10, followed by the requirements of cordless telephones (CT1 and CT2) and digital European cordless telecommunication (DECT) systems. Chapter 10 also deals with a new generation of services known as PCS/PCN; their status in Europe and the U.S. are discussed.

This book is intended for the practicing engineer and senior-year undergraduate or graduate students who want to work in the cellular field. For the practicing engineer, a B.S. degree in electrical engineering and some working knowledge of mobile communications is assumed. For students it is assumed that they have taken senior-level courses in communication engineering.

I would like to express a few words of acknowledgment. First, I would like to express my thanks to TASC for granting permission to publish this book. I extend my appreciation to my colleagues Richard G. Moldt, for helping me review the completed book, and Michael Scheidt, for consulting with me at various stages during the writing of the book. I am extremely grateful to Professor R. Pickholtz of George Washington University, who was instrumental in initiating this project. I also want to thank my graduate students at George Washington University, who have contributed by asking the right questions and, in general, helping me in my thought processes. I also must thank the reviewers of this book for their excellent comments.

Finally, a project of this type can never be completed without the continuous support of one's family. Thanks are due to my wife, Nisha, my daughters, Anuja, Sonia, and Vinita, and my son, Neil. Special thanks to my seventeen-year-old daughter, Sonia, for drawing all the figures and graphs and putting up with this unending project for the last several years.

Chapter 1
Introduction to Cellular Mobile Radio

1.1 A BRIEF HISTORY OF MOBILE COMMUNICATION

The early history of mobile communication is associated with the pioneering work of police. As early as 1921, the Detroit police used mobile telephones in their cars, using a frequency band near 2 MHz. The first two-way mobile telephone system was placed in service about the middle of 1933—also using the 2-MHz band—by the New York City Police Department. In 1934, the FCC authorized four channels in the 30–40 MHz range. By the end of 1940, there were some 10,000 radio transmitters in police cars—all based on amplitude modulation. The pioneering work of Armstrong with FM modulation changed the course of mobile communication. Although FM modulation was first used in the entertainment industry, later, in 1940, the FCC approved it for services other than the entertainment industry. This gave a boost to mobile communication development and started a new era in land mobile communications. So valuable was FM in this type of service that within a period of only six years, virtually all transmitters were converted to FM.

In 1946, Bell Telephone Laboratories (BTL) inaugurated the first mobile system for the public in St. Louis. A "highway" system to serve the corridor between New York and Boston began operating in 1947. This simplex system was in the 35–40 MHz frequency range. In 1955, an 11-channel system around 150 MHz became operational. Due to growing demand, 12 channels were added in 1956 at around 450 MHz. At 150 MHz and 450 MHz, both systems were manual and required operator assistance to set up the call. The first automatic system was put into operation in 1956 at 150 MHz. Subsequently, an automatic system at 450 MHz was built in 1969. In 1971, in response to FCC Docket 18262, the Bell System proposed a new architecture for mobile communication that is presently known as cellular radio. In 1977, the FCC authorized the installation and test of advanced mobile phone service (AMPS). In 1981, the FCC released a 40-MHz bandwidth in the 800–900 MHz range for cellular radio and on June 7, 1982, the FCC started to accept license applications to construct and operate within the top thirty cellular markets in the U.S. The first cellular system went into operation in the Chicago area on October 1983, run by American Telephone and Telegraph (AT&T). In 1990, the FCC issued a Notice of Inquiry (NOI) into the establishment of new personal communication services (PCS). For a fuller discussion, see [1].

In 1945, there were only 18,000 land mobile transmitters in service in the U.S. Since then the number of mobile users has almost tripled every five years. The statistics from France indicate a doubling of the number of users every five years. Present estimates indicate there will be about 20, 15, and 6 million cellular users in the U.S., European, and Asian countries, respectively, by 1995. The estimated users in Figure 1.1 is likely to be adjusted upward, due to growing demand and the improved performance from forthcoming digital cellular systems. A higher capacity system is warranted, based on the congestion in such major metropolitan cities as Los Angeles, New York, London, Tokyo, and Hong Kong. A higher capacity cellular system can be achieved by choosing efficient modulation and speech coding, and by adapting to microcellular communication.

Spectral efficiency of the present analog AMP system is 0.33 bits/sec/Hz, which is increased by about five times by adapting to $\pi/4$ DQPSK and time division multiple access (TDMA) where the present 30-kHz channel will be shared by three users. This is possible by coding the baseband speech at 8 Kbps by vector sum excited linear prediction (VSELP) algorithm [1, Ch. 7]. Similarly, Gaussian minimum shift keying (GMSK) modulation and a TDMA access scheme increases the efficiency of the European GSM system by a factor of two with respect to their present analog FM system. In the future, further increase in spectral efficiency is possible by using CDMA, more efficient modulation, and microcellular architecture. Modulation efficiency greater than 2 bits/sec/Hz has been demonstrated by Sampei for mobile channels by using 16-level quadrature amplitude modulation (16-QAM) [4]. Both the University of California at Davis and Jet Propulsion Laboratory (JPL) have claimed modems with a spectral efficiency greater than 2 bits/sec/Hz [2, 5].

One main contributor to the higher capacity is efficient digital speech coding. Both GSM in Europe and North American digital cellular radio (NADCR) systems use 8-Kbps digital raw speech. In the near future, it is expected that digital speech at 4 Kbps will be possible by using hybrid coders. The hybrid coder describes the performance of a class of algorithms that combine the high-quality potential of waveform coding with the compression efficiency of a model-based vocoder. Figure 1.2 shows a description of quality of speech as a function of data rate. The solid curves are examples of presently realized algorithms. The dotted curve is the objective that is expected to be achievable in the near future. As seen from the figure, the present hybrid coder at 8 Kbps is also based on a hybrid coding technique and its performance quality may fall below good-quality speech (mean opinion score (MOS) below four). However, the quality will still be an improvement over present analog FM cellular systems at about the 18-dB carrier-to-interference level. Table 1.1 summarizes the diagnostic rhyme test (DRT), the diagnostic acceptability

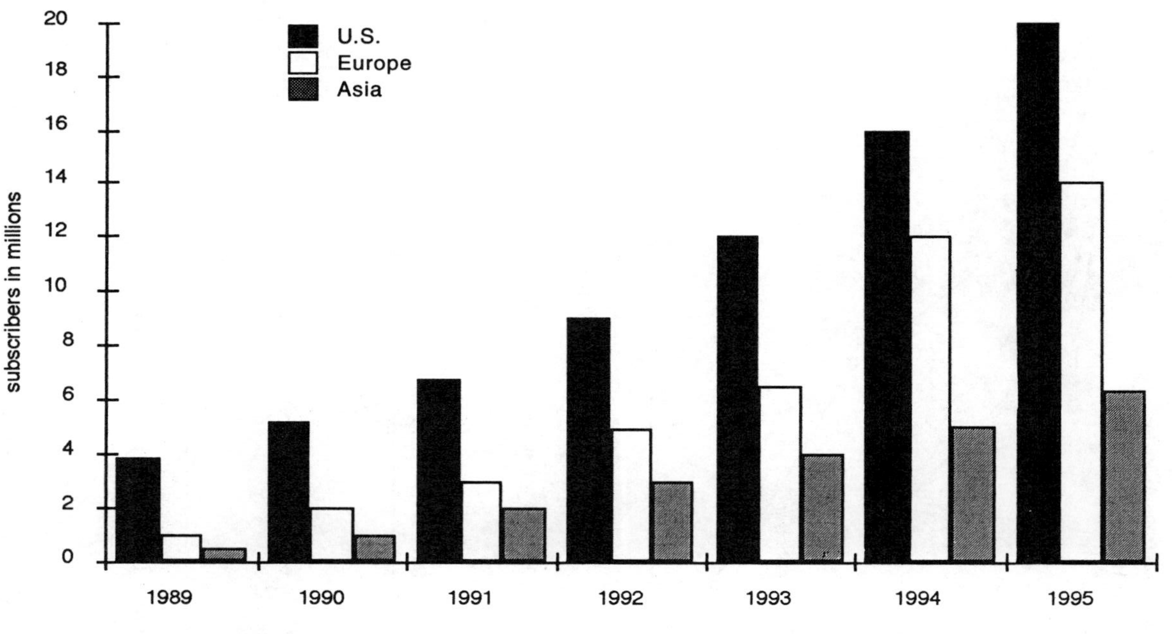

Figure 1.1 Worldwide cellular subscriber forecast. After [7].

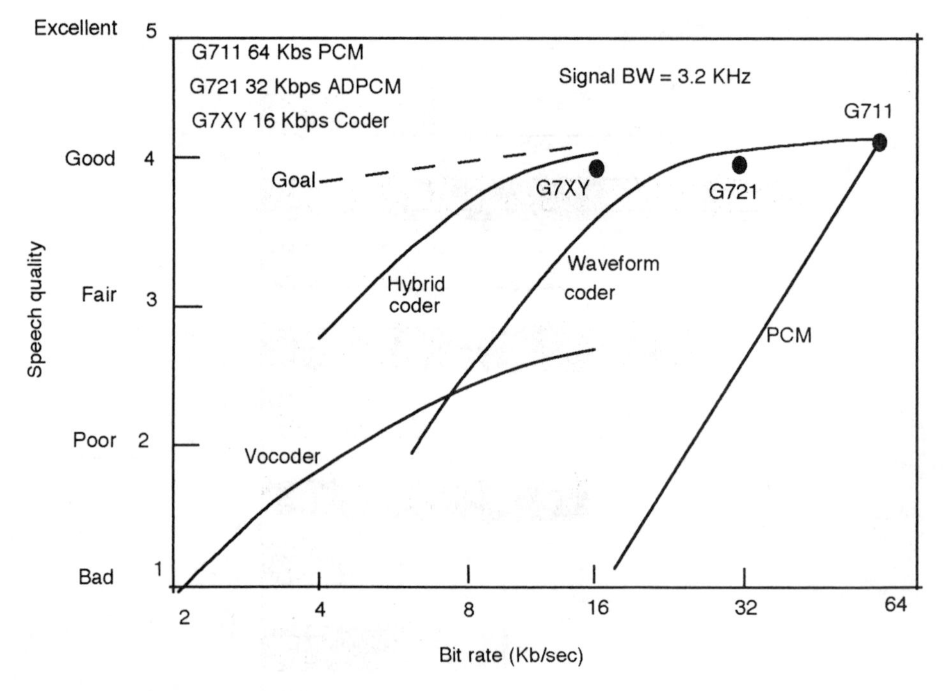

Figure 1.2 Speech quality versus bit rate for speech. Source: [3].

Table 1.1
DRT, DAM, and MOS Scores for Speech Coders

Coder	*DRT*	*DAM*	*MOS*
64-Kbps PCM	95	73	4.3
32-Kbps ADPCM	94	68	4.1
16-Kbps LD-CELP	94*	70*	4.0
8-Kbps CELP	93+	68*	3.7
4.8-Kbps CELP	93−	67−	3.0+
2.4-Kbps LPC	90	54	2.5*

Note: PCM = pulse code modulation; ADPCM = adaptive differential PCM; CELP = code excited linear predictive coding; LD-CELP = low delay CELP; LPC = linear predictive vocoder.
*Estimates-upper bounds, +-lower bounds.

measure (DAM), and the MOS performances of various speech coders in the range of 64 Kbps to 2.4-Kbps speech coders [3]. G711 and G721 are international CCITT standards. Both of these are in used in both public and private speech telecommunications for 64 Kbps PCM and 32 Kbps ADPCM.

DRT is an intelligibility test where the listener is presented with one word from a pair of words—such as "veal-feel," "Joe-go," "jaws-gauze,"—and asked to determine which word was spoken. A correct response from the listener indicates that the speech coding system under consideration preserves the differences (attributes) between two spoken words. Typical DRT values range from 75 to 95. A "good" system would have a DRT score of 90 or above. In DAM, the listener is presented with sentences taken from a Harvard University list of phonetically balanced sentences (e.g., "add the sum to the product of these three") and is asked to rate the speech quality in terms of overall acceptability and individual characteristics. Individual characteristics include signal qualities (interrupted speech, highpass/lowpass filtered speech, peak clipped speech) and background noise (hissing, buzzing, rumbling). Good speech quality has a score above 50%. In MOS, a sentence is passed through two systems, and both sentences are presented to the listener in a random order. The listener rates the system on an absolute scale from 1 to 5. A rating of 5 is considered excellent, while a rating of 1 is bad. For a detailed discussion of these tests, see [6].

Future advancements in the field of mobile communication will really come from PCS. PCS is a network concept that will enable users to establish and receive calls based on a single personal identification number (PIN) across multiple networks from any arbitrary network entry point with service access limited only by the capabilities of the terminal and the type of access point. Various network aspects of PCS/PCN are being studied within CCITT Study Group XVIII, and service aspects are being studied within CCITT Study Group I.

1.2 ORGANIZATION OF THE BOOK

This book deals with the performance aspects of cellular radio. To this regard, cellular channel characterization, propagation, antenna systems, interference, modulation, diversity, and signal processing aspects are considered. These topics are arranged as follows. Chapter 2 deals with the prerequisite background on the relevant parts of probability theory used in cellular communication. A large number of examples are given to provide to the reader a variety of applications.

Chapter 3 discusses channel characterization in terms of fast and slow fading. Spatial distances between two independent fades in terms of wavelength have been arrived at. Receiver threshold determination based on level crossing rate and the fading duration have been mathematically represented. The correlation between electric and magnetic fields have been arrived at, which provides insight to the problem of cellular radio diversity.

Chapter 4 discusses both the analytical and experimental models that can be used for computing the median loss in mobile radio surroundings at 900 MHz. Propagation models by Carey and Okumura are discussed. Propagation behavior inside buildings, tunnels, garages, and so forth have been modeled. Chapter 5 is a study of antenna systems and includes a discussion of cell site and mobile antennas along with different ancillary devices such as transmitter combiners, receiver multicouplers, duplexers, and so forth. The effects of dead spots, antenna mounting, and intermodulation have been extensively treated.

Chapter 6 discusses the problem of cochannel, adjacent channel, and near and-far-end effects. It has been shown that the cochannel problem is of utmost importance in cellular radio. Some techniques of cochannel interference reduction have been proposed. Chapter 7 logically discusses both analog and digital modulations. Emphasis has been given to the requirements of modulation and how far these are satisfied by each modulation type. Cellular radio requires both spectral efficiency and power conservation, and an optimum modulation is the one that will satisfy both. Chapter 8 discusses techniques of diversity and combining both for cell sites and mobiles. It has been shown that scanning combining is the preferable choice for mobiles, and equal-gain combining is the better choice for cell sites.

Chapter 9 deals with the signal processing and data formatting aspects of cellular radio. Baseband signal characterization and filtering, along with applications of PLL as signal processing elements have been discussed. The use of nonlinear channel equalizers and Barker codes for frame synchronization have been discussed. The second half of the chapter deals with different useful block codes along with techniques of majority voting and interleaving to combat the fading effects. Newer speech coding by using RELPC and CELP algorithms are briefly described.

Chapter 10 discusses the requirements for microcellular systems, followed by the requirements for cordless telephones. First and second-generation cordless telephones (CT1 and CT2) and digital European cordless telecommunication (DECT) are extensively treated as are new generations of services, known as PCS/PCN. The status of these in both Europe and the U.S. are discussed.

REFERENCES

[1] Mehrotra, A. *Cellular Radio: Analog and Digital Systems*, Artech House: Norwood, MA, 1994.

[2] Feher, K. " Modems for U.S. Digital Cellular and Emerging Digital Mobile Radio Systems," *IEEE International Communication Conference*, 1991, pp. 19.1.1–19.1.7.

[3] Jayant, N. S. "High Quality Coding of Telephone Speech and Wideband Audio," *IEEE International Communication Conference*, 1990, pp. 322.1.1–322.1.5.

[4] Sampei, S., and T. Sunaga. " Rayleigh Fading Compensation Method for 16-QAM in Digital Land Mobile Channels," *IEEE Vehicular Technology Conference*, 1989,

[5] Divsalar, D., and M. K. Simon. "Multiple Trellis Coded Modulation (MTCM)," *IEEE Trans. on Communication*, April 1988.

[6] Papamichalis, P. E. "Practical Approaches to Speech Coding," Englewood Cliffs, NJ, pp 177–197.

[7] "Digital Communications Systems," Hewlett Packard, Spokane, WA.

Chapter 2
Mathematical Background

2.1 INTRODUCTION

As in any other communication system, the input signal at the cellular receiver is contaminated by noise. This noise has components from both natural and manmade sources. Natural noise is caused by the constant thermal agitation of electrons and is represented by the noise temperature of the receiver. Signal degradation due to thermal noise is computed by knowing the noise figure of the receiver, which is related to its noise temperature. On the other hand, manmade noise is generated by automobile ignition near the receiver as well as other electrical noise caused by power generating equipment, electrical motors, and switching equipment. Automotive traffic density is utilized to estimate the level of manmade noise in a given area. The next most important noise source in the cellular radio is noise generated by the scattering of the signals. When the cell site transmits, it is likely that the signal received by the mobile will vary considerably as the mobile travels from one location to another. It is only in clear areas free from obstruction, when there are no obstructions between the mobile and the cell site, that the received power level will not change if the mobile moves on a locus equidistant from the cell site. In general, due to the presence of buildings, trees, and other obstructions, a multipath situation exists where the received signal can vary as much as 40 dB within a few car lengths [1]. The phenomenon, called fading, is represented by both the amplitude and the width (duration) of the fade. Experimental data in the UHF band indicate that in rural areas the amplitude of the received signal can lie within a certain range with certain degree of confidence. The mobile-received signal can be represented as

$$Y(t) = L(t)R(t)e^{[\omega_c t + \phi(t)]} \tag{2.1}$$

where $\omega_c = 2\pi f_c$ and the nominal value of $f_c = 850$ MHz. The amplitude factor $R(t)$ is Rayleigh distributed and $\phi(t)$, the signal phase, assumes values between 0 to 2π. Thus the amplitude as well as the phase, $\phi(t)$, cannot be assigned specific values and can only be treated in the probabilistic sense. In addition to the fast fading, $R(t)$, the other amplitude factor $L(t)$ changes slowly and, accounts for longterm variation when a mobile changes its location substantially. Similar to the short-term variation of the received signal, which exhibits a Rayleigh distribution, the long-term variation due to $L(t)$ is regarded as a log-normal distribution. The characteristics of these fades have been experimentally verified. In UHF, 1–2 fades per car length of travel occurs. Thus, the fade depth, the rapidity of fade, and the width of the fade are all important. Therefore, for estimating the performance of the cellular receiver in the presence of a complex received signal, one needs to have a good understanding of probability theory and the associated distribution of the received signals.

The FCC filing procedure for cellular radio systems requires a minimum signal strength of 39 dBu (dB above one microvolt) at the boundaries of a cell. The applicant for a license must design a system to provide a signal of 39 dBu or more at 95% of the system locations and for 95% of the time. In other words, the system design and cell layout are based on a probabilistic basis.

We therefore devote this chapter to the necessary background material for statistical analysis. A complete treatment of this is beyond the scope of this chapter. Although the analysis is at an elementary level, some background in probability theory is assumed. For a comprehensive overview of the subject, the reader is referred to [3].

2.2 STATISTICAL THEORY

The objective of this section is to provide the necessary background for probability theory and statistical process that is required for the proper understanding of the theory behind the working of cellular radio. Only those areas of the theory that are directly used in subsequent chapters of this book are discussed.

2.2.1 Definitions and Axioms of Probability Theory

In this section we outline the mathematical techniques for dealing with the results of an experiment whose outcome is not known in advance. An example of this is a communication receiver where the received message has uncertainty due to the random nature of the baseband data because of the noisy nature of the multipath resulting in signal fade, burst noise due to automotive ignition, and/or the shadowing effects. We begin our study of probability theory with some basic fundamentals.

2.2.1.1 Classical Definition

According to the classical definition, the probability $P(A)$ of an event A is determined a priori without actual experimentation. It is given by

$$P(A) = N_A/N \tag{2.2}$$

where N is the number of possible outcomes and N_A is the number of outcomes favorable to the event A.

2.2.1.2 Relative Frequency Definition

Suppose that a random experiment is repeated n times. If event A occurs n_A times, then its probability, $P(A)$, is defined as the limit of the relative frequency n_A/n of the occurrence of A. That is,

$$P(A) = \lim_{n\uparrow\infty} \frac{n_A}{n} \tag{2.3}$$

For example, if a coin is tossed n times and tails shows up n_t times, then the probability of getting tails equals n_t/n as n approaches infinity.

2.2.1.3 Axioms of Probability Theory

We assign to each event A a number, $P(A)$, which we call the probability of even A. This number is so chosen that it satisfies the following conditions:

$$P(A) \geq 0 \tag{2.4a}$$

$$P(S) = 1 \tag{2.4b}$$

$$\text{if } A \cap B = 0, \text{ then } P(A \cup B) = P(A) + P(B) \tag{2.4c}$$

In this case, the outcomes A and B of an experiment are mutually exclusive; that is, the occurrence of one outcome precludes the occurrence of the other. Equation (2.4c) can be extended to more numbers of mutually exclusive events. Thus,

$$P(A_1 \text{ or } A_2 \text{ or } \ldots A_k) = \sum P(A_i) \tag{2.5}$$

2.2.1.4 Some Laws of Probability

The following three relationships can be derived by using (2.4a) through and (2.4c). (1) If $A \cup \overline{A} = S$ and $A \cap \overline{A} = 0$, A and $\overline{A}$ are mutually exclusive. For example, for digital transmission in cellular radio forward and reverse channels, $P(\text{transmitting } 0) = 1 - P(\text{transmitting } 1)$. (2) $(A \cup B) = P(A) + P(B) - P(AB)$.

(3) If $A_1, A_2 \cdots \cup A_n$ are n random events such that $A_i \cap A_j = 0$ for $i \neq j$, mutually exclusive, and $A_1 \cup A_2 \ldots \cup A_n = S$ (A_i's are exhaustive), then

$$\begin{aligned} P(A) &= P(A \cap S) \\ &= P[A \cap (A_1 \cup A_2 \cup A_2 \cdots \cup A_n)] \\ &= P[(A \cap A_1) \cup (A \cap A_2) \cup \cdots \cup (A \cap A_n)] \\ &= P(A \cap A_1) + P(A \cap A_2) + \cdots + P(A \cap A_n) \end{aligned} \quad (2.6)$$

2.2.2 Joint, Marginal, and Conditional Probabilities

Joint, marginal, and conditional probabilities are defined with respect to an experiment E, which in turn can be based on two subexperiments, E_1 and E_2. If E_1 is defined by events $A_1, A_2, \ldots, A_n$ and E_2 is defined by events $B_1, B_2, \ldots, B_n$, then A_iB_j is an event of the total experiment E.

2.2.2.1 Joint Probability

The intersection of events from subexperiments E_1 and E_2 are known as joint probability $P(A_iB_j)$. For example, say the first experiment is the rolling of a fair die, defined as $S_1 = f_1, f_2, \ldots, f_6$. Thus, $P_1(f_i) = 1/6$. If the second experiment is the tossing of a fair coin,

$$S_2 = (h, t) \qquad P_2(h) = P_2(t) = 1/2 \quad (2.7)$$

then the joint probability of tossing tails for the coin and rolling a six for a die is $1/6 \times 1/2 = 1/12$.

2.2.2.2 Marginal Probability

If events $A_1, A_2, \ldots, A_n$ associated with subexperiment E_1 are mutually exclusive and exhaustive, then

$$\begin{aligned} P(B_j) &= P(B_jS) = P[B_j(A_1 \cup A_2 \cdots \cup A_n)] \\ &= \sum P(A_iB_j) \end{aligned} \quad (2.8)$$

where B_J is the associated event with subexperiment E_2.

2.2.2.3 *Conditional Probability*

The conditional probability is expressed in terms of joint and marginal probability as

$$P(B|A) = \frac{P(AB)}{P(A)} \qquad P(A) \neq 0 \tag{2.9}$$

From this we obtain

$$P(AB) = P(A)P(B|A) \tag{2.10}$$

If we do not distinguish between AB and BA, then we can also write

$$P(AB) = P(BA) = P(B)P(A|B) \tag{2.11}$$

From (2.10) and (2.11), we obtain:

$$P(A|B) = \frac{P(AB)}{P(B)} = \frac{P(B|A)P(A)}{P(B)} \qquad P(B) \neq 0 \tag{2.12}$$

In general, if $B_1, B_2, \ldots, B_m$ are a mutually exclusive and exhaustive set of events, then

$$P(A) = \sum_{k=1}^{m} P(A|B_k)P(B_k) \tag{2.13}$$

and

$$P(B_k|A) = \frac{P(A|B_k)P(B_k)}{\sum_{k=1}^{m} P(A|B_k)P(B_k)} \tag{2.14}$$

Independent Events

The two events A and B are called statistically independent if

$$P(AB) = P(A)P(B) \tag{2.15a}$$

or when

$$P(A|B) = P(A) \tag{2.15b}$$

Expressed in words, when two outcomes are independent, the probability of a joint occurrence of the outcomes is the product of the probabilities of the individual outcomes. Alternately, we can also describe the two events to be independent when the conditional probability does not depend on the event on which it is conditioned.

2.2.3 Random Variables, Probability Distribution, and Probability Densities

Given an experiment having a sample space S and elements $\lambda_i \in S$, we define a function $X(\lambda_i)$ whose domain is S and whose range is a set of numbers on the real axis. When $X(\lambda_i)$ assumes only finite discrete values, then it is known as a discrete random variable. An example is the number of telephone calls originated by cellular users in a given time interval, $t_2 - t_1$, $(t_2 > t_1)$. Unlike this example, there are many physical systems that generate continuous outputs. Thermal noise generated in a cellular receiver has a continuous amplitude. Consequently, the sample space S of voltage amplitudes $v \in S$ is continuous and so is the mapping $X(v) = v$. In both the discrete and the continuous case, the possible outcomes of the experiment λ_i need not be numbers. However, the assigned function $X(\lambda_i)$ is such that it maps λ_i into real numbers.

2.2.3.1 Continuous Random Variables

A continuous random variable X is characterized by a probability density function $f_X(x)$, which is related to the distribution function $F_X(x)$ as

$$F_X(a) = P(X \leq a) = \int_{-\infty}^{a} f_X(x)dx \tag{2.16}$$

where $dF_X(x)/dx = f_X(x)$, where $f_X(x)$ is known as the probability density function, which satisfies the following properties:

$$f_X(x) \geq 0 \qquad -\infty < x < \infty \tag{2.17a}$$

$$\int_{-\infty}^{\infty} f_X(x)dx = 1 \tag{2.17b}$$

The symbol $F_X(a)$ is known as the cumulative distribution function. The cumulative distribution function has the following properties:

$$0 \leq F_X(a) \leq 1 \tag{2.18a}$$

$$F_X(-\infty) = 0 \qquad F_X(\infty) = 1 \tag{2.18b}$$

$$F_X(x_1) \leq F_X(x_2) \qquad \text{if } x_2 > x_1 \tag{2.18c}$$

Property (2.18a) follows from the fact that $F_X(\bullet)$ depends on probability. Property (2.18b) follows from the fact that $F_X(-\infty)$ includes only impossible events, while $F_X(\infty)$ includes all events that are certain to occur. Finally, (2.18c) holds since for x_2 and x_1, $F(x_2)$ includes all those outcomes that are contained in $F_X(x_1)$.

By extending the concept of (2.18c), if we consider the outcome of an experiment where $x_1 < X \leq x_2$, then

$$P(x_1 < X \leq x_2) = P(X \leq x_2) - P(X \leq x_1) \tag{2.19}$$

which can also be expressed as:

$$\begin{aligned} P(x_1 < X \leq x_2) &= \int_{-\infty}^{X_2} f_X(x)dx - \int_{-\infty}^{X_1} f_X(x)dx \\ &= \int_{X_1}^{X_2} f_X(x)dx \end{aligned} \tag{2.20}$$

If the random process involves two or more random variables, then we characterize the system by the joint density function, $f_{XY}(x, y)$. From the joint density function, we can obtain the marginal density $f_X(x)$ and the conditional density $f_{XY}(y|x)$ as shown below.

Marginal Density

$$f_X(x) = \int_{-\infty}^{\infty} f_{XY}(x, y)dy \tag{2.21a}$$

and

$$f_Y(y) = \int_{-\infty}^{\infty} f_{XY}(x, y)dx \tag{2.21b}$$

Conditional Density

$$f_{X|Y}(x|y) = \frac{f_{XY}(x, y)}{f_Y(y)} \qquad f_Y(y) > 0 \tag{2.22a}$$

$$F_{Y|X}(y|x) = \frac{f_{XY}(x, y)}{f_X(x)} \qquad f_X(x) > 0 \tag{2.22b}$$

and

$$f_{Y|X}(y|x) = \frac{f_{X|Y}(x|y) f_Y(y)}{\int_{-\infty}^{\infty} f_{X|Y}(x|\alpha) f_Y(\alpha) d\alpha} \qquad f_Y(\alpha) \neq 0 \tag{2.22c}$$

If random variables X and Y are statistically independent, then

$$f_{XY}(x, y) = f_X(x) f_Y(y) \qquad -\infty < x, y < \infty \tag{2.23}$$

The expected value of the random variable X is given by

$$\mu_X = E(X) = \int_{-\infty}^{\infty} x f_X(x) dx \tag{2.24}$$

The variance of the random variable X is given by

$$\sigma_X^2 = E[(X - \mu_X)^2] = \int_{-\infty}^{\infty} (x - \mu_X)^2 f_X(x) dx \tag{2.25}$$

Writing $(X - \mu_X)^2 = X^2 - 2\mu_X x + \mu_X^2$, we obtain

$$\sigma_X^2 = E(X^2) - 2\mu_X E(X) + E(\mu_X^2) = E(X^2) - \mu_X^2 \tag{2.26a}$$

The square root of variance is known as the standard deviation and is the root mean square (rms) value of $(X - \mu_X)$. If the average $\mu_x = 0$, then the above equation is reduced to

$$\sigma_X^2 = E(X^2) \tag{2.26b}$$

The expected value of X^2 is known as the second moment of random variable X. The covariance of random variables X and Y is given by

$$\rho_{XY}(x, y) = \frac{E[(X - \mu_X)(Y - \mu_Y)]}{\sigma_X \sigma_Y} \tag{2.27}$$

If a random variable Z is a function of two random variables X and Y, $Z = g(X, Y)$, then the expected value of the function of a pair of a random variables is defined as

$$E(Z) = E[g(X, Y)] = \int_{-\infty}^{\infty} \int_{-\infty}^{\infty} g(x, y) f_{XY}(x, y) dx dy \tag{2.28}$$

We illustrate this theory by looking at an example of a control channel from the mobile to the cell site.

Example 2.1

In an analog cellular system, a control channel from the mobile to the cell site works on the principle of random access. If more than one user transmits simultaneously on the channel, a collision will occur. If the mobile user does not receive acknowledgment of his or her transmission within a certain time, the message is sent again. Let the parameters of the system be: r = retry time (random variable, assumed to be uniformly distributed between 0 and 2R); $R = E(r)$ = mean retry time; and N = fixed waiting time for acknowledgment; With these parameters, find the message delay and its variance. Mean delay is $\overline{T} = N + R$; mean square delay =

$$\overline{T^2} = \frac{1}{2R} \int_0^{2R} (N + r)^2 \, dr = N^2 + 2NR + \frac{4R^2}{3} \tag{2.29}$$

and

$$\text{Var}(T) = \overline{T^2} - (\overline{T})^2 = N^2 + 2NR + \frac{4}{3} R^2 - (N + R)^2 = \frac{R^2}{3} \tag{2.30}$$

2.2.3.2 Discrete Random Variable

A discrete random variable X is characterized by a set of allowed values x_1, x_2, . . . , x_n and their probabilities. The probability that the random variable X =

$x_i = X(\lambda_i)$ is denoted by $P(X = x_i)$ and is known as the probability mass function. Some important properties of the probability mass function are

$$\sum_{i=1}^{n} P[X = x_i] = 1 \tag{2.31}$$

The cumulative probability distribution function is defined as

$$F_X(x) = P(X \leq x) = \sum_{\text{all } x_i \leq x} P(X = x_i) \tag{2.32}$$

Extending the above concept to two random variables X and Y, which take on values $x_1, x_2, \ldots, x_n$ and $y_1, y_2, \ldots, y_m$. These two random variables can be defined by the joint probability mass function $P[X = x_i, Y = y_k]$, which provides the joint probability that $X = x_i$ and $Y = y_k$.

Using the above definition, the joint distribution function of two random variables can be defined as

$$P(X \leq x, Y \leq y) = F_{XY}(x, y) = \sum_{x_i \leq x} \sum_{y_k \leq y} P(X = x_i, Y = y_k) \tag{2.33}$$

Similarly, the marginal distribution function is given by

$$\begin{aligned} P(X = x_i) &= \sum_{j=1}^{n} P(X = x_i, Y = y_j) \\ &= \sum_{j=1}^{n} P(X = x_i | Y = y_j) P(Y = y_j) \end{aligned} \tag{2.34}$$

The conditional distribution (probability mass) function is given by

$$\begin{aligned} P(X = x_i | Y = y_j) &= \frac{P(X = x_i, Y = y_j)}{P(Y = y_j)} \\ &= \frac{P(Y = y_j | X = x_i) P(x_i)}{\sum_{i=1}^{n} P(Y = y_j | X = x_i) P(x_i)} \qquad P(Y = y_j) \neq 0 \end{aligned} \tag{2.35}$$

Random variables X and Y are independent if

$$P(X = x_i, Y = y_i) = P(X = x_i) P(Y = y_k) \tag{2.36}$$

where $i = 1, 2, \ldots, n$ and $k = 1, 2, \ldots, m$.

Statistical Averages

Two statistical averages that are commonly used for characterizing a discrete random variable, X, are its mean, μ_X, and its variance σ_X^2. These are defined as

$$E(X) = \mu_X = \sum_{i=1}^{n} x_i P(X = x_i) \approx \sum_{i=1}^{n} x_i p(x_i) \Delta x \tag{2.37}$$

The approximation holds as the sample space x is divided into intervals of width Δx centered on x_i such that $P(X = x_i) \approx p(x_i)\Delta x$

$$\begin{aligned} \sigma_X^2 &= E(X = \mu_X)^2 = \sum_{i=1}^{n} (x_i - \mu_x)^2 P(X = x_i) \\ &= \sum_{i=1}^{n} (x - \mu_X)^2 p(x_i) \Delta x \end{aligned} \tag{2.38}$$

where σ_X^2 represents the ac component of the discrete random variable, while μ_X is the dc or the average value. The square root of the variance is known as the standard deviation.

The expected value or statistical average of a function $g(x)$ of a discrete random variable is defined as

$$\begin{aligned} \overline{g(x)} &= E(g(x)) = \sum_{i=1}^{n} g(x_i) P(X = x_i) \\ &\approx \sum_{i=1}^{n} g(x_i) p(x_i) \Delta x \end{aligned} \tag{2.39}$$

Extending this to variable Z, which is a function of two random variables X and Y such that $Z = g(X, Y)$, one can show that

$$\overline{Z} = \sum_{j=1}^{m} \sum_{i=1}^{n} g(x_i, y_j) P(X = x_i, Y = y_j) \tag{2.40}$$

2.2.4 Tchebycheff Inequality

A measure of the concentration of a random variable near its mean is its variance σ. The Tchebycheff inequality states that the probability that random variable (RV) X is outside an arbitrary interval $(\eta - \epsilon, \eta + \epsilon)$ decreases as the ratio σ/ϵ

gets smaller. In other words, as the normalized variance decreases, the probability that the RV X lies outside $(\eta \pm \epsilon)$ decreases. Thus, for any $\epsilon > 0$,

$$P(|X - \eta| \geq \epsilon) \leq \frac{\sigma^2}{\epsilon^2} \tag{2.41}$$

Thus, The probability that the Gaussian random variable lies within $\pm(\eta + \epsilon)$ can be found as shown in the following example.

Example 2.2

$$\begin{aligned} P_{\pm \ell\sigma} &= P(\eta - \ell\sigma \leq X \leq \eta + \ell\sigma) = \int_{\eta - l\sigma}^{\eta + l\sigma} \frac{e^{-(x-\eta)^2/2\sigma^2}}{\sqrt{2\pi}\sigma} dx \\ &= \frac{2}{\sqrt{\pi}} \int_0^{l/\sqrt{2}} e^{-y^2} dy = \operatorname{erf}\left(\frac{l}{\sqrt{2}}\right) \end{aligned} \tag{2.42}$$

2.2.5 Transformation of Random Variables

In communication we are often required to find the density function of the RV at the output of a system from the given density function at the input of the system. Problems of this type can be solved by the following techniques. Given the relationship $y = g(x)$ and the density of x is $f_x(x)$, we are interested in $f_Y(y)$. If x_1, $x_2, \ldots, x_n$ are the real roots of $g(x)$, then $y = g(x_1) = g(x_2) = \cdots = g(x_n)$. The density function of $f(y)$ can be found by

$$f_Y(y) = \frac{f_X(x_1)}{|g'(x_1)|} + \frac{f_X(x_2)}{|g'(x_2)|} + \frac{f_X(x_n)}{|g'(x_n)|} \tag{2.43}$$

where

$$g'(x) = \frac{dg(x)}{dx}$$

For functions of two random variables X, Y, where the joint *PDF* is given by $f_{X,Y}(x, y)$, and $Z = g(x, y)$ with $W = h(x, y)$, then $f_{Z,W}(z, w)$ is arrived at by

$$f_{Z,W}(z, w) = \sum_{i=1}^{n} f_{X,Y}(x_i, y_i)|J_i| \tag{2.44}$$

where J_i is the Jacobian of the transformation, and is given by

$$J_i = \begin{vmatrix} \frac{\partial x_i}{\partial z} & \frac{\partial x_i}{\partial w} \\ \frac{\partial y_i}{\partial z} & \frac{\partial y_i}{\partial w} \end{vmatrix} \tag{2.45}$$

Example 2.3

The Rayleigh density function for the envelope r is given by

$$f(r) = \frac{r}{x_o} e^{-r^2/2x_o} \tag{2.46}$$

where x_0 is the mean signal power.

To find the probability density of signal instantaneous power, $x = 1/2\ r^2$:

$$f_X(x) = \frac{f_r(r)}{|g'(r)|} = \frac{(\sqrt{2x}/x_0)\ e^{-2x/2x_0}}{\sqrt{2x}} = \frac{1}{x_0} e^{-x/x_0} \tag{2.47}$$

Thus, the probability of signal fade (i.e., the signal level below the threshold X_T) is given by

$$P(x \leq x_T) = \int_0^{X_T} \frac{1}{x_0} e^{-x/x_0} dx = 1 - e^{-x_T/x_0} \tag{2.48a}$$

Assuming the signal mean power $x_0 \gg x_T$, then

$$P(x \leq x_T) \approx 1 - [1 - (x_T/x_0)] = x_T/x_0 \tag{2.48b}$$

We shall further discuss this important approximation in Chapter 8.

2.2.6 Some Useful Probability Distributions

In this section we shall discuss those density functions that directly apply to characterizing cellular behavior. Among discrete cases, uniform, gamma, binomial, and Poisson distributions are important. Equally applicable are the uniform, Gaussian, and Rayleigh densities for continuous cases.

2.2.6.1 *Discrete Case*

In this section we provide the summary of those and discuss their probability density function (PDF), cumulative density function (CDF), and moments.

Uniform Density

A discrete random variable, X, is said to be uniformly distributed if

$$P[X = x_i] = 1/n \qquad i = 1, 2, \ldots, n \tag{2.49}$$

Gamma and Erlang Densities

The function $f(x) = Ax^b e^{-cx} U(x)$ is known as the gamma density. Normalization provides the value of A as

$$\int_0^\infty Ax^b e^{-cx} dx = 1 \tag{2.50a}$$

or

$$A = \frac{c^{b+1}}{\Gamma(b + 1)} \tag{2.50b}$$

where

$$\Gamma(b) = \int_0^\infty y^{b-1} e^{-y} dy$$

The integral in (2.50b) is a gamma function. The function $\Gamma(b)$ is also called a generalized factorial because

$$\Gamma(b + 1) = b\Gamma(b)$$

If $b + 1 = n$, an integer, then $\Gamma n = (n - 1)!$ In this case, gamma density is reduced to an erlang density, which is given by

$$f(x) = \frac{c^n}{(n-1)!} x^{n-1} e^{-cx} U(x) \tag{2.50c}$$

A special case of erlang density is the exponential density

$$f(x) = Ce^{-cx} U(x) \tag{2.50d}$$

An application of the exponential density is for characterizing the time delay distribution of the received multipath signal, which is discussed in Chapter 3.

Binomial Distribution

If an experiment is repeated n times, then the probability that event A occurs k times is given by the binomial probability mass function:

$$P_n(X = k) = \binom{n}{k} p^k (1-p)^{n-k} \qquad k = 0, 1, \ldots, n \tag{2.51}$$

where the binomial coefficient is

$$\binom{n}{k} = \frac{n!}{k!(n-k)!}$$

The mean and the variance of the binomial distribution are given by: $\mu_x = np$ and $\sigma_X^2 = np(1-p) = \mu_X(1-p)$, respectively. One of the numerous applications of this density function is in the computation of the probability of error in digital transmission. The probability of having k bits in errors in the transmission length of n bits is given by $P_n(k)$.

Example 2.4

In a binary message transmission, 1,000 bits are transmitted with the probability of an individual bit in error equal to 10^{-3}. To find the probability that the message will be correctly received, we apply (2.51) with $n = 1{,}000$, $p = 10^{-2}$, and $k = 0$ (no error). P_{1000} (correctly receiving the message) is

$$\binom{1000}{0}(10^{-3})^0(1 - 0.001)^{1000} = (0.999)^{1000} \tag{2.52}$$

Poisson Distribution

When burst noise is present at the receiver input, the distribution of noise can be represented as a Poisson distribution. The number of random events that occur in a unit of time often follow a Poisson distribution. Poisson distribution applies particularly to "rare" events, that is, those events that occur infrequently.

When n is very large and p very small, the binomial distribution is difficult to handle. However, if the product np is finite, the binomial distribution can be approximated by the Poisson distribution

$$P(m) = e^{-\overline{m}} \frac{\overline{m}^k}{k!} \tag{2.53}$$

where $\overline{m} = np$ and $\sigma_n^2 = \overline{m}q \approx \overline{m}$ The Poisson distribution can be applied to digital transmission on the forward and reverse setup channels (data, paging, and access channels) between the cell site and the mobile telephone switching office (MTSO). Another application is in the traffic analysis. For example, if there are N calls in the interval (O, T), then the probability of having n calls in a ΔT interval within (O, T) is given by

$$p(n) = e^{-\lambda\Delta T} \frac{(\lambda\Delta T)^n}{n!}, \tag{2.54}$$

where $\lambda = N/T =$ arrival rate, and the mean $m = \lambda\Delta T$.

2.2.6.2 Continuous Density Function

Uniform Density

A random variable x is said to be uniformly distributed (see Figure 2.1) if

$$f_X(x) = 1/(b - a) \qquad a < x < b \tag{2.55}$$

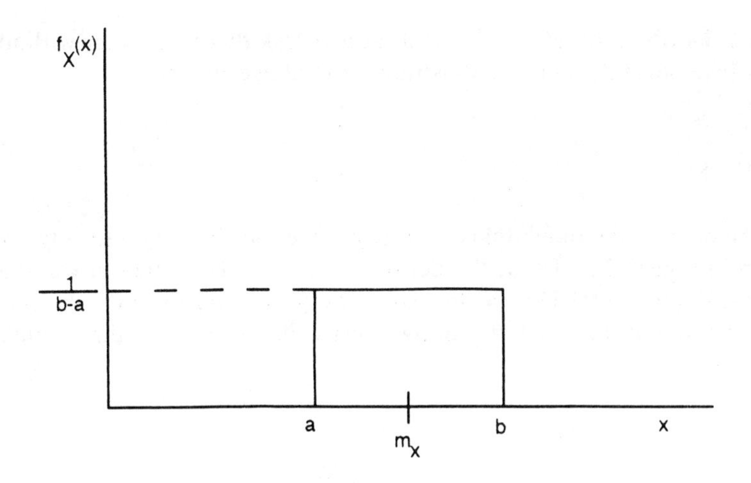

Figure 2.1 Uniform density function.

and zero otherwise. The mean and the variance are given by

$$\mu_x = \frac{(b + a)}{2} \qquad \sigma^2 = \frac{(b - a)^2}{12} \tag{2.56}$$

Thus the mean falls right in the middle of x, where one expects. However, the variance is the function of the range $(b - a)$, and does not depend on its absolute position. One application of this density function in cellular is the phase angle distribution of the received multipath signals.

Gaussian Density

We say that a random variable is normally distributed if its density function is Gaussian and given by

$$f_X(x) = \frac{1}{\sqrt{2\pi\sigma}} e^{-(x-\mu_x)^2/2\sigma^2} \tag{2.57}$$

This function occurs in many applications in communication engineering because of a remarkable phenomenon known as the central limit theorem, which can be stated as follows. If $X_1, X_2, \ldots, X_N$ are independent random variables and the random variable Y is formed as their sum, $Y = \Sigma X_i$, then as N becomes large, the distribution of Y approaches Gaussian distribution. This is independent of the individual distributions of Xs. We will accept this theorem as is, without proof. We now take several examples illustrating the above theory.

Example 2.5

Let us assume that the individual densities are identically and uniformly distributed as shown in Figure 2.2. Then, the density of the first two random variables $X_1 + X_2$ is triangular in shape. This can be proven by performing the convolution between densities $f(x_1)$, and $f(x_2)$. The resulting density due to the first three random vari-

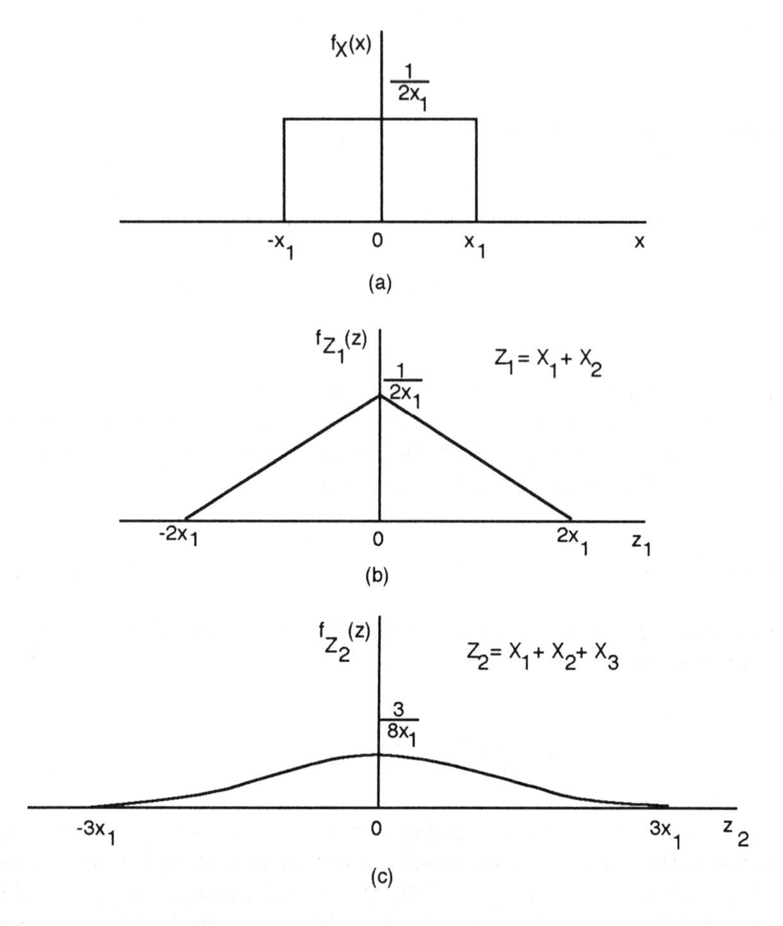

Figure 2.2 (a) A random variable X has a uniform probability density. (b) The probability density of the random variable $X_1 + X_2$. (c) The density of the random variable $X_1 + X_2 + X_3$.

ables, $X_1 + X_2 + X_2$, almost appears to be Gaussian (bell-shaped curve). As more and more terms are added, the resulting density function will approach Gaussian.

Example 2.6

The peak load of a cellular system can be derived by knowing the traffic loads from different business types, which tend to follow a normal random process (large M) with the mean represented as

$$\rho = \sum_{i=0}^{M} \rho_i \tag{2.58a}$$

where, ρ_i is the mean traffic of the ith business group and M = number of traffic sources. The standard deviation of the system is

$$\sigma = \left(\sum_{i=1}^{M} \sigma_i^2\right)^{1/2} \tag{2.58b}$$

where σ_i^2 = variance of the ith business group. Assuming that the traffic sources are uncorrelated, the peak system load can be estimated as

$$\rho_{pk} = \rho + k\sigma = \sum_{i=1}^{M} \rho_i + k\left(\sum_{i=1}^{M} \sigma_i^2\right)^{1/2} \tag{2.58c}$$

where k is a scaling factor, which varies with the averaging interval. A value of 1.5 has been observed as typical for busy-hour estimation in cellular radio.

Example 2.7

Due to a multipath propagation, the received signal at the mobile can be represented as

$$R(w) = \left[\sum_{i=1}^{N} A_i(w)e^{-j\omega\tau_i}\right]s(w) \tag{2.59a}$$

where $s(w)$ is a spectrum of the transmitted signal and τ_i is considered a random variable, which is independent of frequency. Thus, the quantity in the square bracket is the transfer function of the channel. That is,

$$H(w) = \sum_{i=1}^{N} A_i(w)e^{-j\omega\tau_i} \tag{2.59b}$$

or in complex form,

$$H(w) = H_{\text{real}}(w) + jH_{\text{imag}}(w) = \sum_{i=1}^{N} A_i(w)\cos(w\tau_i) - j\sum_{i=1}^{N} A_i(w)\sin(w\tau_i) \tag{2.59c}$$

Assuming the channel characterization to be wide-sense stationary, A_i for each frequency is a random variable. Further, A_i is independent of A_j for $i \neq j$. Similarly, assuming $\{\tau_i\}$ forms a set of independent random variables, then both the real and the imaginary parts of $H(w)$ are each the sum of N independent random variables. Thus, applying the central limit theorem, the distributions of both $H_{\text{real}}(w)$ and $H_{\text{imag}}(w)$ are Gaussian as N increases.

Example 2.8

It is well known that the received signal amplitude variation at the mobile can be described by a mixture of rapid Rayleigh fading, slow fading having a log-normal distribution and a seasonal dependent variation caused by vegetation and so forth. Assuming the standard deviation of these to be: $\sigma_R = 5.6$ dB, $\sigma_L = 2.5$ dB, and $\sigma_S = 2$ dB, and assuming that these processes are uncorrelated, the total standard deviations is: $\sigma_T = \sqrt{\sigma_R^2 + \sigma_L^2 + \sigma_S^2} = \sqrt{(5.6)^2 + (3.5)^2 + 3^2} = 7.3$ dB. Assuming further that prediction errors (σ_ϵ) have Gaussian distributions with a standard deviation of 5 dB, then from the planning point of view, the worst case standard deviation can be $\sigma_P = \sqrt{\sigma_T^2 + \sigma_\epsilon^2} = \sqrt{7.3^2 + 5^2} = 9.0$ dB. In other words, the actual measured value of the deviation, σ_P, can be as high as 9.0 dB.

There are two types of applications of Gaussian density function in cellular radio. In the first application we are looking for error probabilities of data transmission, and in the second application we look for a long-term variation of the UHF signal due to shadowing effects. In the first application, error probabilities are generally expressed in terms of error functions, which is an integral of the Gaussian density function. In the second application, we treat the random variable as log-normally distributed. The Gaussian random variable expressed in decibels is known as a log-normal distribution. The Gaussian density function in this case is expressed as

$$f_X(x) = \frac{1}{\sqrt{2\pi\sigma_x^2}} e^{-[(X-\mu_x)^2/2\sigma_x^2]} \qquad -\infty < x < \infty \tag{2.60a}$$

where the mean and the variance are given by μ_x (dB) and σ_x^2 (dB). As indicated in Figure 2.3, if the value of the function at the mean, μ_x, has a normalized value of 1, then at $\pm\sigma_x$, the value of $f_X(x)$ is approximately 60% of the center value, or 0.6, and at $\pm 2\sigma_x$ the value is reduced to about 13.5% of its center value.

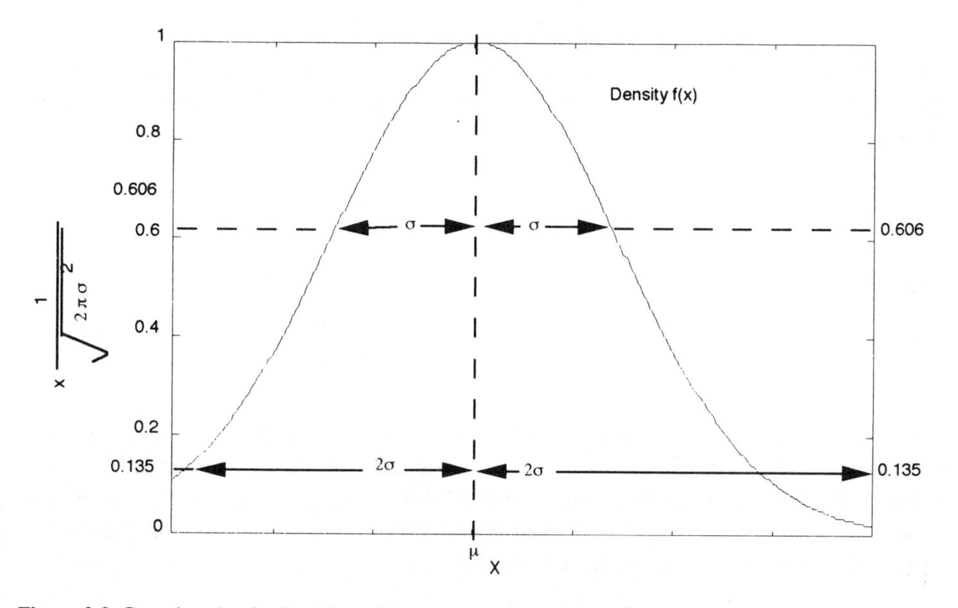

Figure 2.3 Gaussian density function with mean μ_x and variance σ_x^2.

From the density function $f_X(x)$, the probability that $X > a$ can be written as

$$P[X > a] = \int_a^\infty \frac{1}{\sqrt{2\pi\sigma_x^2}} e^{-[(x-\mu_x)^2/2\sigma_x^2]}\, dx \tag{2.60b}$$

The above equation can be expressed in terms of several standard mathematical functions, such as the Marcum Q function, error function, and the complementary error function, which are defined as

$$Q(y) = \frac{1}{\sqrt{2\pi}} \int_y^\infty e^{-Z^2/2} dz \qquad y > 0 \tag{2.60c}$$

$$\mathrm{erf}(y) = \frac{2}{\sqrt{\pi}} \int_0^y e^{-Z^2} dz \tag{2.60d}$$

$$\mathrm{erf}_c(y) = \frac{2}{\sqrt{\pi}} \int_y^\infty e^{-Z^2} dz = 1 - \mathrm{erf}(y) \tag{2.60e}$$

Appendix 2A provides a simplified computer listing and some tabulated values for the error, the complementary error, and the Q functions. It is easy to see from (2.60c) and (2.60e) that

$$\mathrm{erf}_c(y) = 2Q(\sqrt{2}y) \qquad y > 0 \tag{2.60f}$$

and

$$P(X > a) = Q((a - \mu_x)/(\sigma_x)) \tag{2.60g}$$

A similar case of interest is the probability that a measurement will yield an outcome that falls within a certain range about the average, μ_x. Since the "width" of the probability density depends on the standard deviation σ_x, we ask for $P\,[\mu_x - k\sigma_x \leq X \leq \mu_x + k\sigma_x]$; that is, the probability that the random variable X falls in the range of $+k\sigma_x$ around its mean. From (2.42),

$$p_{\pm K\sigma}(\mu_x - K\sigma_x \leq x \leq \mu_x + K\sigma_x) = erf(K/\sqrt{2}) \tag{2.60h}$$

Assuming that the long-term fading signal X is lognormally distributed, then the measured signal strength $10 \log_{10} x$ is in decibels and is normally distributed, and given by

$$f(x) = \frac{1}{\sigma\sqrt{2\pi}} e^{[10\ \log_{10}x - \mu]/2\sigma^2} \tag{2.60i}$$

The values of the mean and variance are also expressed in decibels.

Example 2.9

If the received signal (mean of the received signal, which has a Rayleigh distribution) at a distance R from the cell site is log-normal distributed, then the percentage of area where the threshold level x_0 of the mobile receiver is exceeded can be computed as follows. The probability density of x is

$$p(x) = \frac{1}{\sqrt{2\pi\sigma^2}} e^{-[x-\bar{x}]^2/2\sigma^2} \tag{2.61a}$$

where x, $\bar{x}$, and σ are in decibels. Note that x is the average of the log-normally distributed received signal: here the received signal, x, is itself the short-term average of the Rayleigh distributed signal. Thus:

$$P_X(x_0) = P(X \geq x_0) = \int_{x_0}^{\infty} p(x)dx = 1/2 + 1/2 \operatorname{erf}[(x_0 - \bar{x})/(\sqrt{2}\sigma)] \tag{2.61b}$$

Assume the mean signal strength to be -100 dBm ($\bar{x}$) at a distance R from the cell site, for some specified cell site–radiated power. Let the receiver threshold, x_0, be -125 dBm and the standard deviation of signal be 10 dB. Then we obtain

$$P_X(x_0) = 1/2 + 1/2 \operatorname{erf}[(-125 + 100)/(10\sqrt{2})] = 0.994 \tag{2.61c}$$

Therefore, the percentage of area at a distance R from the cell site where the average signal strength exceeds the threshold value of -125 dBm is 0.994.

Example 2.10

Carey's report [2] for UHF propagation assumes a 14-dB allowance to go from a 50% to a 90% reliability. Find the value of σ required for this case.

$$0.9 = \frac{1}{2} + \frac{1}{2} \operatorname{erf}\left(\frac{x_o - \bar{x}}{\sqrt{2}\sigma}\right)$$

or

$$0.8 = \operatorname{erf}\,(14/\sqrt{2}\sigma)$$

Thus,

$$14/\sqrt{2}\sigma = 0.91 \qquad \sigma = 10.9 \text{ dB}$$

Example 2.11

In cellular radio, the envelope of the received multipath signal follows Rayleigh characteristics. The Rayleigh density function $p(r)$ is given by

$$p(r) = \begin{cases} r/b_0\, e^{-r^2/2b_0} & r \geq 0 \\ 0 & r < 0 \end{cases} \tag{2.62a}$$

Thus, the probability that the envelope will be less than equal to a level R is

$$P(r \leq R) = \int_{-\infty}^{R} p(r)dr = \int_{0}^{R} r/b_0 e^{-r^2/2b_0} dr = 1 - e^{-R^2/2b_0} \tag{2.62b}$$

Assuming a normalized value of the power is to be 10 dB below the channel mean, $(r/\sqrt{2b_0})^2 = 0.1$, then $P\ (r \leq -10\ \text{dB}) = 1 - e^{-0.1} = 9.5\%$. Similarly, $P(r \leq -20\ \text{dB}) = 1 - e^{-0.01} = 1\%$. The graph of the amplitude distribution function is shown in Figure 2.4.

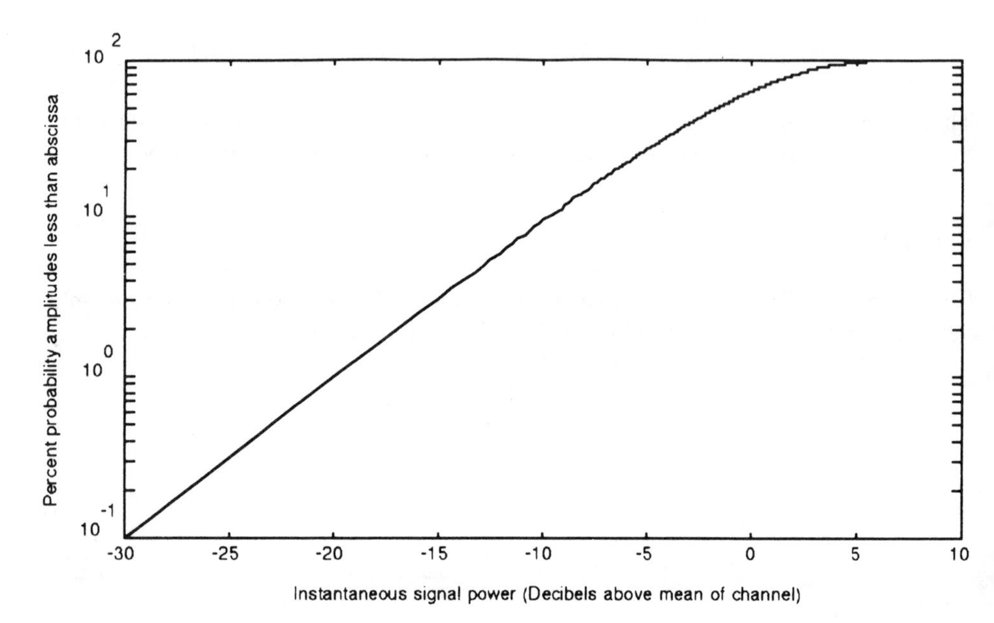

Figure 2.4 Amplitude distribution function.

Example 2.12

As we have seen, signal changes in mobile communication may be described in terms of long and short-term variations. Short-term variations are due to scatterers produce multipath and are directly related to the wavelength involved in the communications processes. On the other hand, long-term variations are associated with the environment and are a result of small changes in the topography.

Although the Rayleigh model has found wide acceptance for urban areas, the Rician model is more general and useful for non-urban environments. A Rician model corresponds to the case where there is a line-of-sight (LOS) or dominant signal component superimposed upon a signal due to scattering. In this case, the density function corresponding to an electric field with intensity E is given by

$$f_R(E) = \frac{E}{\sigma^2} \exp\left(-\frac{E^2 + E_1^2}{2\sigma^2}\right) I_o\left(\frac{EE_1}{\sigma^2}\right) \tag{2.63}$$

where I_o is the modified Bessel function of order zero, E_l is the density of the LOS incoming field, and σ is the standard deviation of the Gaussian-scattered electrical field. The E_l/σ ratio is particularly useful as it is dimensionless and it uniquely defines the Rician distribution function. We shall elaborate further on this equation in Chapter 3.

Before we leave this section, let us take up an example of a mobile interfering with television channels 55–69. Current television receivers are designed to receive in the 806–890 MHz band and have no RF preselection filter; and thus, a mobile using the service can cause interference to television. Our objective is to see the effects of parameters such as traffic density, mobile density, and the distance of the mobile from a television set, on interference.

Assume the interfering area A around a television receiver has a radius of r, as shown in Figure 2.5, then the number mobile located in the radius dr is

$$dr = \overline{v}dt \tag{2.64a}$$

where, $\overline{v}$ is equal to the average mobile speed in miles per hour. Thus, the average number of mobiles, $\overline{m}$, located inside the ring dr is given by

$$\overline{m} = 2\pi r dr D \qquad \overline{m} = 2\pi r D \overline{v} dt \tag{2.64b}$$

where D = average mobile density in mobile per mile^2. Thus, the average number

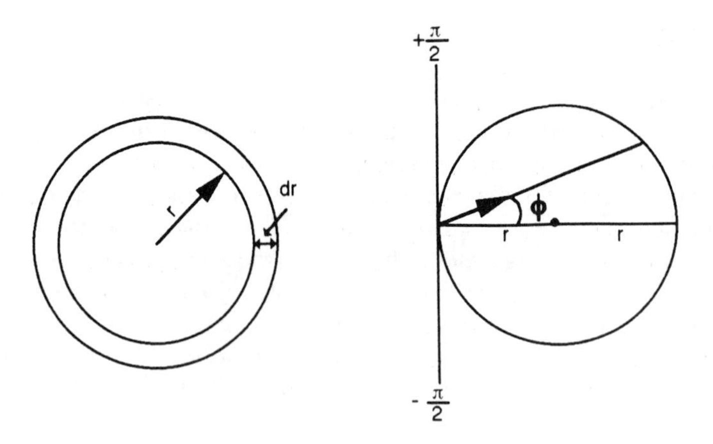

Figure 2.5 Television interference models.

of mobiles, N, crossing area A during the time period T and contributing to the interference is

$$N = 1/2 \int_{t}^{t+T} 2\pi r\bar{v}D dt = \pi r\bar{v}DT \tag{2.64c}$$

The dwell time per mobile for the ith path, t_i, is given by

$$t_i = \frac{2r \cos \phi_i}{\bar{v}} \tag{2.64d}$$

The average dwell time per mobile is given by

$$\bar{t} = \frac{1}{\pi} \int_{-\pi/2}^{\pi/2} \frac{2r \cos \phi}{\bar{v}} d\phi = \frac{4r}{\bar{v}\pi} \tag{2.64e}$$

where ϕ is uniformly distributed between $-\pi/2$ to $+\pi/2$. The average dwell time for all the mobile units during T hours is

$$\bar{t}_T = N\bar{t} = (\pi r\bar{v}DT)(4r/\pi\bar{v}) = 4r^2DT \tag{2.64f}$$

If a mobile generates e erlangs of traffic, then the mobile actually transmits for the time $\bar{t}_h = e\bar{t}_T$. In other words, the duration of interference within the peak one-hour duration ($T = 1$ hr) is

$$\bar{t}_h = 4r^2De \tag{2.64g}$$

However, the mobile channel may not occupy the complete 6-MHz television bandwidth. Thus, if the mobile only occupies the frequency band of K MHz, then the probability of television interference is multiplied by

$$p = 6/K \tag{2.64h}$$

Therefore, the duration of the degradation is given by the product of the fraction of the busy hour when the mobile channel is occupied by all mobiles within area A, multiplied by the probability that the occupied channel lies within the television channel. The duration of degradation, I, of the television signal due to interference from the mobile transmitter is

$$I = 4r^2Dep = 24\,(r^2De)/K \tag{2.64i}$$

2.2.7 Level Crossing Rate and Average Duration of Fades

To evaluate the performance of a receiver, it is essential to know the rate at which the signal level falls below a specified level and how long it remains below that level. The rate at which the stochastic process $x(t)$ decreases below a specified level x is given by Rice as [4]:

$$N(x) = \int_0^\infty \dot{x} f(x, \dot{x}) d\dot{x} \tag{2.65}$$

where $f(x, \dot{x})$ is the joint probability density of x and $\dot{x}$. The physical interpretation of the level crossing rate is shown in Figure 2.6.

From Figure 2.6, $\dot{x} = dx/\tau$, where τ is the time required for a change of level, dx, and $\dot{x}$ is the slope of the variable x. The expected number of signal level crossings for x within $(x, x - dx)$ with a given slope $\dot{x}$, and within the time duration dt is the expected amount of time spent in the dx interval for a specified slope $\dot{x}$ during the interval dt. The time required to cross the interval dx once with $\dot{x}$ slope is:

$$\frac{E(t)}{\tau} = \frac{f(x, \dot{x})dx\, d\dot{x}\, dt}{dx/\dot{x}} = \dot{x} f(x, \dot{x})\, d\dot{x}\, dt$$

Thus, the expected number of crossings during the interval T is

$$\int_0^T \dot{x} f(x, \dot{x})\, d\dot{x}\, dt = T\dot{x} f(x, \dot{x})\, d\dot{x} \tag{2.66}$$

Therefore, the expected number of upward crossings is given by

$$N(X = x) = T \int_0^\infty \dot{x} f(x, \dot{x}) d\dot{x} \tag{2.67}$$

And the expected number of level crossings per second is

$$N(X = x) = \int_0^\infty \dot{x} f(x, \dot{x}) d\dot{x} \tag{2.68}$$

The average duration of fade is found by dividing the cumulative probability distribution by the level crossing rate, or

$$\bar{t}(X = x) = \frac{P(x \leq X)}{N(X = x)} \tag{2.69}$$

Equation (2.69) establishes the relationship between the average duration of fade,

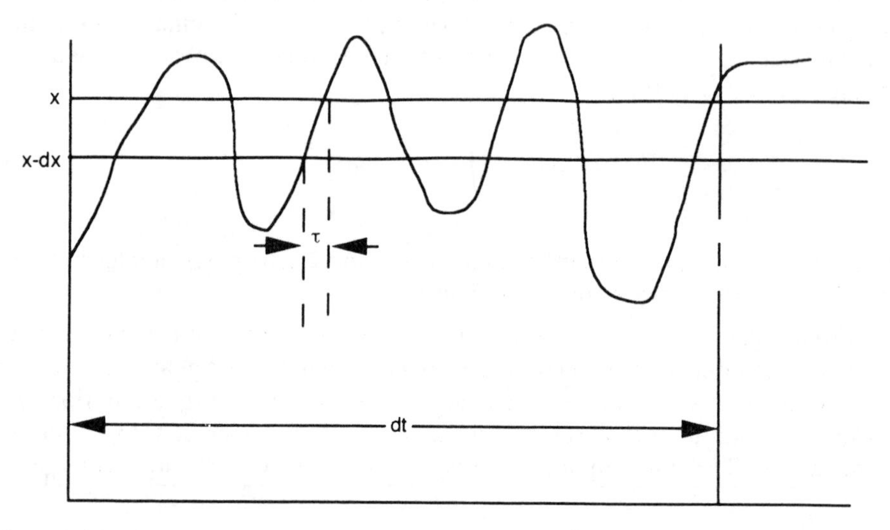

Figure 2.6 Interpretation of level crossing rate.

the cumulative distribution function, and the level crossing rate. Once any two of these parameters are known the third can be found.

2.2.8 Random Process Description

In communication engineering, many random phenomena are a function of time. For example, thermal noise generated inside a receiver is a function of time. Similarly, the signal transmitted from a cellular radio is characterized as a random signal that varies with time. From a system point of view, one is always interested in determining the statistics of the random process. One such statistical process is the multipath effect in the UHF band, for which channel simulators can be built for laboratory use. We will discuss this further in Chapter 3. To determine the statistics of channel noise, one can make repeated measurements of the noise voltage at the output of a single simulator or one can choose to make simultaneous measurements on a large number of channel simulators. If we proceed to make multiple simulators, we call the collection an ensemble. Statistical parameters can be derived by making simultaneous measurements on each of these simulators. Alternately, one may choose a single channel simulator and derive the statistics of the channel by making measurements at different times. In general, it is not likely that the generated statistics by these two approaches will match, but under certain imposed constraints (to be explained later) the statistics may be the same. With this as background, let us now consider a random experiment E with outcome λ and a sample space S. To every outcome $\lambda \in S$ we assign, according to a certain rule, a real-valued function of time $X(t, \lambda)$. We have then created a family of functions, one function for each outcome λ. This family is called a random process.

Having defined the stochastic process $X(t)$ as an ensemble of sample functions, we now consider the values of the process at any set of time instants $t_1 > t_2 > t_2 > \cdots > t_n$, where n is a positive integer. In general, the random variables $X_{ti} = X(t_i)$, $i = 1, 2, \ldots, n$, are characterized statistically by their joint probability density function $p(X_{t1}, X_{t2}, \ldots, X_{tn})$. The joint distribution function can be described by

$$P\{X(t_1) \leq a_1, X(t_2) \leq a_2, \ldots, X(t_n) \leq a_n\} \tag{2.70}$$

2.2.8.1 Statistical Averages

Two statistical averages that are most often used in the description of a random process are the mean, $\mu_X(t)$, and the autocorrelation function, $R_{XX}(t_1, t_2)$. They are defined as $\mu_X(t) = E[X(t)]$ and, for the ensemble average,

$$R_{XX}(t_1, t_2) = E[X(t_1)X(t_2)] \tag{2.71}$$

where the expected values are taken with respect to the appropriate PDFs.

2.2.9 Power Spectral Density

In communication, the basic characteristic one looks for is the frequency content of a signal. In general, a signal can either have a finite or an infinite energy. A finite energy signal can be analyzed for its frequency content from the Fourier transform. Periodic signals have infinite energy and their frequency components can be found by the Fourier series.

A stationary stochastic process is an infinite energy signal and, hence, its Fourier transform does not provide its spectral content. On the other hand, the spectral characteristic of a stochastic signal is obtained by computing the Fourier transform of the associated autocorrelation function. Thus, if

$$\int_{-\infty}^{\infty} |R_{XX}(\tau)| d\tau < \infty \tag{2.72a}$$

then

$$G_X(f) = \int_{-\infty}^{\infty} R_{XX}(\tau) e^{-j2\pi f\tau} d\tau \tag{2.72b}$$

where $G_X(f)$ is the power spectral density of $X(t)$. The inverse Fourier transform relationship is

$$R_{XX}(\tau) = \int_{-\infty}^{\infty} G_X(f) e^{j2\pi f\tau} df \tag{2.72c}$$

We observe that

$$R_{XX}(0) = \int_{-\infty}^{\infty} G_X(f) df = E(X_i^2) \geq 0 \tag{2.72d}$$

Since $R_{XX}(0)$ represents the average power of the stochastic signal, which is the area under $G_X(f)$, then $G_X(f)$ is the distribution of power as a function of frequency and is known as power spectral density.

Example 2.13

The measured value of autocorrelation function of the lognormally distributed local mean signal is reported to be [4].

$$C_{XX} = \Delta \sum_{k=-(L-1)}^{k=(L-1)} W(k)C_{XX}(k)e^{-j2\pi f\Delta k} \qquad 1/2\Delta \leq f < 1/2\Delta \qquad (2.73a)$$

Here, $W(k)$ is the weighing or window function with truncation point M and $L = M/\Delta$ The autocorrelation $C_{XX}(k)$ is given by

$$C_{XX}(k) = \frac{1}{N}\sum_{t=1}^{N-k}(X_t - \overline{X})(X_{t+k} - \overline{X}) \qquad (N - 1) \leq k \leq (N - 1) \quad (2.73b)$$

2.2.10 Linear Time-Invariant Systems

Consider a linear time-invariant system that is characterized by its impulse response $h(t)$ or, equivalently, by its frequency response $H(f)$, where $h(t)$ and $H(f)$ are Fourier transforms of each other. When the input to the system is a stationary random process $X(t)$, the output is also a random process and is arrived at by convolving the input with the impulse response of the system $h(t)$, or

$$Y(t) = \int_{-\infty}^{\infty} X(t - \alpha)h(\alpha)d\alpha = \int_{-\infty}^{\infty} X(\alpha)h(t - \alpha)d\alpha \qquad (2.74)$$

Since convolution is a linear operation preformed on the input signal $x(t)$, the expected value of the integral is equal to the integral of the expected value. In other words,

$$\begin{aligned} E[Y(t)] &= E\int_{-\infty}^{\infty} x(\alpha)h(t - \alpha)d\alpha \\ &= \int_{-\infty}^{\infty} E[(X(\alpha)]h(t - \alpha)d\alpha \qquad (2.75) \\ &= \mu_x \int_{-\infty}^{\infty} h(t - \alpha)d\alpha = \mu_X H(0) \end{aligned}$$

Thus, the expected value of $Y(t)$ does not depend on t. The autocorrelation $Y(t)$ and its Fourier transform is given by

$$R_{YY}(\tau) = R_{XX}(\tau) * h(\tau) * h(-\tau) \qquad (2.76)$$

$$G_Y(f) = G_X(f)|H(f)|^2 \tag{2.77}$$

where * is the convolution. Therefore, we have an important result that the power density spectrum of the output signal is the product of the power density spectrum of the input multiplied by the magnitude squared of the frequency response of the system.

2.2.11 Quadrature Representation of Narrowband Noise

It is often convenient to express narrowband noise in a quadrature form, such as

$$Y(t) = Y_C(t) \cos 2\pi f_c t - Y_s(t) \sin 2\pi f_s t \tag{2.78}$$

This equation can also be expressed in terms of envelope $R(t)$ and phase angle $\theta(t)$ as

$$Y(t) = R(t) \cos[2\pi f_c t + \theta(t)] \tag{2.79}$$

Here, the envelope $R(t)$ and the phase angle $\theta(t)$ are given by

$$R(t) = \sqrt{[Y_c(t)]^2 + [Y_S(t)]^2} \tag{2.80}$$

and

$$\theta(t) = \tan^{-1}\left(\frac{Y_s(t)}{Y_C(t)}\right) \tag{2.81}$$

The PDFs of $R(t)$ and θ are given by $f_R(r) = (r/N_o) \exp - (r^2/2N_o)$, where

$$N_o = E[Y_C^2(t)] = E[Y_S^2(t)] \tag{2.82}$$

$$f_\theta(\theta) = \frac{1}{2\pi} \qquad -\pi < \theta < \pi \tag{2.83}$$

The PDF of R is called the Rayleigh PDF. Before we leave this section, let us illustrate this theory by taking up an example of a cellular radio.

Example 2.14

FM-modulated waves in analog cellular systems can be represented by $Y(t) = \cos[2\pi f_c t + x(t)]$, where

$$x(t) = 2\pi\Delta f \int f(\lambda)d\lambda \tag{2.84a}$$

The signal passing through a multipath channel gets multiplied by a complex coefficient $[a(t), b(t)]$. Thus, the received waveform is given by $R1(\{e^{jx(t)} \cdot [a(t) - jb(t)]\}e^{j2\pi f_c t})$ or

$$\begin{aligned} y(t) &= a(t)\cos(2\pi f_c t + x(t)) + b(t)\sin(2\pi f_c t + x(t)) \\ &= R(t)\cos(2\pi f_c t + \theta(t)) \end{aligned} \tag{2.84b}$$

where

$$\begin{aligned} R(t) &= \sqrt{a(t)^2 + b(t)^2} \\ \theta(t) &= \tan^{-1} \frac{a(t)\sin x(t) - b(t)\cos x(t)}{a(t)\cos x(t) + b(t)\sin x(t)} \end{aligned} \tag{2.84c}$$

The receiver will further add thermal noise in this signal which, as discussed above, which can be represented as

$$\begin{aligned} n(t) &= c(t)\cos 2\pi f_c t + d(t)\sin 2\pi f_c t \\ &= N(t)\cos [2\pi f_c t + \phi(t)] \end{aligned} \tag{2.84d}$$

where

$$\begin{aligned} N(t) &= \sqrt{c(t)^2 + d(t)^2} \\ \phi(t) &= -\tan^{-1} \frac{d(t)}{c(t)} \end{aligned} \tag{2.84e}$$

Thus, the resultant waveform at the discriminator input is

$$Y(t) = R(t)\cos[2\pi f_c t + \theta(t)] + N(t)\cos[2\pi f_c t + \phi(t)] \tag{2.84f}$$

Alternately, this can be written as

$$y(t) = \cos[2\pi f_c t + \theta(t)] + [N(t)/R(t)] \cos[2\pi f_c t + \phi(t)] \tag{2.84g}$$

In this equation, the first term provides the effect of multipath fading while the second term gives the effect of thermal noise. In cellular radio, multipath plays a larger role (i.e., $R(t) \gg N(t)$). Thus, the predominant distortion term is due to $\theta(t)$. We shall elaborate further on this point in Chapter 3.

2.3 CONCLUSIONS

In this chapter we have discussed aspects of probability theory useful in cellular radio. This knowledge will be used in subsequent chapters. In the next chapter we shall characterize the cellular channel.

PROBLEMS

2.1 (a) If

$$h(\tau) = 1/2\pi \int_{-\beta}^{\beta} H(w)e^{jw\tau}dw$$

Show that

$$E[h^2(\tau)] = 1/(2\pi)^2 \int_{-\beta}^{\beta} \int_{-\beta}^{\beta} E[H(w)H(w')]e^{j(w+w')\tau}dwdw'$$

(b) If

$$H(w) = \sum_{i=1}^{N} A_i(w)e^{-jw\tau_i}$$

Show that

$$\begin{aligned} E[H(w)H(w')] &= E\left[\sum_{i=1}^{N} A_i(w)A_i(w')e^{-jw\tau_i}e^{-jw'\tau_i}\right] \\ &= E\left[\sum_{i=1}^{N} A_i(w)A_i(w')e^{-j(w+w')\tau_i}\right] \\ &= E\left[\sum_{i=1}^{N}\sum_{l=1}^{N} A_i(w)A_i(w')e^{-j(w\tau_i+w'\tau_i}\right] \end{aligned}$$

where $l \neq i$.

2.2 If $n(t)$ is a zero-mean white Gaussian noise with two-sided power spectral density $\zeta/2$ and $n_o(T_b)$ is related to $n(t)$ by

$$n_0(T_b) = \int_0^{T_b} n(t)s(t)dt$$

when $s(t) = 0$ for t outside the interval $(0, T_b)$, and

$$\int_0^{T_b} s^2(t)dt = E$$

Show that

$$E[n_o(T_b)] = 0$$

and that

$$E[n_o^2(T_b)] = \zeta E_s/2.$$

2.3 The joint density function of four random variables—r, $\dot{r}$, θ, and $\dot{\theta}$—of the Gaussian process is given by

$$p(r, \dot{r}, \theta\, \dot{\theta}) = \frac{r^2}{4\pi^2 b_0 b_2} e^{-1/2(r^2/b_0 + \dot{r}^2/b_2 + r^2\dot{\theta}^2/b_2)}$$

where $0 \leq \theta \leq 2\pi$ and $-\infty \leq \theta \leq \infty$. Show that

$$p(r, \dot{r}) = \dot{r}/b_0 \exp(-r^2/2b_0) \frac{1}{(2\pi b_2)^{1/2}} e^{-\dot{r}^2/2b_2}$$

What conclusions can you draw by observing $p(r, \dot{r})$?

2.4 The signal distribution at a distance R from a cell site can be regarded as log normally distributed. Assuming the mean signal strength to be -100 dBm and σ to be 6.5 and 8.5 for rural and suburban areas, compute the signal strength required for 90% coverage for both the areas. What conclusions can be drawn by comparing these results?

2.5 Assuming the random variable x to be normally distributed with mean μ and variance σ^2, show that the density functions of $y = e^x$ is given by

$$f_Y(y) = \frac{1}{\sqrt{2\pi\sigma^2 y}} e^{-\ln(y/m)/\sqrt{2}\sigma^2} \qquad y > 0$$
$$= 0 \qquad y < 0$$

and that its distribution function is

$$F_Y(y) = 1/2 + 1/2 \operatorname{erf}\left(\frac{\ln y/m}{\sqrt{2}\sigma}\right) \qquad y > 0$$

where $m = e^\mu$.

2.6 For the transfer function $H(w)$ of the propagation channel in Figure 2.7, show that the channel impulse response $h(\tau)$, is given by $\beta(\sin \beta\tau/2\pi\beta\tau)$.

2.7 Log-normal distribution is a good approximation of automobile-radiated noise power density measured at some distance from a single automobile. Let the measured density be represented as

$$p(y) = \frac{1}{\sqrt{2\pi}\sigma} e^{-(y-\mu)^2/2\sigma^2} \qquad -\infty \leq y \leq \infty$$

where $y = 10 \log x$, with a mean of μ (dB) and a variance of σ^2(dB). Find the density of x. Evaluate its mean and the variance.

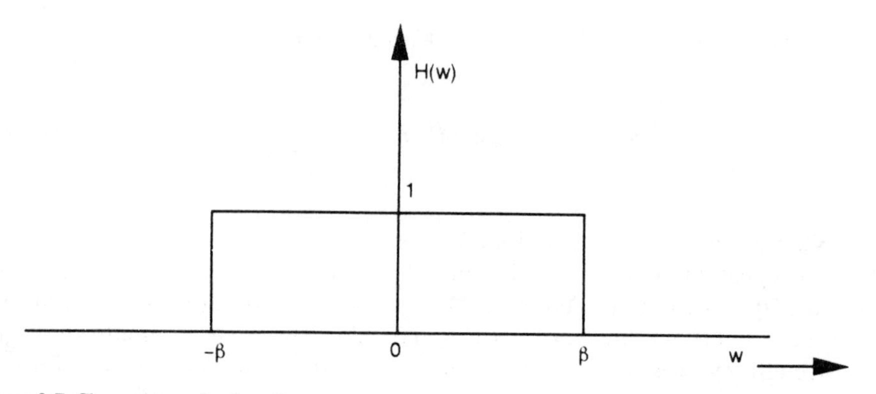

Figure 2.7 Channel transfer function.

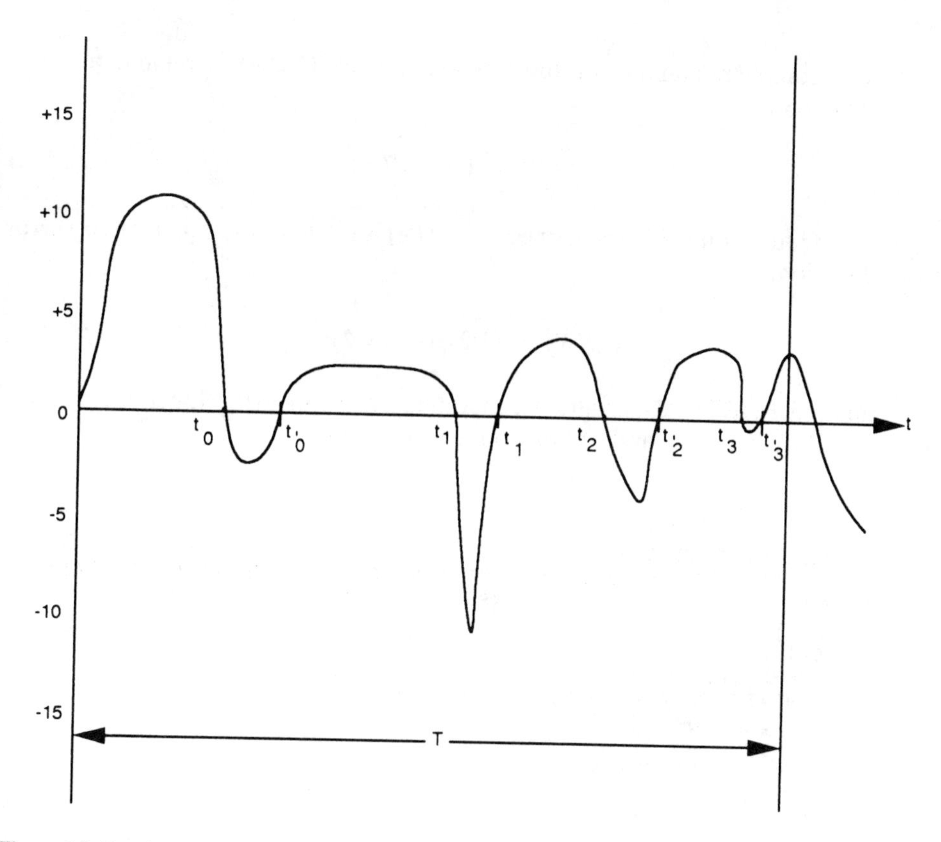

Figure 2.8 Received signal level at a cellular radio.

2.8 For Figure 2.8, find the average duration of fade and the level crossing rate over a period of T seconds.

APPENDIX 2A

The error function is denoted by

$$\operatorname{erf}(X) = \frac{2}{\sqrt{(\pi)}} \int_0^X e^{-y^2} dy \tag{2A.1}$$

The following properties of error function should be noted.

1. $\operatorname{erf}(-X) = -erf(X)$ (i.e., the function is odd symmetric).
2. As X approaches infinity the integral (2A.1) approaches to a value of 1.

3. The complementary error function defined by (2.60e) is related to error function as

$$\mathrm{erf}_c(X) = 1 - \mathrm{erf}(X) \tag{2A.2}$$

4. The Q function, which is defined by (2.60c) is related to complementary error function:

$$Q(X) = (1/2)\mathrm{erf}_c(x/\sqrt{2}) \tag{2A.3}$$

A simplified FORTRAN program generating values of error function, complementary error function, and Q function is given below.

```
      REAL*8 X,ERFC,Q,E,COMP,Y
      OPEN(1, FILE='ERRF.DAT',STATUS='OLD',ACCESS='SEQUENTIAL')
      WRITE(1,*) 'X            ERF,        ERFC,        Q,       '
      DO 10 I = 1,80
      X=0.05*I
      Y=ERFC(X)
      COMP=ERFC(X/(2**0.5))
      Q=1./2.*COMP
      E=1-Y
      WRITE(1,121)X,E,Y,Q
10    CONTINUE
121   FORMAT(4F10.6)
      END
      FUNCTION ERFC(X)
      REAL*8 P,A1,A2,A3,A4,A5,T,X
      P=0.3275911
      A1=0.254829592
      A2=-0.284496736
      A3=1.421413741
      A4=-1.453152027
      A5=1.061405429
      T=1./(1.+P*X)
      ERFC =(EXP(-X**2))*T*(A1+T*(A2+T*(A3+T*(A4+T*A5))))
      RETURN
      END
```

Table 2A.1 on the next page provides the values of the error, complementary error, and Q functions.

Table 2A.1
Error Functions, Complementary Error Function, and Function Values

X	$Erf(x)$	Erf_c	$Q(X)$	X	$Erf(x)$	Erf_c	$Q(X)$
0.050000	0.056372	0.943628	0.480061	2.050000	0.996258	0.003742	0.020182
0.100000	0.112463	0.887537	0.460172	2.100000	0.997020	0.002980	0.017864
0.150000	0.167996	0.832004	0.440382	2.150000	0.997638	0.002362	0.015778
0.200000	0.222702	0.777298	0.420740	2.200000	0.998137	0.001863	0.013903
0.250000	0.276326	0.723674	0.401294	2.250000	0.998537	0.001463	0.012224
0.300000	0.328627	0.671373	0.382089	2.300000	0.998857	0.001143	0.010724
0.350000	0.379382	0.620618	0.363169	2.350000	0.999111	0.000889	0.009387
0.400000	0.428392	0.571608	0.344578	2.400000	0.999311	0.000689	0.008198
0.450000	0.475482	0.524518	0.326355	2.450000	0.999469	0.000531	0.007143
0.500000	0.520500	0.479500	0.308538	2.500000	0.999593	0.000407	0.006210
0.550000	0.563323	0.436677	0.291160	2.550000	0.999689	0.000311	0.005386
0.600000	0.603856	0.396144	0.274253	2.600000	0.999764	0.000236	0.004661
0.650000	0.642029	0.357971	0.257846	2.650000	0.999821	0.000179	0.004025
0.700000	0.677801	0.322199	0.241964	2.700000	0.999866	0.000134	0.003467
0.750000	0.711156	0.288844	0.226627	2.750000	0.999899	0.000101	0.002980
0.800000	0.742101	0.257899	0.211855	2.800000	0.999925	0.000075	0.002555
0.850000	0.770668	0.229332	0.197663	2.850000	0.999944	0.000056	0.002186
0.900000	0.796908	0.203092	0.184060	2.900000	0.999959	0.000041	0.001866
0.950000	0.820891	0.179109	0.171056	2.950000	0.999970	0.000030	0.001589
1.000000	0.842701	0.157299	0.158655	3.000000	0.999978	0.000022	0.001350
1.050000	0.862436	0.137564	0.146859	3.050000	0.999984	0.000016	0.001144
1.100000	0.880205	0.119795	0.135666	3.100000	0.999988	0.000012	0.000968
1.150000	0.896124	0.103876	0.125072	3.150000	0.999992	0.000008	0.000816
1.200000	0.910314	0.089686	0.115070	3.200000	0.999994	0.000006	0.000687
1.250000	0.922900	0.077100	0.105650	3.250000	0.999996	0.000004	0.000577
1.300000	0.934008	0.065992	0.096801	3.300000	0.999997	0.000003	0.000483
1.350000	0.943762	0.056238	0.088508	3.350000	0.999998	0.000002	0.000404
1.400000	0.952285	0.047715	0.080757	3.400000	0.999998	0.000002	0.000337
1.450000	0.959695	0.040305	0.073529	3.450000	0.999999	0.000001	0.000280
1.500000	0.966105	0.033895	0.066807	3.500000	0.999999	0.000001	0.000233
1.550000	0.971623	0.028377	0.060571	3.550000	0.999999	0.000001	0.000193
1.600000	0.976348	0.023652	0.054799	3.600000	1.000000	0.000000	0.000159
1.650000	0.980376	0.019624	0.049471	3.650000	1.000000	0.000000	0.000131
1.700000	0.983790	0.016210	0.044565	3.700000	1.000000	0.000000	0.000108
1.750000	0.986672	0.013328	0.040059	3.750000	1.000000	0.000000	0.000088
1.800000	0.989090	0.010910	0.035930	3.800000	1.000000	0.000000	0.000072
1.850000	0.991111	0.008889	0.032157	3.850000	1.000000	0.000000	0.000059
1.900000	0.992790	0.007210	0.028716	3.900000	1.000000	0.000000	0.000048
1.950000	0.994179	0.005821	0.025588	3.950000	1.000000	0.000000	0.000039
2.000000	0.995322	0.004678	0.022750	4.000000	1.000000	0.000000	0.000032

REFERENCES

[1] Douglas, Reudink 0. "Properties of Mobile Radio Propagation above 400 MHz," *IEEE Trans. on Vehicular Technology*, Vol. VT-22, Nov. 1974, pp. 142–159.
[2] Carey, Roger B. "Technical Factors Affecting the Assignment of Facilities in the Domestic Public Land Mobile Radio Service," *FCC Report No. R-6406,* June 24, 1964.
[3] Papoulis, A. *Probability Random Variables, and Stochastic Processes*, McGraw-Hill, 1984.
[4] Rice, S. O. "Mathematical Analysis of Random Noise," *Bell System Technical Journal* Vol. 22, July, 1944, pp. 282–232.

Chapter 3
Cellular Environment

3.1 INTRODUCTION

In a typical cellular radio surrounding, the communication between the cell site and mobile is not by direct path but via many paths. This is because the direct path between the transmitter and the receiver is obstructed by buildings and other objects. At an ultrahigh frequency (UHF), the mode of propagation of the electromagnetic wave from the transmitter to the receiver is largely by way of scattering, either by reflection from the flat sides of buildings or by diffraction around man-made or natural obstructions. This is shown in Figure 3.1.

In urban areas with a cluster of buildings, the signal received by the mobile at any time would consist of a large number of horizontally traveling plane waves whose amplitude, phase, and angles of arrival relative to the direction of the motion of the vehicle are random. The phase and the spatial angle of arrival of each component of the wave can be regarded as statistically independent. Since the electrical phase of the received signal is random, it can be considered as uniformly distributed between 0 to 2π. The probability density $p(\alpha)$ of the spatial angle α, which provides the probability $p(\alpha)d\alpha$ that the component wave will occur between azimuthal sector α to $\alpha + d\alpha$ cannot be completely specified, since it will be different for different environments and will vary from one region to another within one environment [1–10].

These components of the signals produce a complex standing wave whose signal strength can add or subtract depending upon their relative phase shifts. The resultant field-strength pattern whose minimum and maximum have been experimentally determined to be approximately a quarter-wavelength apart, having fades on the order of 20–30 dB over a few car lengths of travel. The existence of a multitude of propagation paths gives rise to what is commonly known as the multipath phenomenon. This fading signal is not only experienced by the mobile users, but due to reciprocity the cell site also experiences fading, though at different times. Due to the rapidity of signal fades, this phenomenon is known as short-term fading. As the vehicle moves through this fading signal, interruptions of voice

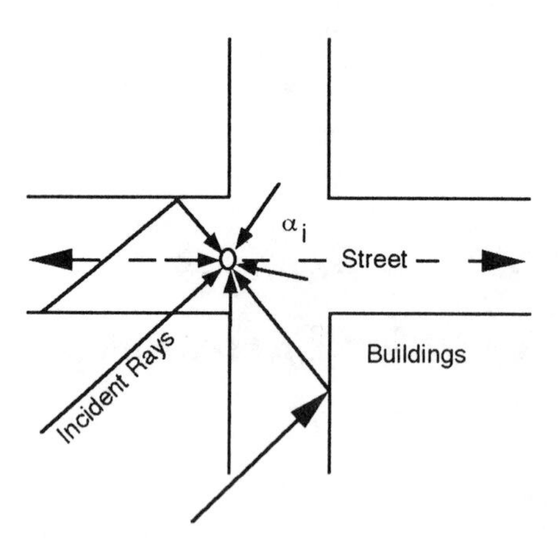

Figure 3.1 A multipath propagation model.

modulation or losses of data bits may occur when noise or interference captures the FM receiver during fades. While the use of linear modulation such as AM is conceptually possible, because of the rapidity and the depth of fades that occur at 800–900 MHz frequency range, frequency modulation has presently been adopted for cellular systems. The model described so far may be termed a "scattered field model" since the energy arrives at the receiver via a number of indirect paths. Another name for the scattered field is the "incoherent field," because its phase is random. Sometimes a significant fraction of the total received energy arrives via a direct line-of-sight path between the transmitter and the receiver. Since the phase of the directly received field is non-random, it is regarded as a "coherent wave." The presence of substantial direct energy is commonly seen in suburban and rural areas, while the signal energy is mostly of the scattered type in urban areas. In addition to short-term fading, the plane wave incident on the mobile receiver changes slowly due to different scatterers and terrain (different surroundings) as the mobile moves from one location to another. Thus, superimposed on the short-term fading are the slow variations in the mean signal strength known as long-term fading. These long-term variations, which are also known as shadowing effects, are generally handled by increasing the radiated power.

Therefore, in summary, there are two types of effects one can expect in cellular surroundings. The first is multipath due to scatters or reflections from buildings and other obstructions, most often within a few hundred feet of the moving vehicle, giving rise to a short-term fading. The second is shadowing of the direct line-of-sight path producing slow variations of the mean received signal strength, providing long-term field strength variation.

Experimental results at 900 MHz in New York City show an excellent match of the receiver envelope variation with Rayleigh distribution provided the sample area is less than about 1,000 square feet [1–3]. The fact that the measured distributions are Rayleigh in urban areas implies that there is no significant directly transmitted component of the electric wave and the field at the receiver is only of the scattered type. Figure 3.2 shows a typical envelope of the received signal at the carrier frequency of 850 MHz [1]. Fading at approximately $\lambda/2$ apart is seen in this figure. In addition to rapid variations of the received signal that follows the Rayleigh distribution, there are long-term changes in the local mean-received signal power as the vehicle moves from one area to another with a change on the order of 5 dB in the mean signal strength for less than one hundred feet of vehicle travel. The measured data, in decibels, indicates that the long-term signal power mean behaves like a log-normal distribution. The variance of this log-normal distribution lies between 6 to 10 dB, with a higher variance generally found in heavily built urban locations. The higher losses are generally caused by high-rise buildings and other obstructions.

The fading envelope directly affects the performance of the mobile receivers. The performance objective of the mobile receivers can only be met so long as the envelope of the received signal power lies above a certain level. Thus, for the evaluation of receivers used for cellular radios, it is necessary to consider the "rate" at which the received signal power envelope crosses a given level and how long it

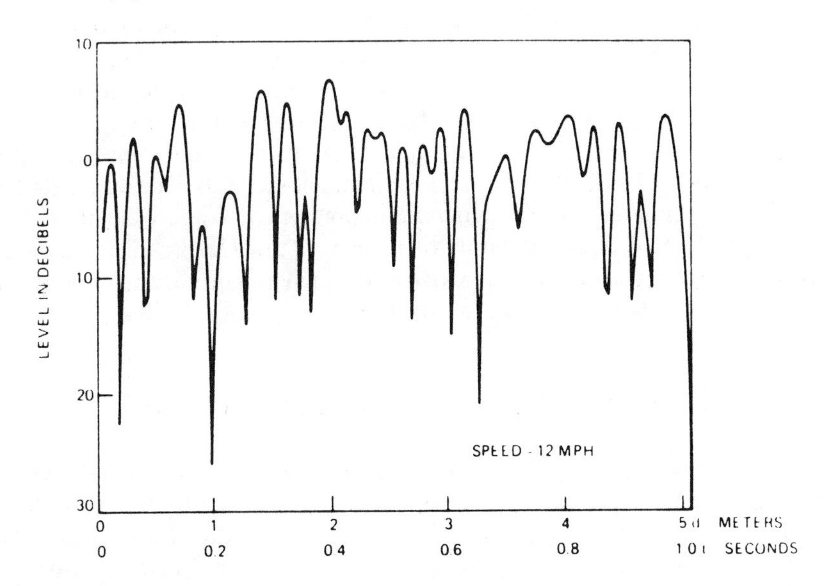

Figure 3.2 Sample of Rayleigh envelope (carrier frequency 850 MHz). Source: [1].

remains below such a level. Equations (2.68) and (2.69) introduced the concept of level crossing rate $N(R = r)$ and the average duration of fade $\bar{t}(R = r)$.

Due to multipath, signals arrive at the receiver at different times giving rise to a spread in channel delay time. Measurements have shown that the signals have a delay spread of the order of 0.5 μs in suburban areas, approximately 3.5 μs in urban areas and a low value of 0.2 μs in rural areas [1]. The resulting distortion provides limitations on the maximum data rate that can be transmitted over the channel. The delay spread related to the coherence bandwidth can be defined as the bandwidth within which a fading signal has a 0.7 or greater correlation. Typical values of the coherence bandwidths lie between 30 kHz for urban surroundings to 1 MHz for suburban areas. Since the signal bandwidth for cellular radio is 25–30 kHz, fading due to multipath delay spread will not significantly impair the signal transmission performance.

Noise caused by other vehicles including the receiver's own automobile also affect the quality of received signal unless these noises have been carefully suppressed. The noise generated by ignition systems of normal production line vehicles will cause data transmission errors at ranges of tens of meters. Complete suppression of noise is possible at the user's vehicle at a reasonable cost at the very beginning when automobile is put into service. However, the noise level tends to increase as the vehicle ages. Therefore, in general it is not possible to control the ignition noise of the automobiles. Therefore, the data transmission through cellular system must be tolerant to ignition noise.

The objective of this chapter is to model the channel, which like other communication system affects the performance of the cellular radio system. To meet this goal, we will describe the statistics of short and long-term fading in Section 3.2. In Section 3.3 we compute the received signal energy at the mobile for different antenna configurations. We will apply the statistics of multipath phenomenon for the level crossing problem in Section 3.4. Sections 3.5, and 3.6 deal with the correlation properties of different electric and magnetic fields. The associated problem of coherence bandwidth for signal combining is considered. Section 3.7 considers the problem of channel simulator design in the UHF range. Sections 3.8 brings out the discussion on other types of noise, such as ignition and environmental noise. Lastly, in Section 3.9 we draw conclusions.

3.2 STATISTICS OF SHORT- AND LONG-TERM FADING

In mobile data communication, extensive errors may be caused by variation of the received signal envelope, which reduces the signal to near the level at which the errors are caused by the front-end noise. We first consider the aspect of signal variation when measured over distances of few tenths of electrical wavelengths where the mean value of the signal remains practically constant. Assuming that the transmitted signal is vertically polarized (i.e., the electric field vector is aligned along the Z axis), the electrical and the magnetic fields described by Clarke [2] at the baseband and at every spatial point are composed of the superposition of N component waves and are given by

$$E_z = E_0 \sum_1^N \exp(j\varphi_n) \tag{3.1}$$

$$H_x = \frac{-E_0}{\eta} \sum_1^N \sin \alpha_n \exp(j\varphi_n) \tag{3.2a}$$

$$\mathrm{H}_y = \frac{E_0}{\eta} \sum_1^N \cos \alpha_n \exp(j\varphi_n) \tag{3.2b}$$

Where E_z represents the electric field vector and H_x and H_y represent the X and Y components of a magnetic field at the spatial point designated as o, as shown in Figure 3.3. The X-Y plane is assumed to be horizontal and the spatial angle to the nth component of the wave is represented as α_n. Since the electric and magnetic fields are 90° apart, the x component of the magnetic field, H_x, contains the term $\sin \alpha_n$ and the y component of the magnetic field, H_y, contains the term $\cos \alpha_n$. The negative sign in (3.2b) is because the vehicle direction is opposite to the

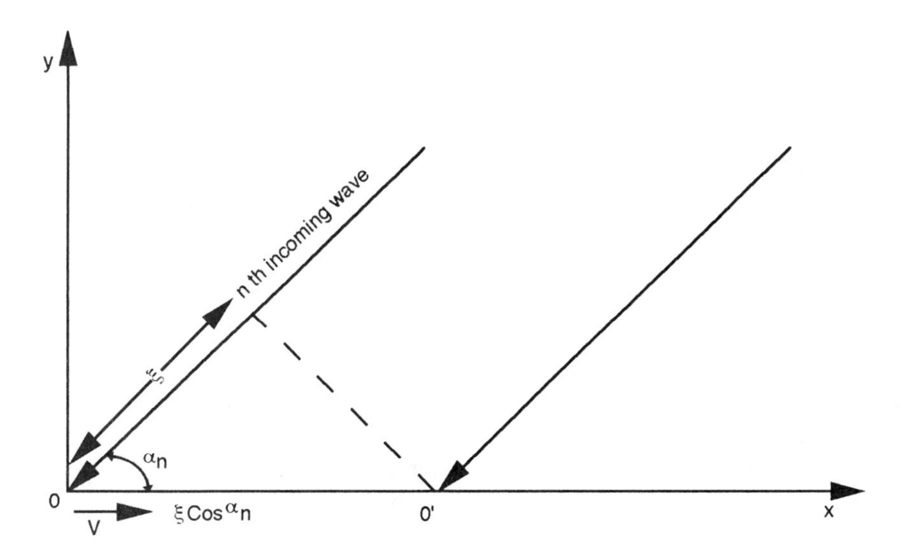

Figure 3.3 Typical component wave incident at 0 and 0′.

direction H_x. Figure 3.4 shows the front and side view of the electric and magnetic fields for a vertically polarized wave traveling over the surface of the earth. In the above equations, E_0 represents the amplitude of the received waves by the mobile in the presence of scatters, η is the intrinsic impedance of free space, and φ_n is the random phase relative to the carrier phase. The above equations are the complex equivalent in the baseband form and the time variation is understood to be of the form $\exp(jw_c t)$. The assumption that at every point exactly N component waves of amplitude E arrive is justified in view of the fact that in urban areas the scatter components are likely to experience a similar path loss.

The electric field E_z and the magnetic field components H_x and H_y are all complex. For the single-frequency excitation at the transmitter, the real and imaginary components of these fields represent the received radio frequency (RF) waves. By applying the central limit theorem and observing that α_n and ϕ_n are independent, it follows that E_z, H_x, and H_y are complex Gaussian random variables. From Figure 3.3, for a vehicle moving in the x direction with constant velocity V, the received carrier is doppler shifted by:

$$f_D = f_m \cos\alpha \tag{3.3}$$

where $f_m = V/\lambda$ = maximum value of doppler frequency f_D at $\alpha = 0$.

This Doppler frequency f_m is directly related to the phase change $\Delta\phi$ caused by the change in path length Δl between signals. We note that the doppler shift is bounded to $\pm f_m$, which in general is far smaller than the carrier frequency f_c. It can be seen that component waves arriving from ahead of the vehicle experience a positive doppler shift, while those arriving from behind the vehicle have a negative doppler shift. Thus, each component of the received signal is shifted by different values of doppler frequency. For example, a vehicle traveling at the speed of 55 miles per hour and receiving signals at a carrier of 850 MHz will introduce a maximum doppler shift of $V/\lambda = 69.3$ Hz. Since the Doppler shift in this case comes out to be approximately 70 Hz, the receiver IF bandwidth must exceed this value also.

3.2.1 Short-Term Statistics

Following example 2.9 and noting that α_n and φ_n are independent random variables, for large N both the real and the imaginary components of E_z, H_x, and H_y are Gaussian random variables. Thus, the envelope of E_z, H_x, and H_y are Rayleigh distributed. We consider below the RF version of (3.1) for the field intensity E_z:

$$E_z = E_0 \sum_{n=1}^{N} \exp[j(w_c t + \varphi_n)] \tag{3.4}$$

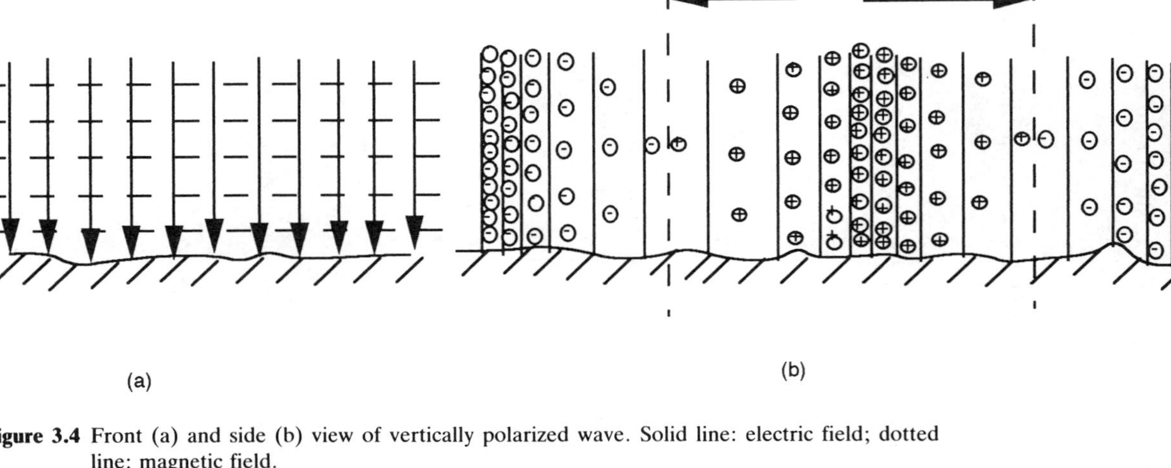

Figure 3.4 Front (a) and side (b) view of vertically polarized wave. Solid line: electric field; dotted line: magnetic field.

The real part of E_z is given by

$$\text{Re}[E_z] = E_0 \sum_{n=1}^{N} \cos \omega_c t \cos \varphi_n - E_0 \sum_{n=1}^{N} \sin \omega_n t \sin \varphi_n$$

$$= T_c \cos \omega_c t - T_s \sin \omega_c t \quad (3.5)$$

where $T_c = \text{E}_0 \Sigma_{n=1}^{N} \cos \varphi_n$ $T_s = E_0 \Sigma_{n=1}^{N} \sin \varphi_n$. Since φ_n is uniformly distributed between 0 to 2π, the mean value of T_c and T_s are:

$$E(T_c) = E(T_s) = E_0 \sum_{n=1}^{N} E(\cos \varphi_n) = E_0 \sum_{n=1}^{E} E(\sin \varphi_n) = 0 \quad (3.6)$$

The mean square value of T_c and T_s are

$$E(T_c^2) = E(T_s^2) = \frac{E_0^2 N}{2} = \sigma^2 \quad (3.7)$$

Therefore, the variance for both components are equal and are given by E_0^2 $N/2$.

Since T_c and T_s are uncorrelated, and therefore independent, we can write $E[T_c\ T_s] = 0$. Therefore, the density of T_c and T_s follows a normal distribution, and the envelope of T_c and T_s is given by:

$$r = (T_c^2 + T_s^2)^{1/2} \quad (3.8)$$

and the phase, θ, is

$$\theta = \arctan (T_s/T_c) \quad (3.9)$$

Thus, the probability density for the signal envelope r is given by

$$p(r) = \frac{r}{\sigma^2} e^{-r^2/2\sigma^2} \quad (3.10)$$

where

$$\sigma^2 = b_0 = \int_{f_c - f_m}^{f_c + f_m} s(f) df$$

which equals mean received power at the mobile $= E_0 N/2$.

The probability that the envelope r is less than equal to a specified signal level R is

$$p(r \leq R) = \int_0^R p(r)dr = 1 - e^{-R^2/2\sigma^2} \tag{3.11}$$

The first and second moments of r are given by

$$E(r) = \langle r \rangle = \int_0^\infty r\, p\,(r)dr = \int_0^\infty \frac{r^2}{\sigma^2} e^{-r^2/2\sigma^2}\, dr = \sqrt{\pi/2}\sigma = \sqrt{\frac{\pi b_0}{2}} \tag{3.12}$$

$$E(r^2) = E(T_c^2) + E(T_s^2) = 2\sigma^2 \tag{3.13}$$

Thus, the variance of r is

$$E\,(r^2) - E^2(r) = \langle r^2 \rangle - \langle r \rangle^2 = (2 - \pi/2)\,\sigma^2 = (2 - \pi/2)\,b_0 \tag{3.14}$$

Assuming the density function $p(\theta)$ to be uniformly distributed between $-\pi$ to π, we can write

$$p(\theta) = \pi/2 \qquad \pi < \theta < \pi \tag{3.15}$$

and 0 elsewhere. It is easy to see that the $E(\theta) = 0$ and $E[\theta^2] = \pi^2/3$. A summary of the different moments for random variables T_c, T_s, r, and θ is shown in Table 3.1.

In addition to the scattered wave, if a direct wave much stronger than the average reflected component reaches the mobile receiver, then Rice's analysis of a sine wave plus a narrow-band Gaussian noise applies [2]. This situation will arise when the mobile is too close to the cell site, where it is assumed that there are

Table 3.1
Moments for Rv, T_c, T_s, r, and θ

Random Variable	*Mean*	*Mean Square*	*Variance*
T_c, T_s	0	$E_0^2\, N/2 = b_0$	$E_0^2\, N/2 = b_0$
r	$\sqrt{\pi b_0/2}$	$E_0^2\, N = 2b_0$	$(2 - \pi/2)b_0$
θ	0	$\pi^2/3$	$\pi^2/3$

practically no obstructions, or when the system is operating in surroundings with fixed scatterers. In this situation, surrounding scatterers produce a random narrow-band Gaussian noise, while the sinusoid is produced by an unobstructed signal. The probability density of the amplitude then takes the form of Rice's distribution and is given by

$$p(r) = r/\sigma^2 \exp\left(\frac{r^2 + Q^2}{2\sigma^2}\right) I_0\left(\frac{rQ}{\sigma^2}\right) \tag{3.16}$$

Where the sine wave component is represented by $Q \cos 2\pi f_c t$ and the scattered wave component is r with mean power σ^2. The frequency f_c includes the Doppler component, if present. The function I_0 is the modified Bessel function of the first kind with a zero index. The general plot of $p(r)$ and its relationship to Rayleigh and Gaussian distribution is shown in Figure 3.5.

A general Rice probability density (3.16) is shown in Figure 3.5, plot i. Rice's density approaches the Gaussian density, function plot II, as $Q^2/2\sigma^2 \gg 1$. If the value of the direct component Q is zero; Rician becomes Rayleigh, as shown in curve III. On the other hand, Rayleigh becomes Rician due to the presence of direct components of the wave.

The experiment performed in New York City by R. W. Young predicts the received signal to be Rayleigh distributed at 900 MHz, provided the sample area is less than about 1,000 square feet. The fact that the measured distributions are Rayleigh signifies that there is no significant energy due to the direct transmitted

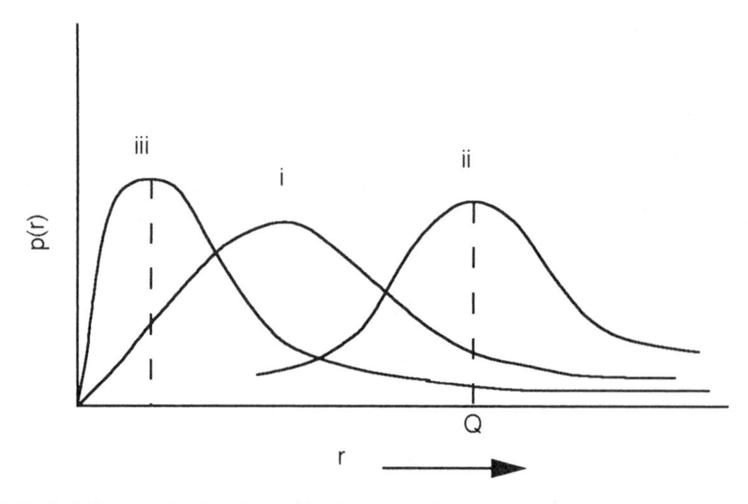

Figure 3.5 Probability density function $p(r)$ of equation (3.16).

component of the field and the fields are only of the scattered type. W. C. Jakes and D. O. Reudink have compared the statistical character of the amplitude of the fluctuating signal at 836 MHz and 11,200 MHz on the same street in a suburban environment at approximately four kilometers from the transmitter [2]. They found the signal variation to follow a Rayleigh distribution at both these frequencies. The experimental amplitude data obtained at 836 MHz is shown in Figure 3.6. The data was obtained with vertical dipole mobile antenna. The speed of the vehicle was 22 feet per second for each of the five frames, which lasted for about one second each. In the figure, the vertical frame is approximately linear in decibels, covering a 70-dB spread with about 7 dB to each vertical division. Statistics of the signal in all the frames except frame four approximates the Rayleigh distribution, while the fourth frame corresponds to the position of a street interaction with one of the intersecting streets pointing in the direction of the transmitter. Thus, a strong component of the transmitted signal is expected to be present, which as shown in frame four raises the average signal level and changes the signal level from Rayleigh to Rice or sometimes even to Gaussian. We now present some examples based on the above discussion.

Example 3.1

For short-term fading in a cellular environment, show that the 1% and 99% points are separated by 26.6 dB. Let r_1 and r_2 be the 1% and 99% points on the Rayleigh distribution curve. Then, from (3.11), we have

$$1 - e^{-r_1^2/2\sigma^2} = P(r \leq r_1) = \int_0^{r_1} p(r)dr = 0.01$$

and

$$1 - e^{-r_2^2/2\sigma^2} = P(r \leq r_2) = \int_0^{r_2} p(r)dr = 0.99$$

Therefore, $-r_1^2/2\sigma^2 = -0.01005$ and $-r_2^2/2\sigma^2 = -3.605$. Thus, $r_1/r_2 = 0.0467$ or $r_1 - r_2 = -26.6$ dB.

Theoretically, the 1% and 99% points are separated by 26 dB. This is also seen from Figure 2.4. However, if the received signal is measured at a large number of widely separated points, all equidistant from the base transmitter, the spread of value will be far greater then 26 dB. Experiments in the New York City area at 150.0 MHz show that the spread of signals is on the order of 36–52 dB between

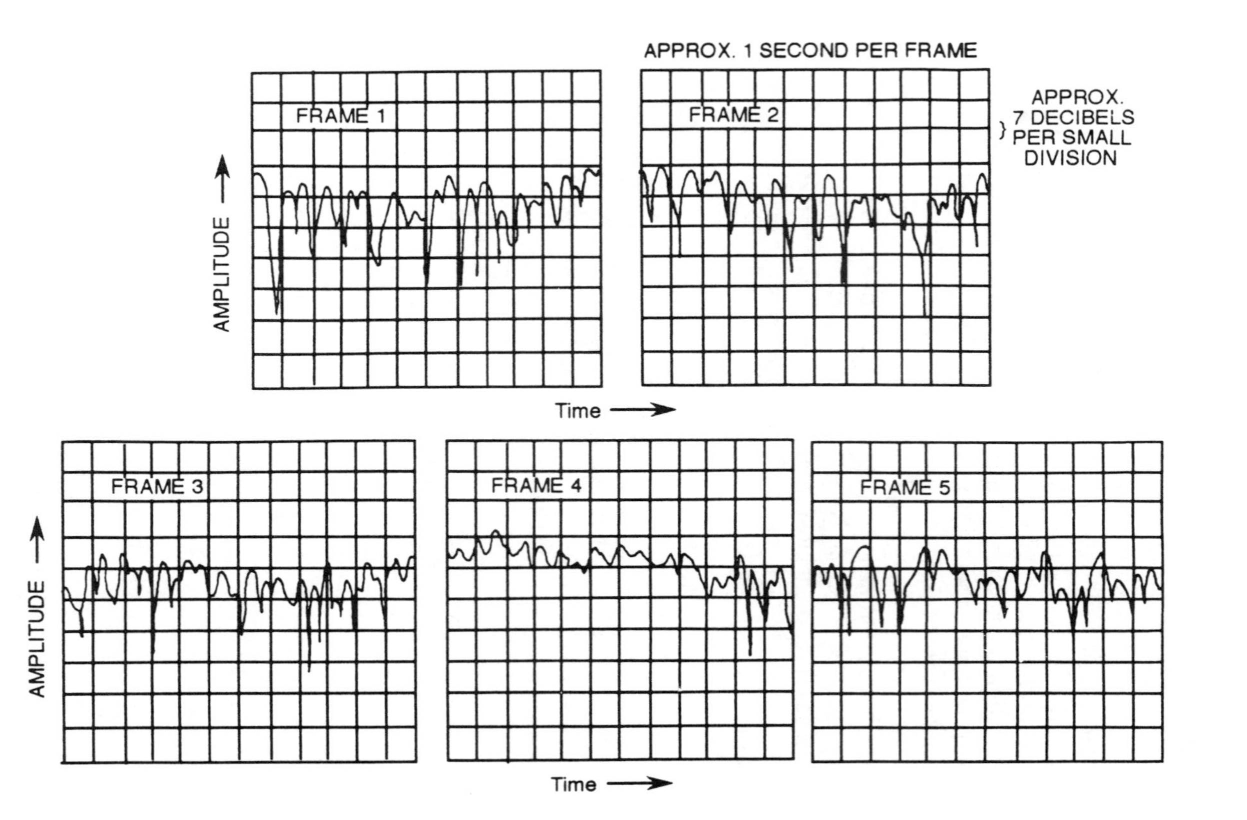

Figure 3.6 Section of a mobile data run showing the variation of signal amplitude with time (one vertical division is 7 dB, one horizontal division is 1 second long). After [2].

the 1% and 99% points. The difference between this and the theoretical Rayleigh distribution is attributed to long-term fading; in other words, due to variations of shadowing losses due to buildings, hills, and so forth [3].

We have discussed the properties of the received signal due to a multipath propagation and have shown that the probability density of the envelope follows a Rayleigh distribution. The signal envelope r has a Rayleigh distribution if measured over distances of a few tenths of a wavelength where the average signal is reasonably constant. The average distance of a few tenths of a wavelength is long compared to the distance between successive minima occurring at about half the wavelength ($\lambda/2$), but short compared to the size of the topographical features. Therefore, the probability of the received envelope r relative to the local mean $\bar{r} = E(r)$ is given by

$$p(r|\bar{r}) = \frac{\pi r}{2\bar{r}^2} \exp\left(\frac{-\pi r^2}{4\bar{r}^2}\right) \qquad r > 0 \tag{3.17}$$

This an alternate form of the Rayleigh distribution, which is obtained by substituting the value of σ in terms of r ($\sigma^2 = 2r^2/\pi$) in (3.10). See Table 3.1, column 2, row 2.

3.2.2 Combined Long-Term and Short-Term Statistics

In a typical cellular surrounding, the received signal will show fading that consists of rapid fluctuations around the mean signal level, which varies slowly as the vehicle moves from one location to another. In Section 3.2.1, we have discussed the short-term fading (rapid fading) of the received signal envelope. In this section we shall derive the probability density function of the signal envelope under Rayleigh fading and log-normal shadowing. However, before we derive this density function we will consider the statistics of the slow variation of the mean level caused by signal shadowing of the radio signal by buildings and hills. Due to the shadowing of the signal, the local mean signal level varies as the vehicle moves past buildings and obstructions. The local mean of the received envelope r varies log-normally with two parameters (m_d, σ) in decibels, where m_d is the mean of the random variable r. Consequently, the probability density of the local mean is

$$p(\bar{r}) = \frac{1}{\sqrt{2\pi\sigma^2}} e^{-\left(\frac{(\bar{r} - m_d)^2}{2\sigma^2}\right)} \tag{3.18}$$

where σ is the standard deviation in decibels of the local mean. The value of σ has been experimentally found to lie between 8–12 dB in U.S. cities and in Japan, while for London this value is somewhat lower [3]. In general, the value is higher in an open area then in urban surroundings. Within an urban area, σ tends to increase with the height of buildings.

Combining the statistics of the short and long-term fading, we can arrive at the probability of fade below a given threshold. For analytical simplicity, in most cases the mean value is assumed to be constant (i.e., the distribution is assumed to be purely Rayleigh). The system performance predicted with this assumption is generally limited to its accuracy [17–19]. Here we shall consider the probability of fade below a given threshold level by superimposing Rayleigh and lognormal fading. Thus, the probability density function of the signal envelope $p(r)$ is

$$p(r) = \int_{-\infty}^{\infty} p(r|\bar{r})\, p(\bar{r})\, d\bar{r} \tag{3.19}$$

Substituting (3.17) and (3.18), we obtain

$$p(r) = \int_{-\infty}^{\infty} \frac{\pi r}{2\bar{r}^2} e^{\frac{-\pi r^2}{4\bar{r}^2}} \frac{1}{\sqrt{2\pi\sigma^2}} e^{\frac{-(\bar{r}-m_d)^2}{2\sigma^2}} d\bar{r} \tag{3.20}$$

The closed-form solution of (3.20) is difficult to obtain, so a numerical integration is performed for its evaluation. In [18], the integration of the above equation was performed within limits of $\pm\ 2.5\sigma$. A plot of r/m_d versus $p(r)$ for different values of σ in decibels is shown in Figure 3.7(a).

For σ equal to zero, the log-normal distribution in the mean is excluded and $p(r)$ is reduced to a Rayleigh density function. With σ equal to 0.01 dB, the density $p(r)$ indeed takes the form of a Rayleigh function. The density function $p(r)$ assumes increasingly higher heights at lower values of r/m_d as the value of σ is increased. From the combined distribution curve, it is clear that as σ attains higher and higher values, the probability that a specified signal level is exceeded is reduced. This is shown in Figure 3.7(b).

3.3 SPECTRAL DENSITY OF THE RECEIVED SIGNAL

Spectral analysis of a narrowband system can be performed by transmitting a continuous wave (CW) signal. For a mobile unit, the signal received from a CW transmission as the mobile moves with a constant velocity can be represented in terms of a complex transmission coefficient. Since the transmitted signal is a single tone, the received signal requires a bandwidth of $2f_m$, where f_m is the maximum doppler shift. The transmission coefficient represents the random amplitude and

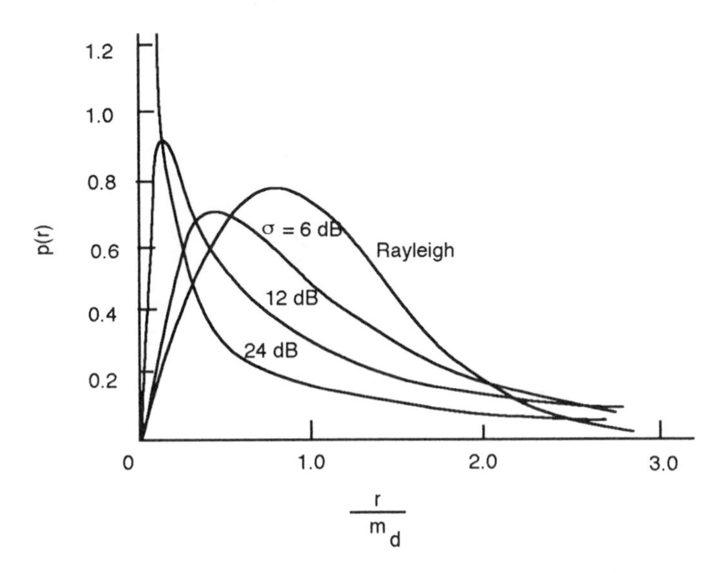

Figure 3.7(a) PDF of signal envelope with fading and shadowing. Source: [16].

phase of the received signal. The statistical properties of this random signal can be determined by finding its moments, which in turn can be easily obtained from its power spectrum. Representing the unit amplitude transmitted signal as:

$$V_T(t) = \text{Re}\,(e^{j\omega t}) = \cos \omega t \tag{3.21}$$

The received signal at the mobile unit can be expressed as

$$V_R(t) = \text{Re}\,[T(f, t)\, e^{j\omega t}] \tag{3.22}$$

where $f = \omega/2\pi$, the frequency of the transmitted signal is in hertz, t is time expressed in seconds, and $T(f, t)$ equals the complex transmission coefficient of the received signal, which represents the channel characteristics.

In cellular radio surroundings, the transmission coefficient is both a function of time and frequency; that is, the mobile-received signal is both time varying and dispersive. The rapid fading as the mobile moves through is represented by the decrease in $T(f, t)$ as the time is varied. Variation of the phase of $T(f, t)$ as a function of time is termed random FM. Variation in both amplitude and phase of $T(f, t)$ as the frequency is varied is known as frequency selective fading and phase distortion of the channel, respectively.

For a given transmitted frequency, say $f = f_c$, the signal $T(f_c, t)$ received by the mobile is the result of many plane waves, each shifted in frequency by a fixed

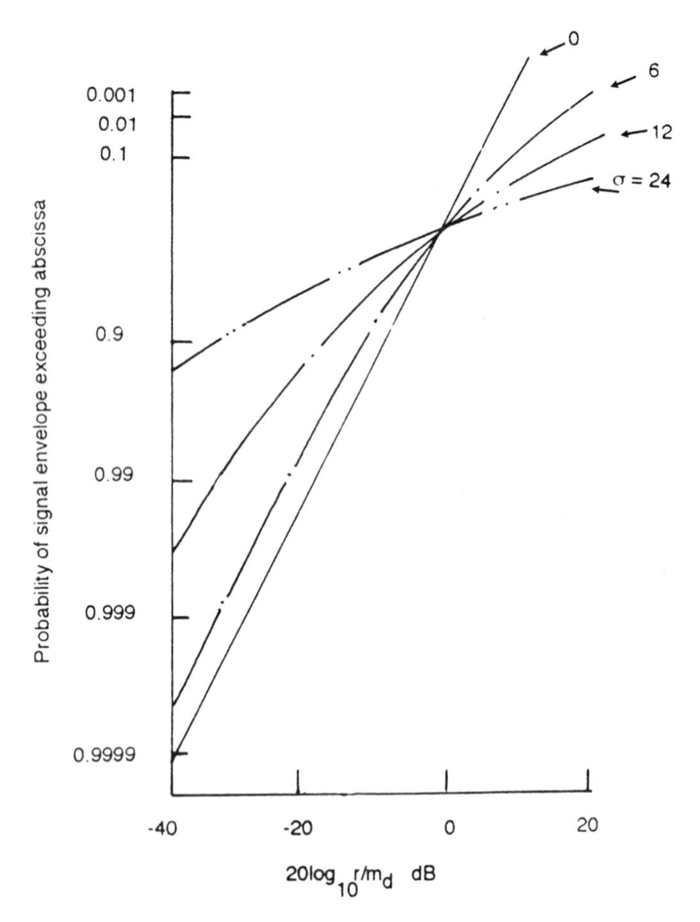

Figure 3.7(b) Distribution of signal envelope with fading and shadowing. Source: [16].

amount of Doppler that is proportional to the relative motion of the mobile with respect to the cell site. Thus, the received signal by the mobile consists of a large number of sinusoids whose amplitude and phase are random, and the frequencies are confined to a Doppler spread around the transmitted frequency f_c. The received signal $T(f_c, t)$ is known as quasi-stationary in this case with respect to f_c and t. By quasi-stationary, we mean the joint density function of $\{T_i\ (f_i,\ t_i)\}$. The elements $i = 1, 2, \ldots, N$ remain unchanged if the components of $\{f_i\}$ are all shifted by the same amount so long as all $\{f_i\}$ remain within the stationary range of Δf, and if the set of $\{t_i\}$ are translated by the same time shift so long as all $\{t_i\}$ remain within the stationary range of Δt. Frequency spread, Δf, is determined by the frequency range

over which the reflective and the diffractive properties of the terrain and buildings remain unchanged. On the other hand, Δt is determined by how long the effects of terrain, other obstructions, and vehicle velocity remain unchanged.

For the transmitted frequency f_c the received signal $T(f_c, t)$ is the result of many plane waves each shifted in frequency by a doppler shift given by (3.3). Assuming that the field is represented by the sum of N plane waves. As $N \rightarrow \infty$ approaches infinity, we can expect the incident power at the receiver antenna between angle α and $\alpha + d\alpha$ to approach a continuous, instead of discrete, distribution. Denoting by $p(\alpha)d\alpha$ the fraction of the total incoming power within $d\alpha$ of the angle α, and also assuming that the receiving antenna is directive in the horizontal plane with the antenna gain of $G(\alpha)$, the fractional received power within $d\alpha$ is given by $G(\alpha)p(\alpha)e_g(\alpha)d\alpha$, where $e_g(\alpha)$ is the mobile antenna polarization. We should note that we are talking here of the normalized power where the actual received power within the angle α is $E_0^2N/2\ G(\alpha)p(\alpha)e_g(\alpha)d\alpha$. Assuming that the polarization angle of the cell-site antenna and the mobile antenna is same, the received power within an angle $d\alpha$ is reduced to $G(\alpha)p(\alpha)d\alpha$. The complete radical spectrum of $T(f_c, t)/\sqrt{2}$ can be expressed in terms of the strength of received signal with a doppler shift. The doppler shift Δf is the frequency away from the carrier frequency f_c. For a signal arriving at an angle α relative to the vehicle motion, the received frequency at the mobile, from (3.3), is given by

$$f(\alpha) = f_m \cos \alpha + f_c \tag{3.23}$$

where $f_m = V/\lambda$ = maximum value of doppler frequency and $\alpha = \cos^{-1} [(f - f_c)/f_m]$.

Equating the differential variation of the received power with angle to that of the differential power with frequency, we obtain

$$S(f)\,|df| = [p(\alpha)\,G(\alpha) + p(-\alpha)\,G(-\alpha)\,|d\alpha| \tag{3.24}$$

Since $f(-\alpha) = f(\alpha)$, and due to the symmetrical nature of $S(f)$, we are justified in writing df and $d\alpha$ in the modulas form.

From (3.23), $df = -f_m \sin \alpha\, d\alpha$; thus, (3.24) is reduced to

$$S(f)\,f_m\,|-\sin \alpha\, d\alpha| = [p(\alpha)\,G(\alpha) + p(-\alpha)\,G(-\alpha)]\,|d\alpha| \tag{3.25}$$

In (3.25), $S(f) = 0$ if $|f - f_c| > f_m$. Substituting the value of sin α from (3.23), we obtain

$$S(f) = \frac{p(\alpha)G(\alpha) + p(-\alpha)G(-\alpha)}{[f_m^2 - (f - f_c)^2]^{1/2}} \tag{3.26}$$

This represents the power spectrum density at the mobile-receive antenna output. We now evaluate the above equation for different antenna types by substituting their respective gains.

Case I: Vertical Whip Antenna. Assuming the transmitted signal to be vertically polarized, the electric field will then be in the Z direction and may be sensed by the mobile vertical whip antenna having a gain of unity. Thus with $G(\alpha) = 1$, (3.26) is reduced to

$$S_{E_z}(f) = \frac{p(\alpha) + p(-\alpha)}{[f_m^2 - (f - f_c)^2]^{1/2}} \tag{3.27}$$

Further, assuming the probability distribution of the signal angle of arrival α to be uniformly distributed or

$$p(\alpha) = \frac{1}{2\pi} \qquad -\pi \leq \alpha \leq \pi \tag{3.28a}$$

Equation (3.27) is reduced to

$$S_{E_z}(f) = \frac{1}{\pi} \frac{1}{[f_m^2 - (f - f_c)^2]^{1/2}} = \frac{2}{\omega_m} \frac{1}{[1 - (f - f_c/f_m)^2]^{1/2}} \tag{3.28b}$$

In the time domain, the effects of the randomly phased and doppler-shifted multipath signals appear as envelope fading. The RF spectrum plot of $S(f)$ is shown in Figure 3.8(a), and the baseband output spectrum is shown in Figure 3.8(b).

Case II. With the vertical $\lambda/4$ monopole having a gain $G(\alpha) = 3/2$ in the azimuth plane, the power spectrum $S(f)$ can easily be arrived at as

$$S(f) = \frac{3}{\omega_m} \frac{1}{[1 - (f - f_c/f_m)^2]^{1/2}} \tag{3.29}$$

Here, $p(\alpha) = 1/2\pi$ as in case I.

The RF spectrum of $S(f)$ is shown in Figure 3.8(c). The baseband spectrum plot is shown in Figure 3.8(d). If there is a dominant component in the arriving signal, this will have substantial influence in the baseband spectrum. If the dominant component arrives at an angle α_o, then the baseband will have spectral peaks at $f(1 \pm \cos \alpha_o)$.

Case III: Vertical Loop Antenna. The vertical loop antenna aligned in a plane perpendicular to the motion operates on the magnetic field and forms the part of

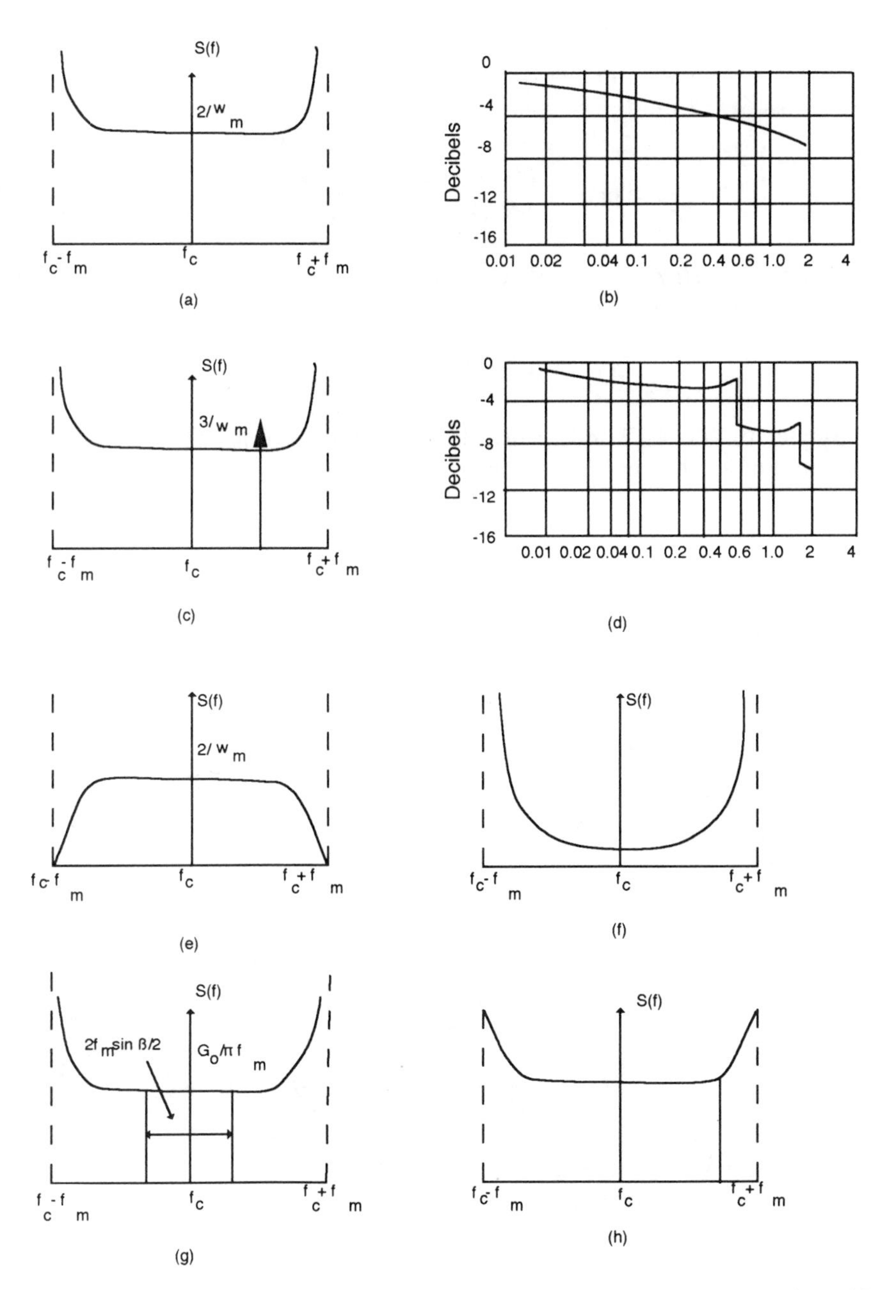

Figure 3.8 (a) Power spectra; (b) vertical whip input and baseband output; $\lambda/4$ monopole input (c) and baseband output (d) having strong spectral peak; vertical loop perpendicular (e) and parrallel (f) to the vehicle motion bean antenna directed perpendicular (g) and parrallel (h) to vehicle motion. After [4].

the Pierce "total field" antenna system. The antenna gain is given by $G(\alpha) = 3/2 \sin^2\alpha$. Thus, $S(f)$ from (3.26) is

$$S(f) = \frac{3}{2} \frac{p(\alpha)\sin^2\alpha + 1p(\alpha)\sin^2\alpha}{f_m \{1 - [(f - f_c)/f_m^2]\}^{1/2}} \tag{3.30}$$

Assuming $p(\alpha) = 1/2\pi$ and substituting the value of $\sin^2\alpha$ from (3.23), the power spectrum density $S(f)$ is

$$S(f) = 3/\omega_m [1 - (f - f_c/f_m)^2]^{1/2} \tag{3.31}$$

Similar to above, if the vertical loop antenna is in a plane parallel to the vertical motion of the vehicle, the antenna gain is

$$G(\alpha) = 3/2 \cos^2\alpha \tag{3.32}$$

and

$$S(f) = \frac{3}{\omega_m} \frac{(f - f_c/f_m)^2}{[1 - (f - f_c/f_m)^2]^{1/2}} \tag{3.33}$$

The spectrum plots for $G(\alpha) = 3/2 \sin^2\alpha$ and $G(\alpha) = 3/2 \cos^2\alpha$ are shown in Figure 3.8(e) and 3.8(f)

Case IV: Beam Antenna For beam antennas directed perpendicular to the motion of the vehicle

$$G(\alpha) = G_0 \tag{3.34}$$

if $|\alpha| < \beta/2$ and equal to zero elsewhere. Thus,

$$S(f) = \frac{G_o}{\pi f_m \sqrt{1 - (f - f_c/f_m)^2}} \tag{3.35}$$

if $|f - f_c/f_m| \geq \sin \beta/2$ and equal to zero elsewhere.

For beam antennas directed along the vehicle motion, the spectrum density $S(f)$ is

$$S(f) = \frac{2G_0}{f_m \sqrt{1 - \left(\frac{f - f_c}{f_m}\right)^2}} \tag{3.36}$$

if $1 > (f - f_c/f_m) \geq 1 - \cos \beta/2$ and equal to zero elsewhere.

The RF spectrum plots for (3.35) and (3.36) are shown in Figure 3.8 (g,h).

As discussed at the beginning of this section, the second important phenomenon connected with $T(f, t)$ is the variation of its phase with time, which is known as random FM. Similar to frequency variation due to Doppler, random FM is also due to the motion of the vehicle. Both the electric and the magnetic fields of the received wave at the mobile exhibit the random FM. Random behavior of the electric field phase, E_Z, is given by the density function $p(\dot{\theta})$ as

$$p(\dot{\theta}) = \frac{1}{\sqrt{2}\omega_n} [1 + 2 (\dot{\theta}/\omega_n)^2]^{3/2} \tag{3.37}$$

The power spectrum of the random FM can be derived by the autocorrelation function of θ, which is rather involved and will not be attempted here. However, the nature of the power spectrum shows a monotonic decrease of spectrum density as the frequency is increased. The plot is shown in Figure 3.9. Inverse proportionality is seen as f increases. Unlike spectrum plots for Doppler at the different receive antennas, the plot of random FM is smooth and has finite probability between any two frequencies. In cellular system the impairment due to random FM can be minimized with speech processing.

3.4 LEVEL CROSSING

As stated in Section 3.2.1, a fade of 20 dB or more occurs less frequently than a fade of 10 dB or more. In general, the shallower the fade, the more often it occurs. The quantitative expression that deals with the phenomena of fading is the level crossing rate N_R, derived in Section 2.2.7, It is the expected rate at which the envelope crosses the specified signal level $r = R$ in the positive direction; that is, N_R is given by the expression

$$N_R = \int_0^\infty \dot{r}\, p(R, \dot{r})\, d\dot{r} \tag{3.38}$$

where $\dot{r}$ is the slope with which the level $r = R$ is crossed. Rice has provided the joint probability function for four random variables r, $\dot{r}$, θ, and $\dot{\theta}$ of a Gaussian process as

$$p(r, \dot{r}, \theta, \dot{\theta}) = \frac{r^2}{4\pi^2 b_0 b_2} e^{-1/2\,(r^2/b_0 + \dot{r}^2/b_2 + r^2\dot{\theta}^2/b_2)} \tag{3.39a}$$

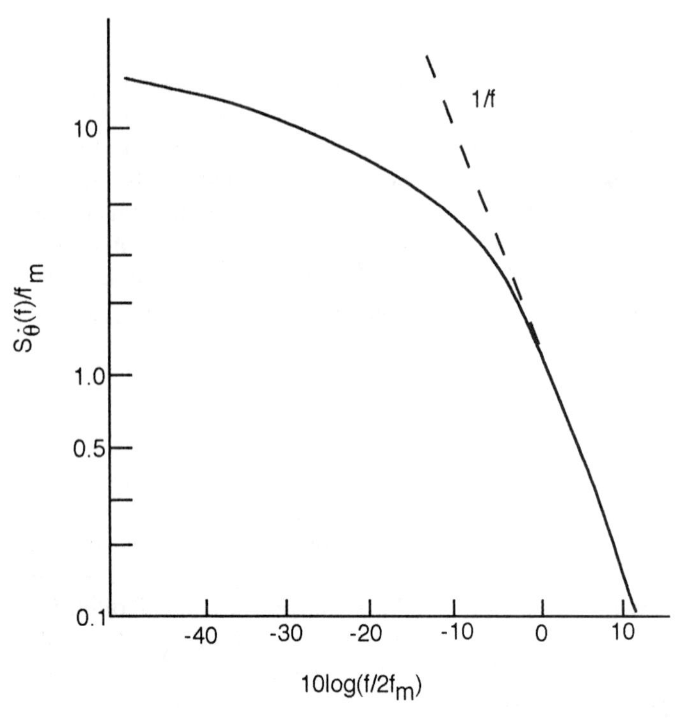

Figure 3.9 Power spectra of random FM.

where $\theta = \tan^{-1} T_s/T_c$ and

$$b_n = (2\pi)^n \int_{f_c - f_m}^{f_c + f_m} S(f)\,(f - f_c)^n\,df \tag{3.39b}$$

To find $p(r, \dot{r})$, required in (4.38) for evaluating the level crossing rate N_R, we integrate the above expression $p(r, \dot{r}, \theta, \dot{\theta})$ for θ between 0 to 2π and $\dot{\theta}$ between $-\infty$ to $+\infty$, or

$$\begin{aligned} p\,(r, \dot{r}) &= \int_{-\infty}^{\infty} \int_{0}^{2\pi} p(r, \dot{r}, \theta, \dot{\theta})\, d\theta\, d\dot{\theta} \\ &= \frac{r^2}{2\pi b_0 b_2} e^{-1/2(r^2/b_0 + \dot{r}^2/b_2)} \int_{-\infty}^{\infty} e^{-1/2\,(r^2\dot{\theta}^2/b_2)}\, d\dot{\theta} \qquad (3.39c) \\ &= \frac{\sqrt{b_2}}{r} k \int_{-\infty}^{\infty} e^{-1/2\eta^2}\, d\eta = \frac{\sqrt{b_2}}{r} k\sqrt{2\pi} \end{aligned}$$

where

$$k = \frac{r^2}{2\pi b_0 b_2} e^{-1/2(r^2/b_0 + \dot{r}^2/b_2)} \qquad \eta = \frac{r}{\sqrt{b_2}} \dot{\theta}$$

and

$$\int_{-\infty}^{\infty} e^{-1/2\eta^2} d\eta = \sqrt{2\pi}$$

Therefore,

$$p(r, \dot{r}) = \left(\frac{r}{b_0} e^{-r^2/2b_0}\right) \left(\frac{1}{\sqrt{2\pi b_2}} e^{-\dot{r}^2/2b_2}\right) = p(r)\, p(\dot{r}) \tag{3.39d}$$

Since the joint density function of Gaussian random variables r and $\dot{r}$ is the product of two density functions, r and $\dot{r}$ are independent and also uncorrelated. This reduces down the expression for level crossing rate N_R (3.38) to

$$p(R) \int_0^{\infty} \dot{r}\, p(\dot{r})\, d\dot{r} = p(R) \int_0^{\infty} \dot{r} \frac{1}{\sqrt{2\pi b_2}} e^{-\dot{r}^2/2b_2} d\dot{r} \tag{3.40a}$$

Substituting $\eta = r/\sqrt{2b_2}$ in the above expression, we obtain the level crossing rate N_R as

$$N_R = \frac{p(R)\,(2b_2)}{\sqrt{2\pi b_2}} \int_0^{\infty} \eta e^{-\eta^2} d\eta = \frac{p(R) 2b_2}{\sqrt{2\pi b_2}} \frac{1}{2} = \left(\frac{b_2}{\pi b_0}\right)^{1/2} \frac{R}{\sqrt{2b_0}} e^{-(R/\sqrt{2b_0})^2}$$
$$= \left[\frac{b_2}{\pi b_0}\right]^{1/2} \rho e^{-\rho^2} \tag{3.40b}$$

where $\rho = R/\sqrt{2b_0} = R/\sqrt{\langle r^2 \rangle} = R/R_{\text{rms}}$. From the above expression, the average duration of fade below $r = R$ can be found as

$$P\,(r \leq R) = \frac{\sum \tau_i}{T} \tag{3.41}$$

Since the level crossing rate is N_R, the average duration of fade (from Section 2.2.7) is

$$\bar{\tau} = \frac{P(r \leq R)}{N_R} = \frac{\sum \tau_i}{TN_R} \tag{3.42}$$

From (3.11)

$$P(r \leq R) = 1 - e^{-(R^2/2\sigma^2)} = 1 - e^{-(R/R_{\text{rms}})^2} = 1 - e^{-\rho^2} \tag{3.43}$$

Substituting (3.43) and (3.40c) in (3.42), we obtain the average duration of fade as

$$\bar{\tau} = \sqrt{\frac{\pi b_0}{b_2}}\,\frac{1 - \rho^{-\rho^2}}{\rho e^{-\rho^2}} = \left(\frac{\pi b_0}{b_2}\right)^{1/2} \frac{1}{\rho}\,(e^{\rho^2} - 1) \tag{3.44}$$

We provide some examples based on (3.44) below.

Example 3.2

For a vertical monopole antenna, find the level crossing rate N_R and the average duration of fade τ. From (3.40c), the level crossing rate is

$$N_R = \left(\frac{b_2}{\pi b_0}\right)^{1/2} \rho e^{-\rho^2}$$

From (3.39b)

$$b_0 = \int_{f_c - f_m}^{f_c + f_m} S(f)df = \frac{2}{\omega_m} \int_{f_c - f_m}^{f_c + f_m} \frac{1}{\sqrt{1 - \left(\dfrac{f - f_c}{f_m}\right)^2}}\, df$$

Substituting $(f - f_c)/f_m = \sin\theta$ in the above equation and integrating, we obtain

$$b_0 = \frac{2}{2\pi f_m} \int_{-\pi/2}^{\pi/2} \frac{f_m \cos\theta}{\cos\theta} d\theta = 1$$

Similarly,

$$b_2 = \frac{2}{\omega_m} 2\pi^2 f_m^2 \int_{-\pi/2}^{\pi/2} \frac{\sin^2\theta \cos\theta}{\cos\theta} d\theta = 2\pi^2 f_m^2$$

Thus, the level crossing rate is

$$N_R = \sqrt{2\pi} f_m \rho e^{-\rho^2}$$

and the average duration of fade from (3.44) is

$$\bar{\tau} = \left(\frac{\pi}{2\pi^2 f_m^2}\right)^{1/2} \left(\frac{e^{\rho^2} - 1}{\rho}\right) = \frac{1}{\sqrt{2\pi^2} f_m} \left(\frac{e^{\rho^2} - 1}{\rho}\right)$$

The plots of N_R/f_m versus ρ and of τ versus ρ are shown in Figure 3.10. The maximum level crossing occurs at $\rho = -3$ dB for vertical whip antennas and decreases on either side of this level. Recall that we have defined the crossing rate for a positive direction, so when the level is set much higher than the rms level, the signal will remain below the set level most of the time. On the other hand, when the level is set much lower than the rms level of the signal, the signal will remain above this level most of the time. Thus, in both the cases the level crossing rate is reduced. For an automobile traveling at a speed of 55 mph and receiving a carrier frequency of 850 MHz, the received Doppler has a maximum value of 69.3 Hz at $\rho = -3$ dB. The computed value of a normalized level crossing rate is 77.5 per second.

3.5 SPATIAL CORRELATION OF ELECTRIC AND MAGNETIC FIELDS

Correlation distance plays an important role in the design of a space diversity system at UHF frequencies. If two or more antennas are placed at distances such that electric and magnetic fields fade independently, signals can be combined that would assure the received signal to maintain a reasonable level and combat the effect of Rayleigh fading. Little difference has been found by A. J. Rustako in the cumulative distributions of the combined amplitudes from four antennas spaced $\lambda/4$, $3\lambda/4$, and $5\lambda/4$ apart [26]. The correlation coefficients of signal amplitudes at

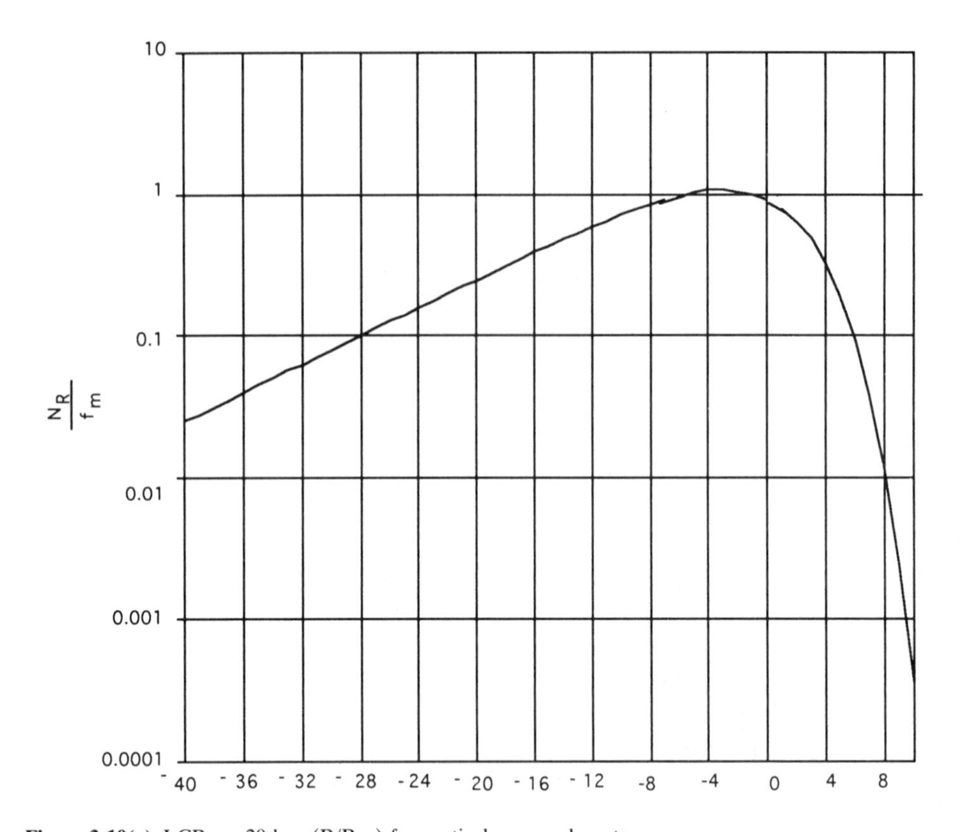

Figure 3.10(a) LCR vs. 20 $\log_{10}(R/R_{rms})$ for vertical monopole antenna.

the terminals of the antenna at these separations has been shown to be 0.25, 0.06, and 0.03 respectively. Thus even at a higher correlation (as high as 0.25), space diversity plays an effective role. We derive below the autocorrelation and cross-correlations of the electric and magnetic fields at points 0 and 0′, separated by a distance ξ as shown in Figure 3.3. The electrical phase for the nth component of the wave at 0 and 0′ will be φ_n and $\varphi_n + k\xi \cos \alpha_n$ respectively, where the value of the free-space phase constant k is $2\pi/\lambda$. This is shown in Figure 3.3. The auto-covariance function of the electric field (3.1) is given by

$$R_{E_z}(\xi) = \langle \overline{E}_Z E'_Z \rangle = \left\langle \left(E_0 \sum_{n=1}^{N} e^{-j\varphi_n} \right) \left(E_0 \sum_{m=1}^{N} e^{j(\varphi_m + k\xi \cos \alpha_n)} \right) \right\rangle \tag{3.45a}$$

where E_Z is the complex conjugate of the electric field at 0 and E'_Z is the value of the electric field at 0′ . The symbol $\langle\ \rangle$ is the ensemble average of φ_n and α_n.

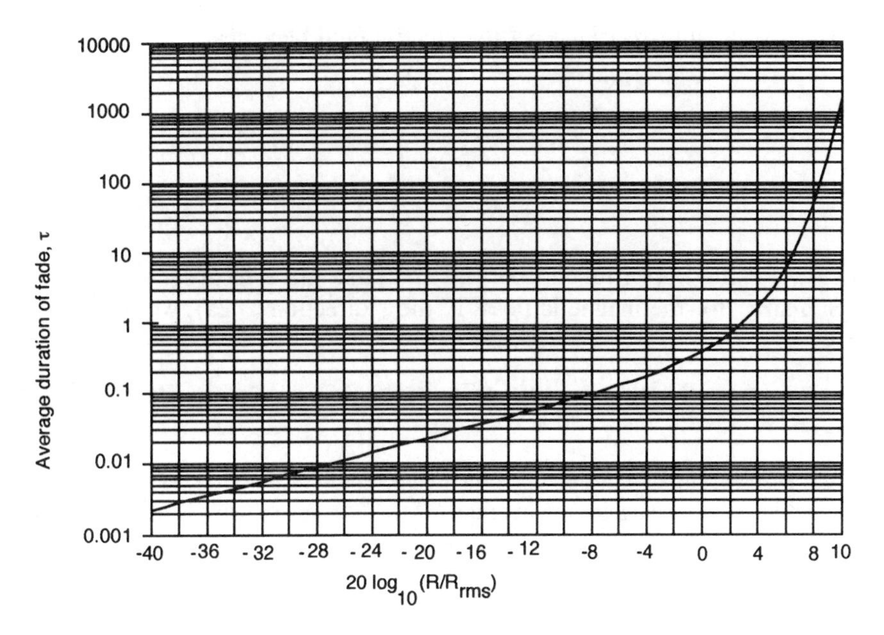

Figure 3.10(b) Average duration of fades for vertical monopole.

Assuming the statistical independence of φ_n and α_n and regrouping the angle terms together, we can rewrite the above equation as

$$R_{E_Z}(\xi) = E_0^2 \sum_{m=1}^{N} \sum_{n=1}^{N} \langle e^{j(\varphi_m - \varphi_n)} \rangle \langle e^{jk\xi\cos\alpha_n} \rangle \tag{3.45b}$$

Since the average of $e^{j(\varphi - \varphi)}$ is zero except when $n = m$, the average is also same for every value of n, so the above equation is reduced to

$$R_{E_Z} = E_0^2 \sum_{n=1}^{N} \langle e^{jk\xi\cos\alpha_n} \rangle = NE_0^2 \int_{-\pi}^{\pi} p(\alpha)\, e^{jk\xi\cos\alpha}\, d\alpha \tag{3.45c}$$

Assume that the waves arrive from any direction with equal probability (isotropic scattering),

$$P(\alpha) = 1/2\pi \qquad -\pi < \alpha < \pi \tag{3.45d}$$

and is equal to zero elsewhere.

The spatial autocovariance of the electric field becomes

$$R_{E_z}(\xi) = NE_0^2 \frac{1}{2\pi} \int_{-\pi}^{\pi} e^{jk\xi\cos\alpha}\, d\alpha \\ = NE_0^2 J_0(k\xi) \tag{3.45e}$$

where J_0 (.) is the Bessel function of the first kind and zero order.

Similarly, for the magnetic fields in the x direction (3.2a), we obtain

$$R_{H_x}(\xi) = \langle \overline{H}_x H_x' \rangle = \frac{E_0^2}{\eta^2} \sum_{n=1}^{N} \sum_{m=1}^{N} \langle e^{j(\varphi_m - \varphi_n)} \rangle \langle \sin\alpha_n, \sin\alpha_m\, e^{j(k\xi\cos\alpha)} \rangle \tag{3.46a}$$

where H_x is the complex conjugate of the electric field at 0 and H_x is the value of the electric field at 0′. Once again, the average $e^{j(\varphi_m - \varphi_n)}$ is zero except when $n = m$. Thus,

$$R_{H_x}(\xi) = \frac{E_0^2}{\eta^2} \sum_{n=1}^{N} \langle \sin^2\alpha_n e^{jk\xi\cos\alpha_n} \rangle \\ = \frac{E_0^2}{2\eta^2} \sum_{n=1}^{N} p(\alpha)\,(1 - \cos 2\alpha)\, e^{jk\xi\cos\alpha} d\alpha \\ = \frac{1}{2} \frac{NE_0^2}{\eta^2} [J_0(k\xi) + J_2(k\xi)] \tag{3.46b}$$

The density function for $p(\alpha)$ has once again been assumed to be uniform between angles of $-\pi$ to $+\pi$. J_0 (.) and J_2 (.) are the Bessel functions of the first kind with zero and second order, respectively. Similarly, the autocovariance function of the y component of the magnetic field can be shown to be

$$R_{H_y}(\xi) = \langle \overline{H}_y H_y' \rangle = \frac{NE_0^2}{2\eta^2} [J_0(k\xi) - J_2(k\xi)] \tag{3.47}$$

The normalized form of autocovariance function $\rho_E(\xi)$, $\rho_H(\xi)$, and $\rho_H(\xi)$ are shown in Figure 3.11. As seen from the plots of autocovariance functions, the magnetic and electric fields decorrelate rapidly to a small value, around $\xi = \lambda/2$. Therefore, antennas placed at $\lambda/2$ or further apart at the mobile can be employed for space diversity at UHF. Obviously, this conclusion is arrived at with an assumption of isotropic scattering, which holds for mobile units that are surrounded by fixed and moving objects producing scattering uniformly along 360°. On the other hand, separations required at the cell site will be a lot larger as the assumption of

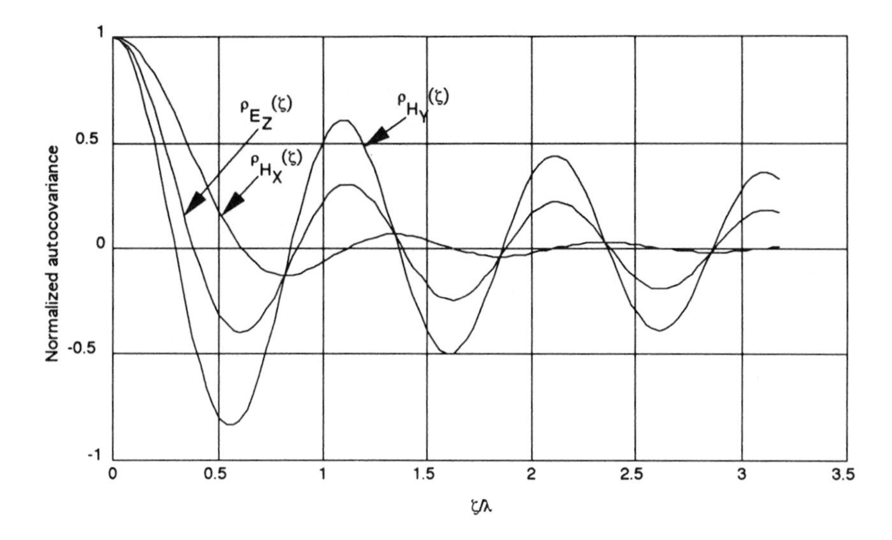

Figure 3.11 The normalized autocovariance function $\rho_E(\zeta)$, $\rho_H(\zeta)$, and $\rho_H(\zeta)$.

uniform scattering does not hold strictly. The cell site is selected such that the receive antenna is high and away from the surrounding clusters.

Proceeding similar to above, the cross-correlation of electric field E_Z and the magnetic field H_x ((3.1), (3.2a)) can be written as

$$R_{E_Z H_x}(\xi) = \langle \overline{E}_Z H_x' \rangle = \frac{E_0^2}{\eta} \sum_{n=1}^{N} \sum_{m=1}^{N} \langle e^{j(\varphi_m - \varphi_n)} \rangle \langle \sin \alpha_m \, e^{jk\xi \cos \alpha_m} \rangle \tag{3.48}$$

Again the independence of φ_n and α_n has been assumed. Assuming a uniform distribution of $p(\alpha)$ (isotropic scattering), the value of the cross-correlation becomes

$$R_{E_z H_x}(\xi) = \left(\frac{NE_0^2}{\eta} \right) \frac{1}{2\pi} \int_{-\pi}^{\pi} \sin \alpha \, e^{-jk\xi \cos \alpha} d\alpha = 0 \tag{3.48b}$$

The value of the integral is zero due to the odd nature of sin α. Proceeding in a similar fashion, it can be shown that the cross-correlation between two components of the magnetic field is also zero. In other words,

$$R_{H_x H_y}(\xi) = \langle \overline{H}_x H_y' \rangle = 0 \tag{3.49}$$

The cross-correlation between electric field E_Z and the H_y component of the magnetic field is given by

$$
\begin{aligned}
R_{E_zH_y}(\xi) &= \langle \overline{E}_z H_y' \rangle \\
&= \frac{E_0^2}{\eta} \sum_{n=1}^{N} \sum_{m=1}^{N} \langle e^{j(\varphi_m - \varphi_n)} \rangle \langle \cos \alpha_m \, e^{k\xi \cos \alpha_m} \rangle \\
&= \frac{E_0^2}{\eta} \sum_{n=1}^{N} \langle \cos \alpha_n \, e^{k\xi \cos \alpha_n} \rangle \\
&= \frac{NE_0^2}{\eta} \int_{-\pi}^{\pi} p(\alpha) \cos \alpha \, e^{k\xi \cos \alpha} \, d\alpha
\end{aligned}
\tag{3.50a}
$$

Assuming the density function $p(\alpha)$ to be uniformly distributed, the above equation can be reduced to

$$
R_{E_zH_y}(\xi) = j \frac{NE_0^2}{2\eta} J_1(k\xi) \tag{3.50b}
$$

where $J_1(.)$ is the Bessel function of first kind and first order. The plot of the normalized function $R_{E_z}H_y(\xi)$ is shown in Figure 3.12.

Since $R_{E_zH_x}(\xi) = R_{H_xH_y}(\xi) = 0$, the fields (E_Z, H_x) and (H_x, H_y) are uncorrelated and independent for all spatial separations. This is a rather surprising result. Thus, the antenna systems, which are separately sensitive to the electric field, and the two components of the magnetic fields can be used for diversity gain in cellular systems.

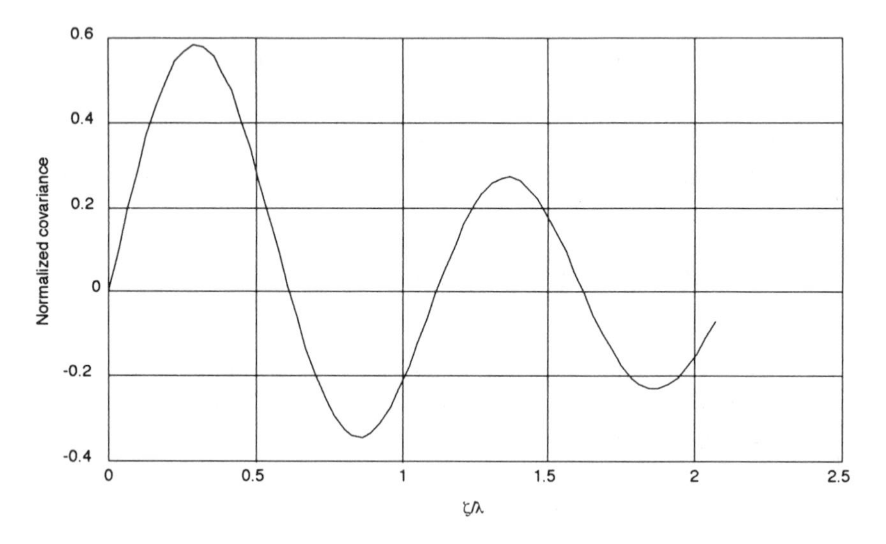

Figure 3.12 The normalized covariance function from (3.50b).

3.6 CORRELATION BETWEEN TWO SIGNALS: COHERENCE BANDWIDTH

The problem of correlation between two signals at cellular radio arises for the determination of channel bandwidth [12–16]. One major consideration, based on which the 30-kHz channel bandwidth has been assigned to cellular radio, is the concept of coherence bandwidth. Both the correlation and the coherence bandwidths are related to each other through the time-delay distribution of the received signals. Different path lengths of the received signal give rise to different propagation time delay. The presence of different time delays in the received waves cause the statistical properties of the two signals at different frequencies to become essentially independent for a large frequency separation. The frequency separation $\Delta\omega$, at which the correlation between signals drops below a certain value (less than 0.7), is known as a coherence bandwidth. If one transmits a signal whose bandwidth exceeds the coherence bandwidth of the channel, it is not possible to correct the amplitude and phase distortions of the received signal by means of a single complex correction factor (equalizer) applied over the whole bandwidth. For digital data transmission having a signal bandwidth greater than the channel bandwidth, the error probabilities is expected to increase rapidly as the ratio of the required signal bandwidth to the channel bandwidth increases. Additionally, since the irreducible error probability for the channel is directly related to the coherence bandwidth, it will not be possible to control errors. We will discuss below the covariance of two signals and prove that it is simply the characteristic function of the time delays suffered by the component plane waves, which compose the cellular radio field.

Let the electric fields at frequencies ω_1 and ω_2 be written as

$$E_1 = E_{01} \sum_{n=1}^{N} e^{j\omega_1(t-t_n)} \tag{3.51a}$$

$$E_2 = E_{02} \sum_{n=1}^{N} e^{j\omega_2(t-t_n)} \tag{3.51b}$$

where E_{01} and E_{02} are the received signals amplitudes of each component E_1 and E_2. The time delays associated with the nth components of electric field are represented by t_n. Thus, the covariance of these two signals are

$$\begin{aligned} R_{12} = \langle \overline{E}_1 \, E_2 \rangle &= \overline{E}_{01} \, E_{02} \left\langle \sum_{n=1}^{N} \sum_{m=1}^{N} e^{-j\omega_1(t-t_n)} e^{j\omega_2(t-t_m)} \right\rangle \\ &= \overline{E}_{01} \, E_{02} \, e^{j(\omega_2-\omega_1)t} \sum_{n=1}^{N} \sum_{m=1}^{N} \langle e^{-j(\omega_2 t_m - \omega_1 t_n)} \rangle \end{aligned} \tag{3.52a}$$

Similar to the last section, assuming that the time delays are independent, the quantity under the summation sign is zero except for $n = m$. In other words,

$$\sum_{n=1}^{N} \sum_{m=1}^{N} \langle e^{-j(\omega_2 t_m - \omega_1 t_n)} \rangle = 0 \tag{3.52b}$$

for $m \neq n$. Thus,

$$R_{12} = \langle \overline{E}_1 E_2 \rangle = \overline{E}_{01} \, \mathrm{E}_{02} \, e^{j(\omega_2 - \omega_1)t} \sum_{n=1}^{N} \langle \, e^{-j(\omega_2 - \omega_1)t_n} \rangle \tag{3.52c}$$

Assuming the difference between two transmitted fundamentals to be $\Delta\omega$, or

$$\Delta\omega = \omega_2 - \omega_1 \tag{3.52d}$$

The difference $\Delta\omega$ is considered here to be small such that it takes approximately the same time to travel from the cell site to the mobile receiver or vice versa. This assumption implies that the propagation along all the paths is via free space waves without dispersion, and the phase changes experienced when reflecting or diffracting objects are independent of frequency. Let us define the time delay associated with the nth component of the received waves to be Δt_n where

$$\Delta t_n = t_n - t_0 \tag{3.52e}$$

where t_n and t_0 are the times of travel for nth and the direct components of the received waves. Thus, Δt_n is the differential time with respect to the reference time of t_0.

Substituting for t_n and $(\omega_2 - \omega_1)$ in terms of Δt and $\Delta\omega$ in (3.52c), we obtain

$$R_{12} \, (\Delta\omega) = N\overline{E}_{01} \, E_{02} \, e^{j\Delta\omega t} \, (e^{-j\Delta\omega t_0}) \, \langle \, e^{-j\Delta\omega\Delta t} \rangle \tag{3.52f}$$

The subscript n has been dropped in (3.52f) because the average is the same for any value of n. Thus, the normalized value of the correlation function is given by

$$\rho_{12} \, (\Delta\omega) = \langle \, e^{-j\Delta\omega\Delta t} \rangle \tag{3.52g}$$

which states that the correlation between signals is simply the characteristic function of random delay Δt. Three different time delay distributions are considered here, namely, exponential, Maxwell, and impulse. Consider first the exponential delay distribution, that is,

$$p(\Delta t) = \frac{K}{T} e^{-\Delta t/T} \qquad 0 \leq \Delta\, t \leq \infty \tag{3.53}$$

where T = standard deviation of time delays.

The correlation function is:

$$\rho_1(\Delta\omega) = \frac{1}{T} \int_0^\infty e^{-\Delta t/T} e^{-j\Delta\omega\Delta t} \, d(\Delta t) = - \frac{1}{1 + j\Delta\omega T} \tag{3.54}$$

or $|\rho_1(\Delta\omega)| = [1 + (\Delta\omega T)^2]^{-1/2}$. Maxwell and impulse distributions have also been assumed for random delay Δt. From Young and Lacy's measurements of impulse response at 450 MHz in New York City, Clark has approximated the delay density to be Maxwell and represented as [2]

$$\begin{aligned} p(\Delta t) &= K\sqrt{2/\pi}\,(3 - 8/\pi)^{3/2} \frac{(\Delta t)^2}{T^3} e^{-[(3 - 8/\pi)\Delta t^2/2T^2]} \qquad \text{if } \Delta t > 0 \\ &= 0 \qquad \text{if } \Delta t < 0 \end{aligned} \tag{3.55}$$

In the most simplistic case, the delay distribution can be considered to concentrate at two values.

$$p(\Delta t) = \frac{K}{2} [\delta(\Delta t) + \delta\,(\Delta t - 2T)] \tag{3.56a}$$

where K is the total power received at the antenna, Δt is the delay relative to a direct path from cell site to mobile, and T is the standard deviation of the delays. A plot of $|\rho_{12}\,(\Delta\omega)|$ based on the exponential density function of Δt is shown in Figure 3.13. The exponential distribution of time delay appears more likely to occur on physical grounds as the shorter delays are more prevalent than longer delays. Figure 3.13 also presents the experimental results of Ossana at 860 MHz in suburban environments for two carriers separated by 0.1, 0.5, 1.0, and 2 MHz [25]. The correlation between experimental points represented by the circles and the theoretical curve of covariance is extremely good with the assumed exponential density of the time-delay spreads.

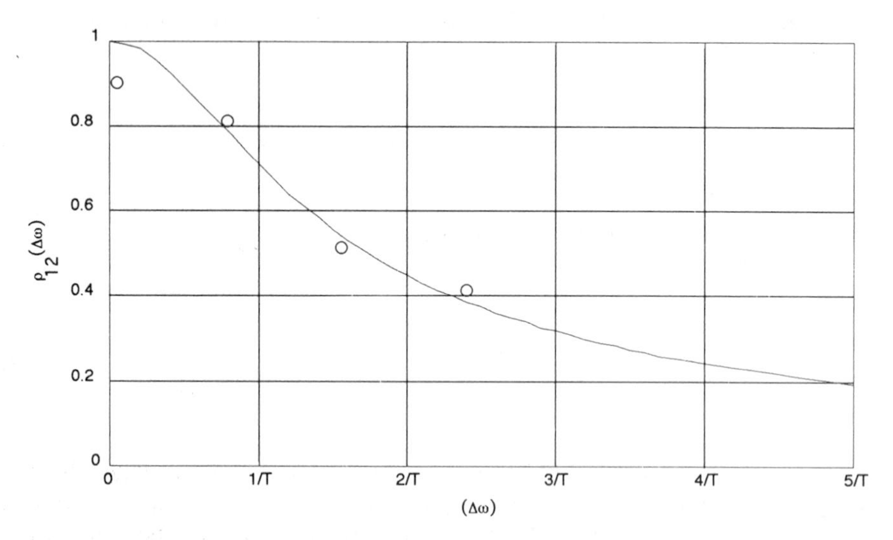

Figure 3.13 Normalized covariance of two signals as a function of their frequency separation assuming exponential distribution of time delay spread T. The circles are ossanna experimental points. After [2].

Normalized correlation for Maxwell and impulse distributions are given by [4]:

$$\rho_2(\Delta\omega) = \exp^{-b^{3/2}} \sqrt{(1 - b^2)^2 + (8/\pi)b^2 \, {}_1F_1^2 \, (-1/2; 3/2; b^2/2)}$$

$$\rho\ (\Delta\omega) = |\cos(\Delta\omega T)| \tag{3.56b}$$

where

$$b = \frac{\Delta\omega T}{\sqrt{3 - (8/\pi)}}$$

and ${}_1F_1$ is the confluent hypergeometric function, which can be expressed in terms of modified Bessel functions of the first type:

$${}_1F_1\ (v + 1/2; 2v + 1, x) = 2^{2v} \Gamma(v + 1)\, x^{-v} \exp\,(x/2)\, I_v\,(x/2) \tag{3.56c}$$

Time-delay distributions and the resultant correlations are shown in Figure 3.14(a,b). Assuming the coherence bandwidth to be the frequency separation at which the normalized value of the correlation coefficient attains a value less than 0.7, then from Figure 3.14(b) it can be seen that the coherent bandwidth can be considered to be independent of the probability density function of time-delay spread. From Figure 3.14, a nominal value of coherence bandwidth (Δf) can be

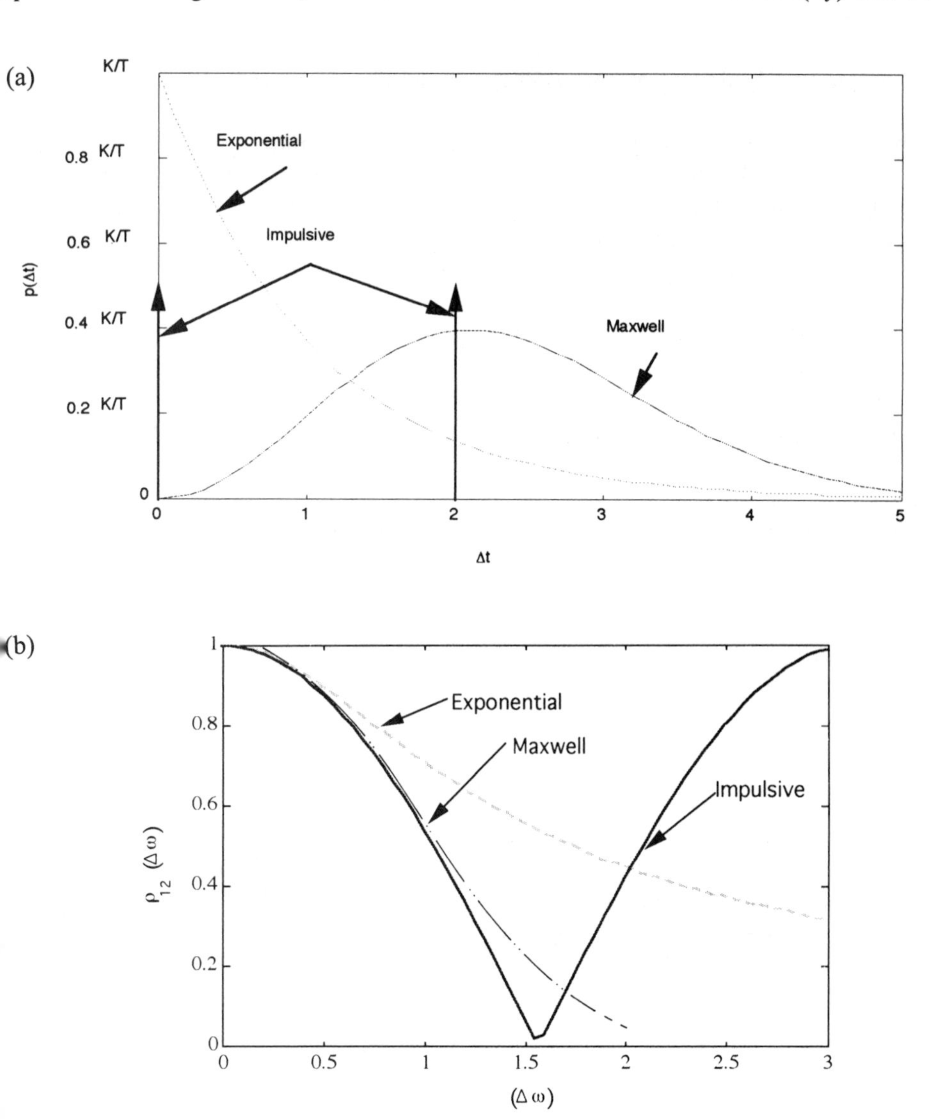

Figure 3.14 (a) Effects of time delays several delay distributions. (b) Autocorrelation versus frequency separation.

assumed to be $1/(2\pi T)$, which corresponds to a correlation value of about 0.7. Thus for New York City, at 900 MHz where $T \approx 0.5\ \mu$sec, the coherence bandwidth is approximately 31 kHz, while in suburban areas with $T \approx 0.25$ msec, the coherence bandwidth is 20 times greater or roughly 0.6 MHz.

3.7 CHANNEL SIMULATOR DESIGN [10, 11, 16, 23–25]

It is mandatory to try out for new developments in cellular radio in the laboratory before using in the field. A hardware simulator having characteristics of real path impairments must therefore be designed so that expensive field trials of the new equipment can be avoided before tests in the laboratory can be done. Here we describe the desired capabilities [16, 22–24]. As described earlier, the PDF of the received envelope at the mobile follows Rayleigh distribution superimposed on a log-normal slow shadowing representing terrain features. In some cases, direct line-of-sight component may also exist along with the fast fading. The doppler frequency shifts representing the relative motion of the mobile with respect to cell site are to be included. This is manifested by a random fluctuation of phase, which is known to be uniformly distributed between 0 and 2π. Channel delay, which generally follows an exponential distribution, has to be a part of simulation. Noise and interference also need to be added to the received signal. Figure 3.15 illustrates the conceptual model of the channel that accounts for all the impairments men-

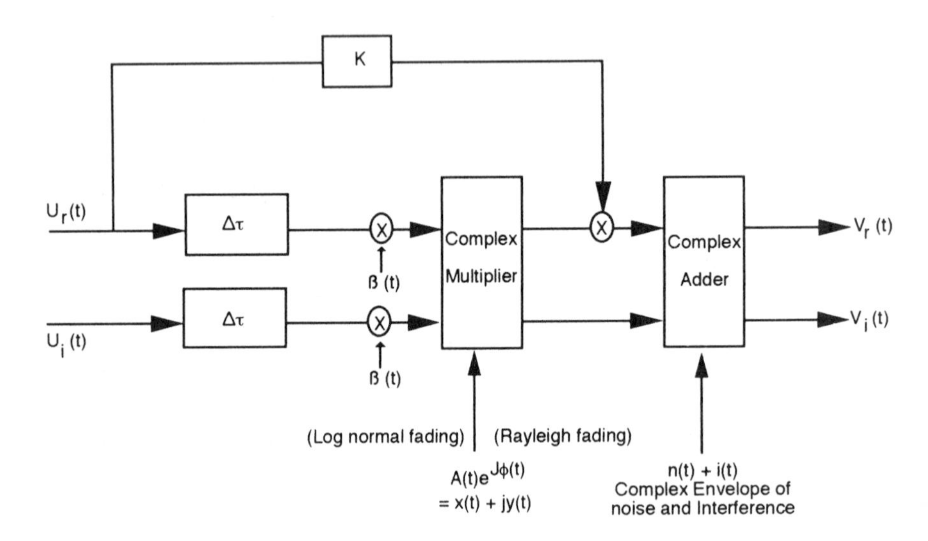

Figure 3.15 Channel model.

tioned above. Here, the inputs to the channel $[u_r(t), u_i(t)]$ are the real and the imaginary parts of the transmitted signal envelope given by

$$S(t) = Rl[(u_r + ju_i)e^{j\omega_c t}] \quad (3.57)$$

where $S(t)$ is the RF transmitted signal, ω_c is the radian carrier frequency, and u_r and u_i are the real and imaginary components of the baseband input signal to the simulator. The channel outputs are represented by (v_r, v_i), the real and the imaginary parts of the simulator output envelope, and can be represented as

$$r(t) = Rl[(v_r + jv_i)\, e^{j\omega_c t}] \quad (3.58)$$

where $r(t)$ is the RF received signal at the simulator output. It should be noted that the simulator transfers the input signal components (u_r, u_i) into output components (v_r, v_i), adding all the channel distortions.

Equal amounts of path delay are added to both u_r and u_i. The delayed components u_r and u_i are multiplied by $\beta(t)$ to represent the slow fading process and are log-normally distributed. The second input to the complex multiplier, $A(t)e^{j\phi(t)}$, represents fast fading and includes the effects of Doppler spread. The scaler K, when multiplied by the real component of the input signal, represents the line-of-sight component of the signal and is simply added at the complex multiplier output. Finally, the additive noise, $n(t)$, and interference, $i(t)$, are summed in a complex adder before the final simulated channels V_r, V_i are generated. Both u_r and u_i are delayed by $\Delta\tau$.

3.8 ENVIRONMENTAL NOISE

The interference environment present at the cellular receiver must be known before it can be predicted whether the system will perform satisfactorily or not. The interference in cellular consists of discrete signals other than the desired signals, broadband impulsive noise, and various other unwanted radiations. The most desired representation for noise is its power spectral density; that is, the noise per unit bandwidth. This parameter must be known as a function of time, location, and frequency, and should be represented in a form that allows it to be combined and compared with other types of noise, such as internal receiver noise. In this section, we shall discuss the source of ignition noise and its representation in frequency domain along with its effects on the data performance of the receiver. It should be noted that ignition noise is the predominant noise source in which the mobile operates. Receiver front-end noise is counted as a receiver noise figure and is in cascade with the external noise. Thus, the total noise is the sum of the external as well as the receiver internal noise. External noise will include galactic and ignition noise, among others.

3.8.1 Ignition Noise

Automotive ignition noise arising out of an individual vehicle generally displays the characteristics of a periodic waveform with highly irregular pulse amplitudes. The noise generated by an individual automobile also does not possess any significant amount of coherence and is regarded as an incidental manmade (unintentionally generated) noise. Also, due to the multiplicity of vehicles in different positions and moving at different speeds, the composite signal trains display random characteristics both in signal amplitude and phase. Often the pulses are clustered together with irregular spacings between clusters. With a group of pulses, individual pulse width can lie anywhere between 1–5 ns [20]. Ignition noise is the predominant form of manmade noise up to about 180 feet from a typical highway in U.S. cities [20] This form of noise also has a high correlation with population density. The automotive ignition noise has been shown to have approximately 16 dB of daily variations between peak and off-peak traffic hours.

As stated in the IEEE Standard No. 263 dated November 1965, the motor-vehicle radio noise arises principally from the ignition circuit of gasoline (or similar) engine-driven vehicle equipment in which steep wavefront electric transients are generated by the high-voltage electric discharges, which occur across the distributor or spark plug gaps [21]. A secondary, but nonetheless troublesome, source of noise is the battery-charging circuit, in which electric transients are generated. Other lesser sources of noise that are occasionally troublesome are due to the fan belt, gauges and instruments, and the generator shaft.

As stated above, the impulsive nature of ignition noise exhibits a wideband frequency spectrum, which is flat in amplitude over 25–1,000 MHz. The voltage spectrum of the noise can be represented as

$$V(\omega) = R(\omega) + jX(\omega) = \int_{-\pi}^{\pi} V(t)\, e^{-j\omega t}\, dt \tag{3.59a}$$

where $\omega = 2\pi f$ and the real and imaginary components of noise are

$$R(\omega) = \int_{-\infty}^{\infty} v(t) \cos \omega t \, dt \tag{3.59b}$$

$$X(\omega) = \int_{-\infty}^{\infty} v(t) \sin \omega t \, dt \tag{3.59c}$$

Thus, the spectrum has amplitude and phase characteristics given by

$$A(\omega) = \sqrt{R^2(\omega) + X^2(\omega)} \tag{3.59d}$$

$$\phi(\omega) = \tan^{-1} \frac{X(\omega)}{R(\omega)} \tag{3.59e}$$

Therefore, the inverse transform of the real component $v(t)$ can be written as

$$v(t) = \frac{1}{\pi} \int_{-\pi}^{\pi} A(\omega) \cos[(\omega t + \phi(\omega)] \, d\omega \tag{3.59f}$$

where $A(w)$, as stated above, is essentially flat over 25–1,000 MHz. The spectrum amplitude can also be expressed in the following terms:

$$S(f) = A(\omega) \neq \pi \quad \text{V/Hz} \tag{3.59g}$$

Some ignition noise measurements effecting data transmission were taken [5] in London at UHF frequencies. The bit error rate curve at 462 MHz for 4,800 bps FM transmission is shown in Figure 3.16. The probability density function approximating the data error rate measurements in the presence of ignition noise is given by

$$p_e(r) = 0.5 \, e^{-4.41 r^{0.27}} \tag{3.60}$$

where $p_e(r)$ is the probability density function of ignition noise. In order to study the effects of ignition noise, the received signal levels were completely eliminated by generating the RF data signal (ignition noise was present in the automobile) in the vehicle itself and feeding this signal directly to the receiver [5]. The base station was turned off, but the input antenna was still connected to the receiver so that the ignition noise from other vehicles could enter into the receiver system. Since the ignition noise measurements on error rate curve were made with a steady-state signal level without fading and shadowing, the experimental curve was modified to include the effects of fading and shadowing. The modification can be done by multiplying the probability of the error curve $P_e(r)$ at a steady signal envelope of r and the probability density function $p(r)$ of that signal level occurring in an environment with fading and shadowing. Thus, the theoretical error rate with shadowing and signal fading is given by

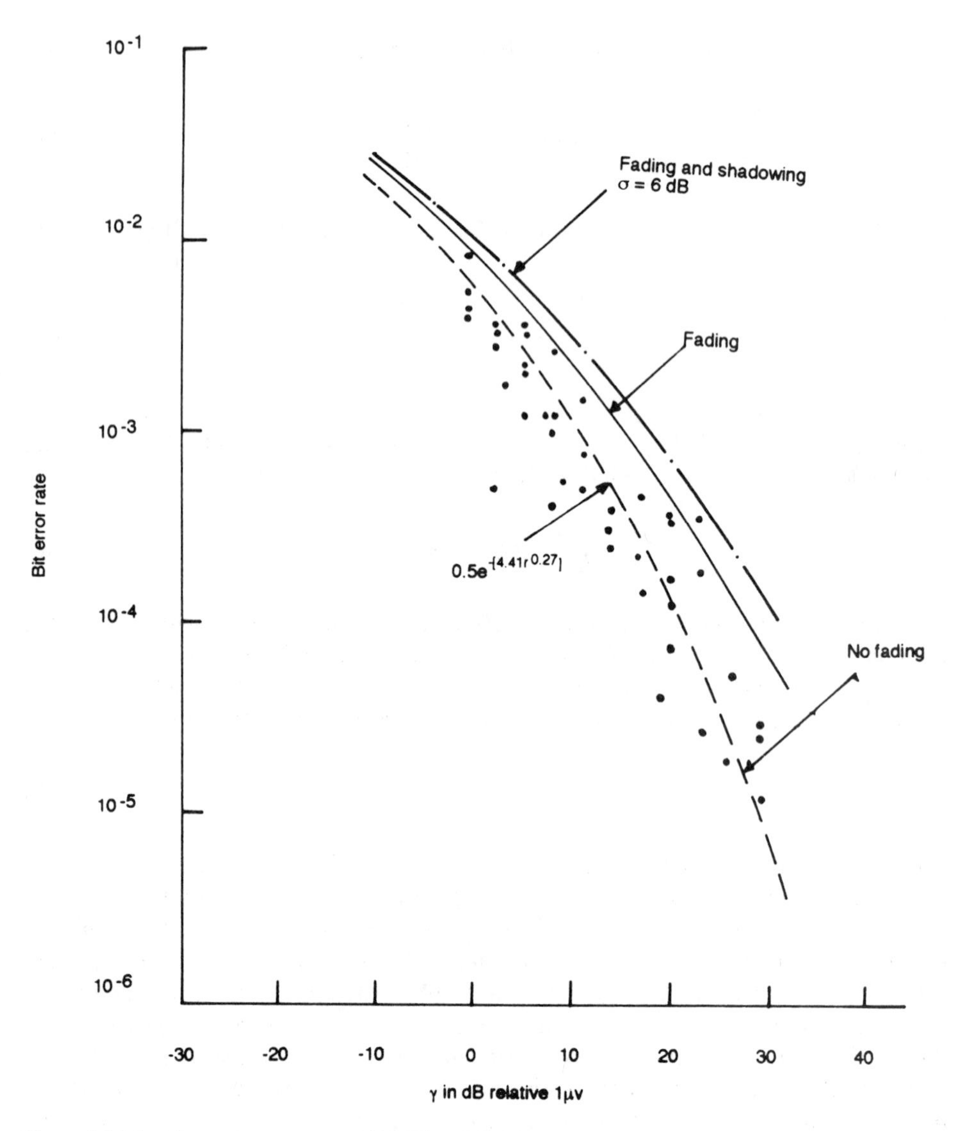

Figure 3.16 Ignition noise errors at 462 MHz, 4800 b/s direct FM and the effects of fading on error rate.

$$P_e(\cdot) = \int_0^\infty p_e(r)p(r)dr$$
$$= \int_0^\infty 0.5e^{-4.41r^{0.27}} p(r)dr \tag{3.61}$$

where the value of $p_e(r)$ is obtained from (3.60). In Figure 3.16, fading degrades the performance by approximately 6 dB and shadowing (σ = 6 dB) adds a further 3 dB in the degradation.

3.8 CONCLUSIONS

In this chapter we have characterized the cellular channel. The channel operates under fast fading (Rayleigh fading), superimposed on the long-term signal variation that follows log-normal distribution. Due to fast fading, the signal level at the receiver fluctuates rapidly. The distance between two fades is of the order of $\lambda/2$ and may cause a sensitivity problem at the receiver. The statistical representation of the problem is done by a level crossing rate and the fade duration. These two are related to each other, as the level crossing rate increases the duration of fade is decreased and vice versa. Electric and magnetic fields received by the mobile are in general uncorrelated except that there is a correlation between the electric field and the y component of the magnetic field. The covariance of two signals follows the characteristics function of random time delay Δt. The density function of time delay can be represented by exponential distribution, as the shorter delays are more prevalent than longer delays. In next chapter we shall take up the study of propagation and its application to cellular radio.

PROBLEMS

3.1 The number of fades $N(R)$ per second is related to fades of depth greater than X by the expression

$$N(R) = \sqrt{2\pi}\,\frac{VX}{\lambda}\,e^{-x^2}$$

Here, the fade depth X is normalized with respect to the rms value of the signal. If the rms level of the signal is 1.6dB above the median level, find the value of X when the fade depth is 20dB below the rms level or 18.4dB below the median level. Find also the number of zero crossings at 900 MHz when the vehicle is traveling at 60 miles per hour.

3.2 The average duration of fade for a vertical monopole is given by

$$\bar{\tau}(R) = \frac{e^{\rho^2 - 1}}{2\pi f_D \rho}$$

where f_D is the Doppler frequency shift, and ρ is the fade depth (signal below the rms value)in decibels. Find the average duration of fade at 836 MHz with a car speed of 30 miles per hour and a signal fade of 10 dB.

3.3 For a 950-MHz carrier frequency, find the distance between two successive dips and peaks of the received signal at the mobile.

3.4 With a 950-MHz cellular carrier frequency and for mobile velocity of 60 miles per hour, find the maximum Doppler frequency in Hz. What should be the receiver bandwidth in this case? Can you expect significant fades between one meter of vehicle travel in this case?

3.5 Show that the power spectra of a beam antenna is given by

$$S(f) = \frac{G_0}{f_m \sqrt{1 - (f/f_m)^2}}$$

if $|f/f_m| < \sin(\beta/2)$ and $\xi = 90°$.

$$S(f) = \frac{2G_0}{f_m \sqrt{1 - (f/f_m)^2}}$$

if $1 > f/f_m > 1 - \cos\beta/2$, where the beam antenna pattern and gain is given by

$$G(\alpha - \xi) = G_0$$

if $|\alpha - \xi| < \beta/2$ and is equal to zero otherwise.

REFERENCES

[1] Arredondo, G. A., J. C. Feggeler, and J. I. Smith. "Voice and Data Transmission," *Bell System Technical Journal*, Vol. 58, No. 1, Jan. 1979.

[2] Clarke, R. H. "A Statistical Theory of Mobile-Radio Reception," *Bell System Technical Journal*, July–Aug. 1968.

[3] Reudink, D. O. "Properties of Mobile Radio Propagation Above MHz," *IEEE Trans. on Vehicular Technology,* Vol. VT-23, Nov. 1973, pp. 143–159.

[4] Gans, M. J. "A Power-Spectral Theory of Propagation in the Mobile-Radio Environment," *IEEE Trans. on Vehicular Technology*, Vol. VT-21, No. 1, Feb. 1972.

[5] Turin, G. L., F. D. Clapp, T. L. Johnston, S. B. Finn, and Lavy, "A Statistical Model of Urban Multipath Propagation," *IEEE Trans. on Vehicular Technology*, Vol. VT-21, No. 1, Feb. 1972.
[6] Zander, J. "Stochastical Model of the Urban UHF Radio Channel," *IEEE Trans. on Vehicular Technology*, Vol. VT-30, No. 4, Nov. 1981.
[7] Schmid, H. F. "A Prediction Model for Multipath Propagation of Pulse Signals at VHF and UHF Over Irregular Terrain," *IEEE Trans. on Antennas and Propagation*, Vol. AP-18, No. 2, March 1970.
[8] Bultitude, R. J. C. "A Study of Coherence Bandwidth Measurements for Frequency Selective Radio Channels."
[9] Engl, J. S. "Effects of Multipath Transmission on the Measured Propagation Delay of an FM Signal," *IEEE Trans. on Vehicular Technology*, Vol. VT-18, No. 1, May 1969.
[10] Capels, E. L., K. E. Massad, and T. R. Minor. "A UHF Channel Simulator for Digital Mobile Radio," *IEEE Conference on Vehicular Technology*, Vol. VT-29, No. 2, May 1980.
[11] Hashhmi, H. "Simulation of the Urban Radio Propagation Channel," *IEEE Trans. on Vehicular Technology*, Vol. VT-28, No. 3, Aug. 1979.
[12] Hubbard, R. W., R. F. Linfield, and W. J. Hartman. "Measuring Characteristics of Microwave Mobile Channels," *IEEE Conference on Vehicular Technology*, 1979.
[13] Cox, D. C., and R. P. Leek, Correla Cox and Robert P. Leek, "Correlation Bandwidth and Delay Spread Multipath Propagation Statistics for 910 MHz Urban Mobile Radio Channel," *IEEE Trans. on Communication*, Vol. COM-23, Nov. 1975, pp. 1271–1280.
[14] Cox, D. C. "910 MHz Urban Mobile Radio Propagation: Multipath Characteristics in New York City," *IEEE Trans. on Communication*, Vol. COM-21, Nov. 1973.
[15] Cox, D. C. "Multipath Delay Spread and Path Loss Correlation for 910-MHz Urban Mobile Radio Propagation," *IEEE Trans. on Vehicular Technology*, Vol. VT26, Nov. 1977.
[16] Neilson, D. L. "Microwave Propagation Measurements for Mobile Digital Radio Application," *IEEE Trans. on Vehicular Technology*, Vol. VT-27, Aug. 1978, pp. 117–131.
[17] Yoshida, F.I.S., T. Takaeuchi, and M. Umehira. "Propagation Factors Controlling Mean Field Strength on Urban Streets," *IEEE Trans. on Antennas and Propagation*, Vol. AP-32, Aug. 1983.
[18] French, R. C. "Error Rate Predictions and Measurements in the Mobile Radio Data Channel, A Statistical Model of Urban Multipath Propagation," *IEEE Trans. on Vehicular Technology*, Vol. VT-21 No. 1, Feb. 1972.
[19] Flemming, H. "Mobile Fading-Rayleigh and Lognormal Superimposed," *IEEE Trans. on Vehicular Technology*, Vol. VT26, No. 4, Nov. 1977.
[20] Skomal, E. N. "The Range and Frequency Dependence of VHF-UHF Man-made Noise in the Above Metropolitan Areas," *IEEE Trans. on Vehicular Technology*, Vol. VT-19, No. 2, May 1970.
[21] IEEE Standard No. 263, "Measurement of Radio Noise, Generated by Motor Vehicles, and Affecting Mobile Communications Receives in the Frequency Range 25 to 1,000 Mc/s," *IEEE Trans. on Vehicular Communications*, Vol. VT-15, No. 2, Oct. 1966.
[22] Lim, M. S., and H. K. Park. "The Implementation of the Mobile Channel Simulator in the Baseband and its Application to the Quadrature Type GMSK Modem Design," *IEEE Conference on Vehicular Technology*, 1990, pp. 496–500.
[23] Kim, J. M., S. C. Lee, and H. K. Park. "Development of Channel Simulator for Mobile Radio Communication," *IEEE Conference on Vehicular Technology*, 1987 , pp. 567–570.
[24] Goubran, R. A., H. M. Hafez, and A.U.H. Sheikh. "Real Time Programmable Land Mobile Channel Simulator," *IEEE Conference on Vehicular Technology*, 1986, pp. 215–218.
[25] Ossana, J. F., Jr. "A Model for Mobile Radio Fading Due to Building Reflections: Theoretical and Experimental Fading Waveform, Power Spectra," *Bell System Technical Journal*, Nov. 1964.
[26] Rustako, A. J., et al., "Performance of Feedback and Switch Space Diversity Reception," *IEEE Trans. on Vehicular Technology*, Vol. VT-25, Aug. 1976, pp. 75–84.

Chapter 4
Propagation

4.1 INTRODUCTION

There are a number of mechanisms by which radio waves can travel from a transmitting to a receiving antenna: direct wave or free-space waves, ground or surface waves, tropospheric waves, and ionospheric waves. The simplest case of radio propagation is the free-space propagation of waves between a transmitter and a receiver. By "free-space," we mean a region whose properties are isotropic (the same properties when measured along any axis) and homogeneous (uniform structure). The other names for free-space radio waves are direct or line-of-sight waves. The wave is propagated in straight lines and thus can be used for satellite or outer space communication, as shown in Figure 4.1(a). Additionally, the definition can also be used for line-of-sight terrestrial transmission (between microwave towers), as shown in Figure 4.1(b). The second transmission mechanism is the ground or the surface-wave mode. As described by Kenneth Bullington of Bell Telephone Laboratories, "ground-wave propagation may be considered to be the sum of three principal terms, namely the direct wave, reflected wave, and the surface wave" [1]. The surface wave, as the name suggests, is guided along the earth's surface. Some energy radiated from the transmitting antenna reaches the receiving antenna directly. Some energy reaches the receiver after reflecting from the earth's surface, and some reaches the receiver via a surface wave. The surface wave is about one wavelength above the ground over land and about 5–10 wavelengths above sea level. Since the ground is not a perfect reflector, some energy is absorbed by the ground also. As energy enters the ground, it sets up the ground currents, or, in other words, the distribution of the electromagnetic field in the region near the surface of the earth is distorted relative to what it would have been over an ideal perfectly reflecting surface. These three components of the surface wave are shown in Figure 4.1(c). The third mode of communication is due to the presence of the troposphere, which extends to about ten miles above the surface of the earth. The troposphere is a nonhomogeneous medium with time-varying properties caused by changing weather conditions. Its refractive index gradually decreases with height.

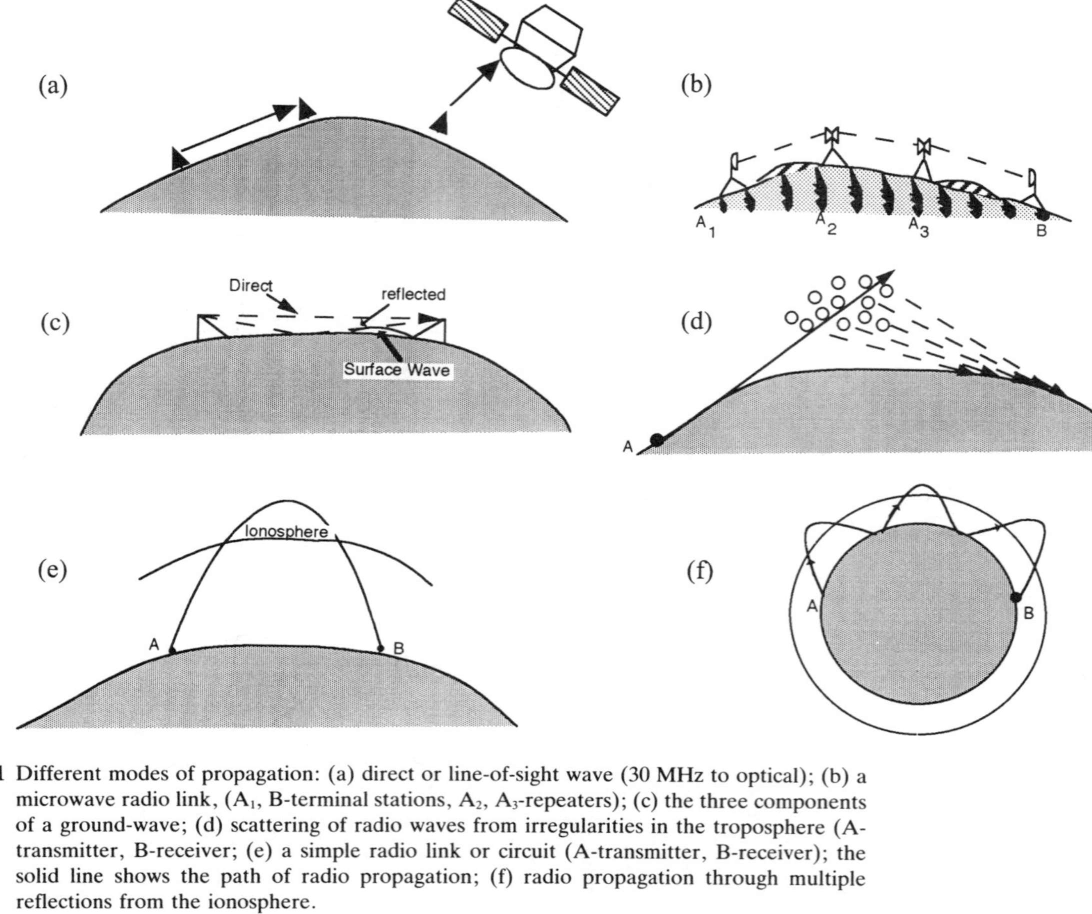

Figure 4.1 Different modes of propagation: (a) direct or line-of-sight wave (30 MHz to optical); (b) a microwave radio link, (A_1, B-terminal stations, A_2, A_3-repeaters); (c) the three components of a ground-wave; (d) scattering of radio waves from irregularities in the troposphere (A-transmitter, B-receiver; (e) a simple radio link or circuit (A-transmitter, B-receiver); the solid line shows the path of radio propagation; (f) radio propagation through multiple reflections from the ionosphere.

This gradual variation of the refractive index results in a bending of paths taken by radio waves, as shown in Figure 4.1(d). The tropospheric mode of operation applies to waves shorter than ten meters. The fourth mode of operation is through the ionosphere. The part of the atmosphere extending from 40 miles to 400 miles is called the ionosphere. For radio waves less than one meter long, the ionosphere acts, as a reflector. The waves reflected from the ionosphere may have one or several hops, as shown in Figure 4.1(e,f). Thus is the propagation through this mode used for long-distance transmission. Apart from reflection, the ionosphere can cause scattering of radio waves due to minute inhomogeneities in the refractive index. Additionally, meteors impinging on the ionosphere may also cause the scattering of radio waves. Like the troposphere, the ionosphere is a medium of continually varying properties. Superimposed on this slow variation is a random fast variation or fluctuation.

There are two reasons why the study of propagation is important for the design of a cellular system. First, it provides the necessary tool for computing the field intensity at different coverage areas of the cell. Since the primary coverage area in most cases lies from 0.5 mile to 10–15 miles maximum, the ground-wave mode of propagation is used. Second, it allows computation of adjacent and co-channel interference, which is limited in distance to less than 50–60 miles from a cell site.

There are two primary approaches used for predicting the radio-wave field strength. The first involves purely theoretical methods, which are suitable for treating isolated features such as hills and other fixed obstructions. However, this approach ignores the irregularity of the earth in its predictions. The other approach is based on measurements in various environments to account for the effects of irregular terrain and man-made obstructions, especially for higher frequencies and low mobile antenna heights prevalent in vehicular communication. A third approach, which is a combination of the above two, is primarily a well-verified model based on measurements and accounts for hills and other obstructions by applying the diffraction theory. In cellular radio design, there are at least two popular propagation models. The first is the FCC-recommended model by Roger B. Carey, which arbitrarily considers the limits of coverage from a cell-site transmitter to be at the 39-dBu contour of the cell, and the calculations of the contour are to be made according to the principles of [2]. The other design model, by Okumura, predicts coverage contours consistent with the experimentally verified results [4].

This chapter is arranged around these two models. After briefly introducing transmission and terrain-related parameters in Section 4.2, we discuss free-space propagation loss where a clear path and no ground reflections are considered. In Section 4.4, we introduce the reflection effects of flat ground. When the distance between the source and the receiver approaches the horizon distance, the curvature of the earth begins to play a part and the spherical smooth earth effect must then be considered. This is discussed in Section 4.5. Bullington's approach for propagation over smooth Earth is discussed next. Since the radio path might be obstructed by natural and man-made obstructions such as hills and buildings, concepts of Fresnel zone and knife-edge modeling are examined in Sections 4.7 and 4.8. Since the surface roughness is a natural characteristic of the terrain, rough surface criteria is discussed in Section 4.9. Section 4.10 is fully devoted to the UHF propagation modes by Carey and Okumura. The accuracy of these models is judged by critically examining test results from measurements conducted at different locations in the United States and around the world. Since there is a need to use the experience gained through these experiments at other locations, different adjustments to the models are considered as a part of this section.

4.2 EXPRESSING PROPAGATION CHARACTERISTICS

In the study of propagation, we are interested in the received signal level for a specified transmitter power. The transmitted signal is reduced due to the intervening propagation path and terrain. This reduction in signal strength is known as transmission loss. Various terms are used to express the transmission loss and define the terrain parameters. This section addresses both of these groups of parameters [6, 4].

4.2.1 Transmission-Related Parameters

The National Institutes of Standards and Technology (NIST) and the theoretical analysis by Bullington express transmission loss in decibels. The Federal Communications Commission (FCC) and the work of Okamura provide propagation in terms of field strength (dBu) for a 1 kW of effective radiated power (ERP). The use of the term field strength is due to the influence of the broadcast industry. In this section we shall define the following transmission-related parameters:

- Transmission loss.
- Definition of a sector.
- Transmission loss distribution.
- Median value of transmission loss.
- Loss deviation.
- Median transmission loss distribution.
- dBu contour.
- Service contour.

Transmission Loss

Transmission loss (TL) is the ratio of the effective radiated power of a transmitting cell-site antenna to the power received at the properly terminated receive mobile antenna minus the gain of the receiving mobile antenna. The ratio, in decibels, is expressed as

$$TL = \underbrace{10 \log P_c + G_c}_{ERP} - \underbrace{(10 \log P_m - G_m)}_{\text{received power}} \tag{4.1}$$

where P_c is the transmitter output power of the cell site, P_m is the received power at the mobile (receiver input), G_c is the transmitting (cell site) antenna gain in decibels, and G_m is the gain of the receiving mobile antenna in decibels. If the receiver uses an isotropic antenna, $G_m = 0$ dB. For a half-wave dipole, $G_m = 2.1$ dB (1.6 numeric).

Example 4.1

Find the net allowable transmission loss, TL, assuming a mobile transmit output power of +36 dBm to a cell-site omniantenna gain of +10 dB. Let the minimum usable receiver input power be −103 dBm and the mobile antenna gain be 3 dB. $TL = 10 \log P_c + G_c - 10 \log P_m + G_m = 36 + 10 + 103 + 3 = 152$ dB. If we use a sector antenna at the cell site with a gain of +17 dB, the total transmission loss can be increased to 159 dB, or for the same channel, receiver input power is increased by 7 dB over an omnidirectional antenna of 10-dB gain.

Definition of a Sector

A sector describes the bounds of a large number (at least four per wavelength) of transmission loss measurement locations. It is usually along the straight path of a vehicle for a short distance of travel, but at times it can be over a small area. Sectors are identified by type corresponding to distance traveled (e.g., 20m, 40m, and 200m in length), which can be along a straight path or a defined boundary. A typical sector is shown in Figure 4.2.

Transmission Loss Distribution

Transmission loss distribution (TLD) is the percentage of TL measurements made in a given sector (area or straight path) that are less than a specific value. The TLD for a sector, in general, is time independent and thus is expressed in terms of the percent of locations.

$$\text{TLD} = P\% \text{ of TL} \tag{4.2}$$

where TL $\leq$ level x/specific sector.

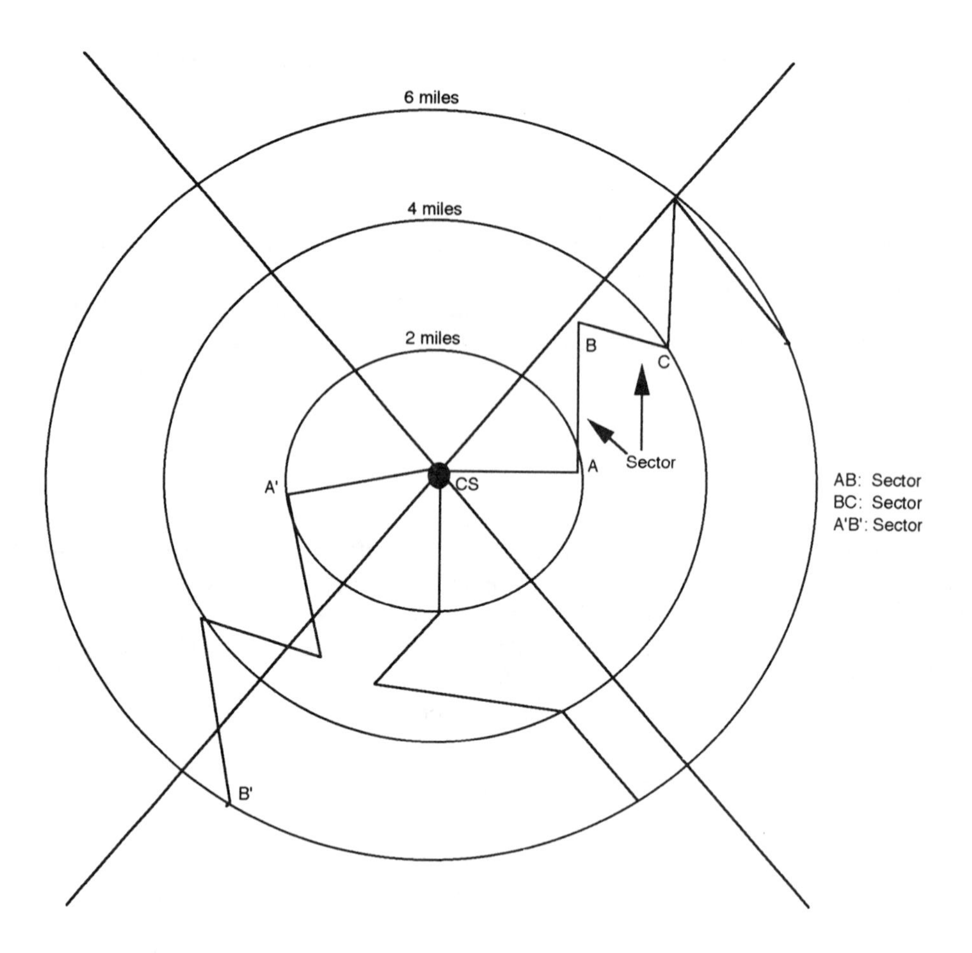

Figure 4.2 Typical measurement locations (sectors).

Median Value of Transmission Loss

Median transmission loss (MTL) is the loss corresponding to 50% of the locations of TLD. This provides the long-term average of the transmission loss distribution, which follows a log-normal distribution. The average transmission loss increases as the distance of the mobile increases from cell site. The deviation is taken to be 8.2 dB and is independent of the distance from the source, as shown in Figure 4.3.

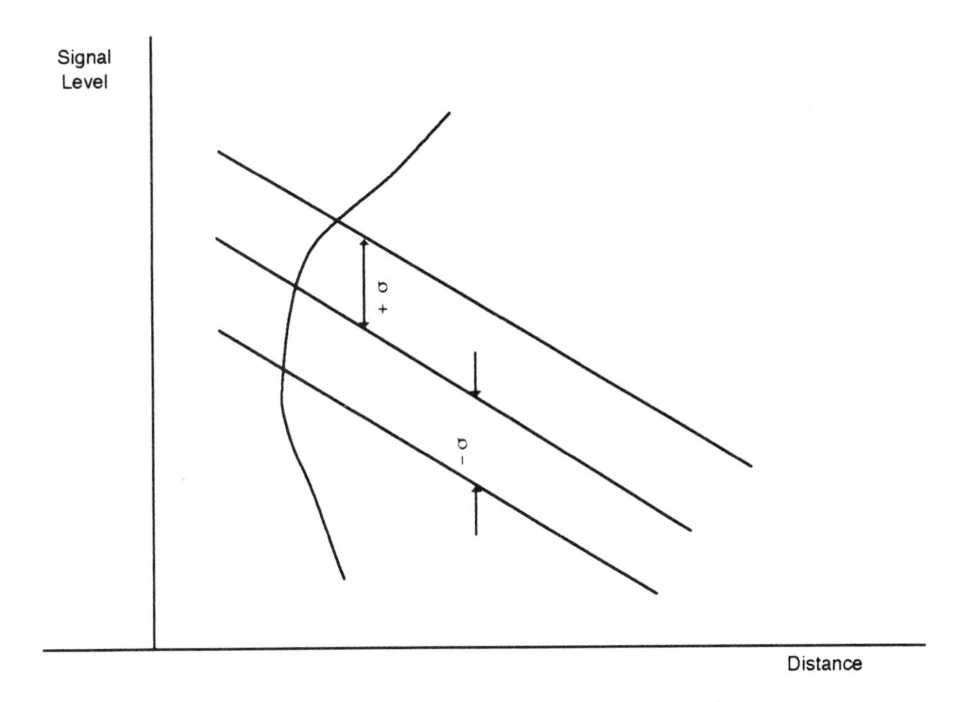

Figure 4.3 Median transmission loss variance.

Loss Deviation

The loss deviation (LD) is the decibel difference in attenuation between the 50% and 90% locations on a TLD curve. In general the value of LD should correspond to the Rayleigh fading loss, which has a theoretical value of 8.2 dB. However, values as high as 30 dB have been observed in some urban surroundings [8, 9].

Median Transmission Loss Distribution

Median transmission loss distribution (MTLD) is the percentage of MTLs for a specified path distance that are less than a given value. The 50% value of the MTLD forms the basis for propagation curves for a specified terrain. The standard deviation for the MTLD is a measure of the variation that can be expected for different types of terrain.

dBu Contour

As stated earlier, both the work of Okumura and the FCC express the received signal intensity in terms of dBu. For a half-wave dipole antenna, the field intensity can be derived as follows [2–4]:

$$P_m = p_m A_m \quad \text{watts} \tag{4.3}$$

where P_m is the mobile receiver input power, p_m is the power density at the mobile receiving antenna, in W/m^2, A_m is the effective area of the mobile antenna, since

$$p_m = E^2 10^{-12}/120\pi \tag{4.4}$$

where E is the field strength in microvolts per meter and 120π (or 377 Ω) is the wave impedance of the free-space wave. The effective area for a half-wave dipole is given by

$$A_m = 1.64\lambda^2/4\pi \tag{4.5}$$

where the factor 1.64 is the antenna gain, and λ is the wavelength in meters and is related to the transmitted frequency f in MHz as

$$\lambda = 300/f_{\text{MHz}} \tag{4.6}$$

With the help of (4.5) and (4.6), it is easy to see that for a half-wave dipole antenna the capture area, A_m, of the mobile is only one quarter at 900 MHz compared to that of 450 MHz. Hence, the 900 MHz half-wave dipole captures only 25% as much power at its antenna terminals in comparison to a 450 MHz antenna. Substituting (4.4), (4.5), and (4.6) in (4.3) we obtain the field intensity as

$$F_m = 20 \log_{10} E = 105 + 10 \log_{10} P_m + 20 \log_{10} f_{\text{MHz}} \tag{4.7}$$

For $F_m = 39$ dBu and $f_{\text{MHz}} = 900$, the received signal power, 10 log P_m, from (4.4) is -125 dBw. This is the median value of received signal power. In UHF service, there is a 90% probability that the signal in a given area will at least be 14 dB below the average median value for the area. Thus, with a 90% reliability, the signal level can only be guaranteed to be -139 dBw. For satisfying the requirement of 39 dBu, the receiver sensitivity must meet or exceed -139 dBW. This corresponds to about 0.8 microvolts across 50 ohms at the receiver input.

Service Contour

The signal contour for a specified receiver sensitivity must be plotted around the cell site to define the coverage area. This contour is a statistical boundary. A typical 95%, one-microvolt service contour is shown in Figure 4.4(a) by the solid line. As the mobile unit travels along this boundary, for 95% of all the locations it is expected to receive signal levels above one microvolt. A typical plot for signal level variation is shown in Figure 4.4(b). The shaded areas along the ordinate represent points along the contour for which the signal is less than one microvolt and represents 5% of the total path length. Signal levels at all other points are higher than one microvolt and can be 15–20 dB stronger. As shown in Figure 4.4(a), the one-microvolt 50% contour covers a larger area. The contour for 0.5 microvolt lies in between the one-microvolt 95% contour and the one-microvolt 50% contour.

Due to the statistical nature of the contour, it is desirable to have some overlap among adjacent cells [3, 8]. This overlap will provide a higher coverage probability,

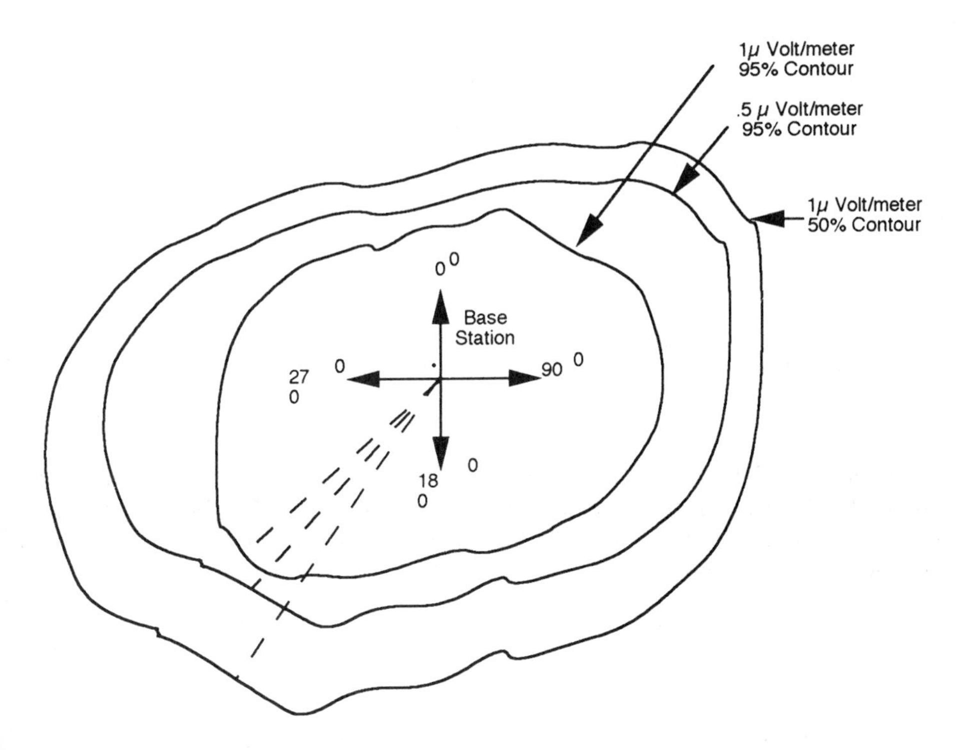

Figure 4.4(a) Typical statistical service contours.

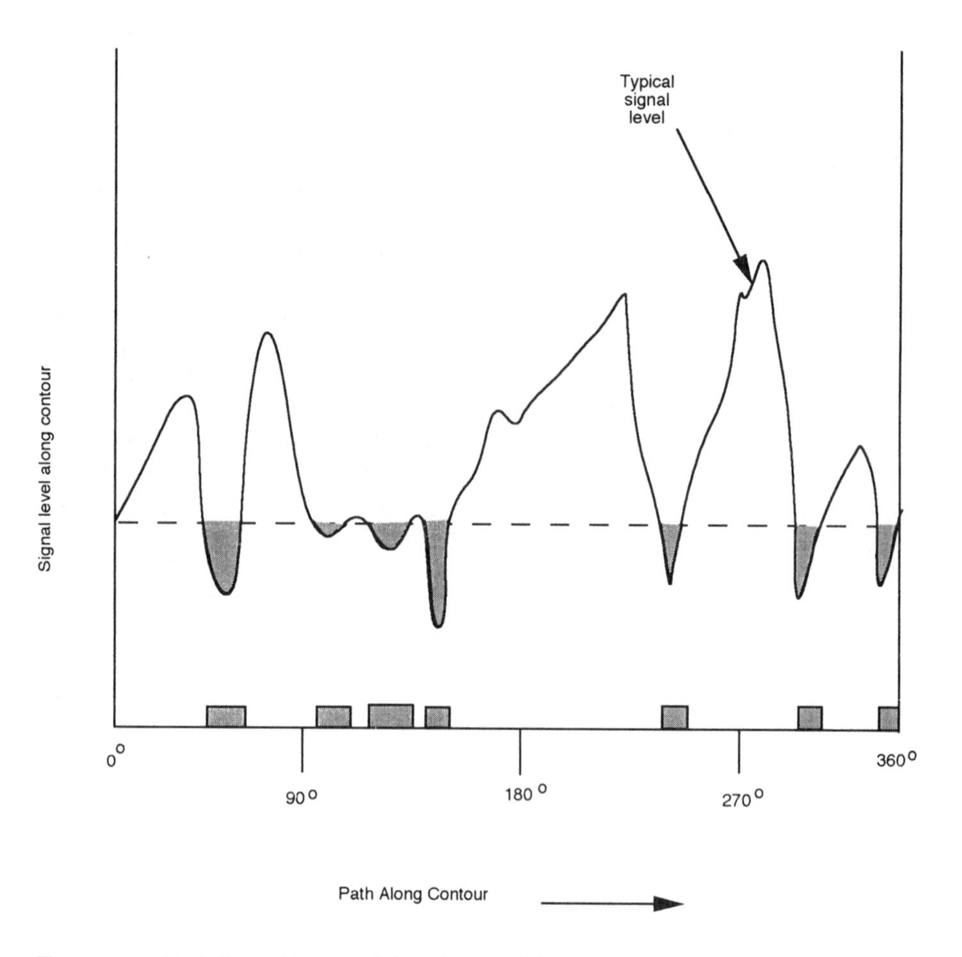

Figure 4.4(b) Typical signal level variation along a 95% contour.

and is shown in Figure 4.5. If two 1-μV 95% contours are for cell site 1 and 2, the probability for the signal to be one microvolt or more at A is 95% for each base station. Thus, the probability that at least one of the signals will be one microvolt or more is the joint probability and is given by the following expression:

$$\begin{aligned} P &= 1 - (1 - P_1)(1 - P_2) \\ &= 1 - (0.05)(0.05) = 0.9975 \end{aligned} \tag{4.8}$$

Therefore, at the intersection points of the two circles, A and A', the probability that the signal will be one microvolt or more is better than 99.45%. Obviously, in

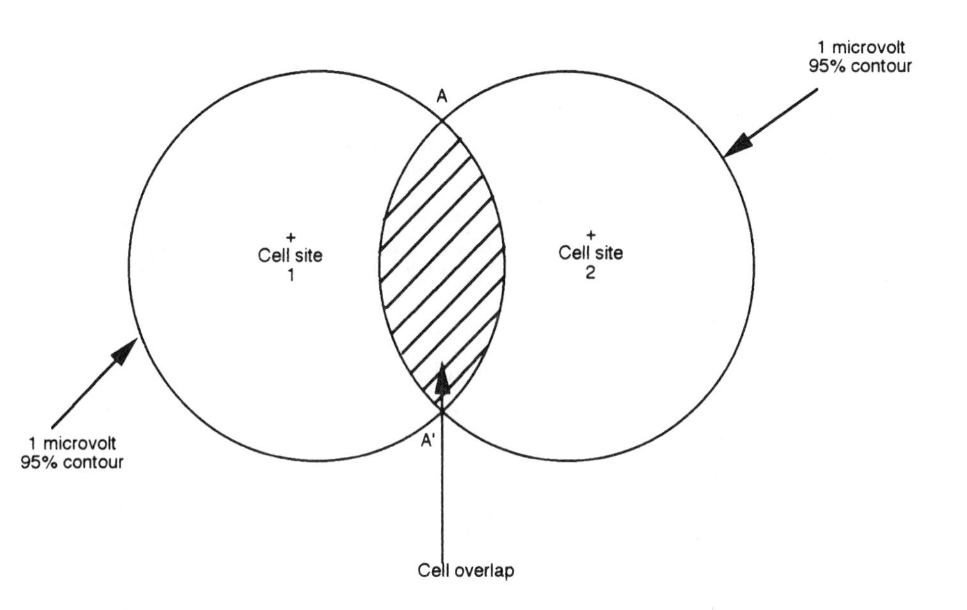

Figure 4.5 Cell overlap geometry and service contour.

the shaded area the probability that the signal level will be greater than 1 μV will be even higher.

4.2.2 Terrain Parameters

As stated earlier, terrain features affect propagation characteristics. Propagation curves have to account for environmental clutter and terrain irregularities correctly so that the field strength predicted by using these terrain irregularities comes close to the actual measured value of the propagation loss. There are an infinite variety of terrain features, but for reference analysis "quasismooth terrain" is assumed. Whether in quasi-smooth or irregular terrain, buildings and trees near the vehicular antenna affect the received field strength in different ways. If we classify obstacles in areas according to building and trees, it will make the prediction of field strength in a given area very difficult. Thus, rather than classifying them into many minute groups, we shall classify the environmental clutter according to the degree of congestion in a geographical area, as shown in Table 4.1 [9].

Environmental clutter of the urban type consists of larger buildings, which can be further subdivided into dense urban, moderate urban, and geographically limited urban. Details of terrain features and their definitions are stated below.

Table 4.1
Classification of Terrain

Category	*Characteristics*
Open area	Very few structures in the propagation path, such as tall buildings and trees: farm land, open fields, and rice fields, for example
Suburban area	Residential, with small one or two-storied houses, area with some obstacles near the vehicular radio, but still not very congested
Urban	Heavily built-up areas with tall buildings and high-rise apartment buildings

As stated earlier, quasi-smooth terrain is defined as a flat terrain where the undulation height of the surface is 20 meters or less with gentle ups and downs: the average level of the ground does not differ very much. A correction factor is applied for terrain undulations greater and less than 20 meters by CCIR [10]. The scatter diagram for the terrain undulation correction factor by Okumura at 922 MHz is shown in Figure 4.6. By knowing the terrain undulation height, Δh, in meters, one can find the correction factor in decibels over Δh of 20 meters.

The terrain undulation height Δh for rolling hill terrain is defined by Okumura as the difference between 10% and 90% of the terrain undulation height within a distance of 10 km in front of a mobile antenna. This is shown in Figure 4.7(a) [4]. The CCIR defines the parameter Δh as the difference between the elevation exceeded by 90% of the terrain and the elevation exceeded by 10% of the terrain

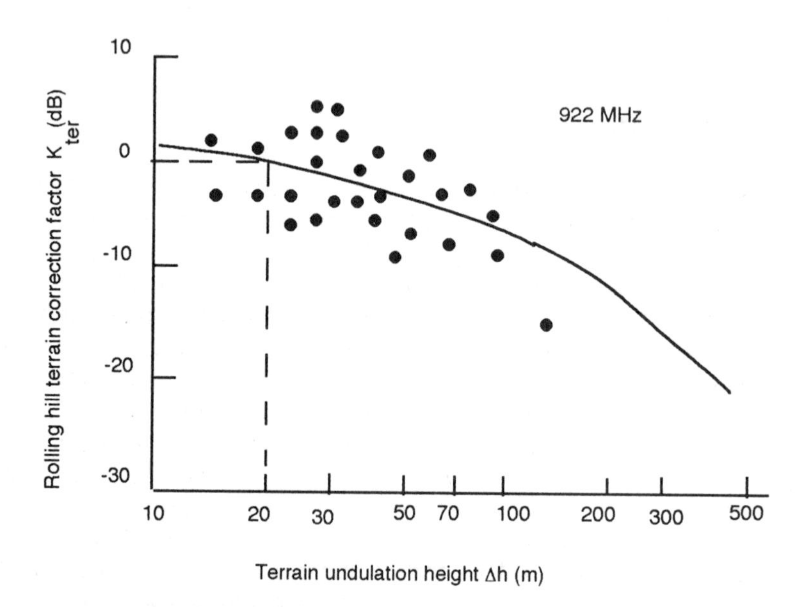

Figure 4.6 Okamura correction factor for terrain undulation height.

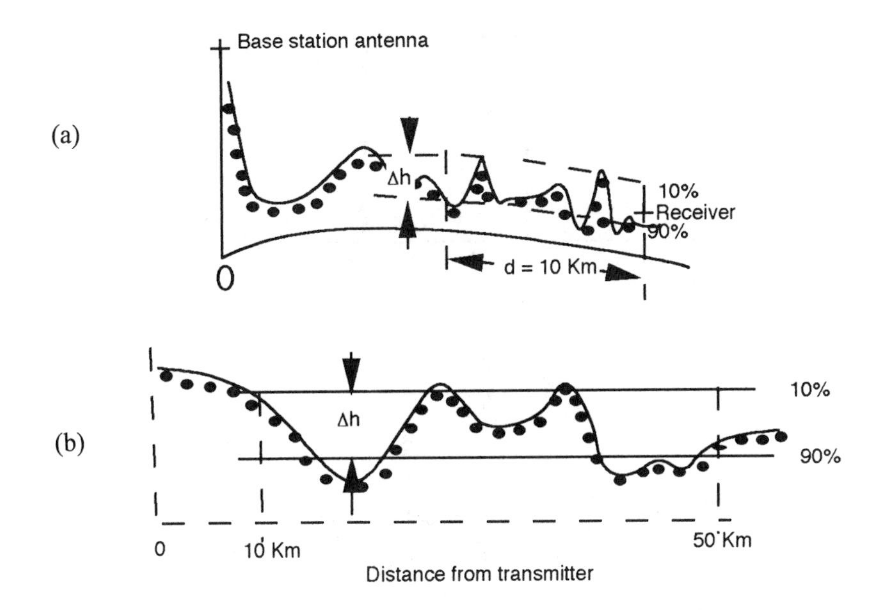

Figure 4.7 (a) Definition of the parameter Δh (terrain undulation height) for rolling hilly terrain; (b) CCIR definition of the parameter Δh.

in the range 10–50 km in front of the transmitter. This is shown in Figure 4.7(b) [10]. CCIR provides correction curves for path loss for Δh up to 500m with Δh = 50m as a reference value of zero-decibel correction. All forms of terrain irregularity are treated via the parameter Δh, with no distinction made between undulations and isolated features. Okumura further considers the terrain undulations in terms of isolated mountains and ridges, the general sloping terrain, and mixed land-sea paths. The isolated mountain has a knife-edge effect at UHF frequencies where the height, h, of the mountain is measured from the average ground level. This is shown in Figure 4.8(a). Okumura's model for an isolated ridge is of a standard height of h = 200m and is shown in Figure 4.8(b). For other heights, proportional attenuation is obtained. General sloping terrain, whether flat or undulated, sloping over a distance of at least 5–10 km is shown in Figure 4.9. The terrain parameter in terms of the average general slope, θ_m, is defined as

$$\theta_m = (h_n - h_m)/d_n \tag{4.9}$$

For a rising hill, the slope θ_m is considered to be positive, and for a falling hill the slope is considered to be negative. For a mixed land-sea path, Okumura defines the distance parameter, β, as the ratio of length covered by sea divided by the total range, as shown in Figure 4.10. It is expected that the signal coverage is greater in mixed configurations of land and sea than for the land case only. Thus,

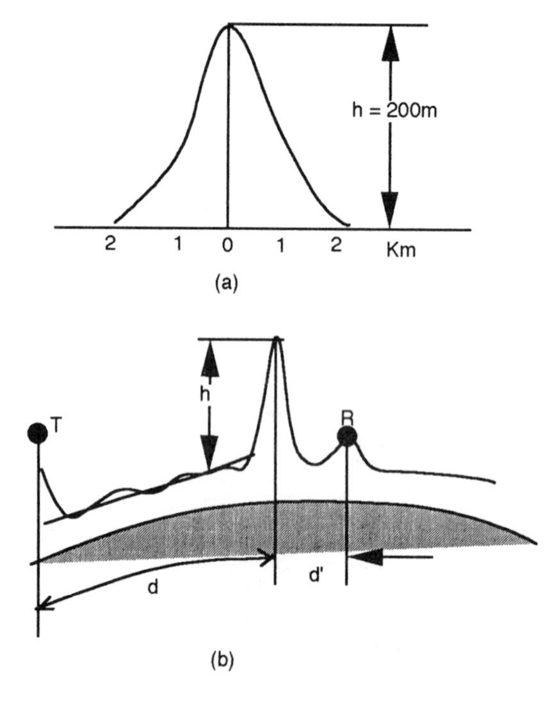

Figure 4.8 The model (a) and geometrical parameter (b) of an isolated ridge.

an exhaustive classification of terrain has been covered by Okumura, while the CCIR classification is general in nature.

4.3 FREE-SPACE PROPAGATION [1, 11–13]

A convenient starting point for the propagation study is the condition of two antennas in free space—an isotropic, homogeneous, and non-absorbing (no atmosphere) medium of dielectric constant unity. We shall consider this below.

4.3.1 Relation Between Transmitted Power and Receive Field Intensity [1, 11]

Assume that a transmitter is radiating an output power of P_c watts from an antenna whose directional gain is G_c and that a receiver is situated at a distance of d meters from the transmitter. Thus, the power flux density or the Poynting vector, S_m, is given by

$$S_m = \frac{P_c G_c}{4\pi d^2} \quad \text{W/m}^2 \tag{4.10}$$

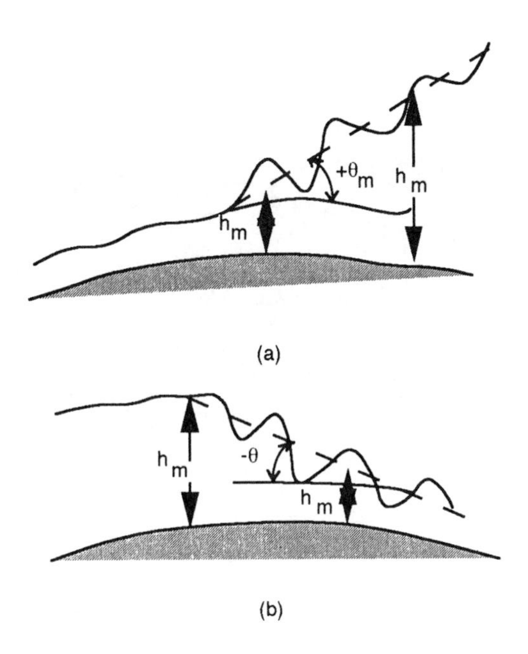

Figure 4.9 Definition of average angle of general terrain slope: (a) positive slope $(+\theta_m)$; (b) negative slope $(-\theta_m)$

If E_{rms} is the electric field strength of a radio wave in volts per meter and H_{rms} is the magnetic field strength of the radio wave, in amperes per meter, then the pointing vector S_m can be expressed as

$$S_m = H_{\text{rms}}E_{\text{rms}} \quad \text{W/m}^2 \tag{4.11}$$

where

$$H_{\text{rms}} = E_{\text{rms}}/120\pi \quad \text{A/m} \tag{4.12}$$

From (4.10), (4.11), and (4.12), we obtain the relationship between the receive field intensity, E_{rms}, and the radiated power, P_cG_c, as

$$E_{\text{rms}} = \sqrt{30P_cG_c}/d \quad \text{V/m} \tag{4.13}$$

Thus the peak field intensity E_m, is given by

$$E_m = \sqrt{60P_cG_c}/d \quad \text{V/m} \tag{4.14}$$

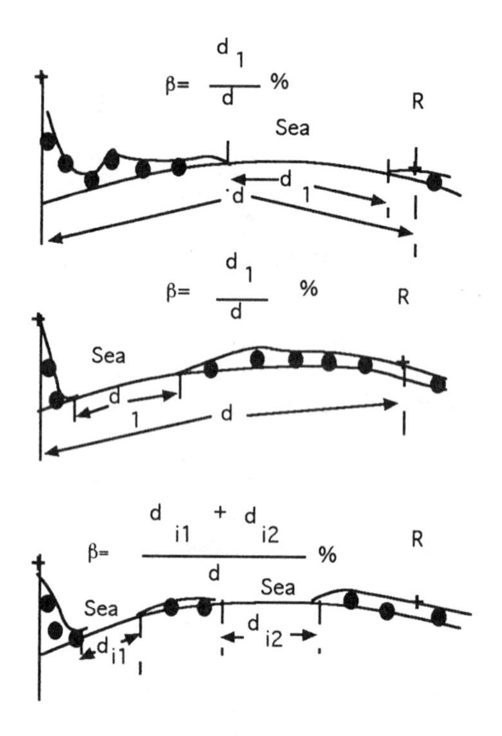

Figure 4.10 Definition of distance parameter β for a mixed land-sea path.

The instantaneous value of the receive field at a distance d is written as

$$E_m = \frac{\sqrt{60P_cG_c}}{d}\cos\omega\,(t - d/c) = \frac{\sqrt{60P_cG_c}}{d}\cos(\omega t - kd)\quad \text{V/m} \tag{4.15}$$

where $k = (2\pi/\lambda)$ is constant for a specified λ and ω is the angular frequency of operation in radians per second.

If the power, P_c, is expressed in kilowatts, $P_{c,kw}$ the distance d in kilometers and the field intensity E in millivolts per meter, then the equation for rms and peak field intensity can be rewritten as

$$E_{\text{rms}} = 173\,\frac{\sqrt{P_{c,\text{kw}}G_c}}{d_{\text{km}}}\quad \text{mV/m} \tag{4.16a}$$

and

$$E_m = 245 \frac{\sqrt{P_{c,\text{kw}} G_c}}{d_{\text{kw}}} \quad \text{mV/m} \tag{4.16b}$$

The above two equations establish the relationship between transmitted power and the value of field intensity at the receiver.

4.3.2 Free-Space Loss

If E is the rms value of field intensity at antenna input, then the received power is given by

$$P_m = \frac{E^2}{120\pi} A_m \quad \text{Watts} \tag{4.17}$$

where A_m is the effective aperture area of the vehicular antenna. Since the aperture, A_m, and the antenna gain, G_m, are related by

$$A_m = G_m \lambda^2 / 4\pi \tag{4.18}$$

we obtain

$$P_m = (E\lambda/2\pi)^2 \, G_m/120 \quad \text{watts} \tag{4.19}$$

From (4.13), the receiver field intensity, E_o (say), is related to the transmitted power, P_c, as

$$P_c = E_0^2 d^2 / 30 G_c \tag{4.20}$$

Since both E and E_0 are the rms values of the receive field intensity, they are equal and hence

$$P_c/P_m = (4\pi d/\lambda)^2 \, 1/G_c G_m \tag{4.21}$$

In radio system design, we use the concept of transmission loss (space loss): The ratio of power radiated from the cell-site antenna to the power available at the terminals of the mobile receive antenna when the system is operating in free space. Therefore,

$$L_{fs} = P_c/P_m = (4\pi d/\lambda)^2 \, 1/G_c G_m \tag{4.22}$$

To make (4.22) independent of antenna gains, basic transmission loss, L_p (or path loss) is defined. This is the transmission loss between two isotropic sources of unity gain, and is given by

$$L_p = (4\pi d/\lambda)^2 \tag{4.23}$$

Since transmission loss may vary over a very wide range, it is usual to express L_{fs} and L_p in decibels and not in watts. Thus,

$$L_{fs,\mathrm{db}} = 10 \log_{10} L_{fs} = 20 \log(4\pi d) - 20 \log \lambda - G_{c,\mathrm{db}} - G_{m,\mathrm{db}} \tag{4.24}$$

and

$$\begin{aligned} L_{p,\mathrm{db}} &= 10 \log L_p = 20 \log(4\pi d) - 20 \log_{10} \lambda \\ &= 32.4 + 20 \log_{10}(f_{\mathrm{MHz}}) + 20 \log_{10}(d_{\mathrm{km}}) \\ &= 37.0 + 20 \log_{10}(f_{\mathrm{MHz}}) + 20 \log_{10}(d_{\mathrm{miles}}) \end{aligned} \tag{4.25}$$

The above formula shows an inverse square law dependence on distance d. Thus, as d is doubled, free-space path loss increases by 6 dB. The formula also shows that as λ is decreased (high f), the path loss increases. In order to compensate for these losses, the transmit and receive antenna gains can be increased.

Example 4.2

Compute the path loss at 900 MHz between the cell site and a mobile located at a radial distance of 15 km from the cell site.

$$\begin{aligned} \lambda &= 300/900 = 0.3333\mathrm{m} \\ L_{p,\mathrm{dB}} &= 20 \log(4\pi \times 15 \times 10^3) - 20 \log(0.3333) \\ &= 115 \text{ dB} \end{aligned} \tag{4.26}$$

4.4 PROPAGATION OVER FLAT EARTH [1, 15]

After arriving at the path loss equation in free space, it is logical to consider real-life propagation between two antennas erected over the flat, imperfectly conducting surface. We assume in our treatment that the earth is perfectly smooth (no diffraction) for the entire length of the propagation path. The problem can be stated as follows. Given the antenna heights h_c and h_m of the cell site and mobile units respectively, separated by a distance d over a smooth earth, find the attenuation function F that describes the received field intensity as

$$E_{\text{rms}} = \frac{\sqrt{30P_cG_c}}{d} F \quad \text{V/m} \tag{4.27}$$

where the transmitted power P_c is in watts, the transmit antenna gain is G_c, and d is in meters. As shown in Figure 4.11(b), the electric field at the mobile antenna h_m consists of a direct wave designated as (a) and a single reflected wave (b). It should be noted that other reflected waves are not captured by the mobile due to point receiving antenna and therefore are not considered in the following analysis. The direct and the reflected instantaneous waves E_1 and E_2 can be defined as

$$E_1 = \frac{\sqrt{60P_cG_c}}{d} e^{j\omega t} \quad \text{V/m} \tag{4.28a}$$

$$E_2 = R\frac{\sqrt{60P_cG_c}}{d + \Delta d} e^{j\left(\omega t - \theta - \frac{2\pi\Delta d}{\lambda}\right)} \quad \text{V/m} \tag{4.28b}$$

where θ is the phase shift due to reflected wave, R is the reflection coefficient, and $(2\pi\Delta d/\lambda)$ is the phase shift due to the path difference, $AC + CB - AB$, as shown in Figure 4.11(c).

If we assume that the antenna heights $h_c \ll d$ and $h_m \ll d$, and the direct path length AB and the reflected wave $AC + CB$ differ only slightly. Thus $d + \Delta d \approx d$ can be used for the denominator of (4.28b). However, the phase shift term $(2\pi\Delta d/\lambda)$ in the exponent cannot be neglected since Δd is not negligible compared to λ. Thus, the resultant electric field at the receiving point B is the vector sum of the direct and the reflected waves represented by (4.28a) and (4.28b).

$$E = E_1 + E_2 = \frac{\sqrt{60P_cG_c}}{d}\left[1 + \text{Re}^{-j(\theta + 2\pi\Delta d/\lambda)}\right] e^{j\omega t} \quad \text{V/m} \tag{4.29}$$

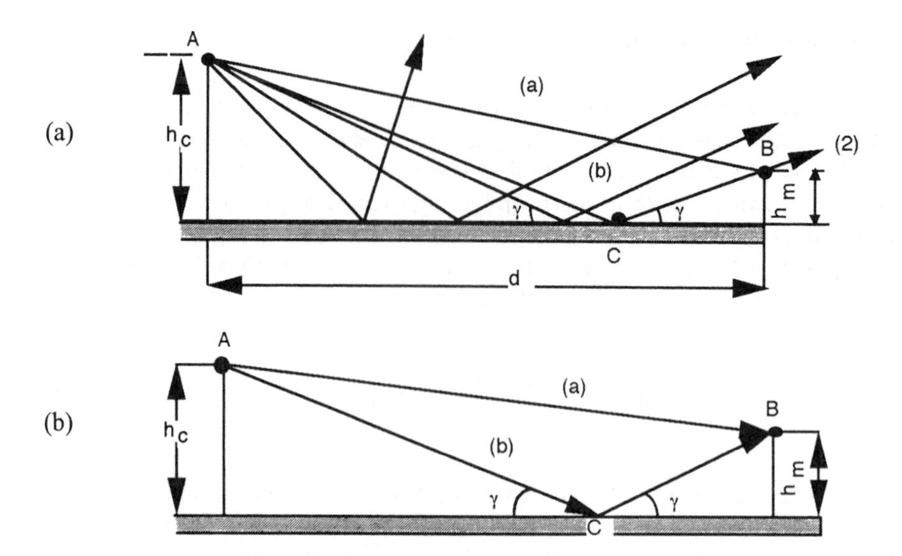

Figure 4.11 The wave field at the receiving antenna: multiple (a) and (b) single reflections.

Expanding the parenthetical quantity $1 + \mathrm{Re}^{-j(\theta + (2\pi \Delta d/\lambda)}$ we obtain

$$\begin{aligned} 1 + \mathrm{Re}^{-j(\theta + (2\pi\Delta d)/\lambda)} &= 1 + \mathrm{Re}^{-j\alpha} \\ &= \sqrt{(1 + R\cos\alpha)^2 + R^2 \sin^2 \alpha}\, e^{-j\varphi} \\ &= \sqrt{1 + R^2 + 2R\cos\alpha}\, e^{-j\varphi} \end{aligned} \tag{4.30}$$

where $\alpha = \theta + (2\pi\Delta d)/\lambda$ and $\varphi = \tan^{-1}(R\sin\alpha)/(1 + R\cos\alpha)$.

Therefore, the resultant value of the field at B is:

$$E = \frac{\sqrt{60 P_c G_c}}{d} \sqrt{1 + R^2 + 2R\cos\left(\theta + \frac{2\pi\Delta d}{\lambda}\right)}\, e^{j(\omega t - \varphi)} \tag{4.31}$$

The equation above contains two unknowns—R, the reflection coefficient, and Δd, the path difference between the reflected and the direct path. In the next section we shall show that the reflection coefficient R is a function of the grazing angle γ, the polarization of the field, the dielectric constant, the conductivity of the earth, and the frequency of transmission. For now let us assume that R is known and our only problem is to find Δd. To find Δd, the geometric construction as shown in

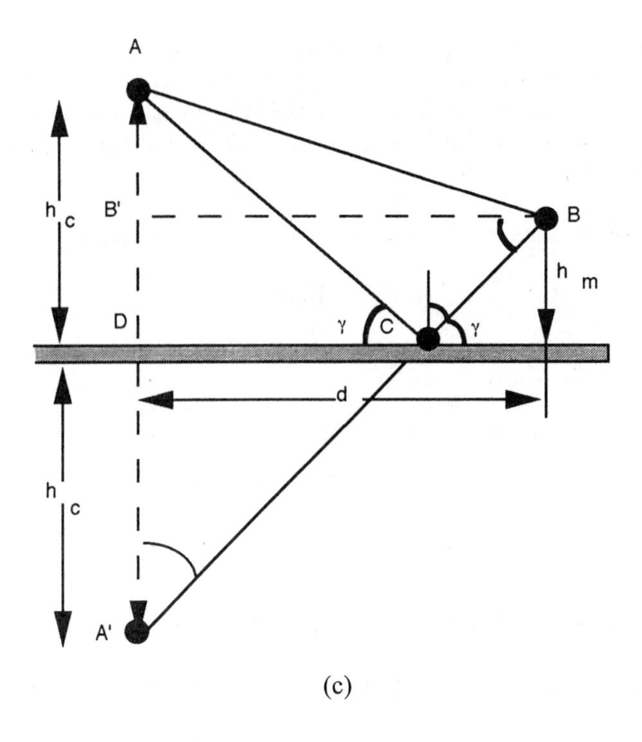

(c)

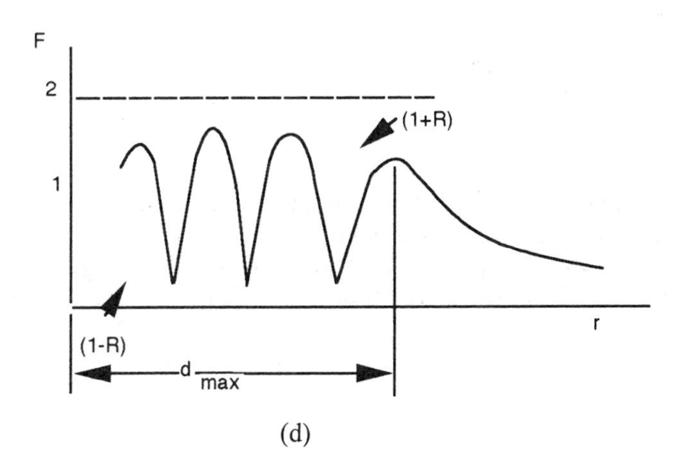

(d)

Figure 4.11 (cont.) (c) Angle of elevation, γ, and the path length, d, difference Δd; (d) attenuation function, F, versus distance.

Figure 4.11(c) is considered. The line BC is extended to intersect the perpendicular dropped from A to the horizontal plane. The intersection point is denoted as A', which is the mirror image of A. The line CD is the right bisector of the triangle ACA'; that is, $AC = A'C$ and $AD = A'D$. From B, let us draw the perpendicular BB' on the ordinate AD. Thus,

$$\begin{aligned} AB &= \sqrt{d^2 + (h_c - h_m)^2} = d\sqrt{1 + [(h_c - h_m)/d]^2} \\ &\approx d + \frac{1}{2d}(h_c^2 + h_m^2 - 2h_ch_m), \end{aligned} \tag{4.32}$$

where $(h_c - h_m)/d \ll 1$, and

$$\begin{aligned} ACB &= A'B = \sqrt{d^2 + (h_c + h_m)^2} \\ &= d\sqrt{1 + [(h_c + h_m)/d]^2} \\ &\approx d + \frac{1}{2d}(h_c^2 + h_m^2 + 2h_ch_m) \end{aligned} \tag{4.33}$$

where $(h_c + h_m)/d \ll 1$. Therefore, the path difference Δd is

$$ACB - AB = \Delta d = A'B - AB = 2(h_ch_m)/d \quad \text{meters} \tag{4.34}$$

Substituting this value of Δd in (4.31), we obtain the resultant value of the electric field as

$$E = \frac{\sqrt{60P_cG_c}}{d}\sqrt{1 + R^2 + 2R\cos\left(\theta + \frac{4\pi h_ch_m}{\lambda d}\right)}\, e^{j(\omega t - \varphi)} \tag{4.35}$$

The above equation can further be simplified if we assume R to be unity (perfect reflection), and the phase angle θ to be 180 deg. As we will see in the next section, this assumption will hold good in general for the small grazing angle γ. Substituting $R = 1$ and $\theta = 180$ deg, the magnitude of (4.35) can be rewritten as

$$E = \frac{\sqrt{60P_cG_c}}{d}\sqrt{2\left(1 - \cos\frac{4\pi h_ch_m}{\lambda d}\right)} = \frac{2\sqrt{60P_cG_c}}{d}\left|\sin\frac{2\pi h_ch_m}{\lambda d}\right| \tag{4.36}$$

or

$$E^2 = 4\,(60P_cG_c/d^2)\,\sin^2(2\pi h_ch_m/\lambda d) \tag{4.37}$$

where E is the peak voltage intensity at the receiver input in volts per meter. From (4.19), the received field intensity E is related to the receiver input power as

$$P_m = (E_{\text{rms}}^2/120\pi)\,(G_m\lambda^2/4\pi) \tag{4.38}$$

and thus,

$$P_m = 4P_c\,(\lambda/4\pi d)^2\,G_cG_m\,\sin^2(2\pi h_ch_m/\lambda d) \tag{4.39}$$

In the above equation, $P_c(\lambda/4\pi d)^2G_cG_m$ can easily be identified as the received power over a free-space path (4.24). Denoting this as P_o, we get

$$P_m = 4P_o\,\sin^2(2\pi h_ch_m/\lambda d) \tag{4.40}$$

For a small value of the argument $\sin(\theta) \cong \theta$. Thus,

$$P_m \approx 4P_o\,(2\pi h_ch_m/\lambda d)^2 \tag{4.41}$$

Substituting, P_o in terms of transmitted power P_c, one can rewrite the above equation as

$$P_m = P_cG_cG_m\,(h_ch_m/d^2)^2 \tag{4.42a}$$

Taking the logarithm on both sides, we can write

$$P_m = 10\log_{10}(P_cG_cG_m) - 40\log d + 20\log h_c + 20\log h_m \quad \text{dBm} \tag{4.42b}$$

Equation (4.42a) shows the fourth power relationship of the received power with the distance from the cell-site antenna and a 6-dB per octave variation on antenna heights. Therefore, 6-dB less attenuation is encountered by doubling the cell-site antenna height. By applying the above equation, a 40% increase in range is seen by doubling the antenna height. The free-space loss also varies as the square of the path length, while for propagation over the flat earth, the received power at the mobile varies as the fourth power of the distance. Thus, the received power decreases by 12 dB when the distance is doubled. Experimentally, the exponent size for ground-wave mode of propagation for urban areas has been shown to be a function of base station antenna height and the distance from the base station. The exponent value varies from a low value of less than 1 to about 4.0. Also the quantity λ, though not explicitly shown in (4.42a), is part of antenna gains G_c and

G_m. Before we leave this section we shall once again explore the behavior of the quantity F in (4.35).

$$F = \sqrt{1 + R^2 + 2R\cos(\theta + 4\pi h_c h_m/\lambda d)} \tag{4.42c}$$

As stated earlier, the magnitude R and the phase angle θ of the reflection coefficient are functions of the reflection point, that is, the distance d, because the grazing angle γ varies with d. As d varies, the function F passes through a number of maxima and minima as $\cos\{\theta + [(4\pi h_c h_m)/\lambda d]\}$ assumes the value of $+1$ and -1, respectively.

At maxima, $F = 1 + R$, and at minima, $F = 1 - R$. Thus, F varies within limits of $+2$ and 0, assuming that R attains the maximum value of 1. The plot of F versus distance d is shown in Figure 4.11 (d). From the figure, the upper broken line represents the maximum value of $F = 1 + R$ and the lower broken line represents the minimum value of $F = 1 - R$. At some distance, d_{max}, where $\theta + [(4\pi h_c h_m)/\lambda d] = 2\pi$, the last maximum of the function F occurs. As d increases further, the argument of the cosine will tend towards π and the grazing angle becomes small. The value of F then decreases monotonically. The area of monotonic decrease of F is shown shaded in Figure 4.11(d). This maxima and minima is also observed from (4.36). The distance to the maxima can be obtained from

$$\frac{2\pi\, h_c\, h_m}{\lambda\, d_{max,n}} = \frac{\pi}{2}(2n + 1) \qquad n = 0, 1, 2, \ldots$$

or

$$d_{max,n} = 4h_c h_m/\lambda(2n + 1) \tag{4.43}$$

The distances to the minima can similarly be found as

$$d_{min,n} = 2h_c h_m/\lambda(n + 1) \tag{4.44a}$$

Thus, the first minima for $n = 0$ occurs at a distance of

$$d_{min,0} = 2h_c h_m/\lambda \quad \text{meters} \tag{4.44b}$$

Example 4.3

From (4.42a), the propagation loss (path loss) over flat earth is

$$L_{direct} = P_m/P_T = (h_c h_m/d^2)^2$$

where P_T is the transmitted power and P_m is the mobile received power. In this mode of transmission, we can see from (4.39) that the free-space loss $\lambda^2/(4\pi d)^2$ is further attenuated by a factor of $(4\pi h_c h_m)^2/(\lambda d)^2$.

4.4.1. Complex Permittivity Equation

In this section we derive the complex permittivity equation for an imperfect ground. We shall be using the results of this in finding the reflected value of energy, R, when an electromagnetic wave strikes the ground at a certain angle. The symbols used are:

- E, E_m,and E_{rms} are the instantaneous, peak, and rms value of the electric field expressed in volts per meter.
- H, H_m, and H_{rms} are the instantaneous, peak, and rms values of the magnetic field expressed in terms of amperes per meter.
- $\mu_o = 4 \times 10^{-7}$, and μ are the permeability of the free space and of the medium in henries per meter.
- μ' = ratio of permeability of the medium to the permeability of free space = μ/μ_o (dimension less).
- $\epsilon_o = (1/36\pi)10^{-9}$, and ϵ are the absolute (free space) and the medium permittivity in farads per meter.
- ϵ' = ratio of permittivity of the medium to the permittivity of free space or the relative permittivity = ϵ/ϵ_o (dimensionless)
- σ = conductivity of the medium in siemens per meter.

From Maxwell's first equation for an imperfect medium, the displacement, conduction, permittivity and total current are related by the differential equation as:

$$\epsilon \frac{\partial \mathbf{E}}{\partial t} + \sigma \mathbf{E} = \text{curl } \mathbf{H} \quad \text{Am/m}^2 \tag{4.45}$$

respectively, where the vector quantities for the instantaneous value of the electric field, E, are given by

$$\mathbf{E} = \mathbf{E}_m \, e^{j\omega t} \quad \text{V/m} \tag{4.46a}$$

or

$$\partial \mathbf{E}/\partial t = j\omega \mathbf{E}_m \, e^{j\omega t} \tag{4.46b}$$

where, E_m is the maximum value of the electric field. Substituting (4.46b) in (4.45), we obtain

$$(\sigma + j\omega\epsilon)\, \mathbf{E}_m e^{j\omega t} = \text{curl } \mathbf{H} \tag{4.46c}$$

Substituting E_m in terms of $\partial\mathbf{E}/\partial t$ from (4.46b), we rewrite the above equation as

$$(\sigma + j\,\omega\epsilon)\left(\frac{1}{j\omega}\frac{\partial \mathbf{E}}{\partial t}\right) = \text{curl } \mathbf{H} \tag{4.47a}$$

or

$$\left(\epsilon - j\frac{\sigma}{\omega}\right)\frac{\partial \boldsymbol{E}}{\partial t} = \text{curl } \mathbf{H} \tag{4.47b}$$

Comparing (4.47) with Maxwell's first equation (4.45) the complex permittivity of the imperfect medium ϵ_c is given by

$$\epsilon_c = \epsilon - j\,\sigma/\omega \tag{4.48}$$

Dividing the above equation by absolute permittivity, ϵ_0, the relative complex permittivity is given by $\epsilon'c = \epsilon' - j\,\sigma/(\omega\epsilon_0) = \epsilon' - j\,(\lambda\sigma 36\pi 10^9)/(2\pi C)$, where C is the velocity of light in vacuum and has a value of 3×10^8 m/s, $\omega = 2\pi f = 2\pi C/\lambda$, and $\epsilon_o = 10^{-9}/36\pi$.
Thus,

$$\begin{aligned}\epsilon_c' &= \epsilon' - j60\lambda\sigma \\ &= \epsilon' - j(18{,}000\sigma)/f_{\text{MHz}}\end{aligned} \tag{4.49}$$

for ground-wave propagation. Electrical characteristics of the ground must be accurately known. Typical values of the relative permittivity ϵ' and the conductivity of the medium, σ, for different types of ground are shown in Table 4.2. We now want to use the knowledge of the above equation to arrive at the reflected value R and the electrical phase shift, Φ_h or Φ_v, shown below in the ground-wave mode of propagation. This is the topic of the next section.

4.4.2 Reflection Coefficient

The ratio of the incident wave to its associated reflected wave is called the reflection coefficient. Neglecting the effect of refraction, the reflection coefficient for horizontal and vertical polarization can be computed by the following relationships:

$$\frac{\mathbf{E}_r}{\mathbf{E}_i} = \frac{\sin\varphi - \sqrt{\epsilon_c' - \cos^2\varphi}}{\sin\varphi + \sqrt{\epsilon_c' - \cos^2\varphi}} = R_h\, e^{j\phi_h} \tag{4.50a}$$

for horizontal polarization and

$$\frac{\mathbf{E}_r}{\mathbf{E}_i} = \frac{\epsilon_c' \sin\varphi - \sqrt{\epsilon_c' - \cos^2\varphi}}{\epsilon_c' \sin\varphi + \sqrt{\epsilon_c' - \cos^2\varphi}} = R_v\, e^{j\phi_v} \tag{4.50b}$$

for vertical polarization. Where φ is the grazing angle to the ground and Φ_v or Φ_h are the electrical phase shifts. For φ close to zero degrees, the ratio $\mathbf{E}_r/\mathbf{E}_i \cong -1$, or in other terms, $R_v = R_h = 1$ and $\Phi_v = \Phi_h = 180$ deg. It should be noted that the complex relative permittivity plays no part in arriving at this result.

Example 4.4

Find the reflection coefficient for a vertically polarized wave at a frequency of 900 MHz over smooth sea water when the grazing angle φ is 4 deg. The value of relative permittivity for sea water is 81 deg and the conductivity = 4.5 mhom/m^2.

$$\epsilon_c' = \epsilon' - j\,\frac{18{,}000\sigma}{f_{\text{MHz}}} = 81.0 - j\,90 = 121.08\, e^{-j48.01}$$

Thus,

$$R_v\, e^{j\phi_v} = \frac{0.1218693 \times 121.08\, e^{-j48.01} - \sqrt{81.0 - j\,90 - 0.9851479}}{0.1218693 \times 121.08\, e^{-j48.01} + \sqrt{81.0 - j\,90 - 0.985179}}$$

Table 4.2
Typical Earth Constants

Type of Surface	*ε′(Average)*	*Average σ (mho/meter)*
Fresh water (lakes and rivers)	81	0.001
Sea water	81	5.0
Good ground	25	0.02
Average ground	15	0.005
Poor ground	4	0.001
Mountains	—	0.00075

$$= \frac{14.75594\, e^{-j48.01} - 10.97387\, e^{-j24.18059}}{14.75594\, e^{-j48.01} + 10.97387\, e^{-j24.18059}}$$

$$= \frac{9.871737 - 10.9675j - 10.01093 + 4.49505j}{9.87173 - 10.9675j + 10.01093 - 4.49505j} = 0.257e^{-53.35j}$$

Thus, $R_v = 0.257$ and $\Phi_v = -53.360$.

4.5 EFFECTS OF THE EARTH'S CURVATURE

We define the radio horizon to be the distance between the elevated base or cell-site antenna and the mobile antenna on the ground. The problem in this case is to find the distance to the visible horizon. Considering the triangle OAB in Figure 4.12(a), we get

$$\cos\theta = \frac{a}{a + h_c} \approx 1 - h_c/a \qquad h_c/a \ll 1 \tag{4.51}$$

where a is the radius of the earth and equal to 6.4×10^6m. In our application, the geocentric angle θ can be considered small, and thus $\cos\theta \approx 1 - \theta^2/2$. Thus, (4.51) can be written as

$$\theta = d_1/a = \sqrt{2h_c/a} \tag{4.52a}$$

Thus,

$$d_1 = \sqrt{2ah_c} \tag{4.52}$$

A generalization for the two elevated antennas having heights h_c and h_m is shown in Figure 4.12(b). Trans-horizon distance $d = d_1 + d_2$ can simply be calculated by extending the above equation

$$d = \sqrt{2ah_c} + \sqrt{2ah_m} \tag{4.53}$$

in meters. Thus for 100′ cell site and 10′ mobile antenna heights, the maximum transhorizon distance is equal to 25.8 km. Therefore, the maximum distance to which the coverage can be extended is about 26 km from the cell site irrespective of the 39-dBu contour distance.

In the previous section we derived the propagation equation between two antennas, h_c and h_m, when they are erected over the flat earth. The flat earth

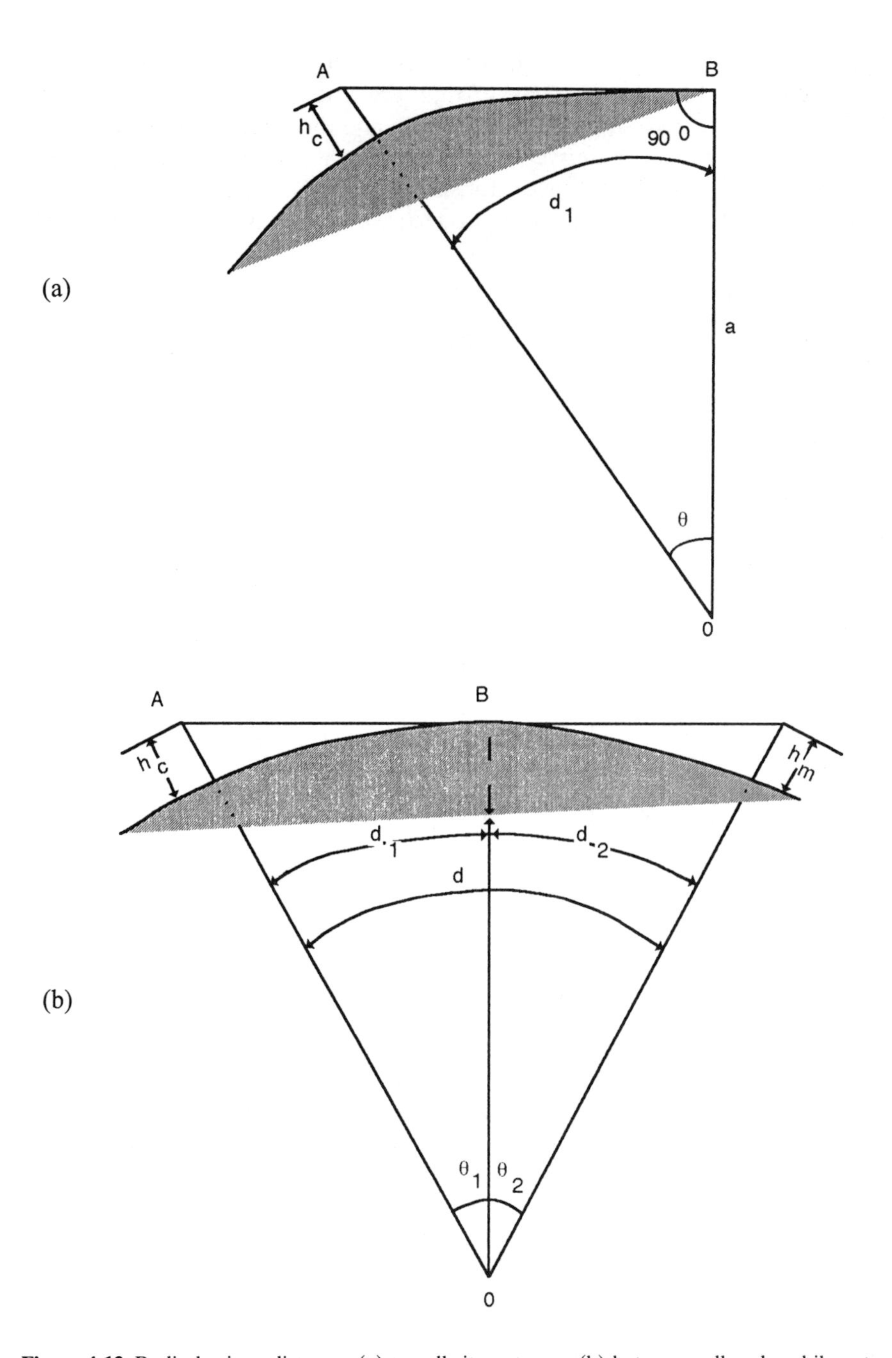

Figure 4.12 Radio horizon distance: (a) to cell-site antenna, (b) between cell and mobile antenna.

assumption is not generally true. The curvature of the earth makes the effective antenna heights h_c' and h_m' appear smaller than the true antenna height h_c and h_m. We shall derive the relationship between the true and the effective antenna heights below so that the previously derived equations for flat earth can be used again. Figure 4.13 shows the configuration of transmit and receive antennas when erected on the curved earth. At the reflection point B, we have drawn the tangent. The

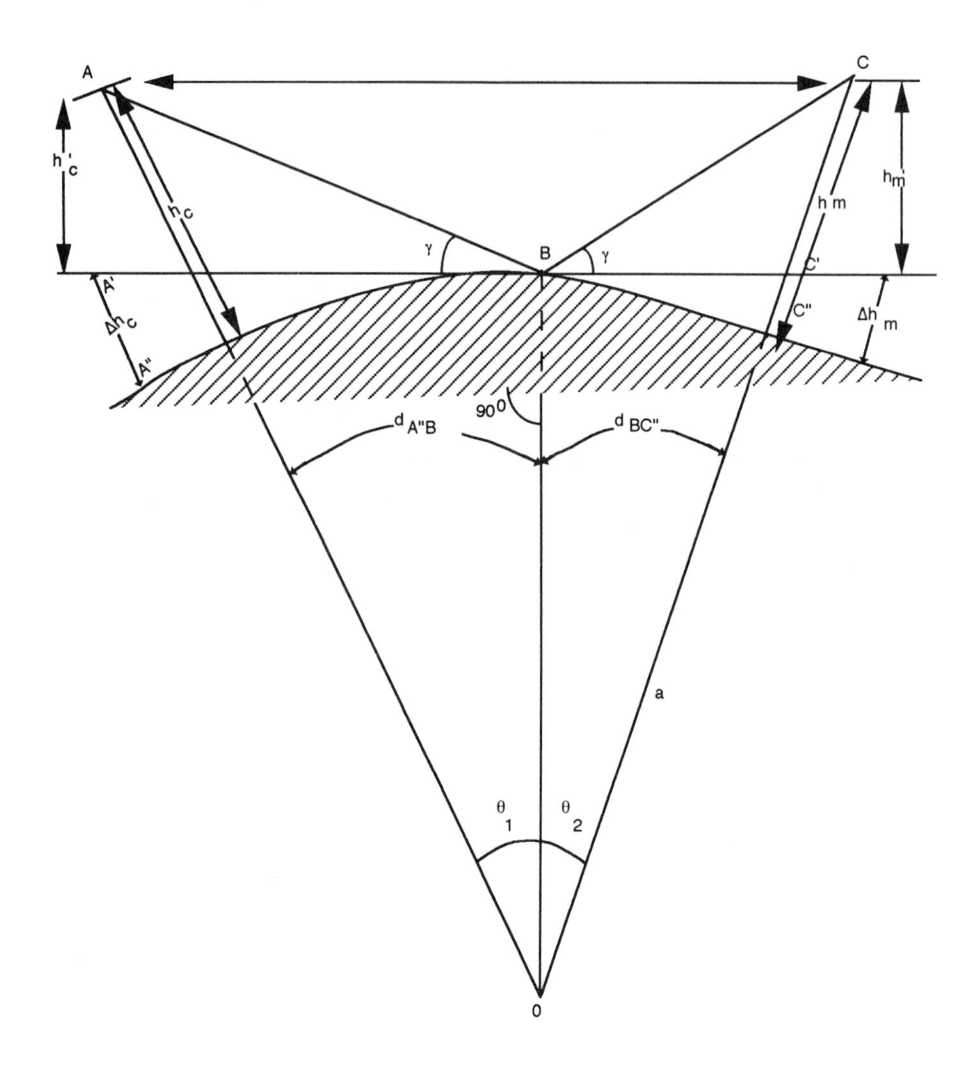

Figure 4.13 Antenna heights for curved earth.

heights of the two antennas measured from the tangent lines $A'BC'$ are the effective antenna heights h_c' and h_m'. The path difference Δd between the direct wave AC and the reflected wave ABC will remain the same since the angle γ remains the same for flat earth as well as for the curved surface. In our derivation below we shall assume that the distance $d_{A''B} + d_{BC''}$ is shorter when compared to the radius of the earth, meaning that the angles θ_1 and θ_2 are small. From Figure 4.13, we get

$$h_c' = h_c - \Delta h_c \qquad h_m' = h_m - \Delta h_m \tag{4.54}$$

where Δh_c and Δh_m represent the straight-line segments $A'A''$ and $C'C''$. Expressing Δh_c and Δh_m in terms of $d_{A''B}$ and $d_{BC''}$ from (4.53), we obtain

$$h_c' = h_c - \frac{d_{A''B}^2}{2a} \qquad h_m' = h_m - \frac{d_{BC''}^2}{2a} \tag{4.55}$$

4.6 CONCEPT OF EQUIVALENT EARTH'S RADIUS

Due to the presence of the earth's atmosphere, whose refractive index decreases with increasing altitude, radio waves do not travel in a straight-line path (geometric path), but rather on a curved path as shown in Figure 4.14. The value of refraction

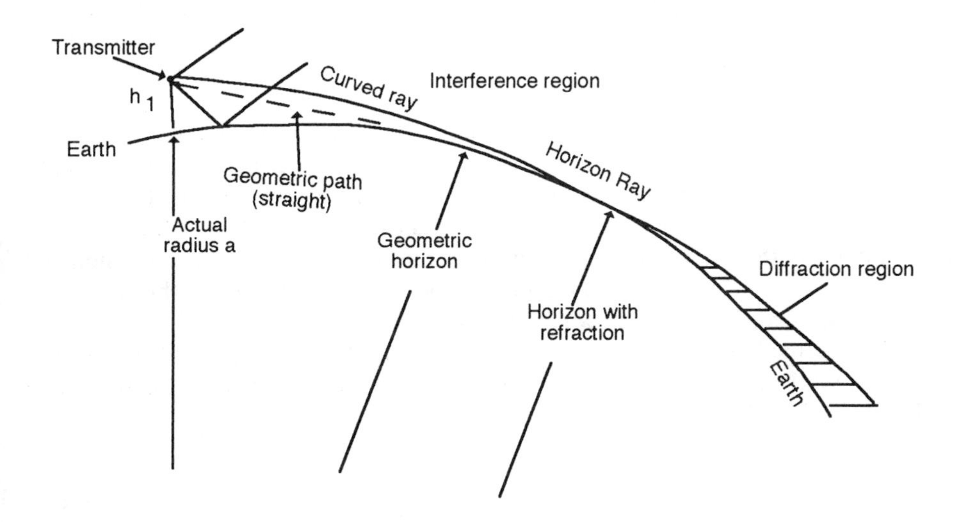

Figure 4.14 Ray curvature over spherical earth of radius a.

is defined as the ratio of propagation velocity in free space to that in the medium of interest. The radio refractive index of air is a function of atmospheric pressure, temperature, and humidity. For a U.S. standard atmosphere, the value of the refractive index n changes from 1.0003 on the surface of the earth to one for free space. The refractive index is always greater than one, since radio energy travels fastest in a vacuum. The refractive index decreases with increasing height. The result is that the velocity of transmission increases with height and the radio energy is bent or refracted toward the earth. The rate of change of the radio refractive index with height determines how rapidly the radio wave is bent. Steep changes in the radio refractive index can be caused by several weather phenomena. Steep changes of the radio refractive index with height allow low-loss transmission over long distances, typically referred to as ducting. For example, one cause for steep changes in the radio refractive index for the Gulf of Mexico is the cool air blown in over the shallow water near land that has been warmed by the sun. As the cool air passes over the warm water, the water evaporates into the air, which increases the humidity, thus increasing the radio refractive index. Due to a small variation of n, a higher value of radio refractivity, $N = (n - 1)10^6$, is defined. The relationship between N and the atmospheric pressure and temperature is given by

$$N = (77.6p)/T + (3.75 \times 10^5\omega)/T^2 \tag{4.56}$$

where p is the atmospheric pressure (millibar), T is the atmospheric temperature in deg/°k, and ω is the partial pressure of water vapor (millibars).

Since the atmospheric pressure decreases faster with height than the atmospheric temperature, there is an overall decrease in radio refractivity with height. The radio horizon of an antenna is formed by the locus of all points where the direct rays from an antenna are tangent to the earth's surface. The surface formed by all such radio horizon lines may be called a radio horizon surface.

Unlike the radio waves, the direct ray (geometric path) travels in a straight line and is tangent to the earth somewhere before the radio horizon. If the earth is replaced by an equivalent earth having a radius $a_e = Ka$ ($K = 4/3$ for normal atmosphere in the middle latitudes and a is the actual Earth's radius), then the radio waves can be considered to travel in straight lines. In this case, the geometric path will be curved. This is shown in Figure 4.15.

The equivalent earth's radius contours for the United States for winter and summer months are shown in Figure 4.16. It can be shown that the effective earth radius K is given by

$$K = \frac{1}{1 + a(dn/dh)} \tag{4.57a}$$

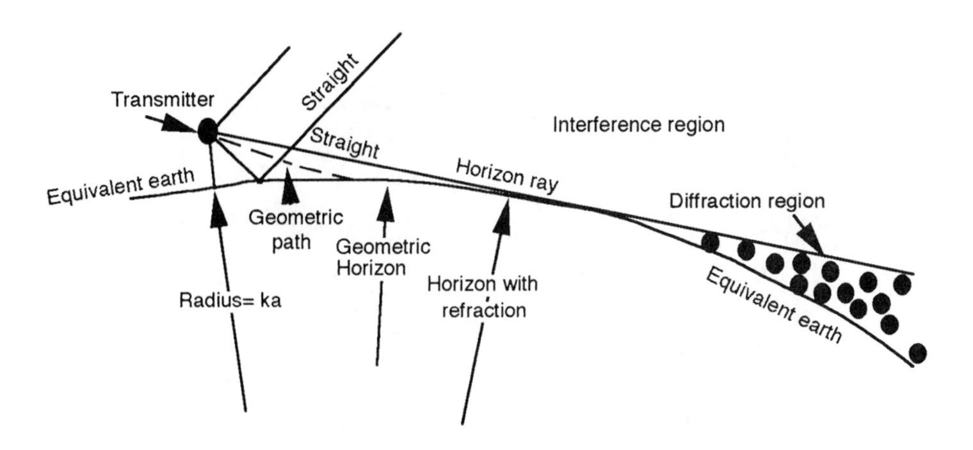

Figure 4.15 Rays when using equivalent earth radius, K_a.

or

$$K = \frac{157}{157 + dN/dh} \tag{4.57b}$$

where dN/dh is the gradient of N per kilometer. The gradient N is negative and constant and has an approximate value of -40 units/km. When this value of dN/dh is substituted in (4.57b), the resulting value of factor K is 4/3, which is used in propagation analysis. From the curves of summer and winter contours of the United States, it is clear that the equivalent earth's radius increases in summer as the atmosphere becomes more homogeneous and thus the value of radio refractive index gradient dn/dh decreases.

4.6.1 Bullington-3 Loss Method over Smooth Spherical Earth

The Bullington 3 loss method provides nomograph accurate to $+2$ dB for predicting path loss over the trans-horizon, as show in Figure 4.17 [10]. The distances d_1 and d_2 can be found from (4.54) as

$$d_{1,2} = \sqrt{2\,aK\,h_{1,2}} = \sqrt{2a_e h_{1,2}} \tag{4.58}$$

where $h_{1,2}$ is the antenna height and a_e is the equivalent earth's radius. In our case,

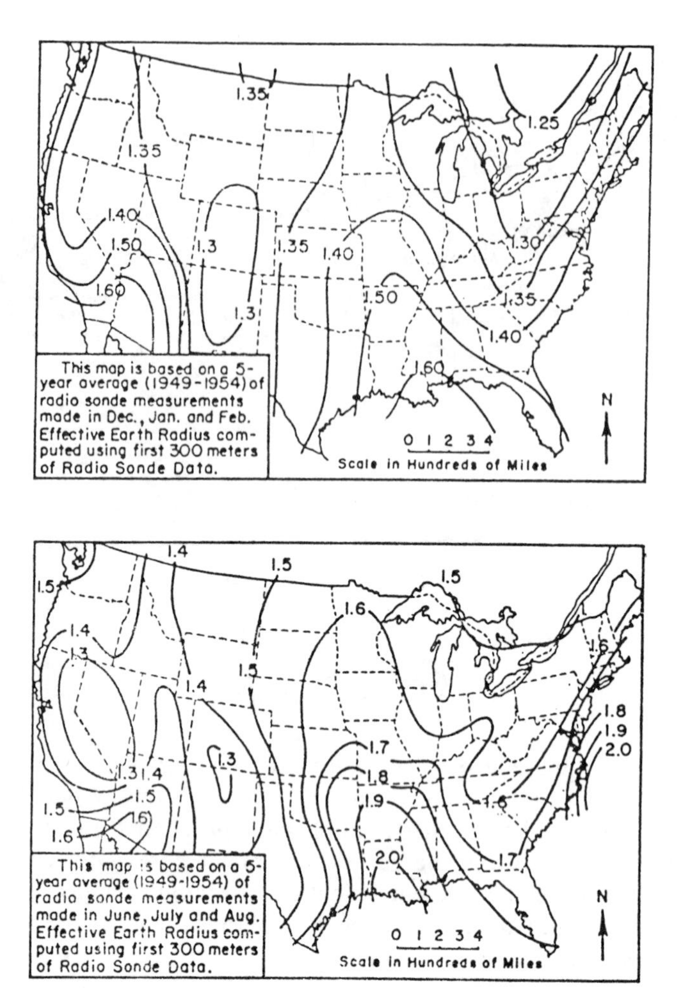

Figure 4.16 Effective earth's radius contours for the U.S.: winter (top), summer (bottom).

h_1 is the cell-site antenna h_c and h_2 is the mobile antenna h_m. An example of the usage of a nomograph is being illustrated.

Example 4.5

Cellular cell design over the Gulf of Mexico requires the subscriber antenna to be 24 feet above sea level, with the cell-site on shore requiring a 450-foot antenna.

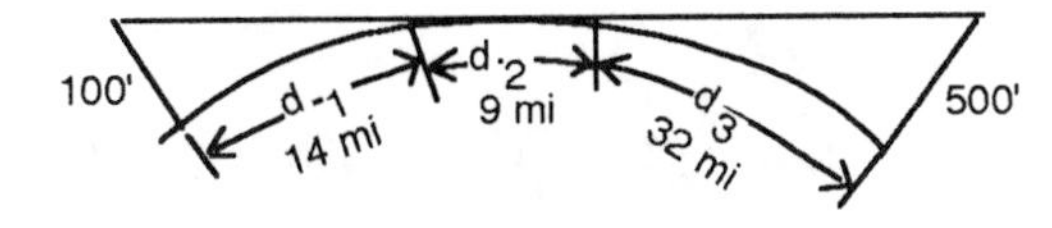

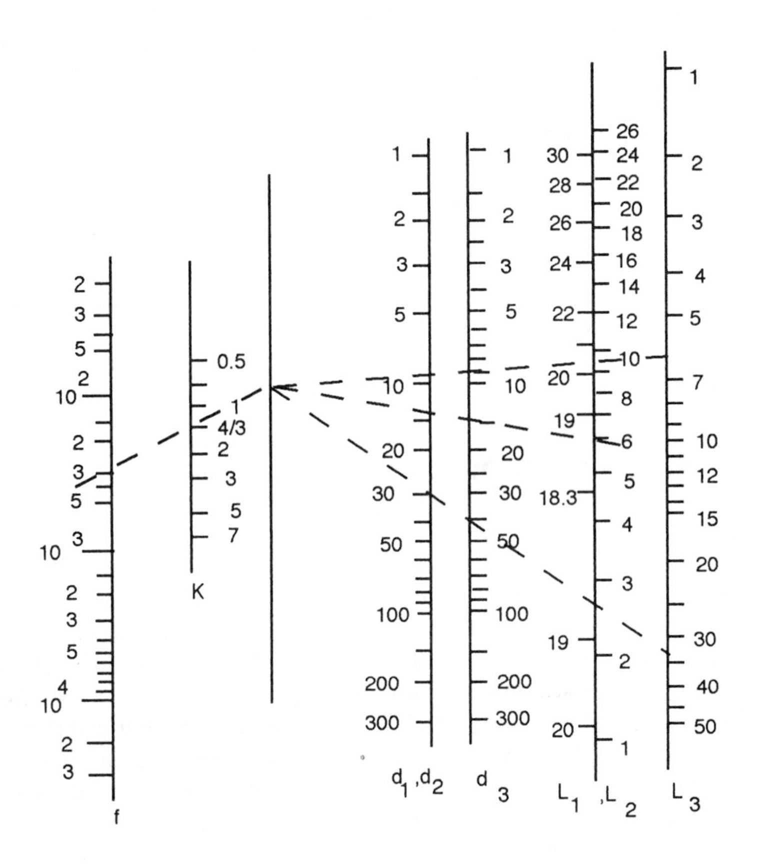

Figure 4.17 Nomograph of signal attenuation from propagation over a smooth spherical Earth. The total loss is $L_1 + L_2 + L_3 = 18.5 + 2.5 + 6.0 = 27$dB. f = frequency in MHz; K = the ratio of effective earth's radius to true earth's radius; d = distance in miles; L = attenuation in dB.

Find the distance d_3 between two horizons d_1 and d_2 such that the design requirement of 39 dBu is satisfied by the mobile receiver. Use UHF frequency of 850 MHz and the cell-site antenna ERP = +20 dBW.

$$d_1 = \sqrt{2 \times (4/3) \times 24} = 8' \qquad d_2 = \sqrt{2 \times (4/3) \times 450} = 34.64'$$

The minimum receive sensitivity corresponding to the median field intensity of 39 dBu, $F(50, 50)$, at 850 MHz is $F_m = 105 + 10 \log_{10} P_m + 20 \log_{10}(850) = 39$ or $10 \log_{10} P_m = 39 - 105 - 58.6 = -124.6$ dBW. Thus, $20 - LP_1 - LP_2 - LP_3 - SL + G_m - 1 = -124.6$ dBW. Space loss is $SL = 36.6 + 20 \log_{10}(850) + 20 \log_{10}(d_1 + d_2 + d_3) = 95.2 + 20 \log_{10}(42.64 + d_3)$. LP_1 and LP_2 from Figure 4.17 are 19.6 and 1.6 respectively.

Therefore, $20.0 - 19.6 - 1.6 - LP_3 - 95.2 - 20 \log_{10}(42.64 + d_3) + G_m - 1 > -124.6$ with $d_3 = 0.0$. $-97.4 - 20 \log_{10}(42.64) + G_m > -124.6$ or $G_m >$ 5.4 dB.

Let us choose the mobile antenna to be 8 dB; then, if $d_3 = 2.5$ miles, $-97.4 - 20 \log(42.64 + 2.5) + 8.0 - 2.0 = -124.5 > -124.6$ dBm. Therefore, d_3 will be positive and the link will close if the mobile has an antenna with a gain 5.4 dB or more. For antennas mounted on ships or boats, this amount of gain is simple to achieve.

4.7 VOLUMES OF RADIO TRANSMISSION: FRESNEL ZONE

There are two ways in which a radio wave can travel from the source to the destination. It either travels along a straight path, as shown in Figure 4.18, or we can imagine that the radio wave occupies a certain finite volume in the space around the axis *TR*. There are two ways in which one can prove this, either experimentally or theoretically. A simple experiment based on Figure 4.18 can resolve the question. Assuming that the window height is similar to that of a camera shutter, then if the radio wave travels in a straight line, point *R* will be unaffected by reducing the height of the window until the window is completely closed. On the other hand, the receiver field intensity will vary as a function of the window opening if the radio waves occupy a finite volume along the axis *TR*. To determine the significant volume, one can start from a large window opening and slowly reduce the size of the opening until the received energy at *R* is significantly reduced. The width of the opening when the receive energy at *R* is significantly affected may provide an answer to a significant volume. The width of the opening may also differ as the window is moved from one location to another along the axis *TR*. This is an experimental method. Alternately, an analytical approach based on Huygen's prin-

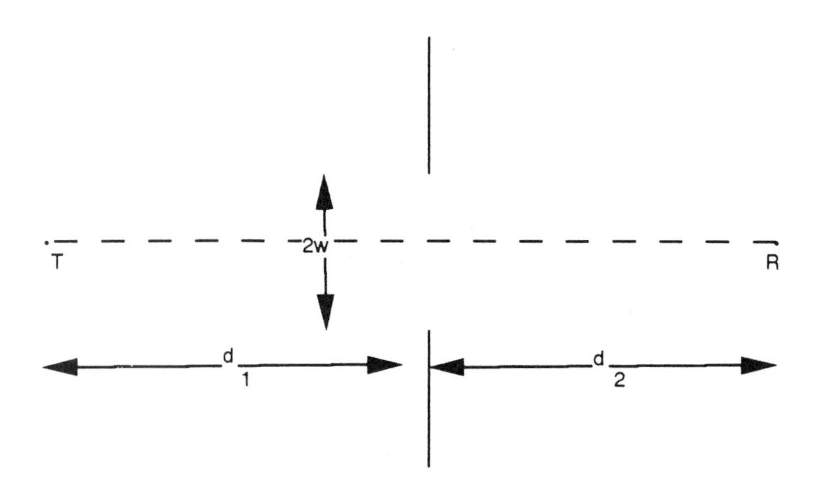

Figure 4.18 Experimental setup.

cipal and the concept of Fresnel zones can be applied to define the significant volume.

In the most elementary form, Huygen's principle states that each point on the front of an advancing wave (produced by some primary source) is the new secondary spherical wave. By applying Huygen's principle, it is possible to compute the field at any point in space from the known field strength on the wave-front of the surface of a sphere, as shown in Figure 4.19.

For a general surface S of an arbitrary shape, Huygen's principle takes the form of Kirchhoff's equation:

$$E_R = -1/4\pi \int_S \left[E_s \frac{\partial}{\partial n}\left(\frac{e^{-ikr}}{r}\right) - \frac{e^{-ikr}}{r}\frac{\partial E_s}{\partial n} \right] ds \tag{4.59}$$

where E_s is the value of field intensity on the surface and $\partial E_s/\partial n$ is the derivative of the field normal to the surface. Thus, in order to find the field intensity at point R from an arbitrary surface one has to know the field intensity on the surface as well as the derivative of the field intensity normal to the surface. Assuming isotropic transmission in both the vertical and horizontal planes, we can construct a sphere around a point source of RF field and can compute the total electric field at R by evaluating the above integral. In order to evaluate this integral, the distance d from R has to be found for each elementary area on the surface, including areas not visible from R. In 1818, the French mathematician Fresnel showed a simplified method of constructing these distances on the sphere. He interpreted the distance

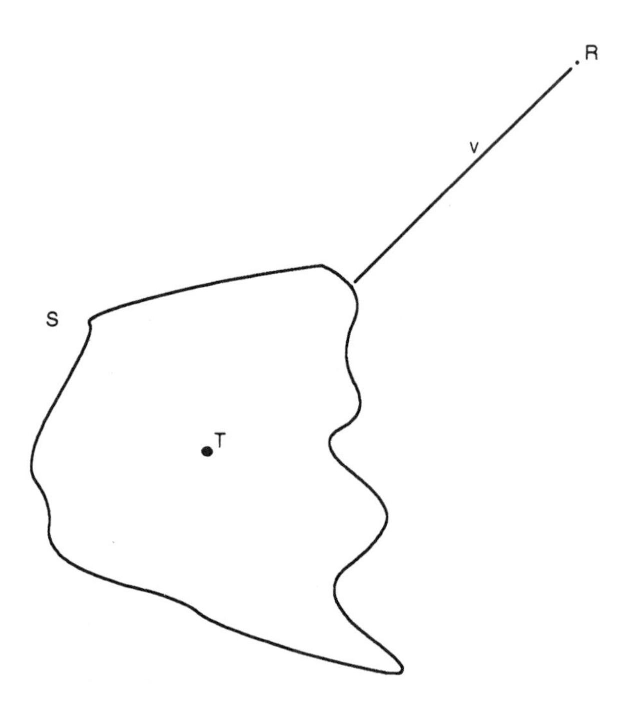

Figure 4.19 Huygen's principle.

RA_1, shown in Figure 4.20, to be equal to $RA_0 + \lambda/2$, or the phase shift at point A_1 is 180 deg with respect to the phase shift at A_0. Similarly, RA_2 is equal to $RA_0 + 2\lambda/2$, and point A_2 is again in phase with A_0. Thus, in general RA_N is equal to $RA_0 + N\lambda/2$. These points of specified phase shifts with respect to A_0 actually form a conical circle on the surface of the sphere. These circles are shown in Figure 4.21. The areas between any two circles are known as Fresnel zones. The first Fresnel zone is the area of the sphere bounded by a circle of radius A_1. Areas between circles A_2 and A_1 are the second Fresnel zone, and so forth. Since each Fresnel zone differs in phase shift by 180 deg, one can perceive that the secondary fields on the surface of the sphere alternate in positive and negative charges. It can easily be proven that the charges on higher order zones cancel each other.

This cancellation is more complete as the order of the Fresnel zones increases. It can also be shown that the sum total of all the contributions to the electric field at a distance R due to second to Nth-order Fresnel zones is equal to one-half of the total electric field. Thus, the volume bounded by the first Fresnel zone effectively contributes half the total energy to the receiving point R and therefore can be regarded as a significant volume. From these discussions, one can conclude that

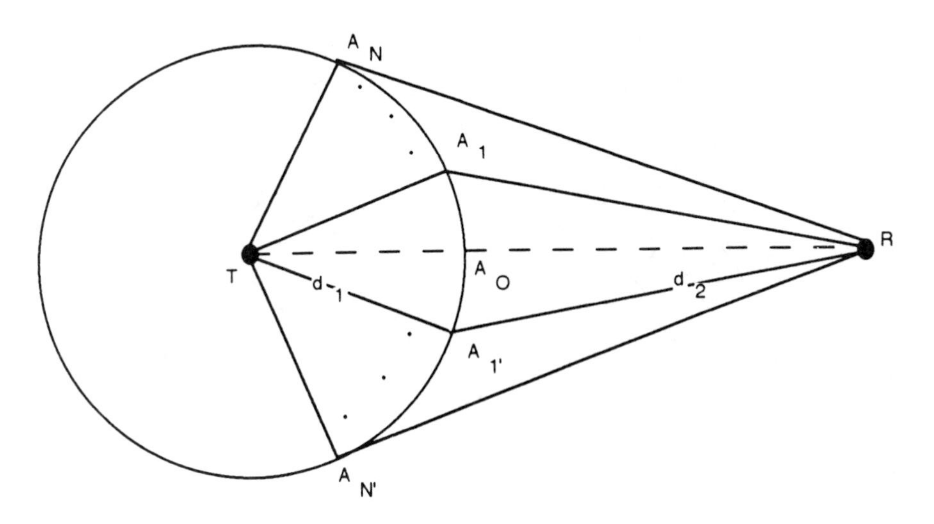

Figure 4.20 Fresnel zones on the surface of a sphere.

if the line joining the transmitter and receiver clears the height of the first Fresnel zone, most of the energy will reach to the receiver.

Fresnel zones can be constructed on surfaces of any arbitrary shape. For this purpose, choose a plane normal to the ray path TR, as shown in Figure 4.22. Then, the radius of the Fresnel zone can simply be derived as follows:

$$TA_NR = TA_N + A_NR = TA_{N0} + A_{N0}R + N\lambda/2 = d_{1N} + d_{2N} + N\lambda/2 \quad (4.60)$$

From the triangle TA_NA_{N0} and $A_{N0}A_NR$ in Figure 4.22 and with the assumption that $b_N \ll TA_{N0}$ and $b_N \ll A_{N0}R$,

$$TA_N = \sqrt{d^2_{1N} + b^2_N} \approx d_{1N} + b^2_N/2d_{1N} \quad (4.61a)$$

and

$$A_NR = \sqrt{d^2_{2N} + b^2_N} \approx d_{2N} + b^2_N/2d_{2N} \quad (4.61b)$$

Substituting these values of TA_N and A_NR in (4.60), we get $d_{1N} + d_{2N} + b_N^2/2d_{1N} + b_N^2/2d_{2N} = d_{1N} + d_{2N} + N\lambda/2$, or

$$b_N = \sqrt{\frac{N\lambda d_{1N}d_{2N}}{(d_{1N} + d_{2N})}} \quad (4.62)$$

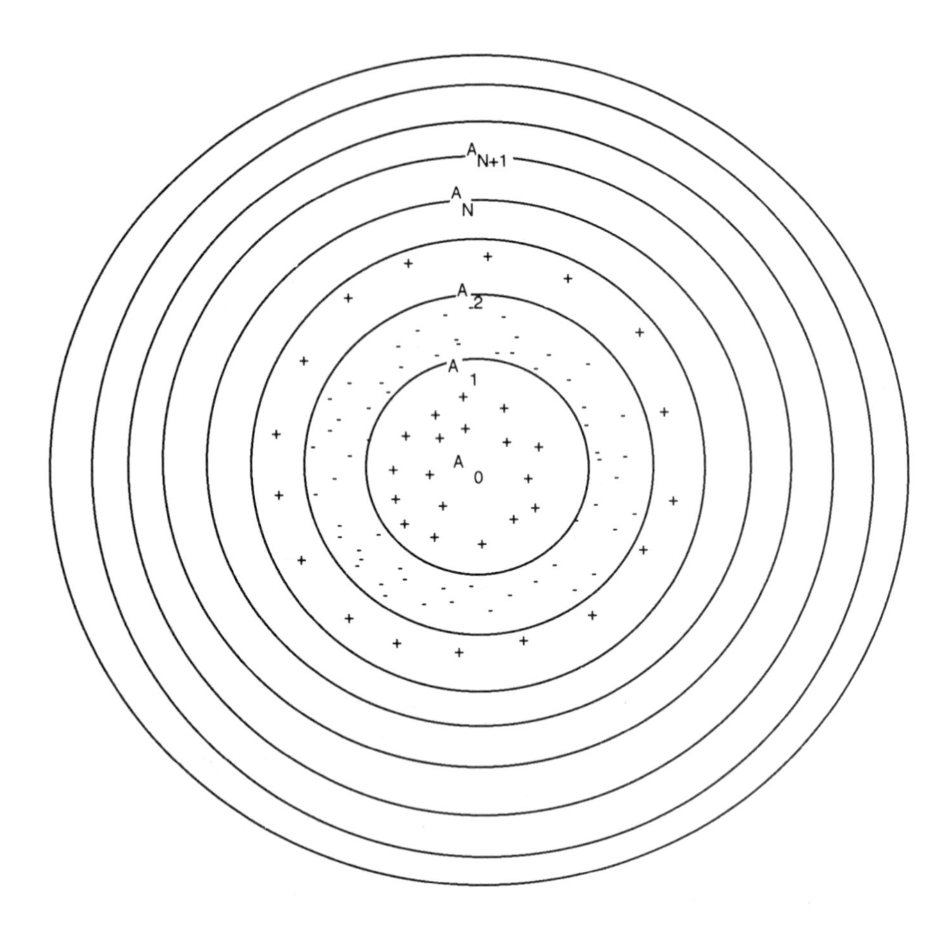

Figure 4.21 Fresnel zones.

Thus, the radius of the first Fresnel zone is given by

$$b_1 = \sqrt{\frac{\lambda d_2 d_2}{d_1 + d_2}} \tag{4.63}$$

The points A_N are on the surface of the ellipsoid of revolution whose foci are at points T and R, as shown in Figure 4.23, and contain significant volume. It should be noted that the radius of the first Fresnel zone is at maximum at the center of the path. We now illustrate this concept with an example.

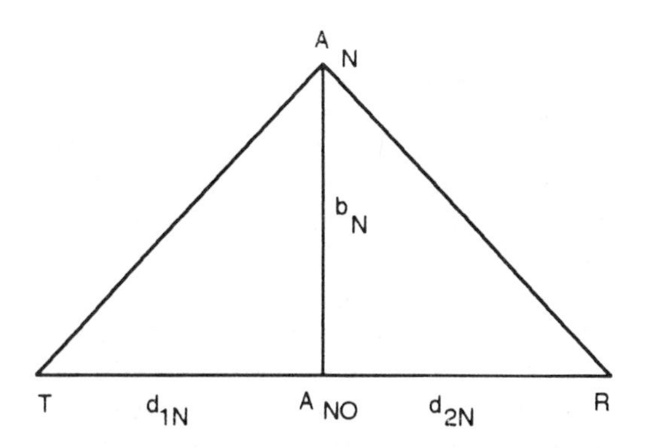

Figure 4.22 Radius of Fresnel zones.

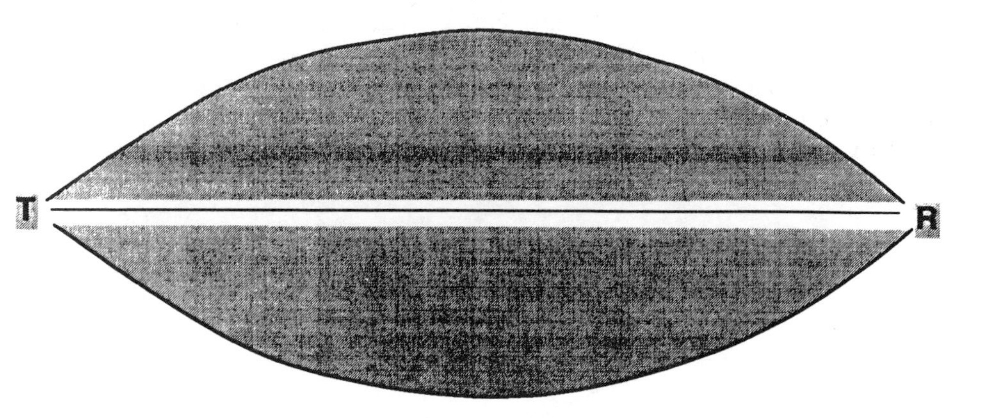

Figure 4.23 Significant volume for RF wave.

Example 4.6

For a cellular system, find the radius of the first Fresnel zone at the center of a 20-km long path for a wave length of 0.3m:

$$b_{\max} = b_1 = \sqrt{\frac{0.3 \times 10 \times 10 \times 10^3}{20}} \text{ meter}$$

$$= 38.7\text{m}$$

If we compute the radius for $\lambda = 30$m and 0.003m, the maximum clearance required will be 274m and 0.274m, respectively. Thus, the same path clearance may be adequate for higher frequency waves than for lower frequency waves. If we compute the radius of the first Fresnel zone at 5-km, 10-km, and 15-km diameters from the cell-site transmitters, we obtain the required Fresnel zone radius to be 33.5m, 38.7m, and 33.5m, respectively. This is shown in Figure 4.24.

The Fresnel zone's loss in decibels relative to free space as a function of Fresnel zone clearance is shown in Figure 4.25. The curves are drawn for three different values of the earth's reflective indices, representing different types of earth. The value of the reflective index R when R equals zero represents a perfect knife edge. For the case when the path clearance is zero (i.e., the observation is just raging the path between the transmitter and the receiver), the loss is 6 dB with respect to the free-space loss. On the other hand, if the clearance is $0.6x$ (first Fresnel zone radius), the total attenuation is zero dB (i.e., the loss in this case corresponds to free-space loss). The value of R equal to -1 represents a perfectly smooth earth. In this case for zero path clearance, the expected value of the loss is about 20 dB. At 0.6 times the first Fresnel zone clearance, there is once again a 0 dB loss. For regular rough ground, values of the reflection coefficient generally lie between -0.3 to -0.4. For R equals -0.3, the curve lies between $R = 0$ and $R = -1.0$. Loss for the path clearance of zero in this case is roughly of the order of 10 dB. Once again, at $0.6x$ (first Fresnel zone radius) the value of the attenuation

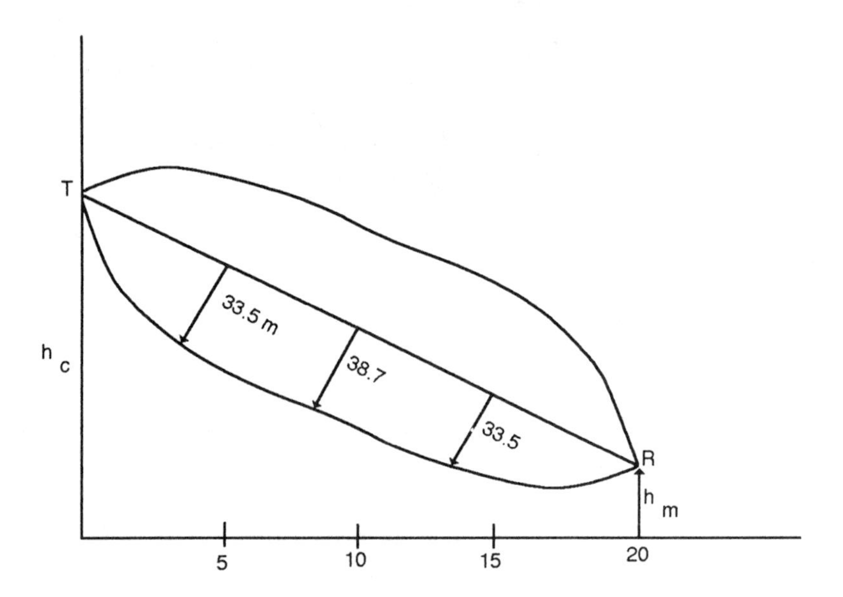

Figure 4.24 Required Fresnel zone clearance for example 4.6.

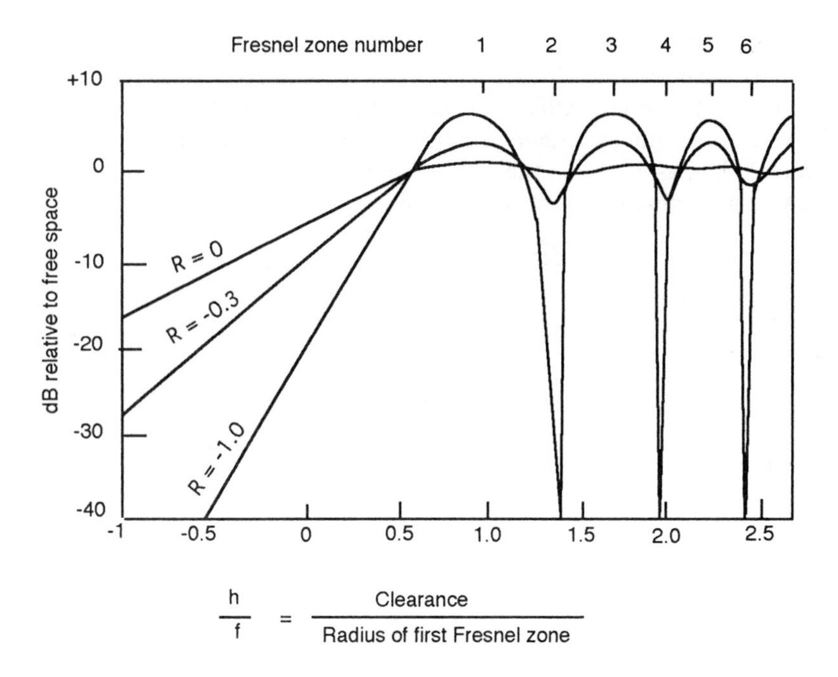

Figure 4.25 Fresnel zones. After [1, p. 303].

is zero dB. The peaks and valleys in the curve are the result of the constructive and destructive interference. At all odd Fresnel zone numbers, the electric field at the receiver due to reflections adds up. At even numbers of Fresnel zone, a marked increase in signal loss is seen. Distance between successive peaks also reduces as the Fresnel zone numbers increase.

4.8 PROPAGATION OVER A KNIFE EDGE

In the ground-wave mode of communication in mountainous regions, direct line-of-sight propagation is often obstructed due to the ridges of the mountains. The radio energy loss due to these obstructions is known as diffraction loss. The diffraction loss is a measure of the height of the obstruction with respect to the antennas used in the system. The height of the obstruction has to be compared to the wavelength of the transmission. The same obstruction height may produce lower diffraction loss at higher wavelengths than at the a lower wavelength. For predicting the path loss, these obstructions are treated as a sharply defined obstacle, known as a "knife edge." The loss may be calculated by a method widely used in physical optics. Figure 4.26 shows two cases of obstructions. In the first case, the line-of-sight path is clear of the obstruction by the height H. In the other case, the obstruction is in the direct path of the radio wave. In the first case, we shall assume the obstruction height as negative, while in the second case we shall treat the obstruction height as positive. The diffraction loss, F, can be calculated by knowing the diffraction parameter v and is given by

$$v = -H\sqrt{2/\lambda(1/d_1 + 1/d_2)} \tag{4.64}$$

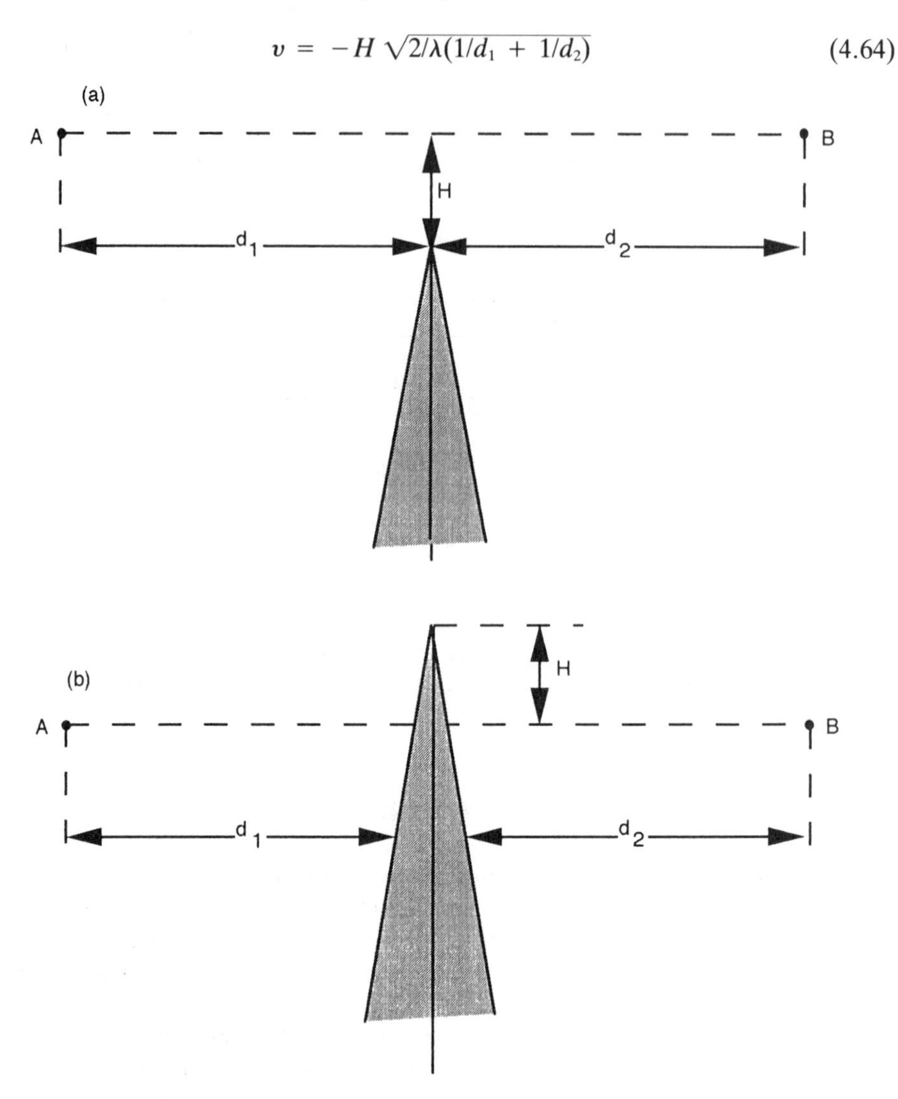

Figure 4.26 Radio propagation across a knife edge, (a) negative height, (b) positive height.

The value of v corresponding to Figure 4.26(a) is positive and the range of F computed from (4.64) is greater than or equal to 0.5.

For Figure 4.26(b), the value of ν is negative and the corresponding computed value of F from (4.48) lies between 0 to 0.5. The approximate values of diffraction loss for different ranges can be computed from the equations given below [12]:

$$\begin{aligned} F &= 0 \text{ dB} \qquad \nu \geq 1 \\ &= 20 \log_{10}(0.5 + 0.62\nu) \qquad 0 \leq \nu < 1 \\ &= 20 \log_{10}(0.5\, e^{0.45\nu}) \qquad -1 \leq \nu \leq 0 \\ &= 20 \log_{10}\left(0.4 - \sqrt{0.1184 - (0.1\nu + 0.38)^2}\right) \qquad -2.4 \leq \nu < -1 \\ &= 20 \log_{10}(-0.225/\nu) \qquad \nu < -2.4 \end{aligned} \tag{4.65}$$

The four-ray model has become popular for computing diffraction loss. The model considers both the direct and the reflected waves from the ground on each side of the obstruction, as shown in Figure 4.27. The field at point B is caused by four waves, each of which has suffered diffraction at the knife edge. To find the resultant field at the receiving antenna B, the field from antenna A, its image A', and the image point B' have to be considered. The four rays which suffer diffraction at M, and are considered for computing the field, are AMB, $A'MB$, $AMB'B$, and $A'MB'B$. The construction as shown in Figure 4.27 is due to an image principle according to which a reflected ray is treated as being due to an image of the actual emitter. In our case, the two images are A' and B' for A and B, respectively.

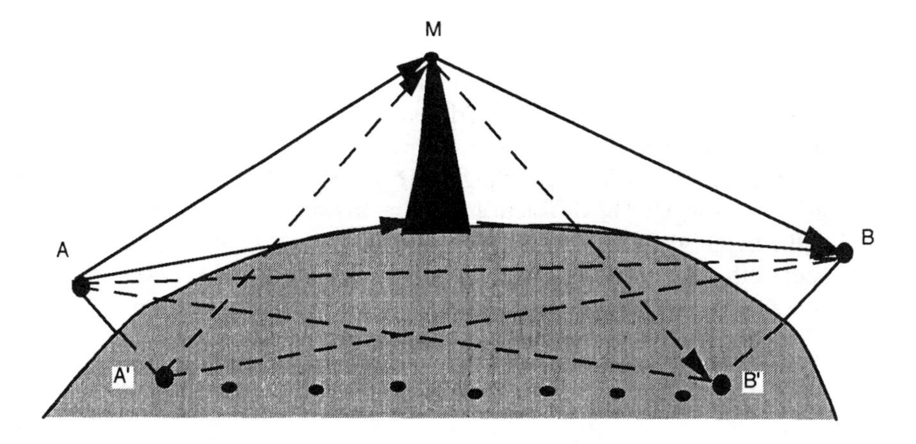

Figure 4.27 The four-ray model of diffraction across a knife edge.

4.8.1 Multiple Knife-Edge Diffraction

Multiple knife-edge diffraction due to multiple obstructions can also be treated through the double integral of the Fresnel form, but the solutions of these integrals are not easy. Due to the mathematical complexity, approximate methods for multiple diffraction have been introduced by different investigators [15–20]. These methods replace the multiple obstructions by a single equivalent obstruction and thus find the loss by the above-discussed technique. We will now describe some popular methods by Bullington, by Epstein-Peterson, and by Japanese Atlas.

In the Bullington equivalent knife-edge method, obstructions at A and B are replaced by an equivalent knife-edge obstruction C at the intersection point of the radio wave [18]. The original obstructions at A and B are now ignored. The height of the obstruction H is read from the Figure 4.28a. The value of the diffraction loss is arrived at by computing ν and then finding F by applying (4.66). The case for multiple knife-edge diffraction is shown in Figure 4.28b. Here, the obstruction B is completely ignored as the intersecting line is above the peak of the obstruction B.

The Epstein-Peterson diffraction loss method calculates the total loss in two different stages and adds the two losses in decibels [16]. In Figure 4.29(a), the loss due to ridge A is first calculated by considering the height of the obstruction H_A above the line joining the transmitter T with the second obstruction at B. Then the diffraction loss due to the second obstruction B is computed by considering the height of the obstruction H_B above AR. The total diffraction loss is the sum of these two losses. This technique produces good results if the obstructions are not too close to each other. Figure 4.29(b) shows a case where the technique may not provide accurate results.

In the Japanese Atlas scheme, the diffraction loss for ridge A is computed by the same technique as in the Epstein-Peterson method [20]. This is shown in Figure 4.30. The path TAB is considered and the height of the obstruction H_A is read. Similar to the other cases discussed above, the value of the diffraction loss is calculated by knowing ν and finding F directly from (4.66). The loss due to obstruction B is calculated by extending the ray AB to the left until it intersects the transmitter axis at T'. The diffraction loss due to obstruction B is now computed by finding the height of the obstruction H_B above $T'R$. Once again, the linear sum of two losses is the total loss. We illustrate below the computing process by taking up an example.

Example 4.7

A cell-site transmitter has an antenna 200-ft long. The terrain has an isolated ridge in the propagation path having a height of 800 ft above the mean sea level (MSL).

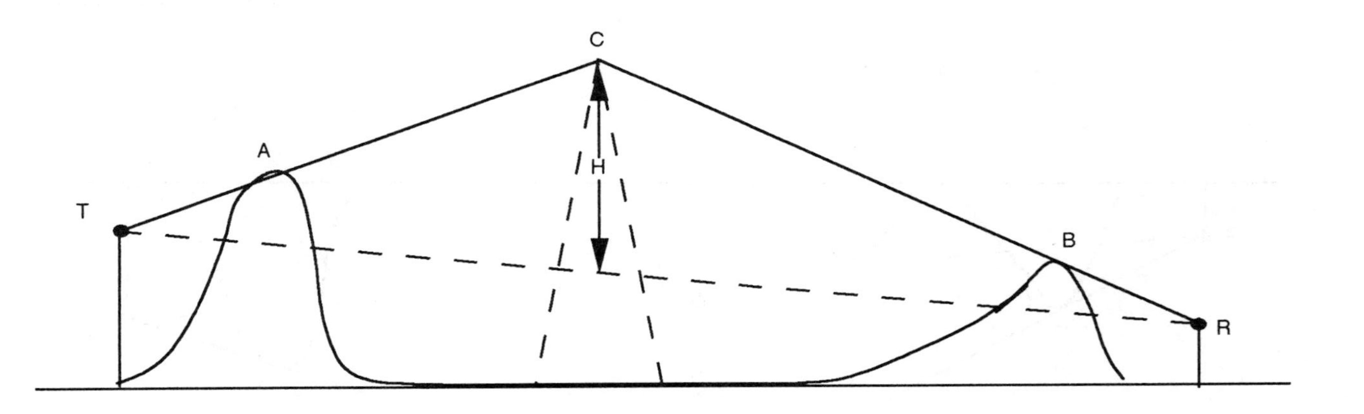

Figure 4.28(a) Bullington equivalent knife-edge method.

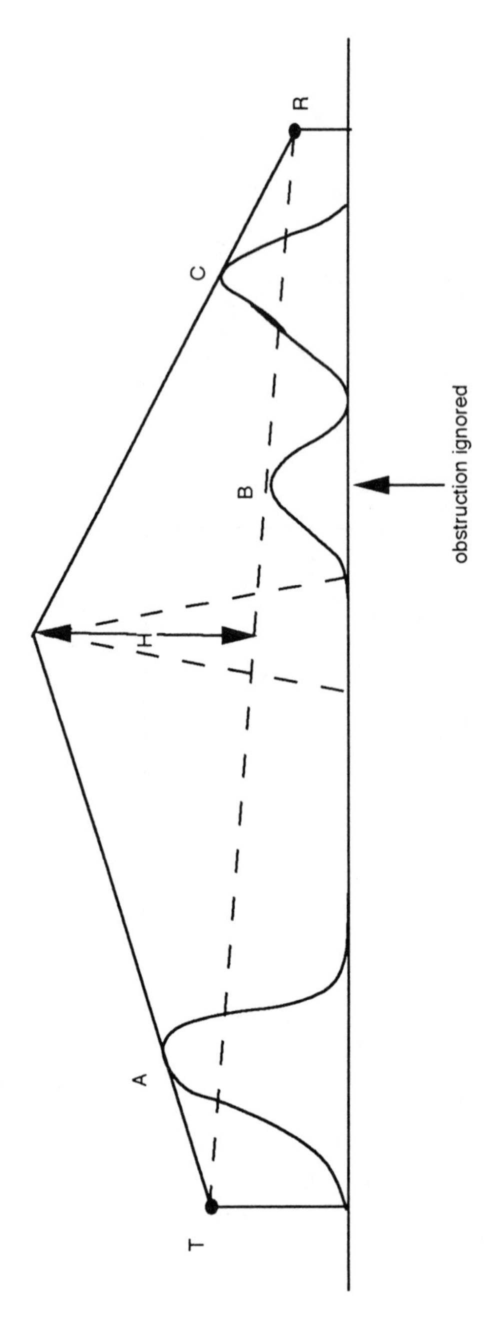

Figure 4.28(b) Bullington equivalent knife-edge for three obstructions.

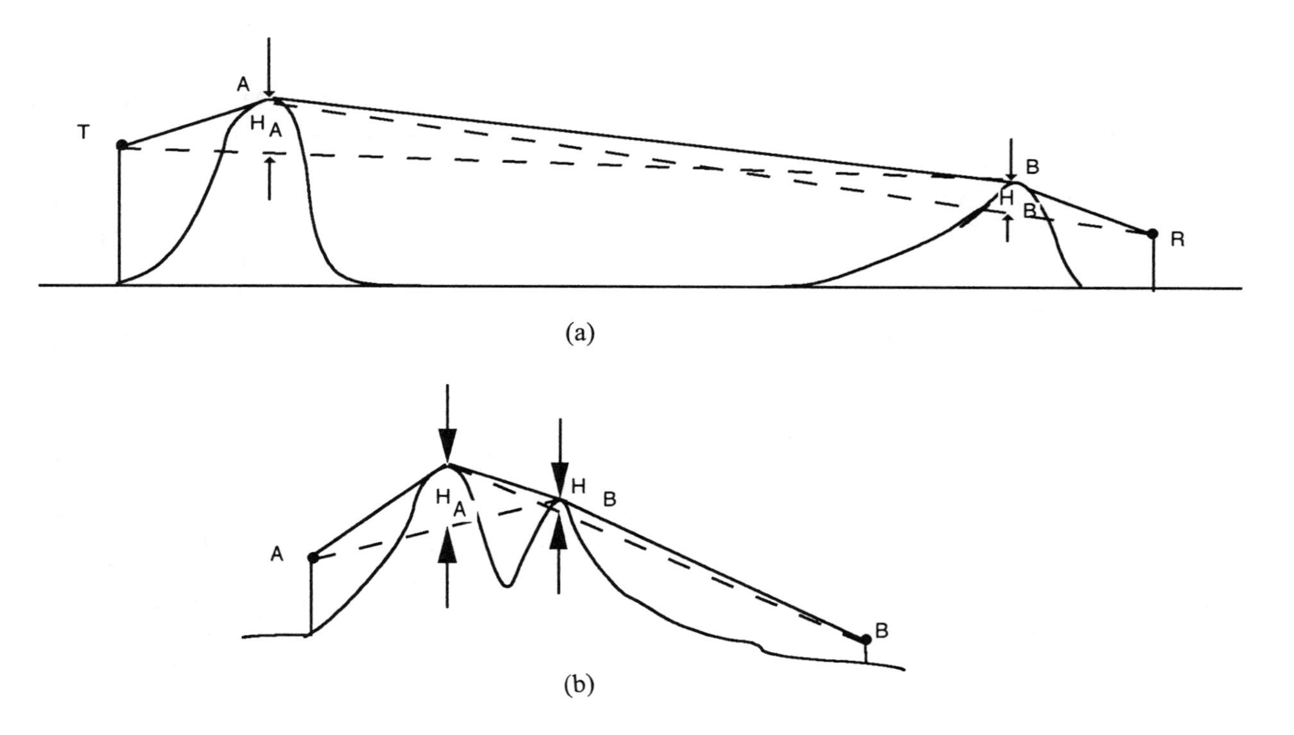

Figure 4.29 (a) Epstein-Peterson knife-edge diffraction; (b) Epstein-Peterson knife-edge diffraction (obstruction close to each other).

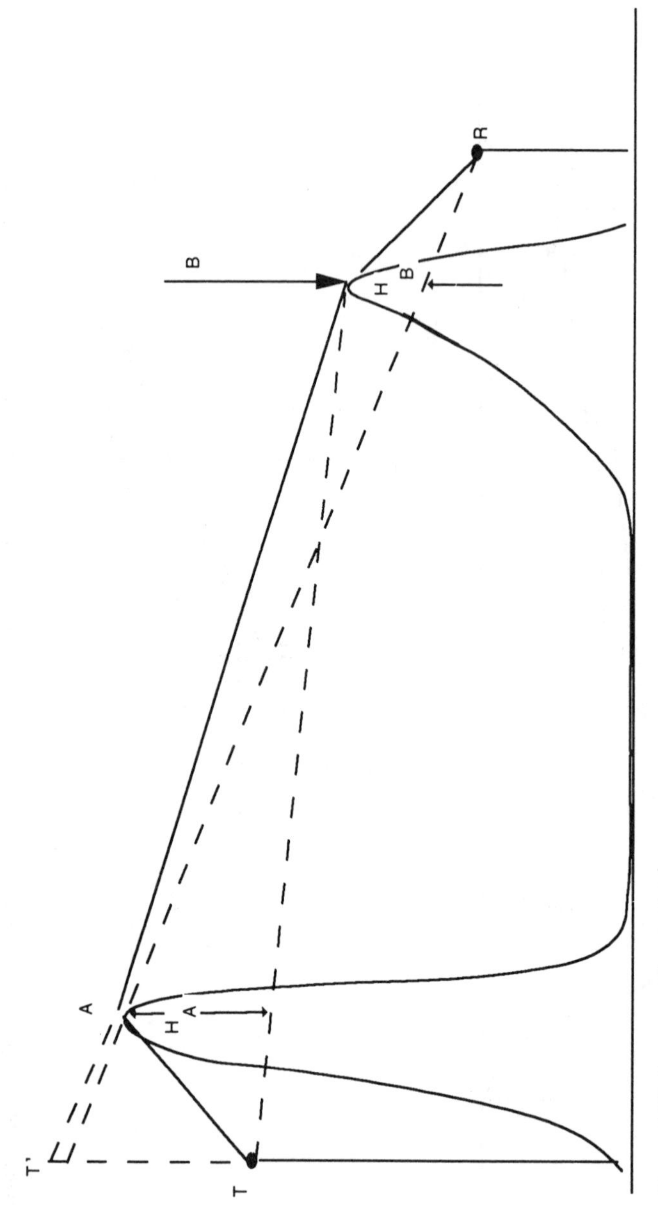

Figure 4.30 Japanese atlas method.

as shown in Figure 4.31. Compute the diffraction loss at a frequency of 900 MHz when the mobile antenna height is 10 ft. Assume the 39-dBu contour to be a distance of 10 miles from the cell site. The MSL at the vehicular location is 50 ft.

Considering the trapezoid *TRMN*, it is easy to see that the height of the mountain, H, above TR is $H' = 800 - (500 - 440 \times 4/10) = 476$ ft. Therefore,

$$\nu = -476 \sqrt{\frac{2}{1.099} (1/4 + 1/6) \frac{1}{5280}} \approx -13.7 \tag{4.66}$$

Thus, F = diffraction loss = $20 \log_{10}(0.225/13.7) = -35.7$ dB.

4.9 ROUGH-SURFACE CRITERIA

In order to predict the propagation loss over an arbitrary surface of the earth, the roughness of the ground must be defined. The higher the roughness of the ground, the less the specular reflection, and therefore the coefficient of reflection is low. The reduction of reflected energy is large in the case of forests, urban areas, or on rough ground. Two planes, one at the top of the surface irregularity and the other without the irregularity, are drawn. Two waves at the same grazing angle θ are considered, as shown in Figure 4.32. The criteria of surface roughness in all the cases is the path difference $r_2 - r_1$ of the ground reflected waves between wavefront A and B without and with surface irregularity.

The criterion generally accepted for defining surface roughness is the Rayleigh criterion. If the path length difference $(r_2 - r_1)$ does not exceed one-quarter of a wavelength, the surface is regarded as electrically smooth. Therefore, the electrical phase difference between the two rays is

$$\Delta\Psi = (2\pi/\lambda)(2H \sin \Theta) \tag{4.67}$$

where H is the height of the obstacle and θ is the grazing angle. The physical distance $(2H \sin \theta)$ contributes to the electrical phase difference of $\Delta\Psi$. Equating the path length difference to $\lambda/4$, the electrical phase shift between the top and bottom waves becomes $\pi/2$. Therefore, the Rayleigh height for the surface roughness can be defined as

$$H_R = \lambda/(8 \sin \Theta) \cong \lambda/8\Theta \tag{4.68}$$

for small grazing angle Θ. An observation of the above equation states that for a small grazing angle, Θ, the physical size of the irregularity can be relatively larger without destroying the electrical smoothness of the reflecting surface. As an exam-

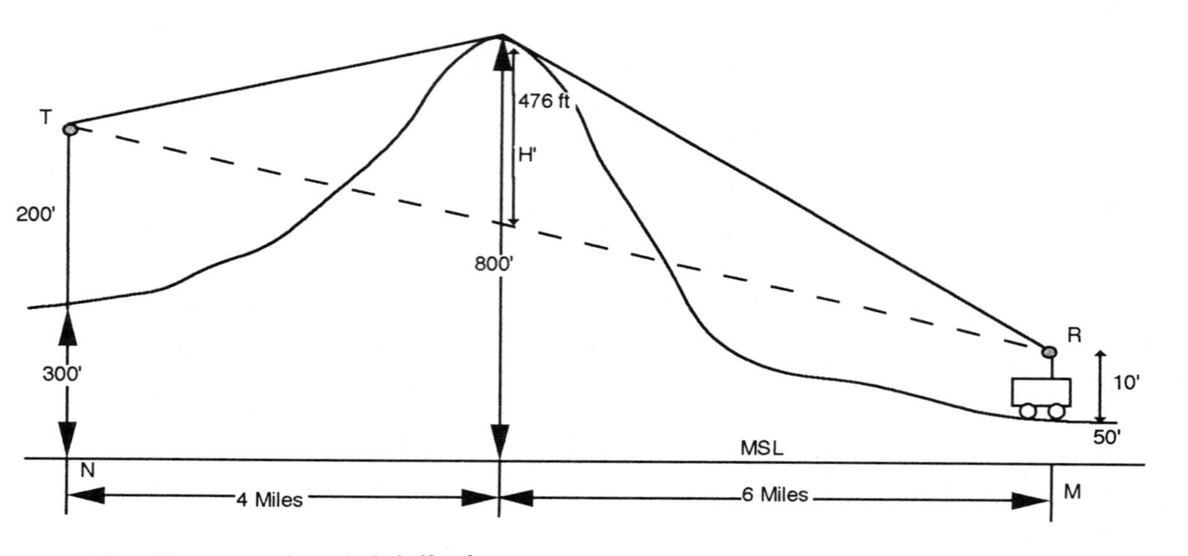

Figure 4.31 Diffraction loss for a single knife edge.

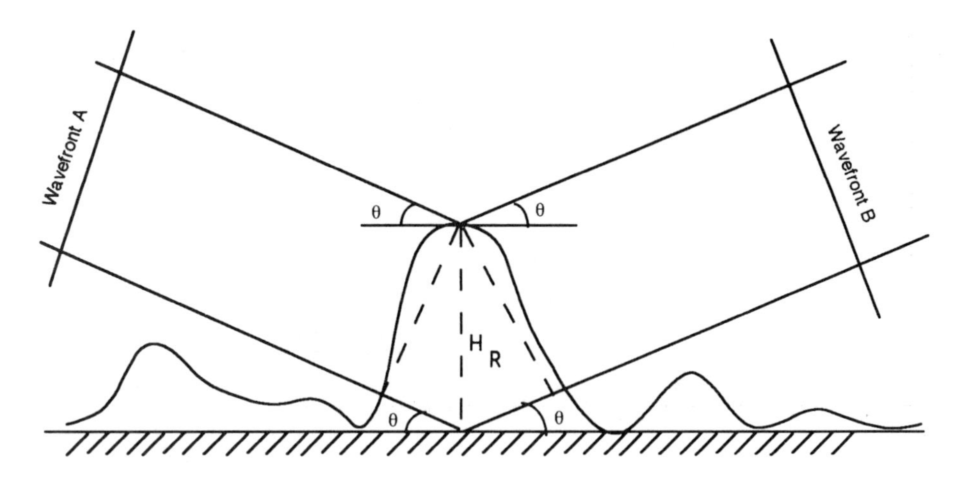

Figure 4.32 Path length difference due to surface roughness.

ple, at 900 MHz and for grazing angle Θ of 1 deg, the computed value for Rayleigh height H_R is 2.38m. Unless the irregularity height exceeds 2.38m, the surface is regarded as smooth. At very long wavelengths, a terrain with hills of the order of a few hundred feet of height may be regarded as smooth surface, but for millimeter wavelengths even hundredths of an inch of surface irregularity may be enough to cause diffraction.

If the surface is considered to be normally distributed with a mean of μ_Z and a variance of σ_Z^2, then the probability density function of the surface irregularity is given by

$$p_z(z) = \frac{1}{\sqrt{2\pi}\sigma_z} e^{-(z - \mu_z)^2/2\sigma_x^2} \tag{4.69}$$

Surface undulation versus distance, x, plot is shown in Figure 4.33.

The Rayleigh criterion for this case is given by [9]:

$$C = 4\pi\sigma_z\Theta/4\lambda \tag{4.70}$$

where the angle Θ is measured in radian from the horizontal plane, λ is the wavelength, and σ_z is the standard deviation of surface irregularities relative to the mean height of the surface. The range of the values of C for different roughness are

$$C < 0.1 \qquad C > 10.0 \tag{4.71}$$

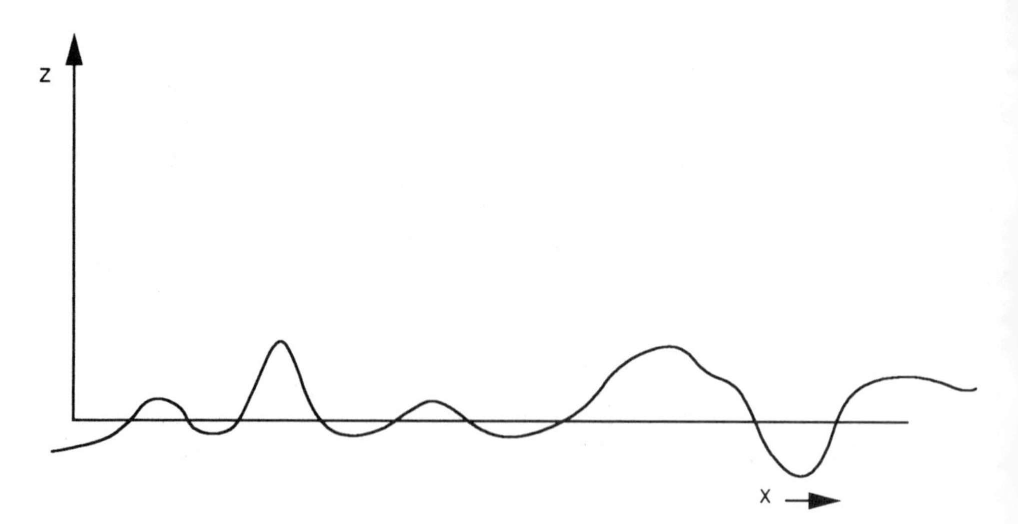

Figure 4.33 Surface undulation versus distance.

for smooth surfaces and rough surfaces, respectively. For values of $C < 0.1$, specular reflection results. For $C > 10$, the coefficient of the reflected wave is very small. Bullington [1] has found experimentally that for most microwave applications the ground is relatively rough with a reflection coefficient in the range of 0.2 to 0.4.

4.10 UHF PROPAGATION MODELS

In this section, we first outline considerations that seem appropriate in both the selection of propagation models and their application in planning and evaluating mobile communications systems performance. Following this we shall describe the various popular UHF models that can be used for cellular mobile design. Various characteristics of concern in the selection of a propagation model include:

- Functional design.
- Accuracy and validity.
- Simplicity and ease of use.
- Computational resource requirements.
- Data requirements.
- Intermodel relationship.
- Recognition and acceptability.
- Development cost.
- Operating cost.
- Generality.

By functional design, we refer to the ability of the model to recognize special physical and operational aspects to be studied. Common functional features will include: ground-wave mode of propagation, communication between two ground points, and other factors, such as atmospheric absorption, vegetation, channel characteristics, and other urban effects. Accuracy requirements for propagation models should be consistent with the expected accuracy from other components of the planning process and with the accuracy of the input data to the model. Models based on measurements should be examined with great care, as the statistical data gathering is costly and the accuracy is based on measurement precautions.

Simple models are usually easier to use, particularly for manual computation. For machine computation, simple models are easier to program and take less memory space, and therefore can be installed in a variety of computers. However, one should also be careful with the simpler model, as often they provide less realistic results.

Since the growing number of applications require computer automation, storage requirements, computational time, and computer size requirements are important before a model is selected. Complex models usually require more computer storage, both high-speed random access memory for computation and large memory for storage of input databases, intermediate results, and so forth. Speed of computation may be important also for those cases where many links are to be evaluated for "snap shots" or event intervals during the design phase. In many cases, propagation model will require computational power found only in machines of a larger size. This may require considerations of the word and memory size of the computer. Propagation models require topographic data, and thus require a supporting database for its operation. A good choice of a database can significantly reduce the cost of the planning process and on computational requirements.

The relationships of the propagation model to other modeling in the study, such as antenna patterns, demographic considerations, receiver signal processing, and so forth are to be considered. Benefits from a high degree of detail in one field of analysis will be negated with lesser detail or details with less confidence in other areas.

Most important of all the requirements is the acceptance of the model by those who will be affected by analysis outputs. Recognition is important when resolving differences between two conflicting results by competing studies. The most obvious example of this is the Carey model, recommended by FCC even though the model does not provide the best technical results in most cases.

Both the developmental and the operating costs can be important considerations for model selection. Even a minor change in an existing model can be very costly if the change is to be quantitatively validated. Operating costs are always important, especially for the larger volume of calculations required for a high degree of confidence. The operating cost often is influenced by the size and complexity of the environmental databases associated with the computations. Traditionally, extracting terrain profiles consumes maximum time and thus leads the cost figure. Lastly, the model selected should be such that it produces consistent results for both specific and general conclusions.

There are two prominent propagation models used in mobile communication: the Carey model and the Okamura model. In this section we shall first define and apply the Carey and Okamura models for the design of a UHF system and then compare these two models based on test results. Following this, an area test model based on [12] is described. This method makes it possible to predict path loss in certain surroundings based on measurements of similar surroundings.

Carey Model

The Carey model is derived from the 1963 CCIR curve for frequencies in the range of 450–460 MHz with a transmit antenna height of 150m and a receive antenna height of 10m. The basic curves that consider a receive antenna height of 10m were adjusted downward by 9 dB to reflect the 6-ft receiving antenna height prevalent in mobile service. An additional factor of 6 dB of loss can be added to the curves shown in Figure 4.34 to account for the conversion from the 450–460 MHz range to the 900 MHz range. The curves are for $F(50, 50)$ for different transmit antenna heights and for a fixed mobile antenna height of 6 ft. The curve for $F(50, 50)$ represents loss for 50% of the time and for 50% of the locations. The field strength is plotted against the ordinate in terms of dBu for a transmitted power of 1 kW through a $\lambda/2$ dipole antenna. Curves for fixed transmit antenna heights of 100, 200, 500, 1,000, 2,000, and 5,000 ft are shown. The FCC recommends that coverage for other intermediate antenna heights be obtained through linear interpolation, while by (4.41) the adjustment should actually be proportional to the square of the antenna height. For example, if the actual antenna height is 125 ft and the transmitted power is 1 kW, then the coverage to the 39-dBu field intensity contour is 15 miles using the curve corresponding to 100-ft antenna height. The predicted 39-dBu contour distance as per FCC recommendation (due to a 125-ft antenna height) will be $15 + 10 \log 125/100 = 16$ mi. However, according to (4.41), the distance to a 39-dBu contour should be $15 + 20 \log 125/100 = 16.9$ mi. The Carey curves as shown in Figure 4.34 do not provide loss below 5 mi. We shall now illustrate the use of the Carey curves for the design of a single cell system.

Example 4.8

Assuming a cell site ERP of 100W through a $\lambda/2$ dipole, find the cell radius assuming the terrain to be perfectly flat all around, including the cell site. The cell-site antenna height is assumed to be 232 ft and the mobiles have antennas 6 ft above the ground plane. Since the radiated power is 10-dB below 1 kW and the nearest antenna

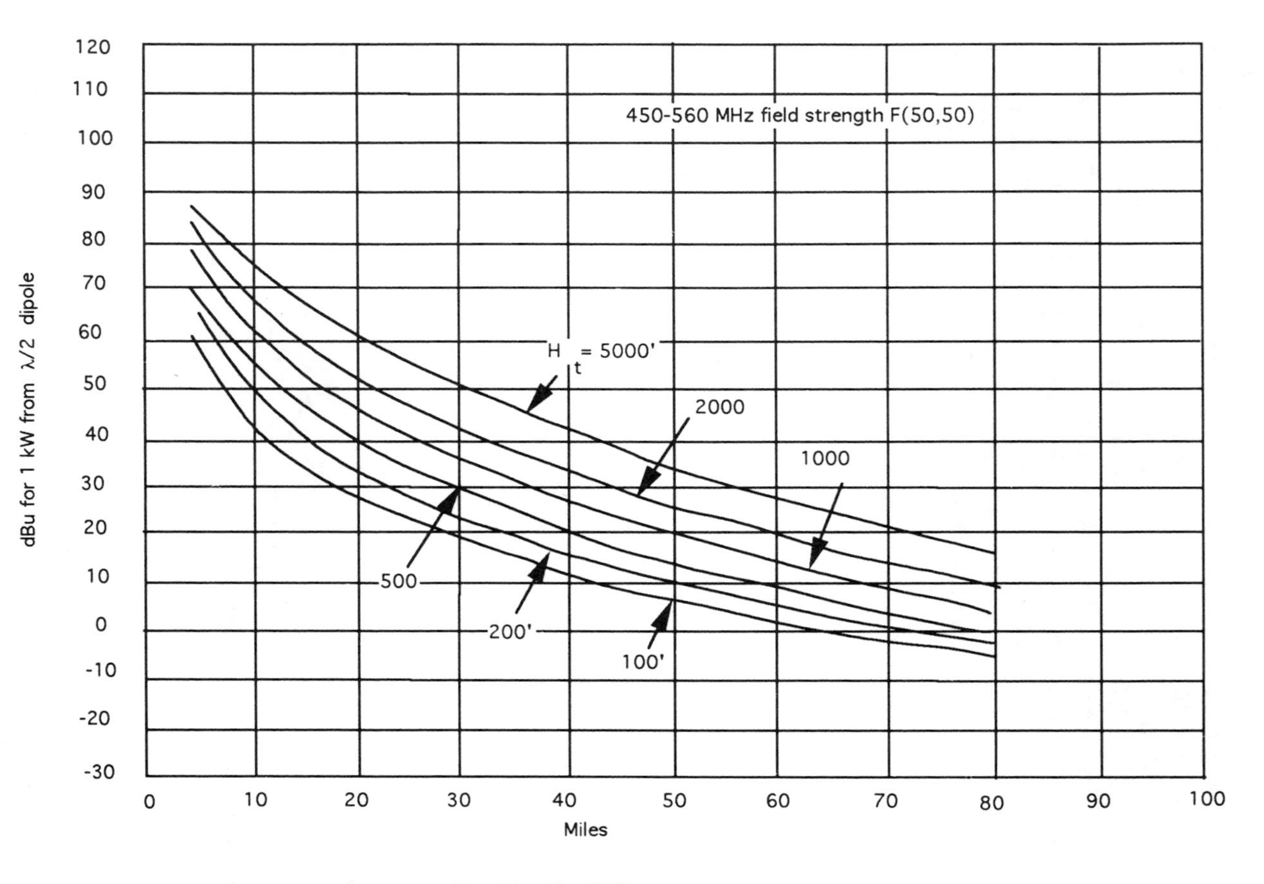

Figure 4.34 A 450–460 MHz field strength F(50, 50). After [35].

height is 200 ft, the field strength curve for 200 ft should be shifted downward by 10 dB before one can see the coverage to 39 dBu. This is approximately equivalent to determining the coverage distance corresponding to 39 dBu + 10 dB = 49 dBu on the curve for the 200-ft antenna height. From the Carey curve, the cell radius r is 11 mi. Making corrections for antenna height, the adjusted cell radius r = 11 + 10 log 232/200 = 11.6 mi. This is shown in Figure 4.35. It should be noted that this is an oversimplified case as the ground surrounding the cell sites is assumed here to be flat uniformly (around 360 deg) with respect to MSL. Also, it should be noted that this calculation is valid for 450 MHz. For 900 MHz, an additional loss of 6 dB is expected, which will reduce the coverage distance 39 dBu further. It should be noted that we have applied linear correction here, as recommended by the FCC.

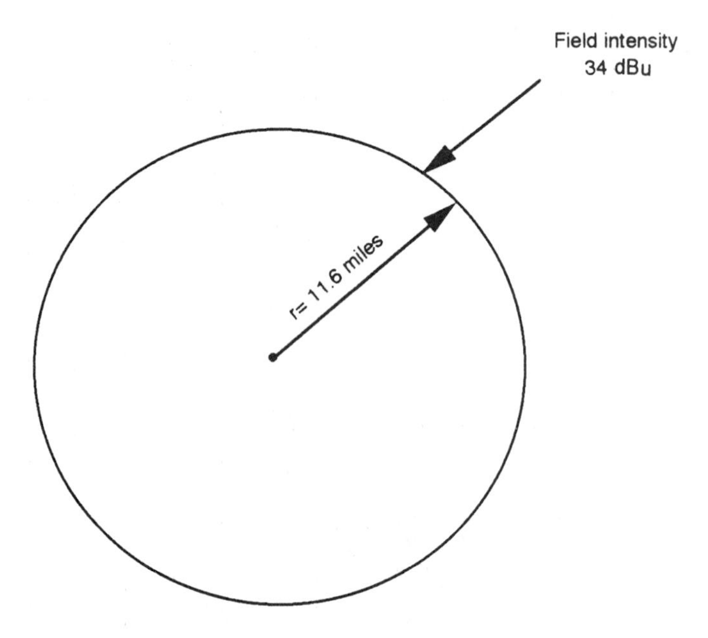

Figure 4.35 Design of a single cell system by Carey curve.

Okamura Model

For prediction of the median power received by a mobile unit from a base station, (4.72) and the corresponding curves are used for a quasi-smooth surface in an urban environment. The equation is based on extensive measurements conducted by Okumura in and around Tokyo.

$$P_p = P_0 - A_m(f, d) + H_b(h_b, d) + H_m(h_m, f) \tag{4.72}$$

where P_p is the the value of predicted receiver power; P_o is the power received under free-space transmission = P_T − free-space loss = $P_T - L_{p,dB}$; $A_{mu}(f, d)$ is the median attenuation relative to free space in an urban area over quasi-smooth terrain with a base antenna height of 200m and a mobile antenna height of 3m (see Figure 4.36); $H_b(h_b, d)$ is the height gain factor from Figure 4.37 if the base antenna height is other than 200 meters; and as shown in Figure 4.38, $H_m(h_m, f)$ is the

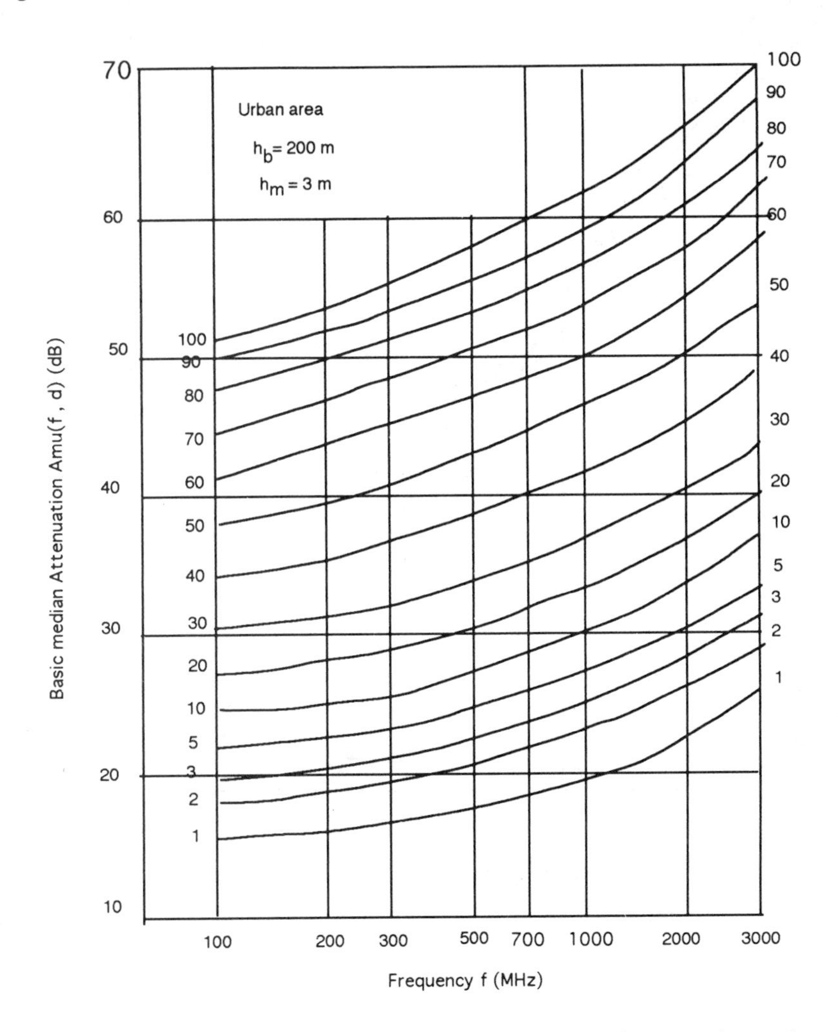

Figure 4.36 Prediction curve for basic median attenuation relative to free space in an urban area over quasi-smooth, referred to h_b = 200m, h_m = 3m. After [9].

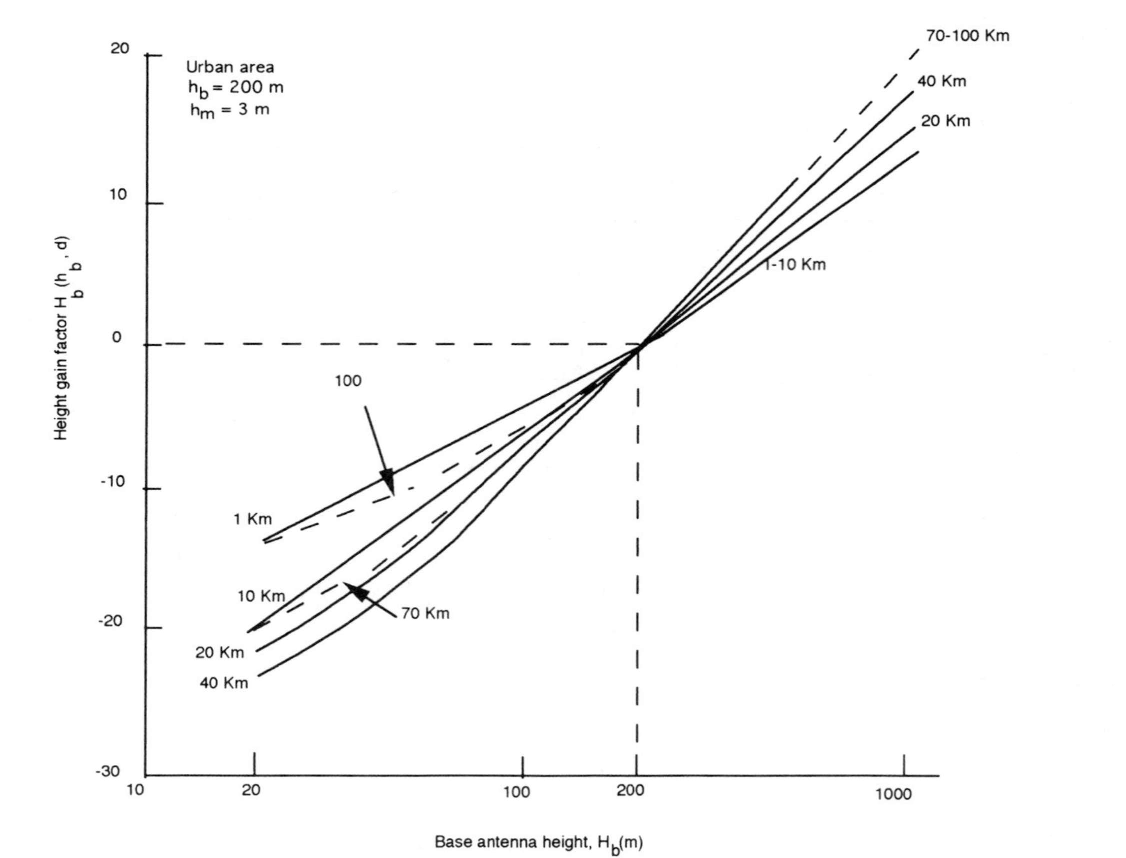

Figure 4.37 Prediction curves for base station antenna height gain factor referred to h_b = 200m as a function of distance. After [9]

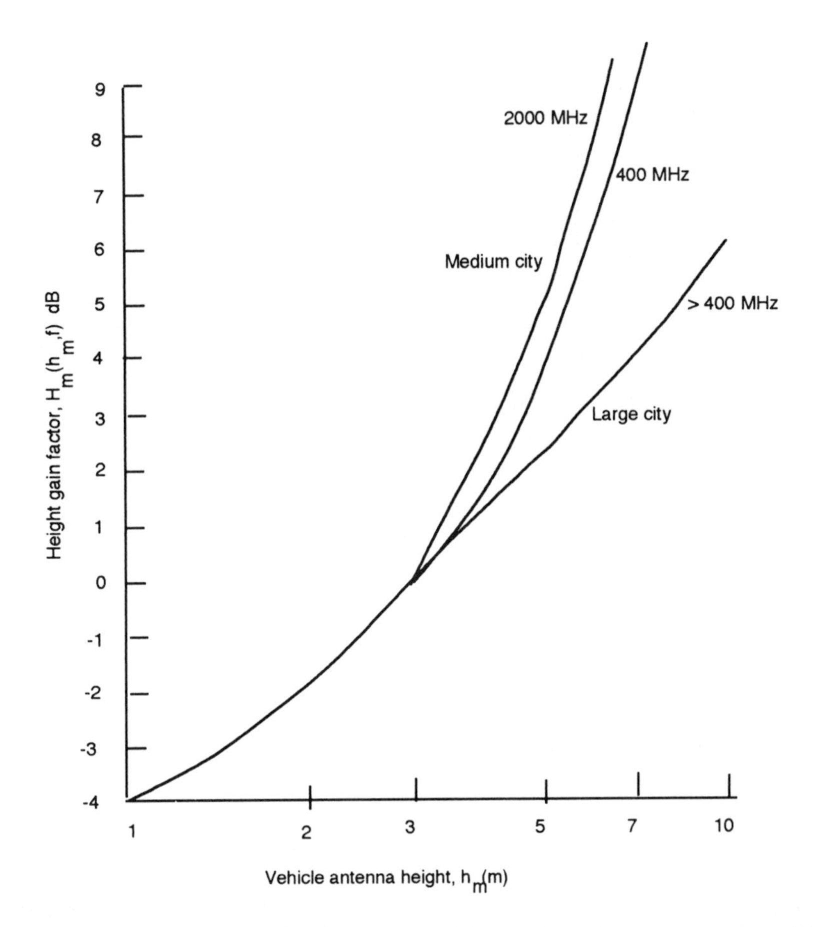

Figure 4.38 Prediction curves for vehicular antenna height gain factor in urban area. After [9].

vehicular antenna height gain adjustment factor in an urban area. Correction applies if the antenna height is different than 3m.

If the propagation is for a different environment, that is, if the surroundings are not urban and the ground is not quasi-smooth, the following adjustments are made on P_p:

$$P_c = P_p + K_{so} + K_{\text{ter}} + K_{sp} + K_{is} \tag{4.73}$$

where k_{so} = correction factor for suburban or open areas, as desired. This is shown in Figure 4.39. As shown in Figure 4.40, k_{ter} = rolling hill terrain correction factor. The definition of Δh has been covered in Section 4.3.2. As shown in Figure 4.41,

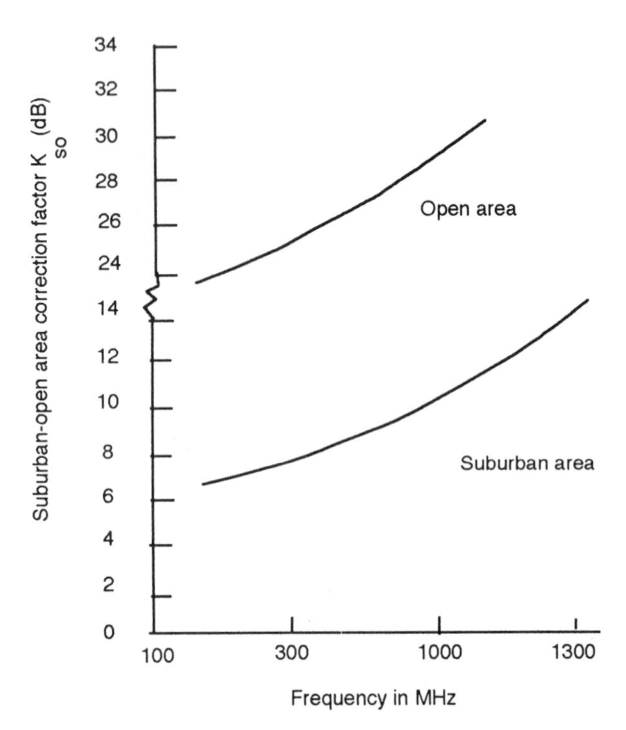

Figure 4.39 Prediction curves for suburban and open area correction factor k_{so}. After [9].

k_{sp} = correction factor for sloping terrain. The definition of terrain slope has been covered in Section 4.3.2. As shown in Figure 4.42, k_{ls} = land-sea correction factor. The definition of mixed land-sea parameter has been discussed in Section 4.3.2.

Additionally, the effects of isolated ridges, street orientation relative to the base station, the presence or the absence of foliage, effects of atmosphere, and (in the case of undulating terrain) the position of the vehicle relative to median height have to be accounted for. We shall describe these factors below, but before we consider these corrections we shall describe Hata's equations for Okumura model.

The propagation loss computed by the Okumura technique matches quite well in practice. However, it is rather cumbersome to implement this in a computer because the data is available in graphical form. Thus, for computer implementation data has to be entered in the computer memory point-to-point and the interpolation routines have to be written for intermediate calculation. In order to make the Okumura technique suitable for computer implementation, Hata has developed an analytic expression for medium path loss, L_p, for urban, suburban, and open areas [28]. The path loss expression is given in Table 4.3.

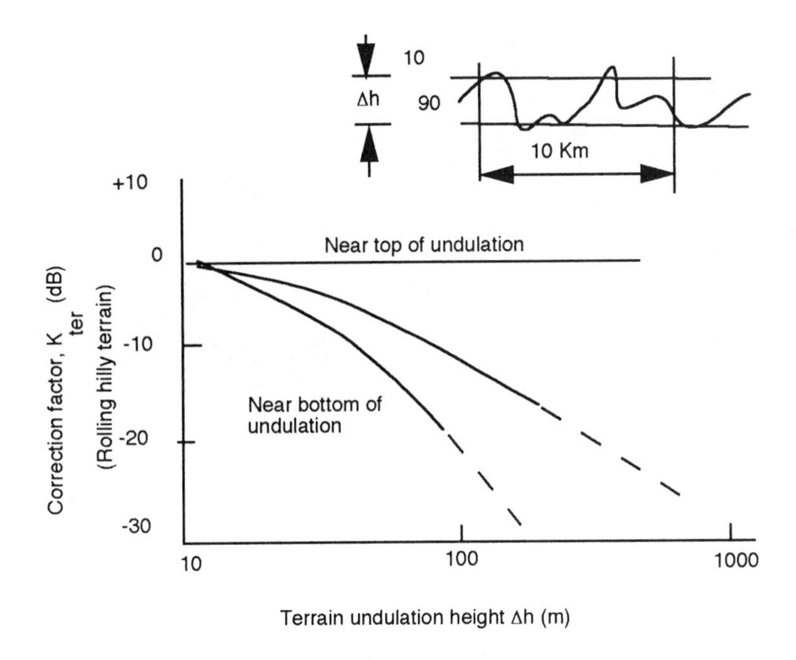

Figure 4.40 Rolling hill terrain correction factor k_{ter}. After [9].

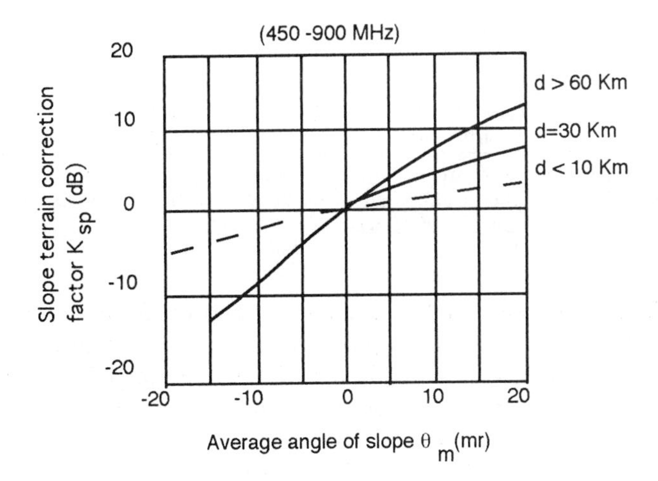

Figure 4.41 Measured value and prediction curves for slope terrain correction factor. After [9].

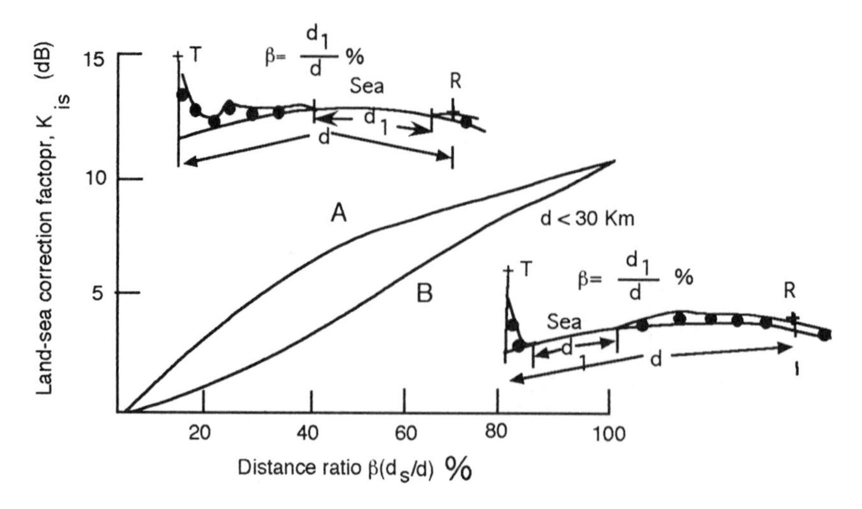

Figure 4.42 Prediction curves for land-sea correction factor. After [9].

Table 4.3
Analytic Approximation for Median Path Loss (Hata Approximation to Okumura Model)

Area	*Approximation*
Urban	$L_p = 69.55 + 26.16 \log_{10}(f_c) - 13.82 \log_{10}(h_b) - a(h_m)$ $+ [44.9 - 6.55 \log_{10}(h_b)] \log_{10}(r)$ (dB)
Medium or large	$a(h_m) = [(1.1)\log_{10}(f_c) - 0.7]h_m - [(1.56)\log_{10}(f_c) - 0.8]$
Large	$a(h_m) = 8.29[\log_{10}(1.54h_m)]^2 - 1.1\ f_c \leq 200$ MHz
	$= 3.2[\log_{10}(11.75h_m]^2 - 4.94\ f_c$ 400 MHz
Suburban	$L_{ps} = L_p(\text{urban area}) - 2[\log_{10}(f_c/28)]^2 - 5.4$ (dB)
Open area	$L_{po} = L_p(\text{urban area}) - 4.78[\log_{10}(f_c)]^2 + (18.33)\log_{10}(f_c) - 40.94$ dB)

Note: L_p = median path loss; u = urban; s = suburban; o = open; a = antenna; f_c = frequency, 150 ~ 1,500 (MHz); h_b = base station effective antenna height, 30–200 (meters); h_m = mobile antenna height, 1~10 (meters); r = distance, 1–20 (Km).

Corrections have been made for suburban and open areas. The correction factor for mobile antenna height $a(h_m)$ for medium and large cities is also made. A large city is understood to be heavily built with relatively large buildings averaging more than four stories. A city is considered medium or small if it has a low average building height, less than four stories, such as Rochester, New York. The reader is advised at this point to refer to Table 4.1 for correlating this definition with general terrain classification. Plots of medium path loss at 900 MHz for the base antenna height of 200m and the mobile antenna of 3m is shown in Figure 4.43.

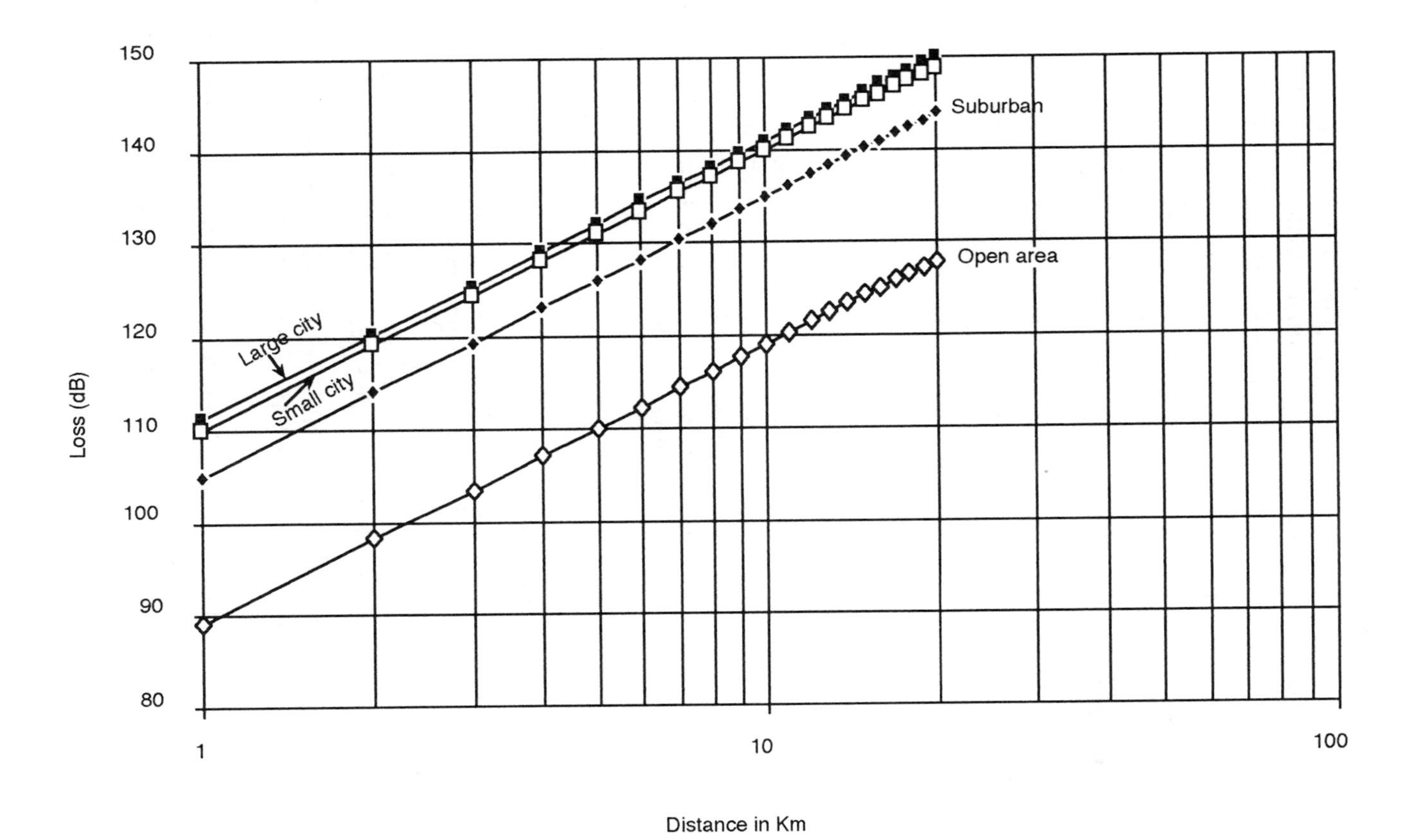

Figure 4.43 Medium path loss at 900 MHz.

For other antenna heights corrections can be applied from Figures 4.36 and 4.37. We shall demonstrate the use of these curves by looking at some examples based on the test results conducted in Philadelphia, Manhattan, and the Bronx. However, before we turn to the examples, let us discuss a general area-to-area model by Lee [12].

4.10.1 General Test Models at UHF

For the test and prediction of received signal power at different radials, measurements are made at a certain fixed distance from the base station. The usual reference distance is taken as one mile. The most general form of the equation relating the received signal power with respect to reference power measured at one mile can be written as

$$P_r = P_{r1}\,(r/r_o)^{-\gamma}\,(f/f_o)^{-n}\,\alpha_o \tag{4.74}$$

where P_{r1} is the received signal power at the reference point of one mile, γ is the slope of the signal power curve, r is the distance from the base, r_1 is the reference distance of one mile from base, f is the measurement frequency, f_o is the reference frequency of measurement, and α_0 is the signal adjustment factor. The above equation can alternately be expressed interns of path loss as

$$P_T - P_r = P_T - P_{r1} + \gamma\log(r/r_1) + n\log(f/f_o) - \alpha_o \quad \text{dB} \tag{4.75}$$

There are at least two advantages to this approach:

1. From the set of measurements in one type of surroundings, predictions for path loss can be made for similar but other surroundings (one suburban to other suburban).
2. By proper adjustment of the model, one can adopt this model for other dissimilar surroundings (urban to rural and vice versa).

Let us demonstrate the utility of this approach through measured results for the suburb of Philadelphia-Camden. Tests were made under the following conditions:

- A base station transmits power of 10W and has an antenna gain of 6 dB with respect to a $\lambda/2$ dipole. The antenna height is 100 ft above ground. The measurement frequency f and the reference frequency f_o are both at 900 MHz.
- The mobile antenna is a $\lambda/4$ whip and is 10 ft above ground. Its antenna gain is zero dB with respect to the gain of dipole antenna.

The measured results under the above conditions are:

- The local mean was found to be log-normally distributed with a standard deviation σ of 8 dB.
- The measured power intercept at the one-mile point $= -61.7$ dBm.
- The slope is 38.4 dB per decade or the numeric value of γ is 3.84.

The measured data is plotted in Figure 4.44. By applying (4.76) the path loss, P_L, is computed as

$$\begin{aligned} P_L &= 40.0 + 61.4 + 38.4 \log(r) + n \log(900/900) + \alpha_0 \text{ dB} \\ &= 101.7 + 38.4 \log(r) + \alpha_o \end{aligned} \tag{4.76}$$

The adjustment factor accounts for base and mobile heights and the transmitted signal power, and can be written as

$$\alpha_0 = (adj)_1\,(adj)_2(adj)_3\,(adj)_4 = \alpha_1\alpha_2\alpha_3\alpha_4$$

where $(adj)_1 = \alpha_1 =$ (the new base antenna height/100)^2; $(adj)_2 = \alpha_2 =$ (the new value of transmitted power in watts)/10W; $(adj)_3 =$ (the new cell-station antenna gain with $\lambda/2$ dipole)/4; and $(adj)_4 =$ (the new mobile antenna height)/10^v.

The adjustment α_1 should be proportional to the square of the antenna height from (4.41). The adjustment to α_2 should be proportional to the new transmitted power from (4.39), and the gain of the cell-site antenna (α_3) should be proportional to the new value of antenna gain, also from (4.39). Factor 4 is due to cell-site antenna gain of 6 dB. The value of v is 2 for mobile antenna heights exceeding 30 ft and 1 for mobile antenna heights below 10 ft.

Thus, the path loss can be expressed in the general form as

$$101.7 + 38.4 \log r + n \log(f/900.0) - \alpha_o \tag{4.77}$$

From (4.77), path loss for suburban surroundings similar to the Philadelphia-Camden area (same γ) can easily be estimated. The measured value P_n and exponent γ for Newark, Tokyo, and Philadelphia (urban, free space, and openarea conditions) are given below in Table 4.4 along with their path loss equations. In all cases the base antenna height is 100 ft, the base transmit power is 10W, and the reference frequency f_o is 900 MHz. Thus, in these cities the path loss can be found easily by knowing r, f, and correction factors α_o, which are equal to $\alpha_1\alpha_2\alpha_3\alpha_4$.

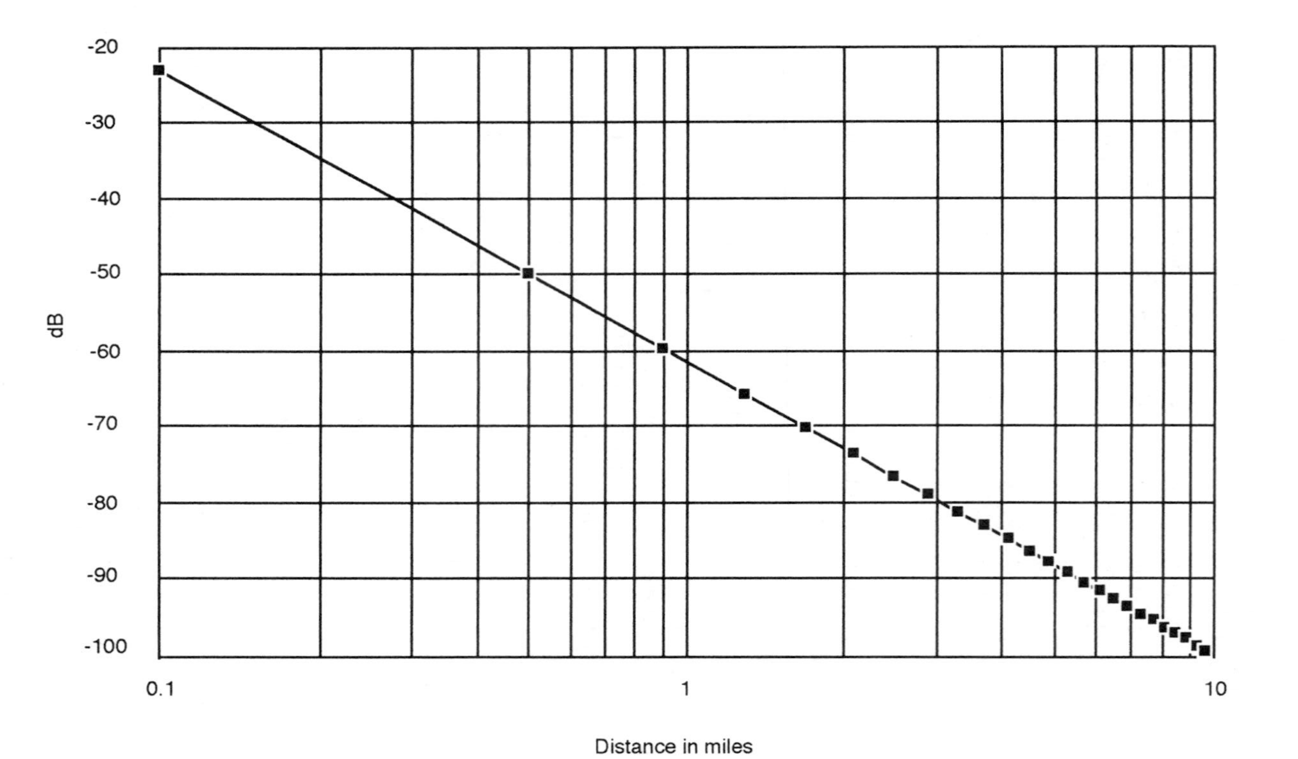

Figure 4.44 Path loss in a typical suburban area at 900 MHz.

Table 4.4
Path Loss Equation For Different Cities

Area of Interest	γ	P_{ri} *(dBW)*	*Path Loss,* $P_T - P_r$
Newark	4.31	−94.0	$104.0 + 43.1 \log_{10}(r) + n \log_{10}(f/900) - \alpha_o$
Tokyo	3.05	−114	$124.0 + 30.5 \log_{10}(r) + n \log_{10}(f/900) - \alpha_o$
Philadelphia:	3.68	−100.0	$110.0 + 36.8 \log_{10}(r) + n \log_{10}(f/900) - \alpha_o$
Urban			
Free space	2.0	−75.0	$85.0 + 20 \log_{10}(r) + n \log_{10}(f/900) - \alpha_o$
Open	4.35	−79.0	$89.0 + 43.5 \log_{10}(r) + n \log_{10}(f/900) - \alpha_o$

4.10.2 Test Results

In this section, we summarize the test results conducted in various cities in the United States and compare them with the Okamura and Carey models [22–24]. We present the reported test results on the Philadelphia urban and suburban areas and Motorola experiments in Bethesda and Patuxtent, MD (near Washington, D.C.).

Philadelphia Urban and Suburban Results

The measurements for the Philadelphia urban area were conducted at a frequency of 820 MHz with a transmit antenna height of 20m and a receive antenna height of 2.4m. These measurements formed the basis for the Bell System's developmental cellular system design. The measured results are shown in Table 4.5 along with

Table 4.5
Measured Results of Philadelphia Urban Area

	Measured Data (dB)			*Okumura (dB)*			*Carey (dB)*	
Distance, Km (mi)	*L*	*ΔL/1*	*ΔL/8*	*L*	*ΔL/1*	*ΔL/8*	*L*	*ΔL/8*
1 (0.6)	122.0	—	—	123.8	—	—	—	—
2 (1.25)	134.0	12.0	—	134.8	11.0	—	—	—
3 (1.87)	140.5	18.5	—	141.2	17.4	—	—	—
4 (2.5)	145.0	23.0	—	145.7	21.9	—	—	—
6 (3.75)	151.5	29.5	—	152.1	28.3	—	—	—
8 (5.0)	156.0	34.0	—	156.7	32.9	—	134.5	—
10 (6.25)	159.5	37.5	3.5	160.2	36.4	3.5	142.8	5.3
12 (7.5)	163.0	41.0	4.0	163.1	39.3	6.4	147.0	9.5
14 (8.75)	165.0	43.0	9.0	165.5	41.7	8.8	150.6	13.1
16 (10.0)	167.5	45.0	11.0	167.6	43.8	10.9	153.8	16.3

Note: Measurement conditions: $h_t = 64'$, $h_r = 9'$, $f = 820$ MHz. Measured data is adjusted to half-wave dipole receiving antennas with 1.5m height. $\Delta L/1$ and $\Delta L/8$ represent path loss slope with respect to path loss at 1 km and 8 km, respectively.

computed values by Okumura and Carey. The measured data in this case has been converted to a 1.5m high mobile antenna so that the comparison can be made on a common basis. The slope of the curve or the propagation characteristic as a function of distance from the base station is illustrated by computing $\Delta L/1$ and $\Delta L/8$, where $\Delta L/1$ and $\Delta L/8$ are the changes in the path loss relative to the path loss at 1 km and 8 km, respectively. For the Carey model, $\Delta L/8$ only is computed, as the values of propagation loss below 5 mi (8 km) are not provided for by Carey curves. As shown by the Okumura, curve values for both $\Delta L/1$ and $\Delta L/8$ are in close agreement with the measured values of these parameters. The measured value of path loss always shows to be higher than the predicted values of loss using the Carey curves, while the path loss changes when referred to an 8-km reference is always higher by Carey curve than the experimental $\Delta L/8$. This is attributed to a constant exponent of 5.5 in the case of the Carey model. Similar measurements were made in the suburban Philadelphia area with a 200-ft transmitting antenna height at a frequency of 820 MHz. The measured data shows a path loss slope of 38.4 dB/decade between 1–10 mi and the exponent value is 3.84. Okumura predicts the propagation loss slope to be 33.23 dB/decade, which is approximately 4-dB lower than the measured value over the whole range. This discrepancy can be attributed to the fact that while the measured data was for a suburban setting, the Okumura prediction is for urban surroundings. Once the correction is made for suburban surroundings, Okumura's predictions are quite close to the measured values of the propagation loss. There is another set of measured data in Philadelphia urban at 836 MHz with a 500ft transmitter antenna height. The data provides an exponent value of -3 up to 4 mi from the base station while the Okumura predicted exponent value is -3.2. This further assures that Okumura follows the measured loss much more closely.

Motorola Experiment in Bethesda and Patuxent, MD

The measured values of the path loss versus distance from 1–20 mi are shown in Figure 4.45. The test at Bethesda was with a 459-ft high transmit antenna and a 5-ft high receive antenna. As seen from the path loss curve, Okumura constantly shows approximately a 4-dB higher loss then the measured curve, but the two curves run parallel to each other. Thus, test results once again indicate that the exponent for Okumura matches well with the measured propagation loss behavior with distances. On the other hand, the calculated values from Carey show a marked disparity with the measured values at lesser distances from the cell site then at higher distances. The curve matches closely to the experimental values at distances exceeding 15 mi. However, one can observe that due to the higher slope of the Carey curve, the computed values of propagation loss will exceed the measured values beyond 20 mi. The constant difference of about 4–8 dB on the mea-

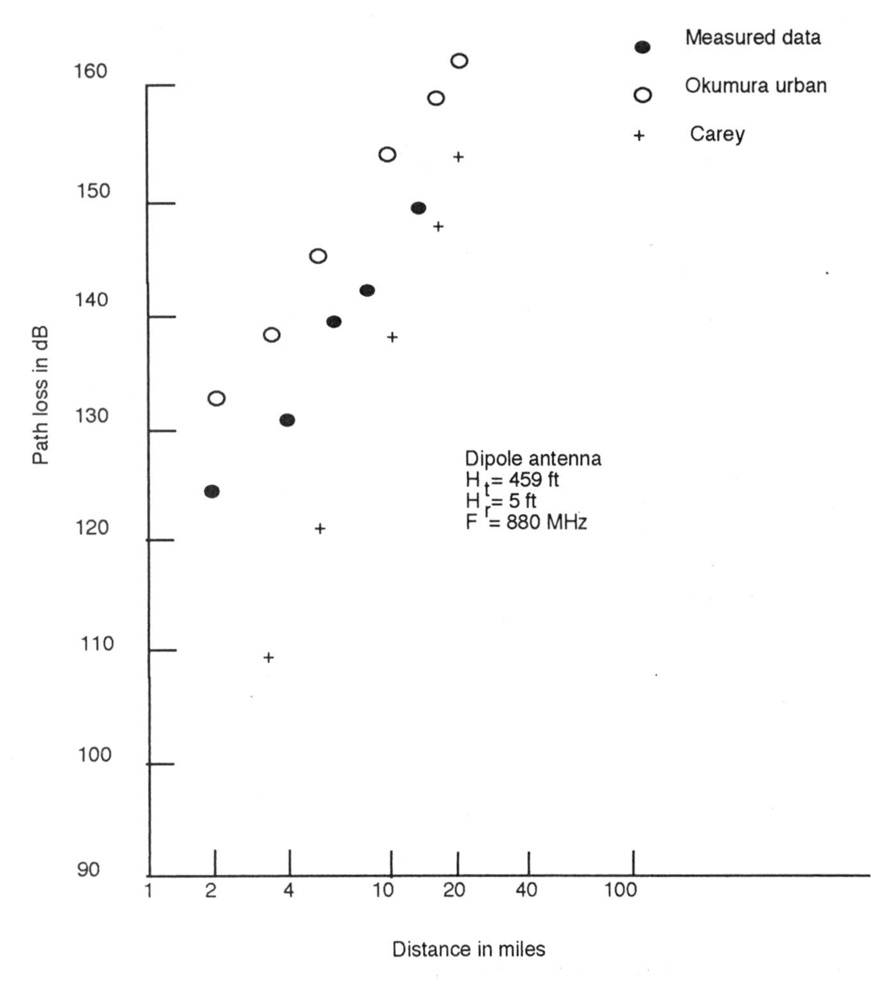

Figure 4.45(a) Path loss at Bethesda, MD site (Washington, D.C.).

sured values of propagation loss and Okumura can be attributed to the Bethesda site not being fully urban. In other words, buildings are not so tall and not so much in a cluster to classify Bethesda as truly urban in Okumura's model. The measured and the computed values by Okumura and Carey for Patuxent are shown in Figure 4.45(b). Actually, the measured and the computed values by Okumura are in complete agreement here, while the Carey curve projects the path loss to be consistently lower up to 20 mi. Beyond 20 mi, the Carey curve shows a trend of higher loss than the measured value.

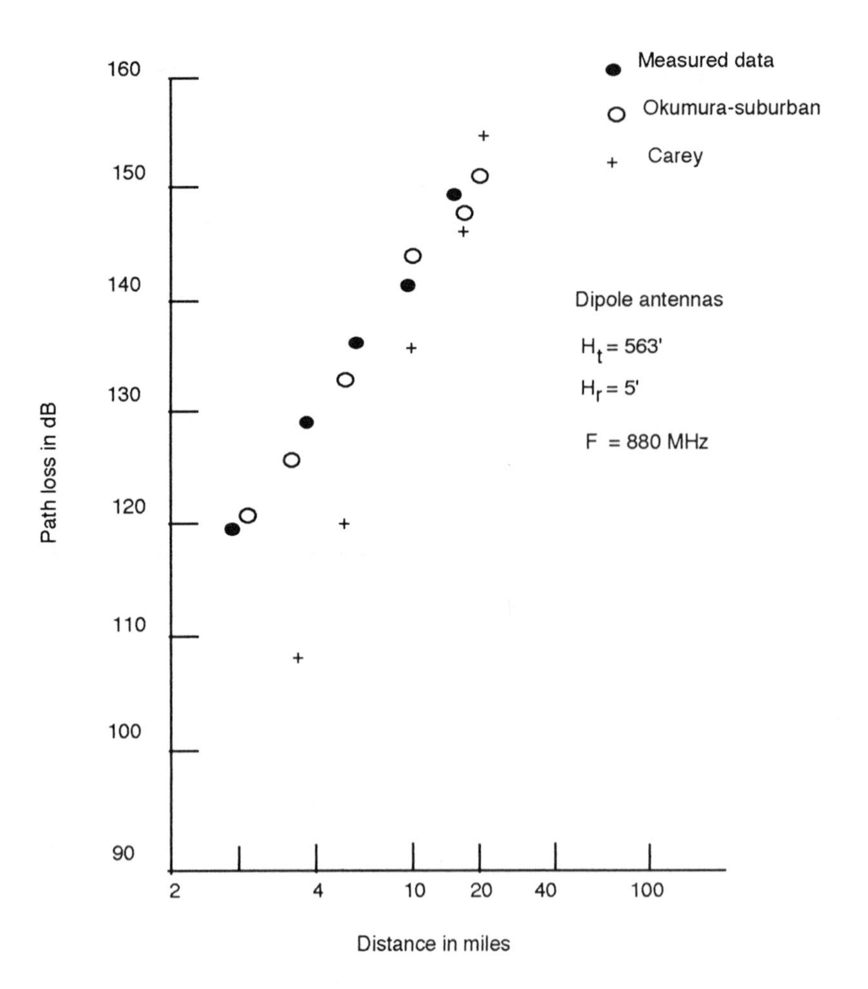

Figure 4.45(b) Path loss at Patuxent, MD (suburban).

Conclusions

From the different test data and the discussion above, one can draw the following conclusions regarding the Okumura and Carey models.

- The Okumura model is reasonably accurate for most urban and suburban areas. Obviously, the propagation model by Okumura accounts for diffraction losses of radio waves due to buildings and trees, while the Carey model does not.

- The Okumura model in some cases shows a constant difference between measured and predicted data. In cases like Philadelphia and Washington, the Okumura model overestimated the path loss by 4.5 dB.
- Distance-dependent behavior of the Okumura model is quite in agreement with the measured values while, on the other hand, Carey provides a constant exponent independent of distance (exponent = − 5.5 for Philadelphia). Thus, the errors of the Carey model are not correctable. Carey curves are an averaging of all the conditions. They do not distinguish between cities and rural areas and they do not account for the increased fading effects at high frequencies. The predicted loss by Carey curve below 5 mi (8 km) has also to be extrapolated. The excess loss due to diffraction over isolated hills is also not accounted for. The distance-dependence property of the field strength is the most important characteristic of the wave. Okumura established a distance-dependent relationship in urban environments with the exponent n related to antenna height, as shown in Figure 4.45.
- Losses predicted by Carey are usually underestimated for distance lower than 15 mi. For distances exceeding 20 mi, the computed values of loss by Carey will in most cases exceed the measured loss.

Example 4.9

Verify the entry in Table 4.5 for path loss at a distance of 10 km. From Figure 4.43, the value of median path loss at 10 km is approximately 140 dB (considering Philadelphia to be a small city). Corrections for base and mobile antenna heights of 20m and 1.5m, from Figures 4.37 and 4.38, are approximately 20 dB and +3 dB, respectively. Also, the correction factors for 820 MHz and for dipole antenna gain will be 20 log(820/900) = −0.8 dB and −2.1 dB. Accounting for all these factors the adjusted path loss is 140 + 20.0 + 3.0 − 0.8 − 2.1 = 160.1 dB, which is close to the 160.18-dB loss shown in Table 4.5.

Example 4.10

The measured path loss on Manhatten at a distance of 8 Km is 144 dB. The cell site transmit antenna is 136m. Verify the result by Okumura model. The frequency of test is 900 MHz. From Figure 4.43, the loss at a distance of 8 Km is roughly 139.0 dB. Adjustment for the base antenna height of 136m from Figure 4.37 is about 3.0. Thus, the total loss is 142.0 dB, a difference of 2 dB from the measured loss.

4.10.3 Adjustment to Models

In this section we shall discuss the different adjustments necessary due to base and mobile antenna heights, and adjustments due to building losses, street orientation, and so forth.

Effects of Base and Mobile Antennas

The effect of base antenna height was observed by Okumura, who found that the effects remain the same within the frequency range of 200–2,000 MHz. For distances less than 10 km, received signal power varies as the square of the antenna height; that is, doubling or reducing the antenna height to half its original length increases or decreases the gain by 6 dB. For high base antennas and for distances exceeding 30 km, the received signal power varies as the cube of the antenna height; that is, doubling or reducing the antenna height to half increases or decreases the gain by 9 dB. For a fixed mobile station height of 3m in an urban environment, the height gain factor as a function of base antenna height is shown in Figure 4.37. The reference antenna height is 200m, which corresponds to a gain of zero dB. Plots are shown for separation distances of 1–10 km, 20 km, 40 km, and 40–100 km.

The effects of mobile antenna height variation have also been predicted by Okumura [4]. A 3-dB increase in gain has been observed by changing the antenna height from 1.5m to 3m. Unlike the case of a mobile antenna, the gain variation of 6 dB was observed for base antenna height variations for distances less than 10 km. For a mobile antenna height exceeding 3m, antenna gain variation is the function of both frequency and the environment. The gain at 2,000 MHz for a medium-size city can increase as much 10 dB by doubling the antenna height. However, for large cities, doubling the antenna height can only increase the gain by about 4 dB for frequencies below 1,000 MHz and for antenna heights exceeding 5m. This gain variation as a function of vehicular antenna height is shown in Figure 4.38.

Building Loss

The use of indoor portable service in cellular systems requires that an estimate of building loss be made. Unfortunately, the building loss is the function of outer wall construction (steel framed, glass, brick, and so on), floor elevation, building orientation with respect to cell sites, percentage of window area, and so forth. Due to the multiplicity of variables and based on the surroundings of the building, the losses can at best be predicted statistically. Our description here is based on tests conducted in Chicago, IL, and Newark, NJ [22]. Some general conclusions that can be drawn from these tests are:

- The average penetration loss in a building situated in an urban area is greater than that for a suburban or rural area.
- A windowed area generally has less loss than a non-windowed area.
- Loss in an open area inside the building is less than if the area has hallways and other walls.
- The average penetration loss decreases as the floor number increases.
- Sheetrock with aluminum backing has a higher attenuation than sheetrock without aluminum backing.
- Buildings with insulation only in the ceiling show less attenuation than those with insulation both in the ceiling and in the interior walls.

The average penetration loss as a function of floor number for measurements made in Chicago is shown in Figure 4.46. The average shown by the solid line is the mean loss in decibels. The building penetration loss for a given floor area is defined to be the difference between the average measurements inside the building and the average measurements made outside the building. As seen from Figure 4.46, the average first-floor penetration loss is 10.4 dB. The slope of the loss line is -1.9 dB per floor. This slope remains about the same for the first several floors till the floor height exceeds the local skyline, beyond which the attenuation gradually tapers off. Measurements were conducted in Chicago in both urban and suburban areas. The penetration loss for urban areas was on the average about 5-dB higher than suburban areas when measured at both locations on the first floor. The absolute numbers of losses and the standard deviations are shown in Table 4.6. The losses and their standard deviation data were gathered from ten suburban buildings and three urban buildings.

Measurements made at a specific floor indicate that the loss characteristic inside the buildings can be classified as a lossy wave guide with attenuation measured in dB per foot of penetration. A typical run along a corridor perpendicular to an outside window in Chicago area is shown in Figure 4.47. Results of about 54 runs of the same type show that the attenuation constant is approximately equal 0.12 dB/foot or 0.4 dB/meter.

Coverage Inside a Tunnel

Radio waves at microwave frequencies inside a tunnel are generally unsatisfactory because of the excessive loss occurring at obstructions, bends and at corners inside the tunnel. Several authors have studied the propagation inside a tunnel by modeling the tunnel as a smooth-walled, lossy, and homogeneous straight-wave guide [29–31]. The various theories conclude that in a straight, smooth, uniform tunnel,

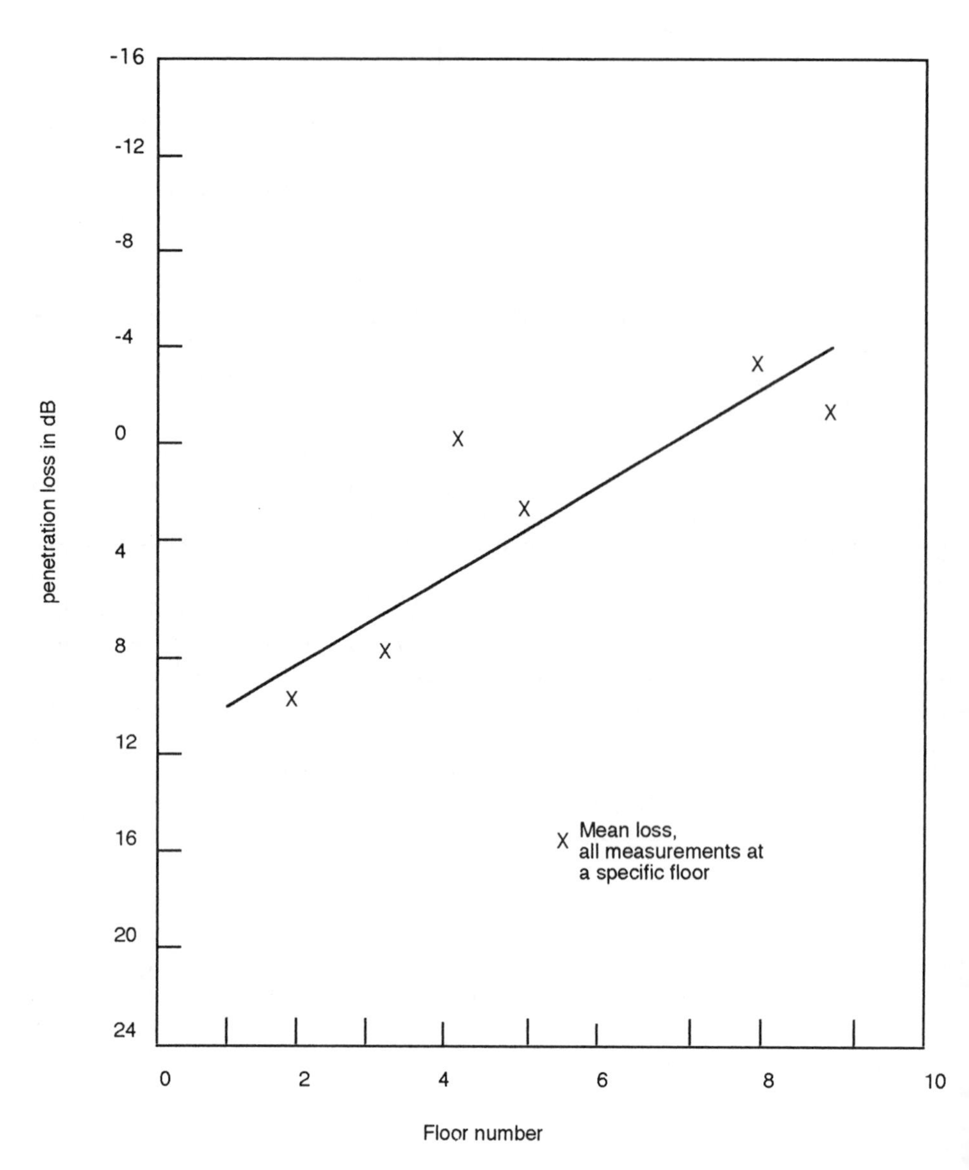

Figure 4.46 Penetration loss measurements. All measurements are at a specific floor. Source: [36, p. 279].

Table 4.6
Building Mean Loss and Its Standard Deviation

Location	*Loss (dB)*	*Standard Deviation (dB)*	*Number of Buildings*
Suburban	13.1	9.5	10
Urban	18.0	7.7	3

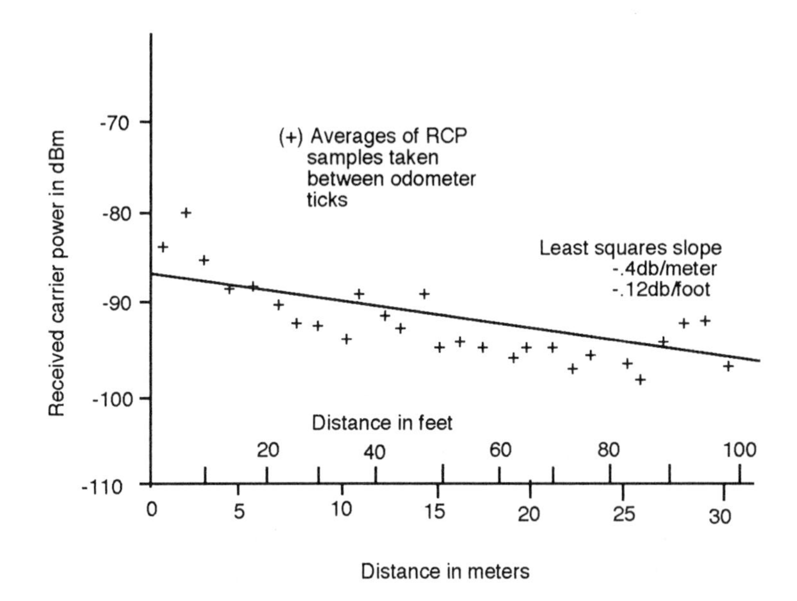

Figure 4.47 Received carrier power relative to distance from outside window. After [29, p. 35].

attenuation varies as an inverse square of the frequency presuming the wall constants are really constant with frequency, and that the wall roughness will increase the attenuation, but the relationship of this effect with frequency is uncertain.

Results of propagation inside the Glenfield tunnel in England and the Lincoln Tunnel, which connects midtown Manhattan to New Jersey, suggest a decrease of propagation loss at a specified distance as frequency is increased. The curve of attenuation versus frequency for the Glenfield tunnel is shown in Figure 4.48(a). The least-square fit for the curve above 438 MHz is given by

$$\alpha = 1.7\lambda^{0.95} \tag{4.78}$$

where α is the attenuation dB/km and the wavelength λ is in cm. Exponential attenuation behavior in decibels is seen below about 2 GHz. Above 2 GHz, the

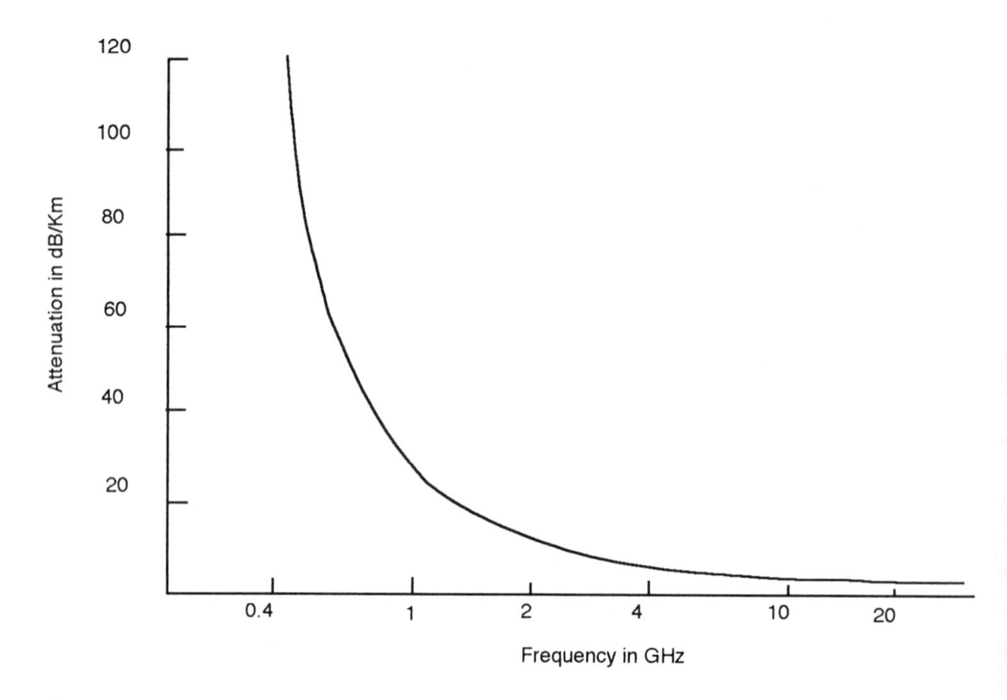

Figure 4.48(a) Attenuation in Glenfield tunnel as a function of frequency. After [29].

attenuation is almost directly proportional to frequency. Between 2 GHz and 24 GHz, attenuation is approximately represented by the expression $A = 1.2\lambda$. Loss of signal strength as a function of antenna separation at 980 MHz for the Lincoln Tunnel test is shown in Figure 4.48(b). At 980 MHz, attenuation inside the tunnel varies inversely as the fourth power; that is, as the distance between two antennas is doubled, attenuation increases by about 12 dB.

Effects of Street Orientation

Street and building layout have pronounced effects on the signal arriving at the mobile antenna. Signals arriving on streets parallel to the transmitter show less loss compared to signals on the perpendicular roads. Signals along the road parallel to the transmitter show channeling effects. A typical test reported in urban surroundings at 854 MHz shows an additional mean loss of the order of 6 dB on streets that are perpendicular to the base station antenna [23]. Graphically, the variation of attenuation along roads designated as 011–013 and 101–103 are shown in Figure 4.49(b) and 4.49(c) along with the layout of the typical area in Figure 4.49(a). In this case, the mean value of the clutter loss at the street level was 31 dB where

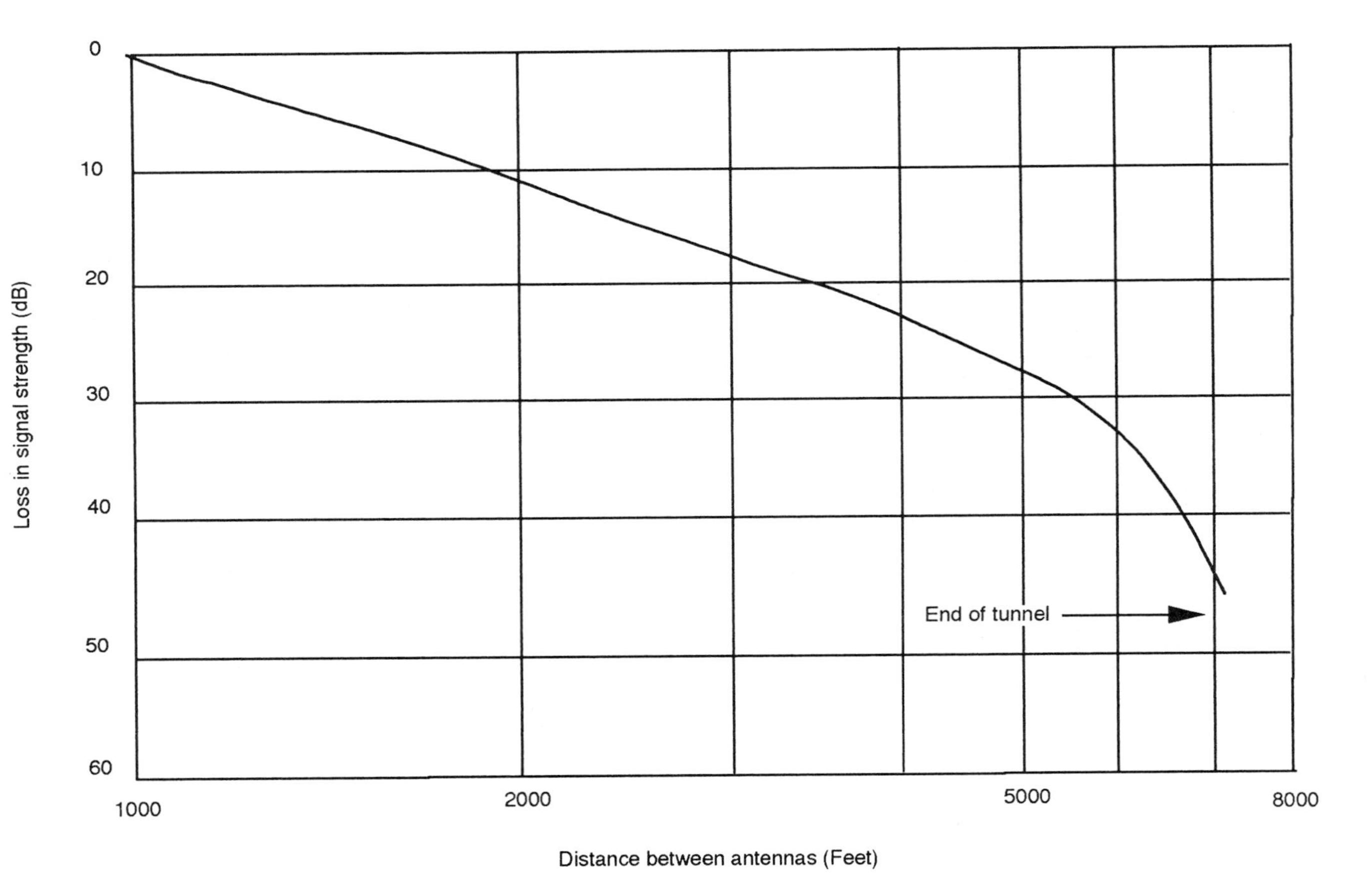

Figure 4.48(b) Signal loss versus antenna separation at 980 MHz. After [9].

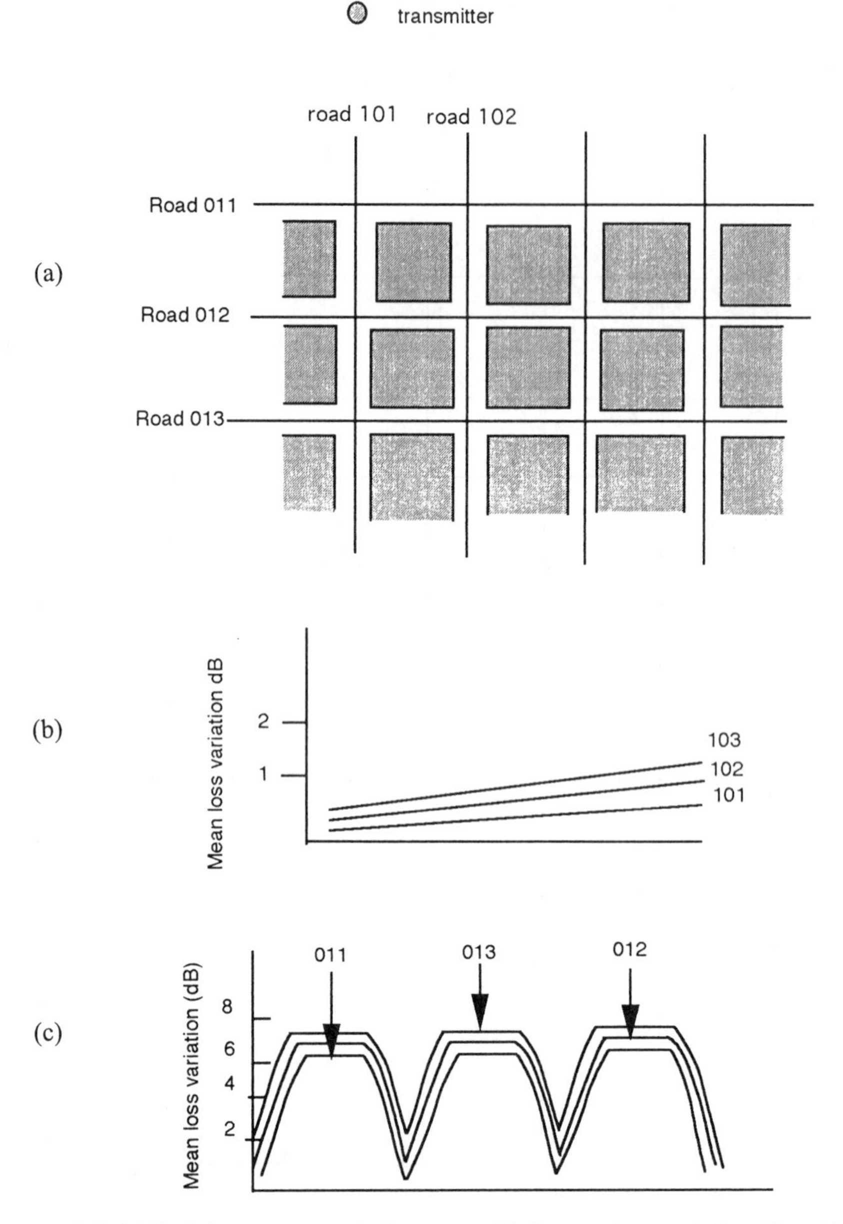

Figure 4.49 (a) Shaded areas represent built-up areas; (b) distance along roads 101–103; and (c) distance along roads 011–013. After [37].

the streets and the buildings were oriented along the direction of the transmitting antenna, whereas the loss was 34 dB for roads perpendicular to the base site location. These losses are with respect to the plane-earth model.

Effects of Foliage

The effects of foliage are considerable at the UHF band. Measurements were made at 836 MHz during both the summer and winter months in suburban Holmdel, NJ, and comparisons of signal loss were made. The average losses in summer months were about 10-dB higher than in winter due to the trees being full of leaves. The exact prediction of signal loss due to foliage is difficult to arrive at due to the type, shape, water content, density of leaves, and uniformity of tree heights. Signal loss also varies seasonally. Lagrone has reported the result of an experimental study behind a grove of oak and hackberry trees in Texas at frequencies in the range of 0.5–3 GHz [32]. The measurement results in this case were correlated with the theoretical loss based on diffraction theory. Some general conclusions of the study based on the work of Tamir and others are [33]:

- Vegetation produces a constant loss independent of distance so long as the T_x to R_x distance exceeds 1 km.
- Measurement by Tamir shows that loss increases as the fourth power of the frequency [33]. These measurements were made at a distance of 1 km with both T_x and R_x antennas at tree tops.
- Studies have shown that signal loss for vertical polarization is higher than that of the horizontal polarization.

Coverage in Underground Parking Facilities

Coverage in underground parking facilities is important in major cosmopolitan cities. Tests were conducted in downtown Pittsburgh in an underground parking facility adjacent to the Hilton Hotel to determine the reliability of the vehicles parked there receiving a usable signal [34]. Test results, as shown in Figure 4.50, indicate that the signal drops as a function of the parking sublevel. On the third sub-level, the test data indicate that the signal became totally unusable although the 39-dBu point was quite rapidly reached as the automobile came up from the bottom of the third sublevel.

4.11 CONCLUSIONS

This chapter deals with both the analytical and the experimental models that can be used to compute the median loss in mobile radio surroundings at 900 MHz.

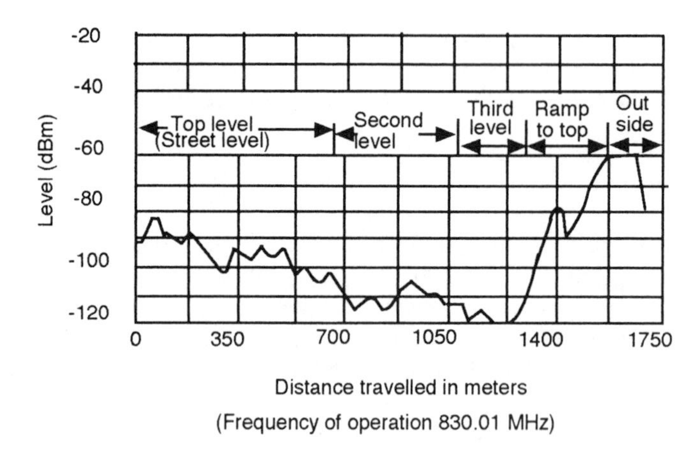

Figure 4.50 Received signal, underground parking area. After [34, p. 53].

Among the widely used propagation models are those by Carey, and Okumura. Coverage inside tunnels and buildings and the effects of foliage on propagation have also been considered.

PROBLEMS

4.1 Find receiver sensitivity in dBW for the field strength of 0.235 μV/m for 20 dB of noise quieting (dBQ).

4.2 Detector noise in urban and semiurban areas are often characterized by 0.5μV and 0.3 μV across 50 ohms. Express the noise floor in dBm.

4.3 Prove that at 800 MHz, half-wave dipole makes available only 31% as much power at its antenna terminal as the comparable 450-MHz antenna.

4.4 For propagation over plane flat earth, show that the median field strength versus propagation distance *d* is given by:

$$E_{\text{plane}} = |E_o| \frac{4\pi h_t h_r}{\lambda d}$$

where E_o is the field intensity in the free space, and h_t and h_r are the transmitting and the receiving antenna height. Plot the value of the median field strength at 900 MHz versus distance in kilometers when the transmit and receive antenna heights are 100m and 3m, respectively.

4.5 Find the net allowable transmission loss for portable cellular radio if the mobile ERP is +28 dBm to an omnidirectional cell-site antenna gain of +10 dB. Let the minimum receiver usable power be −102 dBm(for 18-dB pre-detection SNR). Repeat the calculation if the omniantenna is replaced by a sectored antenna having a gain of +17 dB. Assume the transmission loss at the cell site to be 3 dB and the base diversity gain to be 3 dB.

4.6 Assuming that a mobile receiver is located 10 mi from the base station and the communication is satisfactory. The antenna heights h_b and h_m are 100 ft and 10 ft above the ground. Now, if the base antenna height is reduced to 50% of its original height, what will be its effects in coverage distance. What other modification can one do to mobile antenna height h_m so that satisfactory communication can be maintained for the original distance.

4.7 Assuming the test results of the Glenfield tunnel, compute the one-mile tunnel loss at 900 MHz assuming that a cell transmitter is located at one entrance to the tunnel. Suggest techniques to enhance the field intensity.

4.8 Find the time delay between the direct path and the reflected path where the direct path length is ten miles and the height of the cell site and mobile antennas are 100 ft and 10 ft, respectively.

4.9 Compute the field intensity at 2 km, 5 km, and 10 km from the cell site in μV/m under free-space transmission for transmitted output power of 100W and a dipole gain of 2.15 dB.

4.10 Find the free-space transmission loss at a distance of 20 km for λ of 0.33m (900 MHz).

4.11 Find the reflection coefficient for a vertically polarized wave at a frequency of 900 MHz over smooth sea water when the grazing angle is 7 deg. The value of the relative permittivity ϵ' for sea water is 81 and the conductivity σ is 4.5 mhom/m^2.

4.12 Find the phase and the magnitude of the reflection coefficient for a damp soil having relative permittivity of 10 and the conductivity of 0.01 S/m. Assume λ =0.33m, a grazing angle of 5 deg, and the polarization to be vertical.

4.13 For the configuration shown below, compute the total diffraction loss between cell site and the vehicle receiver by applying the Epstein-Peterson method, assuming the frequency of operation to be 800 MHz.

4.14 Cellular cell design over the Gulf of Mexico requires the boat antenna to be 24-ft above sea level, and the cell-site antenna on the shore to be 750 ft. Find the loss L_{d1} and L_{d2} using Bullington 3 loss method, assuming the ratio of earth effective radius to true radius as 4/3. If the cell site *ERP* is 20 dBW, the boat antenna gain is +3 dB, and the coaxial cable loss between the antenna and the receiver is 2 dB, show that the distance d_3 between two horizons, d_1 and d_2, is zero. Assume the required field intensity to be 39 dBu and the frequency of operation to be 850 MHz.

4.15 Find the maximum line-of-sight range corresponding to the transmit antenna height h_t = 400 ft and the mobile antenna height h_m at 20 ft above the ground.
4.16 Using the Carey curve, find the radial distance in miles for an antenna height of 100 ft when the ERP is 20 dBW. How will the radial distance change if the antenna height is changed to 150 ft?
4.17 What are the causes of dropouts and dead spots in cellular mobile radio system? Can dead spots occur in mountains, large cities, or a flat rice field? Can reflection from the buildings cause dead spots?
4.18 Assume that a 39-dBu signal level criteria is satisfied up to a distance of 10 mi with a cell antenna height of 100 ft and a mobile antenna height of 10 ft. Now, if the cell antenna height is increased by 25 ft, what will be the new 39-dBu contour? Can we reduce the height of mobile antenna so that the original 10-mi coverage is maintained?

REFERENCES

[1] Bullington, K. "Radio Propagation for Vehicular Communications," *IEEE Trans. on Vehicular Technology,* Vol. VT26, Nov. 1977, pp. 295–308.
[2] FCC Code of Federal Regulations 44, § 22-502-22.504, Oct. 1982.
[3] Carey, R. B. "Technical Factors Affecting the Assignment of Facilities in the Domestic Public Land Mobile Radio Service, Report No. R-6406, June 1964.
[4] Okamura, Y., E. Ohmori, T. Kawano, and K. Fukuda. "Field Strength and its Variability in VHF and UHF Land-Mobile Radio Service," Rev. Elec. Communication Lab., Vol. 16, Sept.–Oct. 1968, pp. 825–843.
[5] Longley, A. G., and R. L. Rice. "Prediction of Tropospheric Radio Transmission Loss Over Irregular Terrain," ERL49-ITS64, July 1968.
[6] Shepherd, N. H. "Radio Wave Loss Deviation and Shadow Loss at 900 MHz," *IEEE Trans. on Vehicular Technology*, Vol. VT-26, Nov. 1977, pp. 309–313.
[7] Shepherd, N. H. "UHF Radio Wave Propagation in Dallas, Texas Base to Mobile Station for Vertical Polarization," General Electric Report No. R45-MRD-1, March 1975.
[8] Kessler, W. J., and M. J. Wiggins. "A Simplified Method for Calculations UHF Base-to-Mobile Statistical Coverage Contours Over Irregular Terrain," *IEEE Trans. on Vehicular Technology*, Vol. VT-1977.
[9] Reudink, D. O. "Properties of Mobile Radio Propagation Above 400 MHz," *IEEE Trans. on Vehicular Technology*, Vol. VT23, Nov. 1977, pp. 143–159.
[10] CCIR Report No. 340-4, "VHF and UHF Propagation Curves for the Frequency Range From 30 MHz to 1000 MHz.
[11] Bodson, D., G. F. McClure, and S. R. McConoughey, eds. *Land-Mobile Communications Engineering*, IEEE Press, 1984.
[12] Lee, W.C.Y. *Mobile Communications Engineering*, McGraw-Hill Book Co.: New York, NY, 1982.
[13] Holbeche, R. J., ed. *Land Mobile radio System*, Peter Pergerinus Ltd., 1985.
[14] Ippolito L. J., Jr. *Radiowave Propagation in Satellite Communications*, Van Nostrand Reinhold Co.: New York, NY, 1986.
[15] Epstein, J., and G. W. Peterson. "An Experimental Study of Wave Propagation at 850 MHz," *IRE Proc.*, Vol. 41, pp. 595–611.

[16] Wilkerson R. E., "Approximation of the Double Knife-Edge Attenuation Coefficient," *Radio Science*, Vol. 1, No. 12, Dec. 1966, pp. 1439–1443.
[17] Millington, G., R. Hewitt, and F. S. Immirzi. "Double Knife-Edge Diffraction in Field Strength Predictions," *IEEE Mon.* 504E, March 1962, pp. 419–429.
[18] Bullington, K. "Radio Propagation at Frequencies Above 30 Mc/s, *IRE Proc.*, Vol. 34, 1977, pp. 1122–1136.
[19] "Atlas of Radio Wave Propagation Curve for Frequencies Between 30 and 10,000 Mc/s," *Radio Research Laboratory*, Ministry of Postal Services, Tokyo, Japan, 1954, pp. 142–149.
[20] Lee, W. Y. "Studies of Base-Station Antenna Height Effects on Mobile Radio," *IEEE Trans. on Vehicular Technology*, Vol. VT-29, No.2, May 1980.
[21] Kelly, K. K. II. "First Suburban Area Propagation at 800 MHz," *IEEE Trans. on Vehicular Technology*, Vol. 24, Nov. 1978.
[22] Wells, P. I. "The Attenuation of UHF Radio Signals by Houses," *IEEE Trans. on Vehicular Technology*, Vol. VT-26, No. 4 Nov. 1977.
[23] Compson, B., and S. C. Bhattacharya. "Mobile radio Communication in 800 MHz Band in Metropolitan Toronto," *IEEE Vehicular Technology Conference*, 1982.
[24] Longley, A. G. "Radio Propagation in Urban Areas," *IEEE Vehicular Technology Conference*, 1978.
[25] Jensen, R. "900 MHz Mobile Radio Propagation in Copenhagen Area," *IEEE Trans.*, Vol. 26, No. 4, Nov. 1977.
[26] Ott, G. D., and A. Plitkins. "Urban Path-Loss Characteristics at 820 MHz," *IEEE Trans. on Vehicular Technology*, Vol., No. 4, Nov. 1978.
[27] Gaziano, V. "Propagation Models for 900 MHz Cellular Systems Using Portable and Mobiles," *IEEE Vehicular Technology Conference*, 1981.
[28] Hata, M. "Empirical Formula for Propagation Loss in Land Mobile Radio Service," *IEEE Trans. on Vehicular Technology*, Vol. 29, No. 3 Aug. 1980.
[29] Davis, Q. V., D. J. R. Martin, and R. W. Haining. "Microwave Radio in Mines and Tunnel," *IEEE Conference on Vehicular Technology*, 1984.
[30] Isberg, R. A., and R. L. Chufu. "Passive Reflectors as a Means for Extending UHF Signals Down Intersecting Cross Cuts in Mines or Large Corridors," *IEEE Conference on Vehicular Technology*, 1979.
[31] Deryck, L. "Natural Propagation of Electronic Waves in Tunnels," *IEEE Trans. on Vehicular Technology*, Vol. 24, No. 3, Aug. 1978.
[32] LaGrone, A. H., and C. W. Chapman. "Some Propagation Characteristics of High UHF Signals in the Immediate Vicinity of Trees," *IEEE Trans. on Antennas and Propagation*, Vol. AP-9, 1961.
[33] Tamir, T. "On Radio-Wave Propagation in Forest Environment," *IEEE Trans. on Antennas and Propagation*, Vol. AP15, Nov. 1967.
[34] Forrest, R. T. "800 MHz Mobile Radio Propagation Test in the Pittsburgh Area," *IEEE Conference on Vehicular Technology*, 1984.
[35] U.S. National Archieves, Code of Federal Regulations, Part 47, §22.504, Telecommunication, p. 164.
[36] Walker, E. H. "Penetration of Radio Signals into Buildings in the Cellular Radio Environment," *Bell Systems Technical Journal*, Vol. 62, No. 9, Nov. 1983, p. 2730.
[37] Compson, B., and S. C. Bhattacharya "Mobile Radio Communication in 800 MHz Band in Metropolitan Toronto," *IEEE Vehicular Technology Conference*, 1982, p. 104.

Chapter 5
Antenna Systems

5.1 INTRODUCTION

As in any RF broadcast communication system, the interface to the transmission media is the antenna subsystem. The peculiarities of a cellular radio system requiring reliable communication from a fixed base to a moving terrestrial vehicle mandate unique solutions to the antenna problem. Since the cellular radio is a duplex system, the ideal situation is to provide the same performance in both the transmit and in the receive direction. The types of antennas, their gain and coverage patterns, the power available to drive them, the use of simple or multiple antenna configurations, and antenna polarization are all factors that effect the system performance and are controlled by the system designer. Factors not under the control of the system designer include the topography between cell-site antennas, speed and the direction of the vehicle using cellular system, and where the antenna or antennas are mounted on the vehicle. Each of these factors affect system performance.

Due to a multiplicity of design parameters, often a compromise is sought between antenna gain, vertical beamwidth, and antenna height. For cell-site antennas with heights up to 600 ft, a high-gain omniantenna with a gain of 7–10 dB is selected. For heights of 600–1,000 ft, a medium-gain antenna with 5–7 dBi gain is appropriate. For heights above 1,000 ft, a low-gain antenna with a gain of 0–3 dBi produces excellent results.

Due to the proximity of the transmitter, the receiver has to withstand the out-of-channel interference. The interference caused by co-sited transmitters can basically be categorized into two groups. The first group of interference is essentially caused due to the radiation of equipment installed at the same site. Items falling into this category are power supplies, interactions among antennas, feeders, and so forth. Interference in this group is mostly non-signal related. The second type of interference is the reflective interference, which is caused by corroded joints and contacts from metal structures. The characteristics of these are nonlinear, resulting in unwanted spurious when illuminated by the transmitter power. Unlike the first type, the interference in this group is signal-related. The other interference related to signal is the cochannel interference, but it is not a related to antenna.

In this chapter we shall consider the system aspects of the cellular antenna based on the information in Figure 5.1. On the transmit side, the system consists of transceiver, power amplifier, combiner, and the transmit antenna. On the receive side, it consists of receive antenna, receiver multicoupler, and the receiver. Thus we first discuss the basic antenna parameters relevant to cellular systems, followed by a detailed description of the cell site and mobile antennas in Sections 5.3 and 5.4, respectively. Section 5.5 describes ancillary devices associated with the design of a typical cellular system. Discussion on ancillary devices includes the combiner, receiver multicoupler, duplexer, and the transmission line. Problems with antenna systems are discussed in Section 5.6. The discussion includes the dead spot problem, cell-site antenna mounting problem, and the required isolation between transmit and receive antenna at the cell site for minimizing near-end crosstalk (NEXT). Section 5.7 outlines the standard techniques of computing the signal-to-thermal noise ratio at the receiver. Lastly, the problem of intermodulation in high-power amplifiers is discussed, along with a simple technique of channel assignment in a cell site to avoid intermodulation products of one channel falling at the center frequency of another channel.

5.2 ANTENNA GAIN

Gain is one of the most significant parameters for the design of an antenna system. A high-gain antenna is achieved by increasing the physical size, and thus the aperture or the capture area of the antenna. The concept of the aperture can be introduced by considering a receiving horn antenna, as shown in Figure 5.2. If the Poynting vector, or the power density impinging on the mouth of area A, (in square meters) is S W/m^2, then the total power captured is

$$P = SA \tag{5.1}$$

in watts. Thus, the horn antenna may be regarded as an aperture of area that captures the electromagnetic wave impinging at its mouth. Antenna gain can be increased by increasing the horn mouth or the capture area A. However, increasing the aperture decreases the vertical beamwidth. A high-gain antenna also introduces the problem of nearby dead spots due to minor lobe nulls, but the signal levels at the far sight increase.

One way of increasing the gain of an antenna is by the vertical staking of multiple radiators and electrically feeding them so that their radiation adds. For a high-gain antenna, it is important that they be properly mounted. For example, a vertical antenna mounted at a slight angle may produce a significant degradation

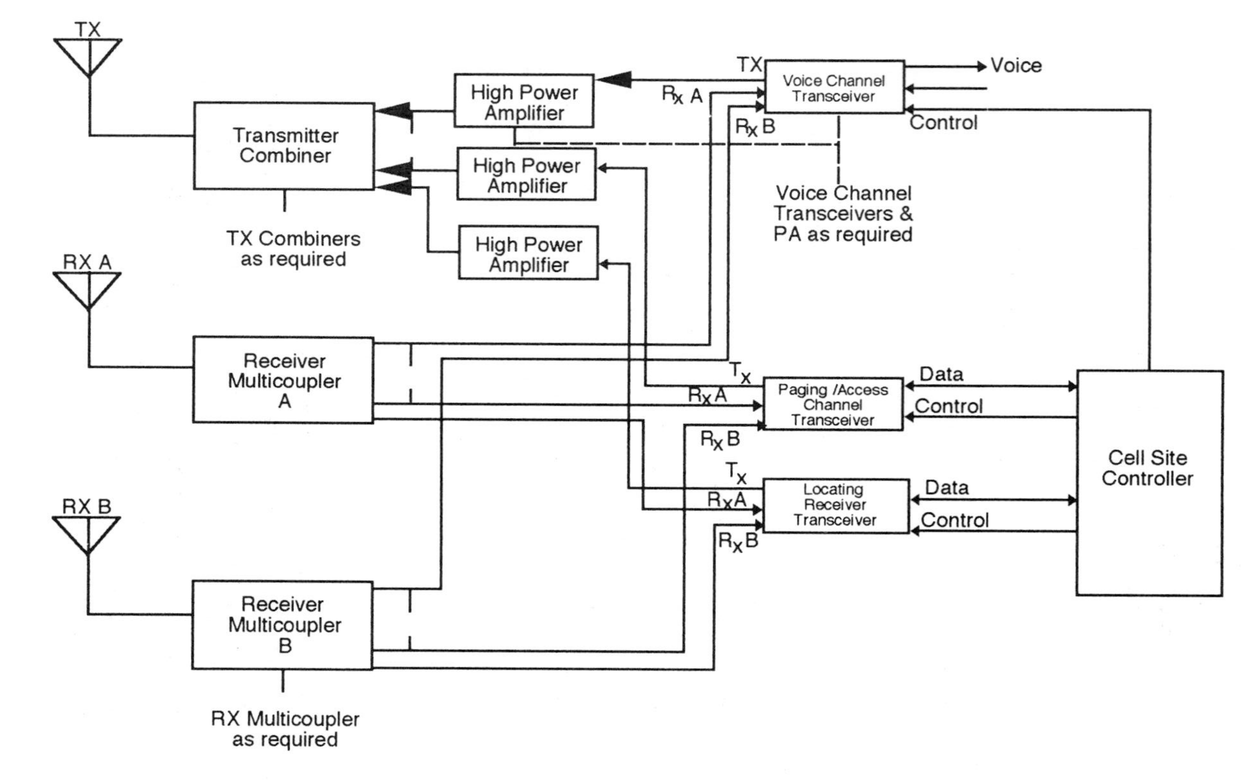

Figure 5.1 Typical cellular cell-site equipment configuration.

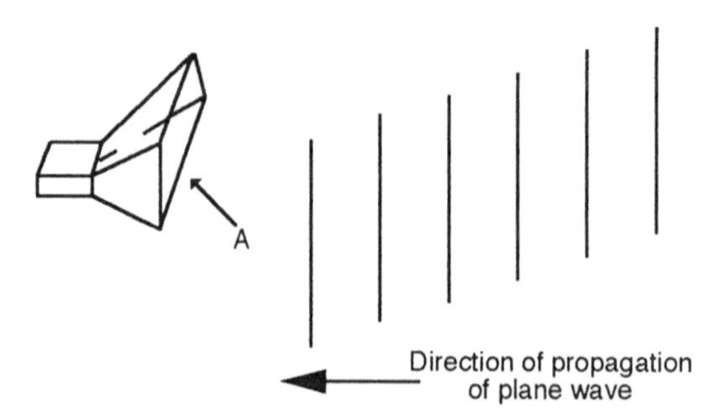

Figure 5.2 Plane wave impinging on horn antenna.

in the desired direction. The gain will also be reduced if the antenna swings under ice or wind loading.

There are two types of antenna patterns commonly used at cell sites, 360 deg or omnidirectional antennas, and sectored antennas of 120 deg, 90 deg, or 60 deg, which provide higher gain compared to omnidirectional antennas (60 and 120 deg are common types). Mobile antennas are low gain (3 dB), collinear mounted on the car roof, cowl, trunk, or on the glass.

The type of pattern utilized is determined by the cell-site geographical coverage area and shadows caused by terrain and other fixed structures. Geographical constraints, such as bodies of water and mountain ranges, are other factors that have an impact on the antenna pattern. We derive below the basic equation for antenna gain and establish its relationship with antenna beamwidth.

Gain Equation

Antenna gain is defined either with respect to isotropic antenna or with respect to the half-wave dipole. An isotropic radiator is a fictitious radiator that radiates equally in all directions. The gain of the antenna in a particular direction is the ratio of the intensity produced by it in that direction divided by the ratio of the radiation intensity that would be produced by an isotropic radiator. The term dBi is used to reference the antenna gain with respect to an isotropic radiator. When referenced to a half-wave dipole, the term used is dBd (0 dBd = 2.1 dBi). The normalized directive gain is given by

$$P_t = \int_S G_{cs}(\theta, \phi)ds \tag{5.2}$$

where the antenna having a normalized directive gain of $G_{cs}(\theta, \phi)$ is located at the center of a large sphere. The directive gain $G_{cs}(\theta, \phi)$ is expressed as power per unit area, or the Poynting vector. The poynting vector is related to the normalized field pattern by

$$G_{cs}(\theta, \phi) = \left[\frac{E_{\theta n}^2(\theta, \phi) + E_{\phi n}^2(\theta, \phi)}{Z_o} \right] \tag{5.3}$$

in W/m^2, where

$$\begin{aligned} E_{\theta n}(\theta, \phi) &= \frac{E_\theta(\theta, \phi)}{E_\theta(\theta, \phi)_{\max}} \\ E_{\phi n}(\theta, \phi) &= \frac{E_\phi(\theta, \phi)}{E_\phi(\theta, \phi)_{\max}} \end{aligned} \tag{5.4}$$

where Z_o = intrinsic impedance of space = 375.752Ω, and $E_\theta(\theta,\phi)_{\max}$ and $E_\phi(\theta, \phi)_{\max}$ are the maximum values of field intensity in θ and ϕ directions.

The E_θ and E_ϕ components of the field pattern at a distance from the antenna location are show in Figure 5.3(a). The pattern has its main-lobe maximum in the Z direction where $\theta = 0$ deg with minor lobes (side and back) in other directions (± 180 deg). Between these minor lobes are nulls, where the antenna has zero gains. The antenna pattern G_{cs} represented in (r, θ) coordinates is shown in Figure 5.3(b). Figure 5.3(c) is the gain representation G_{cs} in rectangular coordinates. Figure 5.3(b,c) shows the vertical cut along the ZY plane. Usually both the vertical and horizontal cuts are required to completely represent an antenna pattern unless the antenna gain $G_{cs}(\theta, \phi)$ is symmetrical along both the xy and zy planes. In this case, one cut is sufficient to characterize the gain pattern [20].

The integral S in (5.2) is performed over the surface of a sphere, as shown in Figure 5.4. The elementary hatched area is given by $r^2 \sin \theta d\theta d\phi$. Thus,

$$P_t = r^2 \int \int G_{cs}(\theta, \phi) \sin \theta d\theta d\phi \tag{5.5}$$

where r is the mobile distance from the cell site.

If the cell-site antenna is isotropic, the antenna gain $G_{cs}(\theta, \phi)$ is independent of angles θ and ϕ, and thus the above equation becomes

$$P_t = G_{cs}(\text{isotropic}) \int_0^{2\pi} \int_0^{\pi} r^2 \sin \theta d\theta d\phi \tag{5.6a}$$

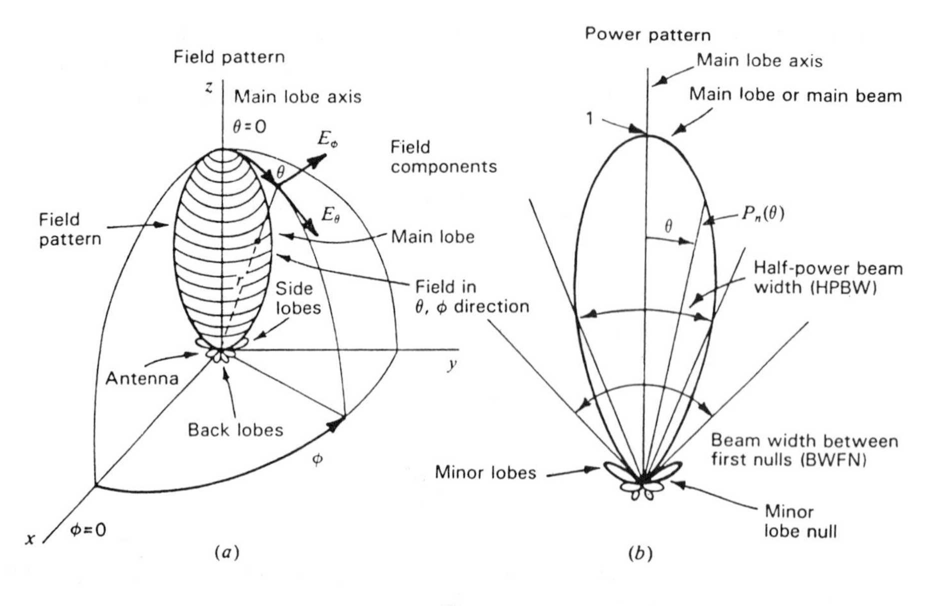

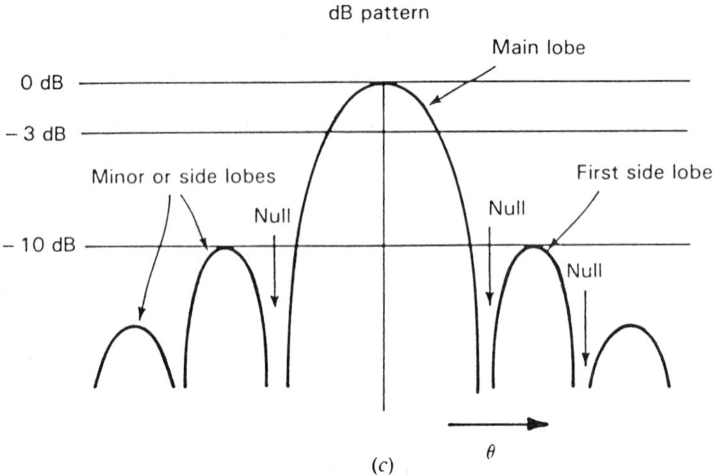

Figure 5.3 (a) Antenna field pattern with coordinate system. (b) Antenna power pattern in polar coordinates (linear scale). (c) Antenna pattern in rectangular coordinates and in decibel (logarithmic) scale (patterns (b) and (c) are the same). Source: [20, p. 21].

Thus,

$$G_{cs}(\text{isotropic}) = \frac{P_t}{4\pi r^2} \tag{5.6b}$$

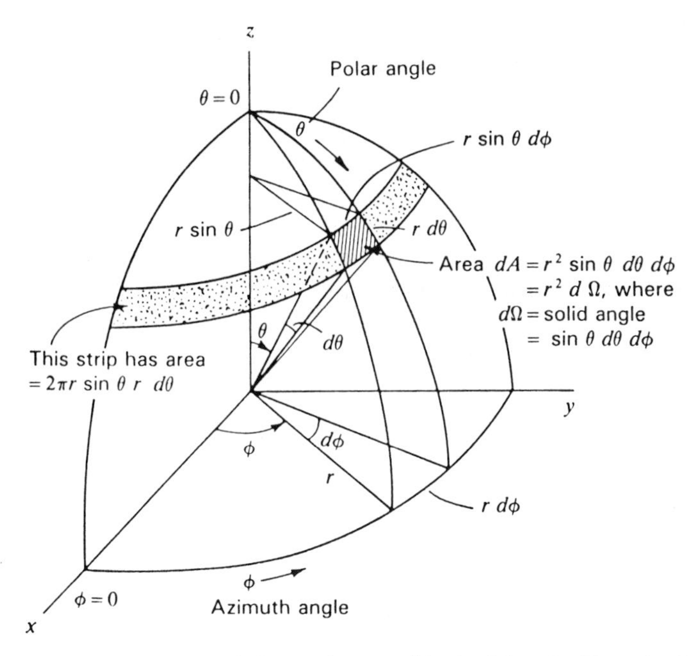

Figure 5.4 Spherical coordinates in relation to the area dA of solid angle $d\Omega\ =\ \sin\ \theta d\theta d\eta$. Source: [20, p. 24].

If the transmitter at the cell site has a directive gain of g_t, then the above expression can be modified to include the directive gain towards mobile antenna.

$$G_{cs} = \frac{P_t g_t}{4\pi r^2} \tag{5.7}$$

If the capture area of the receive antenna is A_m, then the power received by the mobile is

$$P_m = \frac{P_t g_t}{4\pi r^2} A_m \tag{5.8}$$

Since the antenna gain is related to its capture area by

$$g_m = \frac{4\pi A_m}{\lambda^2} \tag{5.9}$$

From (5.7) and (5.8), the signal power received by the mobile can be expressed in terms of the mobile antenna gain g_m.

$$P_m = \frac{P_t g_t g_m \lambda^2}{(4\pi r)^2} \tag{5.10}$$

Let $L_p = (4\pi r/\lambda)^2$. The factor L_p is the loss associated with the propagation of electromagnetic waves from the transmitter to the receiver. Note that L_p depends on both carrier frequency and the distance r. This loss is always present. When expressed in decibels, (5.10) is

$$(P_m)_{\text{dB}} = (\text{ERP})_{\text{dB}} + (L_p)_{\text{dB}} + (g_m)_{\text{dB}} \tag{5.11}$$

where $(\text{ERP})_{\text{dB}} = 10 \log_{10}(P_t g_t)$. The propagation loss L_p expressed in decibels is

$$(L_p)_{\text{dB}} = -36.6 - 20 \log(r_{\text{miles}} f_{\text{MHz}}) \tag{5.12}$$

The expression of mobile antenna gain, g_m, can be related to its half-power beamwidth (HPBW) as

$$g_m = \frac{4\pi}{\theta^o_{HP}\phi^o_{HP}} \tag{5.13}$$

where θ^o_{HP} and ϕ^o_{HP} are half-power beamwidths in the θ and ϕ planes. The factor 4π is the solid angle subtended by a sphere in steradians (square radians).

The 4π steradians $= 4\pi \times (180/\pi)^2 \approx 41{,}250 \text{ deg}^2 =$ solid angle in a sphere. Therefore, the mobile antenna gain g_m in (5.13) is

$$g_m = 41{,}250/\theta^o_{HP}\phi^o_{HP} \tag{5.14}$$

Assuming that within the 3-dB beamwidths of θ and ϕ the radiation intensity is uniform and outside this sector it is zero, the gain g_m can also be expressed as

$$g_m = 41{,}250/\theta\phi \tag{5.15}$$

where θ and ϕ are in degrees and contained in an angle such that $g_m \leq 3$ dB. This formula is rather unrealistic as it does not account for the side lobes. Considering the side-lobe effects, (5.15) becomes

$$g_m = 32{,}600/\theta\phi \tag{5.16}$$

We consider below an example to compute antenna gain g_m.

Example 5.1

For an antenna having half-power beamwidths of 60 deg in both θ and ϕ directions (sectored antenna with vertical and horizontal beamwidths of 60 deg), the directive gain from (5.16) is

$$g_m = 32{,}600/(60 \times 60) = 9.055 \simeq 9.6 \text{ dBi} = 7.5 \text{ dBd} \tag{5.17}$$

This means that the antenna radiates about nine times more power than an omnidirectional antenna of unity gain.

5.2.1 Composite Gain of Two Antennas

Often cell-site antennas are directive and mounted at a separation S. Directive gain is added to produce the desired antenna pattern which for the startup of a cellular system can be omnidirectional. This will be discussed in the next section. Here we arrive at the power gain equation G for the composite antenna, which of course will depend on the mutual impedance between antennas. We shall consider the simplest two antenna configurations, as shown in Figure 5.5. However, this configuration can easily be extended of cases more than two antennas [1].

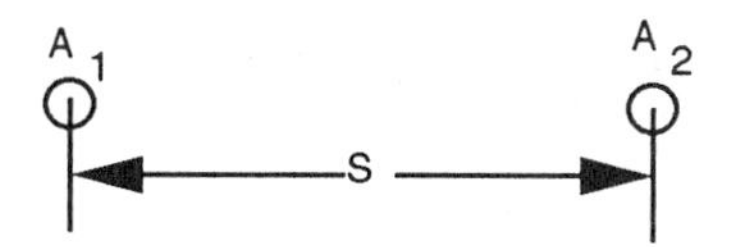

Figure 5.5 Two antennas spaced a distance S apart.

Assume that two identical antennas A_1 and A_2 are carrying currents I_1 and I_2 spaced at a distance S apart. The antennas are assumed to be oriented in the same direction. The relationship between the voltages and currents in the two antennas is given by

$$\begin{aligned} V_1 &= Z_{11}I_1 + Z_{12}I_2 \\ V_2 &= Z_{12}I_1 + Z_{11}I_2 \end{aligned} \tag{5.18}$$

where Z_{11} and Z_{12} are self and mutual impedances for identical antennas. Let the input power to the two antennas A_1 and A_2 be P_1 and P_2. Then,

$$\begin{aligned} P_1 &= 1/2(V_1I_1^* + V_1^*I_1) \\ P_2 &= 1/2(V_2I_2^* + V_2^*I_2) \end{aligned} \tag{5.19}$$

The total power P is then obtained as

$$\begin{aligned} P_0 &= P_1 + P_2 = 1/2(V_1I_1^* + V_1^*I_1) + 1/2(V_2I_2^* + V_2^*I_2) \\ &= R_{11}(I_1I_1^* + I_2I_2^*) + R_{12}(I_2^*I_1 + I_1^*I_2) \end{aligned} \tag{5.20}$$

where R_{11} and R_{12} are the self and mutual resistances. Assuming two currents to have a phase shift of ϕ, we can express I_2 in terms of I_1 as

$$I_2 = AI_1e^{j\phi} \tag{5.21a}$$

where A and ϕ are the same positive constants. There are two special cases of interest: the end-fire array case with $\phi = 180$ deg and the broadside array for $\phi = 0$ deg. The antenna patterns for two isotropic elements of equal amplitudes spaced at a distance of $S = \lambda/2$ with $\phi = 180$ deg and $\phi = 0$ deg are shown in Figure 5.6.

Substituting I_2 in terms of I_1, (5.20) is reduced to

$$P_o = [R_{11}(1 + A^2) + 2R_{12}A\cos\phi]I_1I_1^* \tag{5.21b}$$

If the quantity P_3 is defined such that

$$P_3 = [(I_1I_1^*)^{1/2} + (I_2I_2^*)^{1/2}]^2 \tag{5.22}$$

(a)

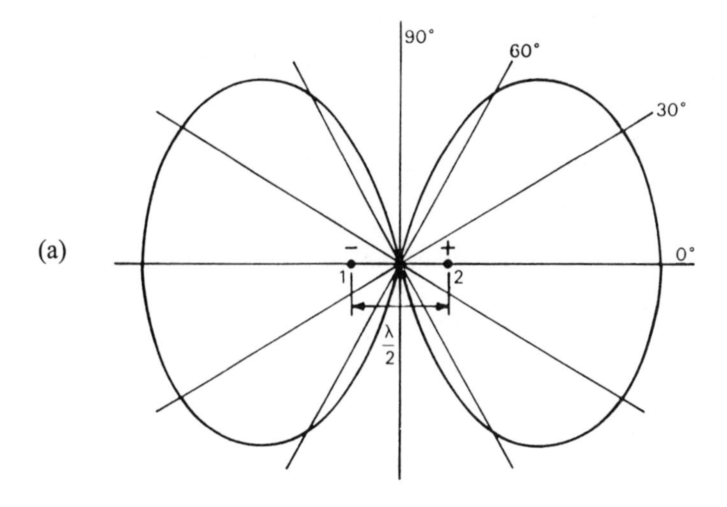

(b)

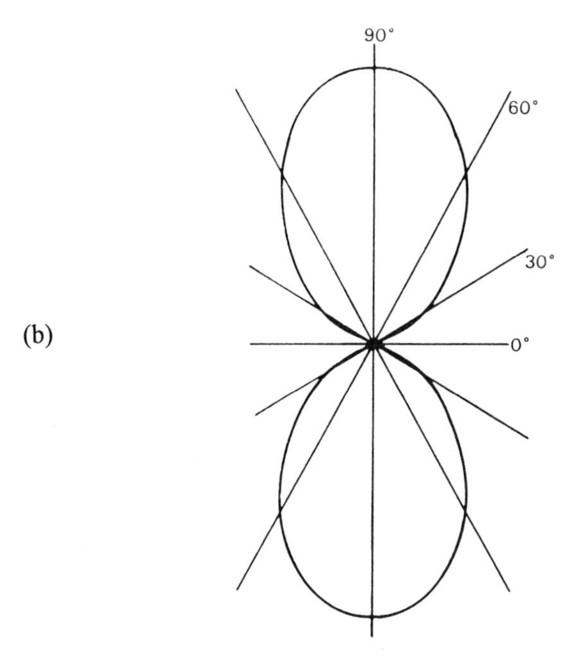

Figure 5.6 Patterns for end-fire array of two isotropic sources: (a) with equal and out-of-phase currents with spacing $d = \lambda/2$; (b) with equal and in-phase currents with spacing $d = \lambda/2$. After [20, pp. 123, 119].

where P_3 is the measure of power flux density at a large distance from the transmitting antennas A_1 and A_2. At large distances, signals add inphase and from (5.21a) and (5.21b) we obtain

$$P_3 = \frac{(1 + A)^2 P_o}{R_{11}(1 + A^2) + 2R_{12}A \cos \phi} \tag{5.23}$$

If $A = 0$, then there is only one antenna and gain P_3 becomes P_3' and is given by

$$P_3' = P_o/R_{11} \tag{5.24}$$

We define the power gain of the two antennas at far field as the ratio of composite gain P_3 of two antennas with respect to a single-element antenna gain (5.24). Thus,

$$G = \frac{(1 + A)^2}{(1 + A^2) + 2A \cos \phi \, (R_{12}/R_{11})} \tag{5.25}$$

When the amplitudes of two currents are equal, and equal to unity, (5.25) becomes

$$G = \frac{2}{1 + \cos \phi \, (R_{12}/R_{11})} \tag{5.26}$$

We shall now investigate the gain, G, for collinear omnidirectional antennas. When $\phi = 0$, the currents in two antennas are in phase and the gain G is reduced to

$$G = \frac{2R_{11}}{R_{11} + R_{12}} \tag{5.27}$$

If the mutual impedance R_{12} is made zero, the composite gain of two antennas is 3 dB. For a half-wave dipole, this occurs when the center-to-center spacing is approximately equal to $\approx 0.55\lambda$. However, for a slightly larger spacing the mutual impedance becomes negative and a gain slightly greater that 3 dB is obtained. An interesting case with $\phi = \pi/2$ is seen from (5.26) when the gain G is 3 dB and is independent of mutual impedance. In this case, a Cardiod pattern is obtained with spacing of $\lambda/4$ between antennas. A large mutual impedance between elements does not hurt the gain, G. This property can be used to an advantage in some cases. The concept of phase shifting in one element with respect to another so that the overall gain can be increased, is used in mobile antennas and is discussed in Section 5.4.

5.2.2 Polarization

An important property of the electromagnetic wave is its polarization—the quantity describing the orientation of the electric field E. A plane wave traveling out of the paper, as shown in Figure 5.7, with its electric field always aligned to the y axis is known as a linearly polarized wave. As a function of time and distance along the z axis, the electric field equation is given by

$$E_y = E_2 \sin(\omega t - \beta z) \tag{5.28}$$

In general, an electric field traveling in the z direction will have both the horizontal and the vertical components of the electric field and the wave will be regarded as elliptically polarized. At a fixed point on the z axis, the electric vector rotates as a function of time and the tip of the vector describes an ellipse known as polarization ellipse. The ratio of the major to minor axis is known as the axial ratio (AR). Two extreme cases of elliptical polarization are circular and linear polarizations. For circular polarization, the major and the minor axes are equal ($E_1 = E_2$) and for linear polarization the minor axis $E_1 = 0$ and the $AR = \infty$. An elliptically polarized wave can be decomposed into vertical and horizontal field components in y and x directions. The electric field components in x and y directions are represented as

$$\begin{aligned} E_x &= E_1 \sin(\omega t - \beta z) \\ E_y &= E_2 \sin(\omega t - \beta z + \delta) \end{aligned} \tag{5.29}$$

where E_1 = amplitude of the linearized wave in x direction; E_2 = amplitude of the linearized wave in y direction; $\beta = 2\pi/\lambda$, radians/meter; and δ = time-phase angle by which E_y leads E_x.

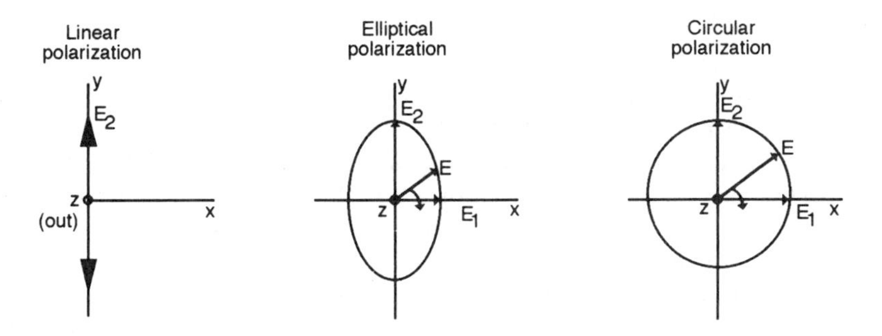

Figure 5.7 Linear, elliptical, and circular polarization (wave propagation is out of page).

Combining (5.28) and (5.29), the resultant vector field E can be expressed as

$$E = \hat{X}E_1 \sin(\omega t - \beta z) + \hat{Y}E_2 \sin(\omega t - \beta z + \delta) \tag{5.30}$$

where $\hat{X}$ and $\hat{Y}$ are the unit vectors in the horizontal and vertical directions. At $z = 0$, on paper $E_x = E_1 \sin(\omega t)$ and $E_y = E_2 \sin(\omega t + \delta)$. From the relation for E_x, $\sin(\omega t) = E_x/E_1$. Thus, $\cos(\omega t) = \sqrt{1 - (E_x/E_1)^2}$. Cellular radio antennas normally are vertically polarized. Cell site and mobile antenna polarization should not be mixed.

5.2.3 Bandwidth

The bandwidth of an antenna can be considered to be the range of frequencies on either side of the center frequency where the antenna characteristics, such as beamwidth, gain, input impedance, polarization, and side-lobe level, are within an acceptable value of those at the center frequency. The use of narrow and wideband antenna is dictated by the type of frequency coverage. For broadband antennas, the bandwidth is usually expressed as the ratio of the upper-to-lower frequencies of acceptable operation. For example, a 5:1 antenna bandwidth indicates that the upper frequency for acceptable operation is five times greater than the lower frequency. Generally, a wideband antenna is best suited for most applications since it can be used for both transmitting and receiving. Additionally, wideband antennas detune less than narrowband antennas under ice and wind loading. For cellular application, AMP in the United States, the bandwidth for a wideband antenna must be 70 MHz. For wideband antennas, the ratio of the antenna diameter to its length must also be large. A simple quarter-wave ground plane antenna has a narrow bandwidth. Even the available collinear antennas were designed for gain rather than bandwidth. We shall discuss these in Section 5.3.

5.3 CELL-SITE ANTENNAS [5–7, 12, 14]

Cell-site antennas are chosen to provide reliable coverage to the geographic design limits of each cell's projected service area. The composite locus of the external cell service area boundaries should conform to the cellular geographic service area (CGSA) design as closely as possible. Since few if any CGSA boundaries exhibit any symmetry and the topography of most service areas is not uniform, omnidirectional coverage from a cell site is usually not possible, nor is it necessary or even desirable. Interference criteria, especially in cell overlap areas where both cochannel and adjacent channel may simultaneously be present, must be considered. Radiation patterns with precise coverage can preclude degradation of system performance. With this in mind, the objective of this and the next section is to expose the reader to different types of cell sites and the mobile antennas that are normally employed. We do not intend to provide complete details of the design as there are books on antennas that cover these exhaustively [20].

There are three main types of antennas which are used for cell site applications: collinear, corner reflector, and Yagi. Among these three, the most important type is the collinear design, which provides an omnidirectional radiation pattern. The azimuth and the vertical pattern of an omnidirectional 6-dB azimuth gain is shown in Figure 5.8(a). The vertical beamwidth is 20 deg, with sidelobes at least 13-dB below the peak gain of the antenna.

Corner reflector and Yagi antennas are of the directional type. Azimuthal and vertical patterns of a 120-deg sectored antenna are shown in Figure 5.8(b) and 5.8(c). A directive antenna in multipath propagation environments has the effect of reducing the fading, elongation of fading period, and gain in mean signal strength.

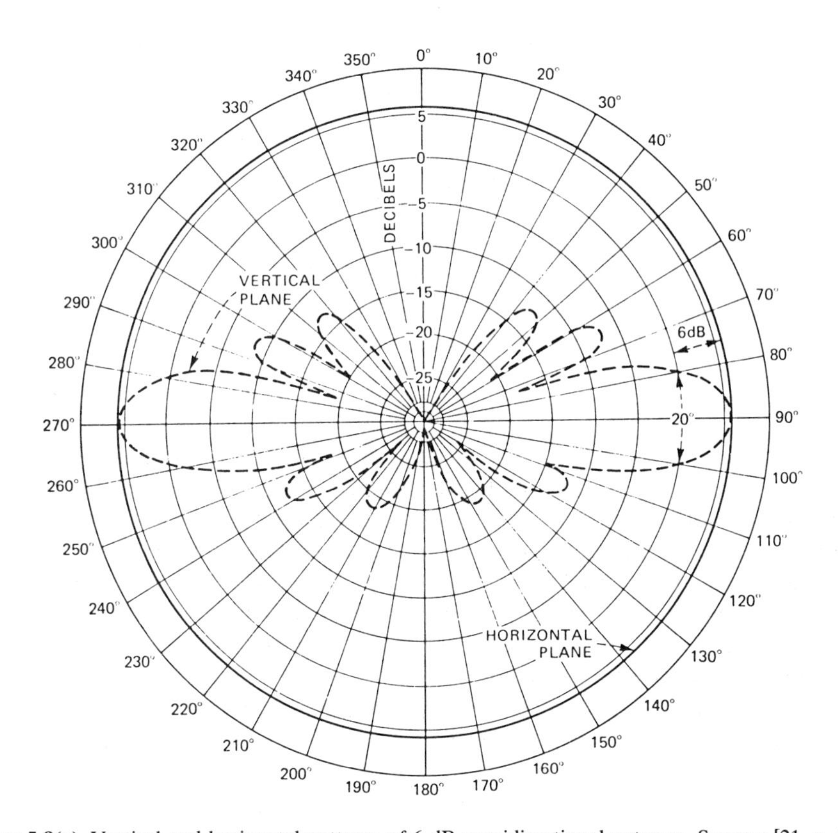

Figure 5.8(a) Vertical and horizontal patterns of 6-dB omnidirectional antenna. Source: [21, p. 224].

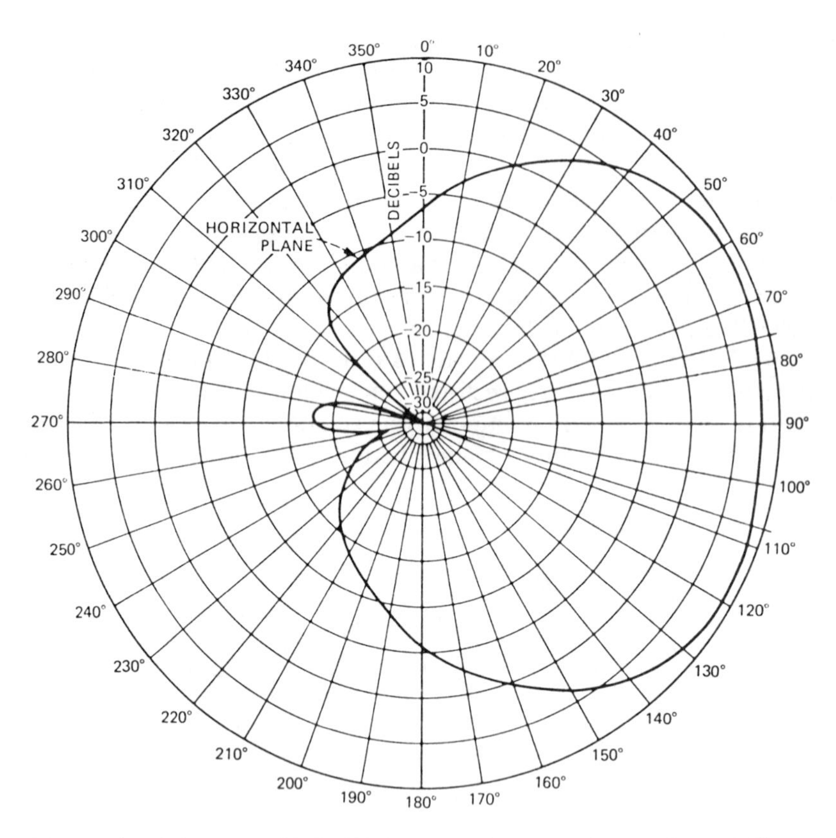

Figure 5.8(b) Azimuthal pattern of 8-dB 120-deg sectored antenna. Source: [21, p. 223].

Also, multipath delay spread is reduced, as long-delayed waves often arrive in opposite direction which is blocked due to reflector in the back of the antenna. In applications where a single antenna cannot be mounted to provide reasonable circular coverage, multiple directional antennas can be used through a power splitter and phasing the system such that their resultant azimuth coverage is omnidirectional. As shown in Table 5.1, collinear antennas mostly have a rating of 500W. However, in some cases like Sinclair Yagi, a power rating as low as 40W is available. The gain lies between 5 and 10 dB. The minimum beamwidth of the antenna is 5.5 deg. It is often desirable to have higher beamwidth so that, for a slight tilt, gain can be maintained. The bandwidth of these antennas lies between 35 MHz and 65 MHz.

There are two factors to be considered when planning the coverage pattern for a given area: the antenna pattern itself and the terrain characteristics around

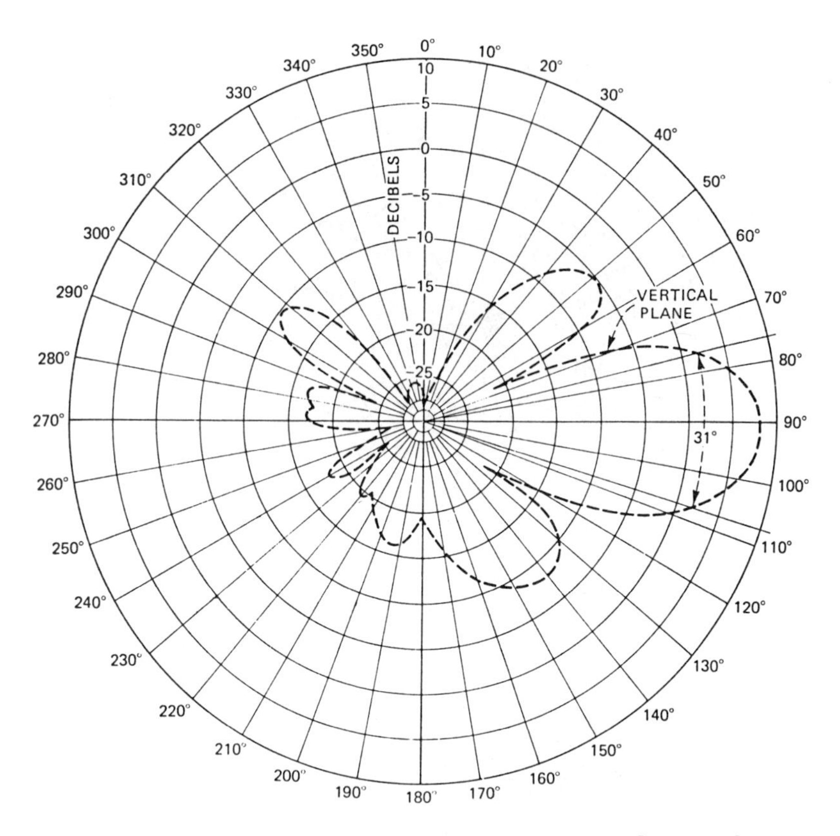

Figure 5.8(c) Vertical pattern of 8-dB 120-deg sectored antenna. Source: [21, p. 225].

the antenna. If an omnidirectional antenna is used, the distortion of that pattern by the average terrain characteristics in the coverage area will probably make the resultant coverage pattern unrecognizable to its source pattern. The use of a "lobed" omnidirectional pattern is achieved by an antenna array as shown in Figure 5.9(a). This will allow the system designer some degree of freedom in shaping the final coverage pattern. Three antennas used to form the composite array pattern are shown in Figure 5.9(b).

The antenna array coverage pattern is accomplished by calculating the height gain factor in each critical direction of interest, which is arrived at by knowing the average terrain in each of the directions of interest and the antenna height itself. Once the effect of height gain (which is nonuniform from radial to radial) is known, the antenna array can be rotated to increase or decrease gain in a critical direction.

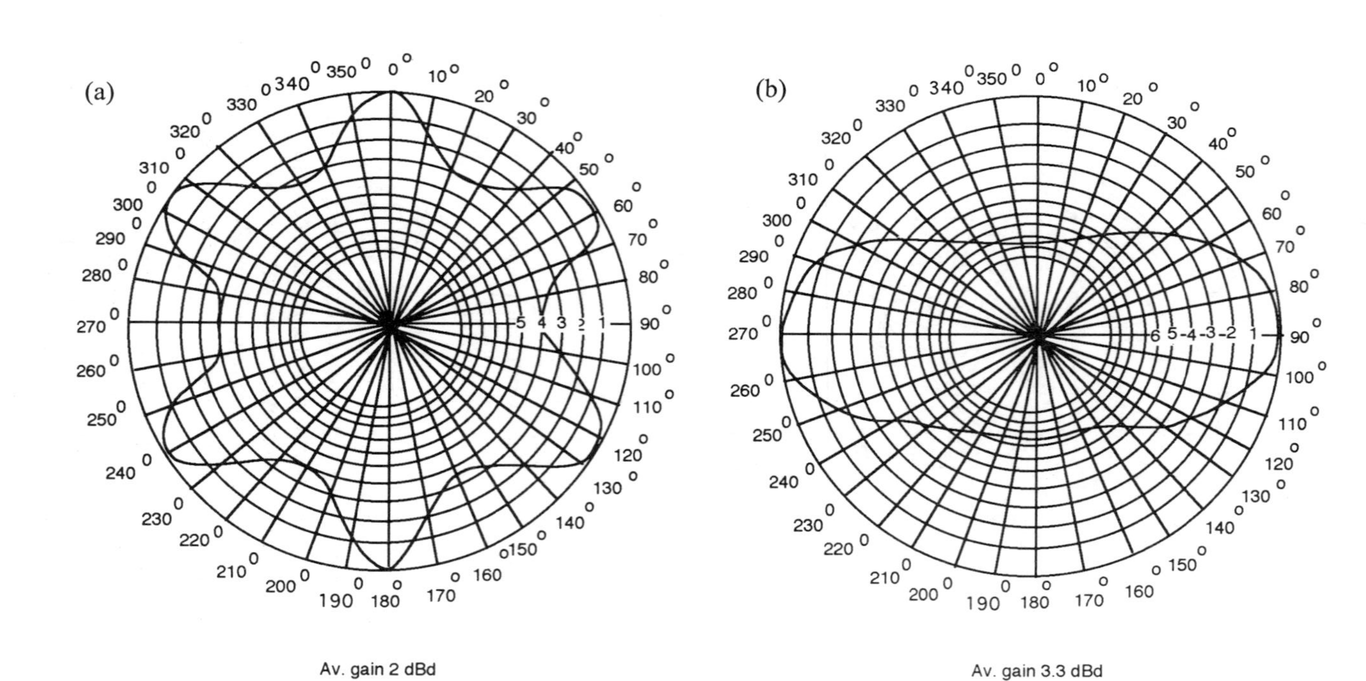

Figure 5.9 (a) "Lobed" omnidirectional pattern (b) individual antenna pattern KATHREIN A 63 30 61 to form individual pattern in (e). After [22].

Table 5.1
Power Rating of Different Cell-Site Antennas

Type	*Model*	*Gain (dB)*	*Vertical Bandwidth (deg)*	*Frequency Bandwidth (MHz)*	*Power Rating (W)*
Yagi	DB 492	6	—	64	100
	PD-1612	10	—	60	150
	SRL-406	10	40	60	40
Corner reflector	ASP-950	9	56	60	100
	DB494	9.0	60	64	150
	PD1124	9.0	54	60	100
Collinear	ASP-950	7.5	8.5	60	500
	DB480	7.5	14	64	500
	PD-10017	10	5.5	35	500

While this technique is not a cure in all cases, it has been a valuable tool in the design of CGSAs.

In some cases, where a cell site may be located in close proximity to a CGSA boundary, a radiation pattern that restricts radiation across the CGSA boundary but shapes the pattern within the CGSA is desirable. The rotation of an antenna array to conform with the topography and the desired coverage is the first choice to minimize the coverage in an unwanted area.

5.4 MOBILE ANTENNAS

Vehicular antennas are limited to a zero-dB gain, 3-dB collinear, elevated feed, 5-dB collinear, and collinear on-glass. A quarter-wave antenna has a broad pattern and unity gain. It works well in densely populated areas where buildings and terrain features block a line of sight between the mobile and cell-site antennas. A collinear design with a $\lambda/2$ upper radiator and a $\lambda/4$ lower radiator separated by an air-wound phasing coil is generally known in the industry as 3-dB gain antenna. Adding one phasing coil and one element section to a quarter-wave antenna doubles the power (i.e., a 3-dB gain). The second collinear design is an elevated feed-point antenna with the radiator above the feed point being a 3-dB antenna, and a broadband $\lambda/4$ skirt below the feed point and enclosed in a polycarbonate tube. For each doubling of the aperture, the antenna element's overall length minus the length of the coils increases the gain by an additional 3 dB. A 5-dB gain antenna consists of two phasing elements with an overall length of 24 inches. Lastly, on-glass design uses a similar structure to that of the 3-dB collinear, except it has a longer lower section. These five antenna types are shown in Figure 5.10. Staking elements increase the antenna gain, but both the practical size and material strength limit

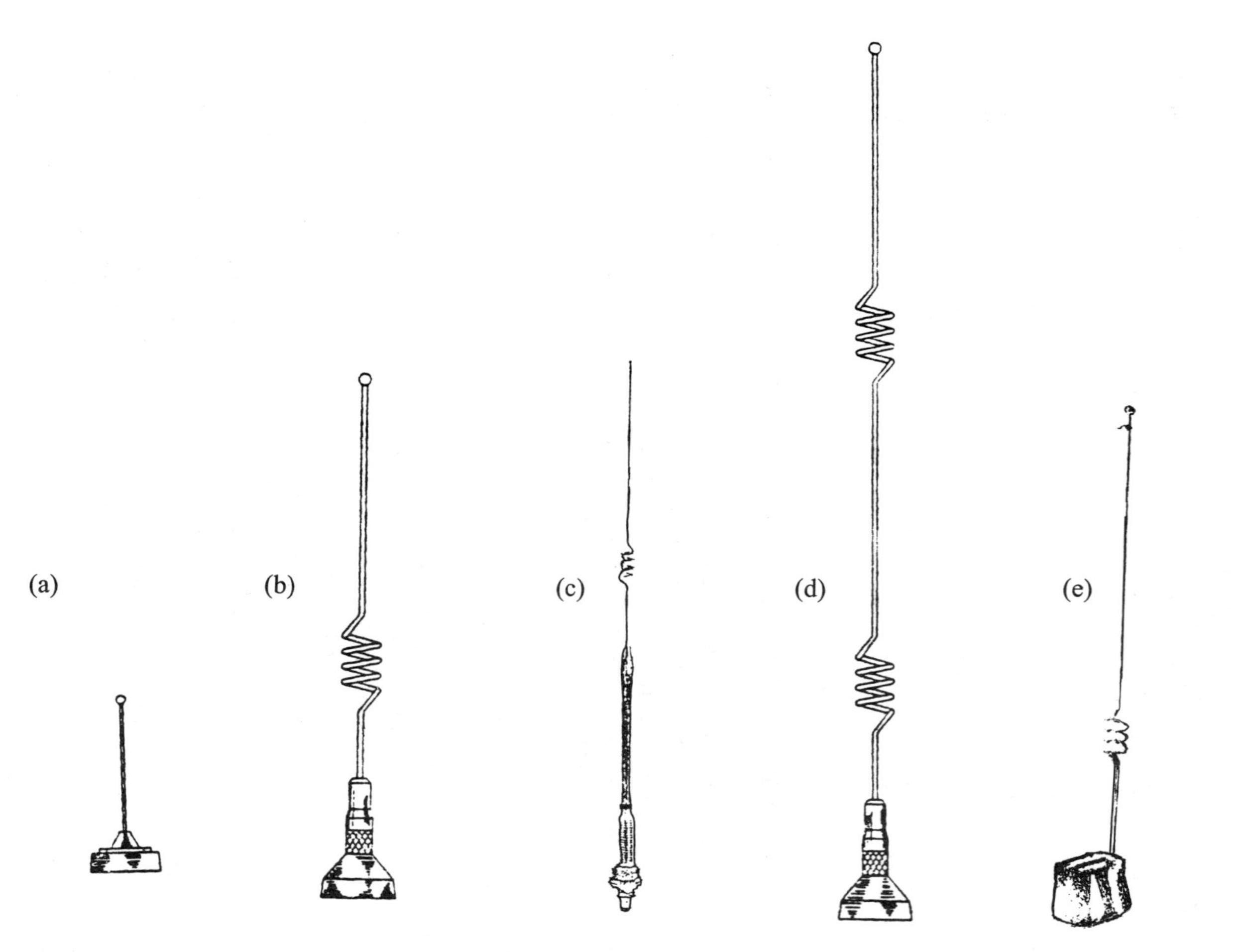

Figure 5.10 Five types of common mobile antennas: (a) unity-gain; (b) 3-dB collinear; (c) elevated feed; (d) 5-dB gain; and (e) on-glass.

an indefinite stacking of elements. As a result, the manufacturer currently produces the following antennas in the 800-MHz frequency range. The unity-gain antenna is about 3.5 in long, the two-element 3-dB collinear gain antenna is about 14 in tall, and the three-element 5-dB gain antenna is about 24 in tall.

As stated above, the quarter-wave antenna is most suitable in densely populated areas where tall buildings may reflect signals and cause radio shadows that make reception difficult. The antenna shown in Figure 5.11 has most of its energy in the vertical plane with angles less than 65-deg above horizon. Less than half of its power is sent at angles greater than 65-deg.

A typical mechanical structure and specifications for a 3-dB collinear roof-mounted antenna by Antenna Specialist is shown in Figure 5.12. The radiator is in one piece, including the phasing coil. The $5\lambda/8$ or $\lambda/2$ top radiator plus the $\lambda/4$ lower section produces a 3-dB omnidirectional gain at horizontal plane. The whip is of stainless steel construction, which effectively reduces the skin effect losses.

The current distribution on a $5\lambda/4$ monopole is shown in Figure 5.13(a). As seen from the figure, the current in the bottom $\lambda/4$ section is in phase with the top $\lambda/2$ section, but the current in between $\lambda/4$ through $3\lambda/4$ is out to phase. Due to out-of-phase current, the gain decreases and causes pattern breakup. The usual practice is to replace this section of the antenna by a helical resonator that matches the impedance of the lower and upper section.

A 3-dB collinear gain antenna may not provide full gain in all cases. In general, a 3-dB gain really performs closer to 2.0 dB. The vertical plane patterns measured at 835 MHz for ASP920 are shown in Figure 5.14.

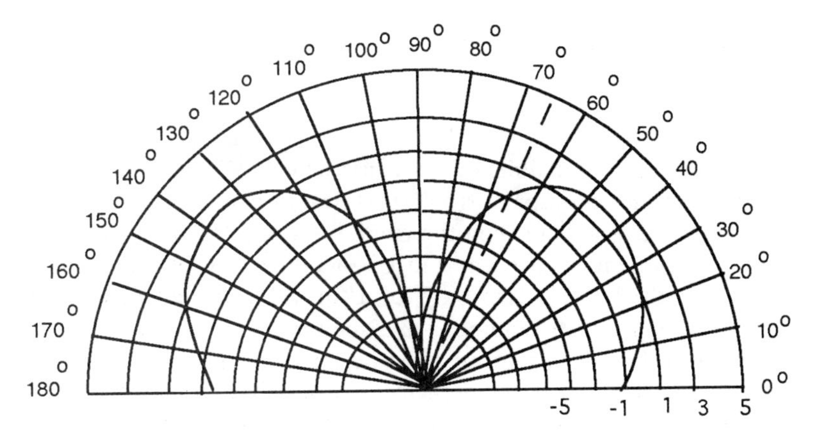

Figure 5.11 Unity-gain antenna with less than half its power above 65-deg. After [13, p. 46].

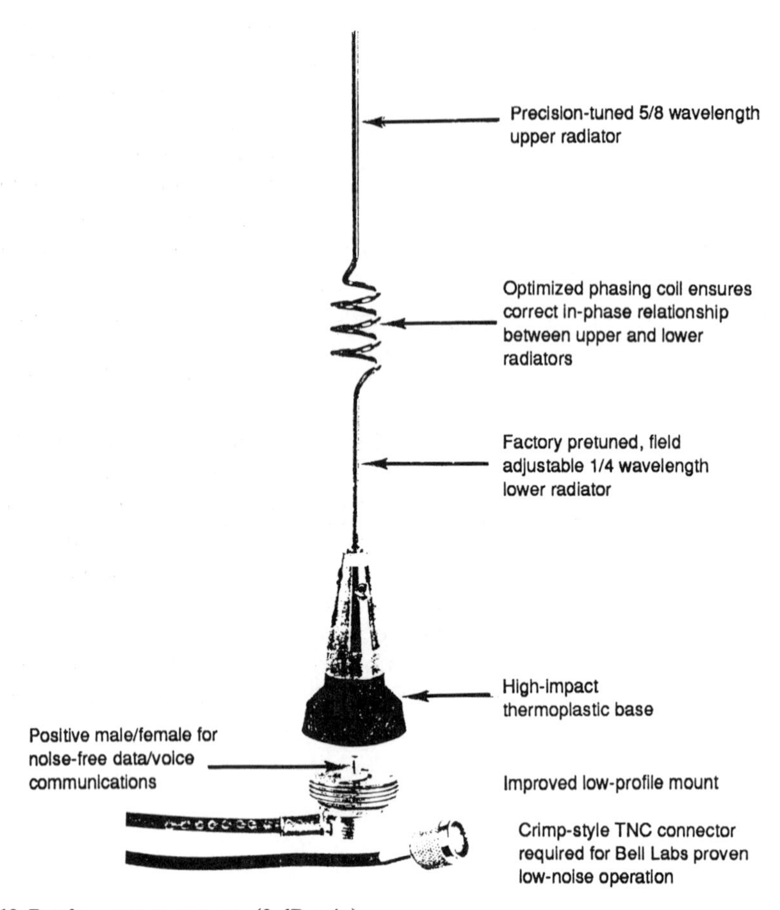

Figure 5.12 Rooftop-mount antenna (3-dB gain).

A one-piece whip of stainless steel construction, which requires only one-hole installation on the automobile trunk, is shown in Figure 5.15. Sometimes the antenna can also be clipped between the trunk body and the trunk opening. Here the radiator is above the feed point and the antenna gain is 3 dB. In order to achieve a bandwidth 70 MHz (coverage of a full transmit and receive band) in cellular radio, the quarter-wave antenna must be thick; that is, the diameter of the antenna at the wavelength of interest must be large. In this case, the $\lambda/4$ skirt is below the feedpoint and is generally enclosed in a polycarbonate tube. Since the elevated collinear design has its own skirt or counterpoise, it operates equally well without any ground plane.

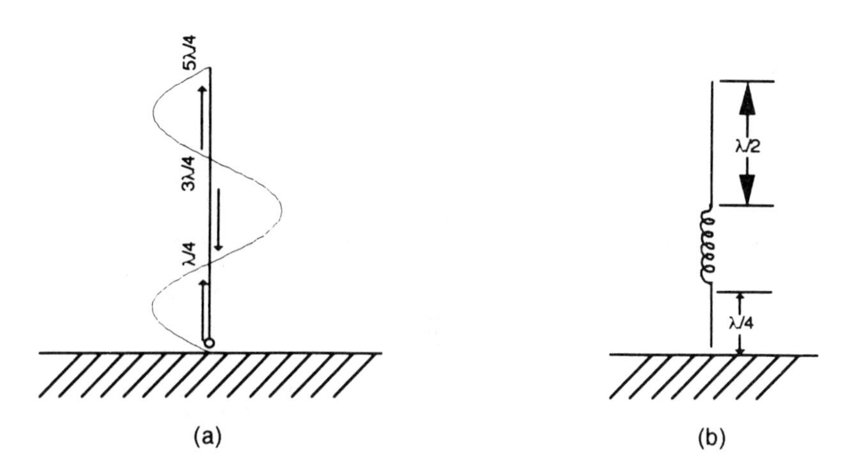

Figure 5.13 (a) Current distribution on a 5λ/4 monopole; (b) a practical collinear for mobile applications.

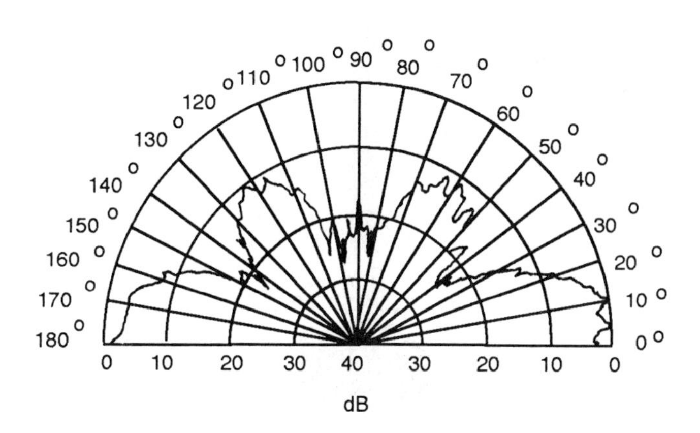

Figure 5.14 Vertical plane pattern for ASP-920.

For rural applications, a 5-dB collinear, three-element trilinear antenna offers the best choice as most of its energy is contained in the line-of-sight horizontal plane. See Figure 5.16.

Consumer preference in cellular radio has indicated that hole in the car body are totally unacceptable. On-glass antenna technology solves this problem. On-glass means that the radiator element is mounted on the glass windshield of a vehicle in such a manner that the RF energy is coupled through the glass without any modification to any part of the vehicle. A common misconception about glass-

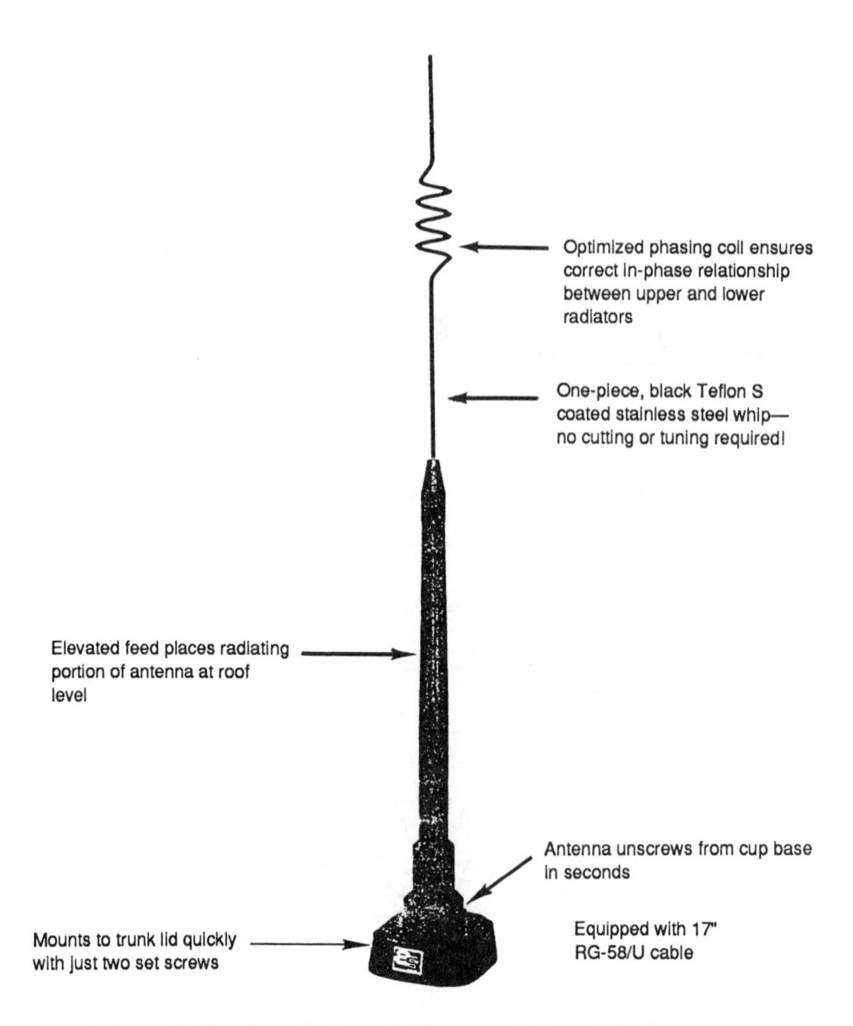

Figure 5.15 ASP-912 3-dB gain no-hole trunk-lid mounted elevated feed.

mount antennas is that the energy they transfer from the coaxial cable through the glass to the antenna base is substantially reduced in the process. Instead, the glass is a transmission medium. With a well-designed glass-mount antenna, a negligible amount of RF energy is lost as it passes through glass. Feeding the element on-glass radiator requires a low-loss, relatively broadband coupling circuit, through dielectric of the glass. The coupling network reactively couples the radiator element to the transceiver and provides a 50Ω input impedance. The mounting of these antennas can be done instantly.

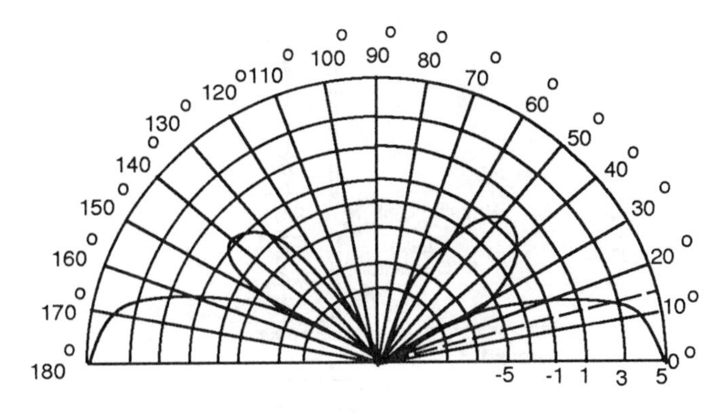

Figure 5.16 A 5-dB collinear, trilinear antenna with most energy within 15 deg in vertical plane. After [13, p. 46].

Voltage-fed and current-fed are two types of glass-mount antennas. The names refer to the distribution of energy on the antenna and the characteristic of that energy at the antenna base. Voltage-fed antennas usually have smaller mounting feet than current-fed antennas because metal plates within the feet form capacitors that transfer the RF energy through glass. Current-fed antennas, on the other hand, require larger capacitors than voltage-fed antennas.

A popular type voltage-fed mobile antenna is the 3-dB glass-mount type shown in Figure 5.17(a). RF energy is fed via the coaxial cable though an impedance matching tap on the L-C tuned circuit. The capacitor is variable, which is tuned to the middle of the cellular band. If not tuned properly, some of the available transmitter power may not be available at the antenna. A problem of voltage-fed antennas is their sensitivity to environment. Snow, ice, or slush buildup on the glass may reduce a voltage-fed antenna's performance due to detuning the L-C loading circuit alignment. The mobile user should be made aware of this.

Figure 5.17(b) shows the current distribution on a typical current-fed 3-dB gain mobile antenna. Here, the currents flowing on the conductors on the inside of the glass oppose each other. It should be noted that the antenna's vertical element is of a different length ($5\lambda/8$) than a voltage-fed antenna as this provides the best non-fading range. As the electrical current is maximum at the base and the voltage is minimum, it performs well under difficult conditions. The antenna is less sensitive to snow, ice, and dirt than voltage-fed antennas.

The azimuth pattern for an on-glass antenna along with $\lambda/4$ roof mounted is shown in Figure 5.18. On-glass antennas show better performance all along the azimuth plane [7, 12–18].

In the U.S. market, both unity gain and 5-dB gain antennas will be required. Unity-gain antennas may not be able to replace high-gain antennas completely. Unlike the United Kingdom, the United States has vast open areas without dense

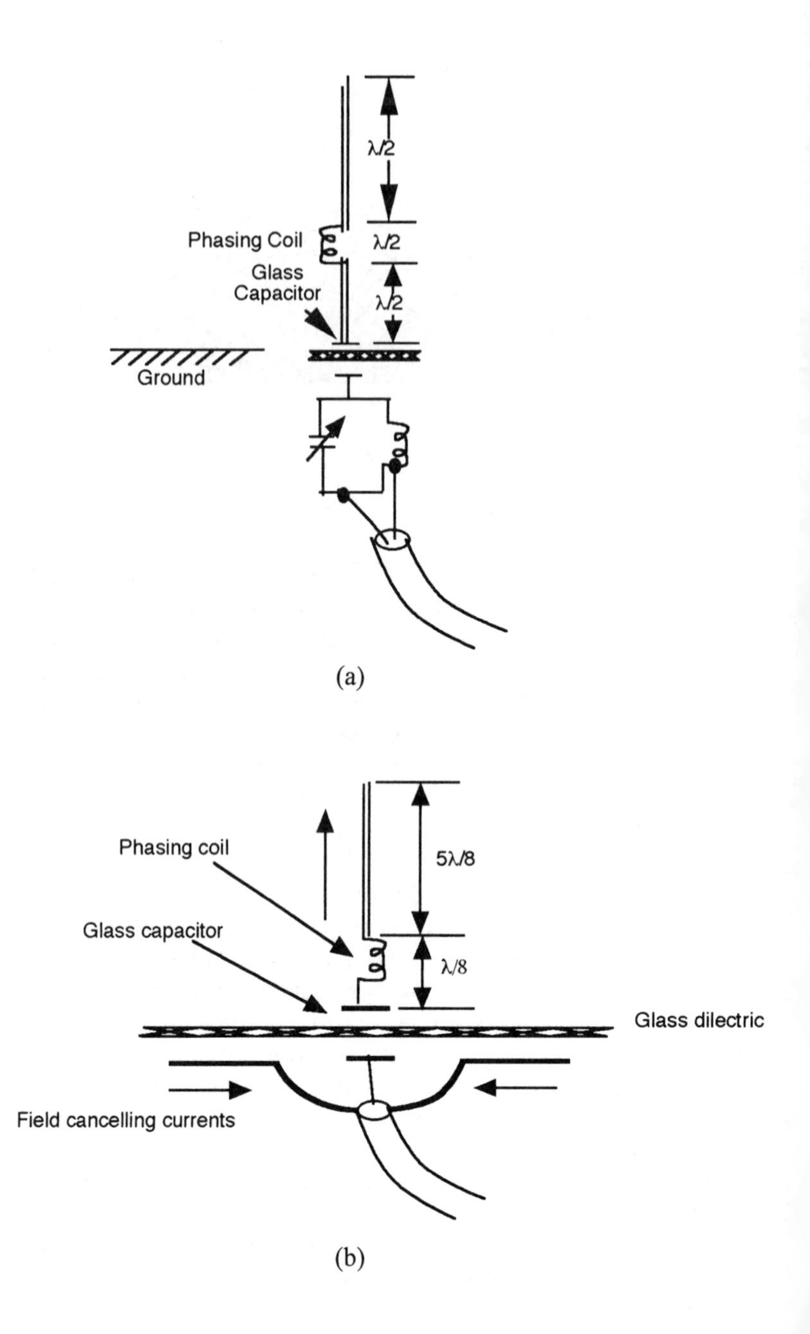

Figure 5.17 (a) A voltage-fed 3-dB glass antenna and (b) a 3-dB gain current-fed antenna. [After 14, pp. 26, 30].

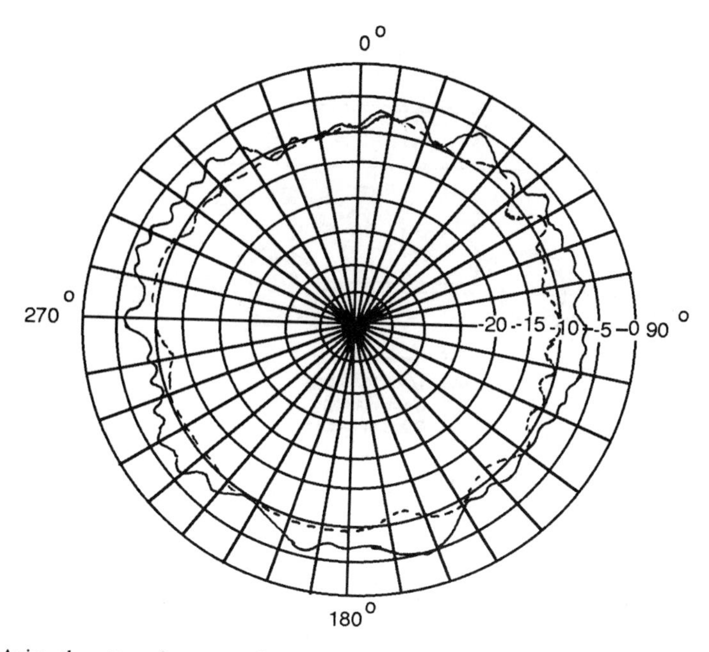

Figure 5.18 Azimuth pattern for an on-glass antenna (solid line) and roof-mounted $\lambda/4$ (dotted line).

cell-site coverage where high-gain antennas will be suitably used. Thus, the U.S. market will be divided between 5-dB high-gain antennas in open rural areas and unity-gain antennas in metropolitan areas.

Subscriber performance as to the location where the antenna should be mounted on the vehicle often has deleterious effects on the cellular system performance. History has shown that the roof location gives the most omnidirectional coverage, but in reality customers, and to a lesser extent some vehicles, do not permit antennas and the associated transmission lines to be installed in the roof.

Although customer dislike for roof mounting may be an overriding consideration, technically it is the most desired location. Being the highest location on the automobile, it enables signals emitted from the antenna or coming into the antenna to be less obstructed by nearby objects. Secondly, the roof of the car provides an ideal ground plane for the launching of radio waves. Usually ,a ground plane having a radius of $\lambda/4$ is required for efficient operation of the antenna. A roof of a car provides sufficient surface area for a 900-MHz cellular operation.

Figure 5.19 provides mounting locations for antennas. In order to realize the full potential of a radio system at 800 MHz it is important to understand how the placement of the antenna plays a part in its performance. A properly designed mobile antenna with adequate bandwidth and matched impedance cannot make

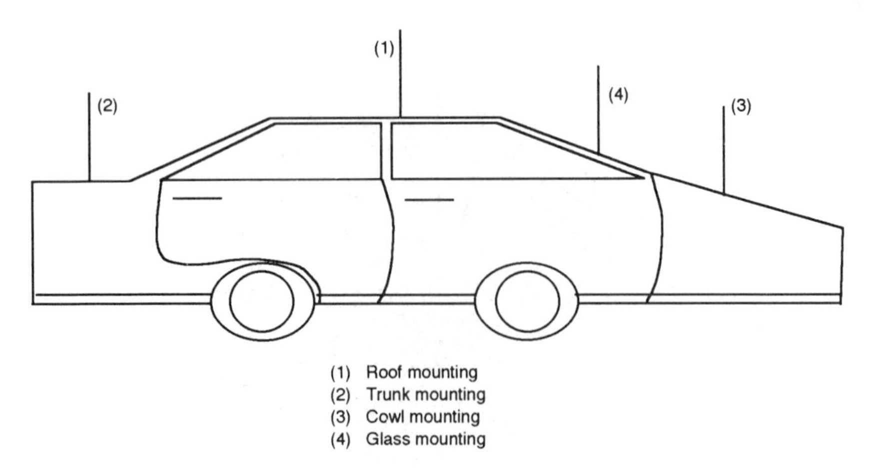

Figure 5.19 Vehicle dimensions and mounting locations.

up for poor mounting locations. Whether body mounted or glass mounted, an antenna is sensitive to its environment. Thus, the vehicle shape and the antenna's mounting position on the vehicle does to a certain degree affect the RF emission pattern. For example, identical antennas mounted on different cars may produce different patterns. More than a 3-dB additional gain can be realized by moving a quarter-wave antenna from the trunk lid to the middle of the roof. Commonly, glass-mount antennas affixed near the glass edge along the rooftop outperform body-mount antennas placed lower on the car body. The simple rule is to mount the antenna high for 360-deg coverage.

5.5 ANCILLARY DEVICES

Different ancillary devices are included in an antenna system to enhance the overall performance. This equipment includes, but is not limited to, a transmitter combiner, a receiver muticoupler, and a duplexer. Transmitter combiners and receiver multicoupler are standard cell-site equipment that allow up to a certain number of transceivers to be connected to a single antenna. A duplexer allows a single antenna to be used for simultaneous transmitting and receiving. The decision to use individual transmit and receive antennas or a single antenna with a duplexer depends on the overall system gain and economy of space limitations on the tower and in the equipment frame. Typical cell-site setups of three-antenna configurations and a two-antenna with duplexer configuration are shown in Figure 5.20. In both cases, N channels are combined in the transmit direction while the receiver multicoupler

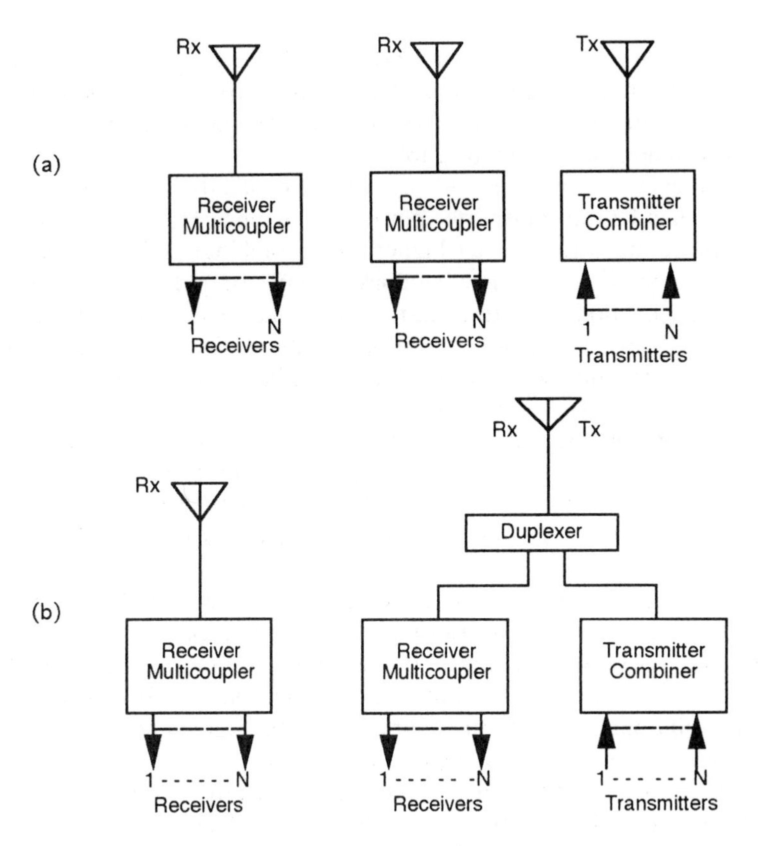

Figure 5.20 (a) Cell-site three-antenna configuration. (b) Antenna and duplexer configurations (up to 16 channels).

provides N outputs, which are connected to N receivers. We discuss each of these devices below. The design numbers provided in this discussion directly apply to U.S. systems.

5.5.1 Combiner [23]

It has always been difficult to combine transmitters that are closely spaced in frequency. Two principal techniques that are used in combining are cavity combining hybrid combining. Among the important characteristics of cavity combining are the number of channels to be combined, isolation between channels, insertion loss, and the power rating of the combiner. The channels to be combined must have a minimum frequency separation, which depends on insertion loss and the frequency stability of the cavities. In a 900-MHz cellular system cavity, combining has been used for signals as close as 0.4 MHz. Some typical manufacturer values for channel-to-channel isolation, insertion loss, and the power ratings of five and ten-channel combiners are shown in Table 5.2.

Power rating is an important and easy characteristic to deal with. The device must be able to handle the transmitter power per channel. If there is a thought to move up in power, one must choose a higher rating combiner at the very beginning.

In cavity combining, a high-Q bandpass cavity's outputs are combined in a transmission line junction. The transmitted signal passes through the cavity and through the length of cable from the output of the cavity to the junction. The length of the cable is adjusted to place a high impedance at the junction point for other channel frequencies except its own.

A typical eight-channel configuration of a cavity combiner is shown in Figure 5.21. Unloaded Q of these cavities are of the order of 10,000. The frequency response of the cavity is shown in Figure 5.22. Isolation between channels is generally low and circulators are required to attain acceptable decoupling between transmitters.

For combining transmitters that are separated by less than 0.4 MHz, hybrid combining can be used. When two transmitters are combined using this technique, half of the power is lost into the matched load. Furthermore, for each doubling of the number of transmitters to be combined, an additional 3 dB of insertion loss occurs due to additional stages in the hybrid. Also, the signal encounters less than 1/10 dB of insertion loss for every stage of the hybrid.

Configurations for combining two and eight transmitters are shown in Figure 5.23. Based on above discussion, the transmitter signal loss for two- and eight-transmitter configurations are little more than 3.3 and 9.9 dB. This includes an insertion loss of 1/3 dB for each of the hybrids. Insertion loss increases by about

Table 5.2
Five- and Ten-Channel Cavity Combiner Specifications

Manufacturer and Model Number	*Number of Channels*	*Transmitter–Transmitter Isolation (dB)*	*Transmitter Antenna Loss (dB)*	*Power Rating (W)*
Microwave Associates				
Tcx 800-5x 7-3A	5	40	2.7	60/150
Decibel Products				
DB-4469	5	70	2.9	150
DB-4458-2	10	70	3.3	150
Phelps Dodge				
TJT800-5	5	85	3.3	125
TJD800-10T	10	60	3.5	125

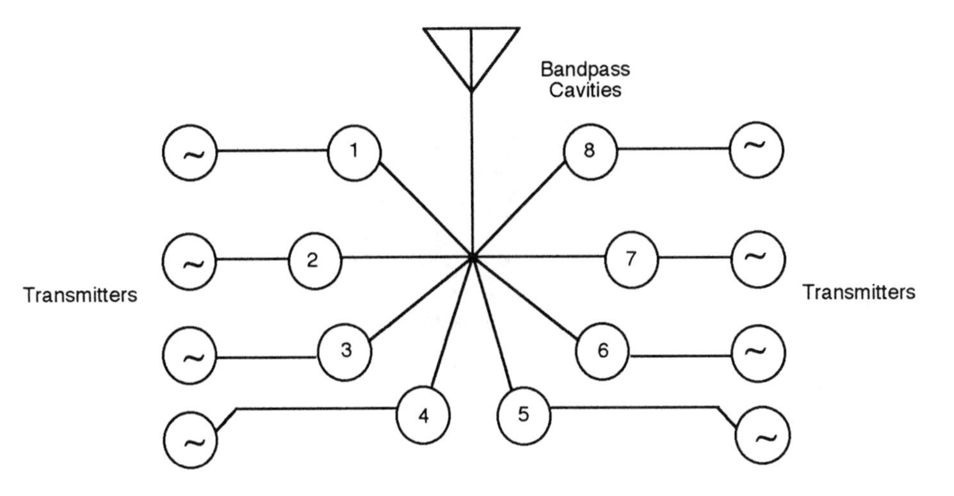

Figure 5.21 Eight-channel combiner using cavities and a junction.

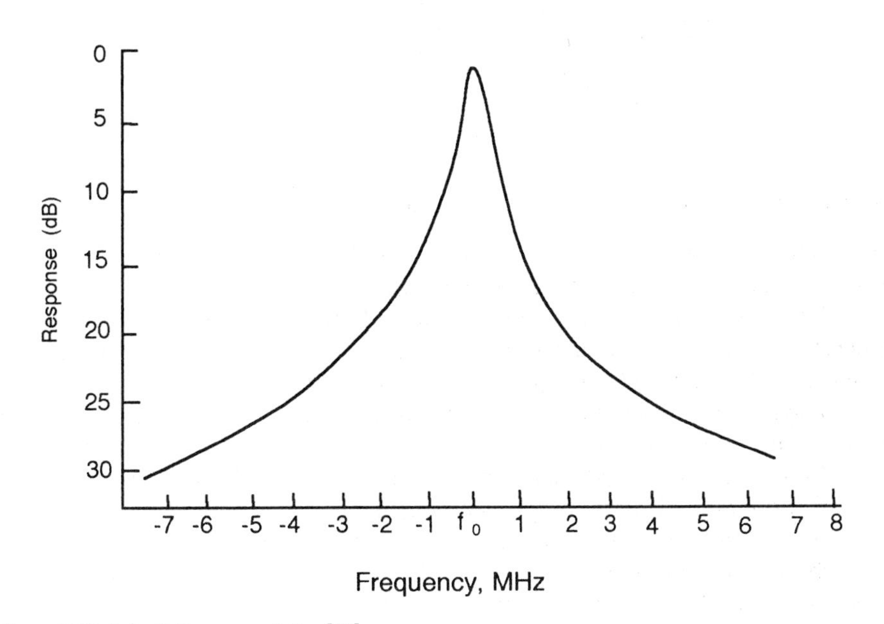

Figure 5.22 Selectivity curve. After [23].

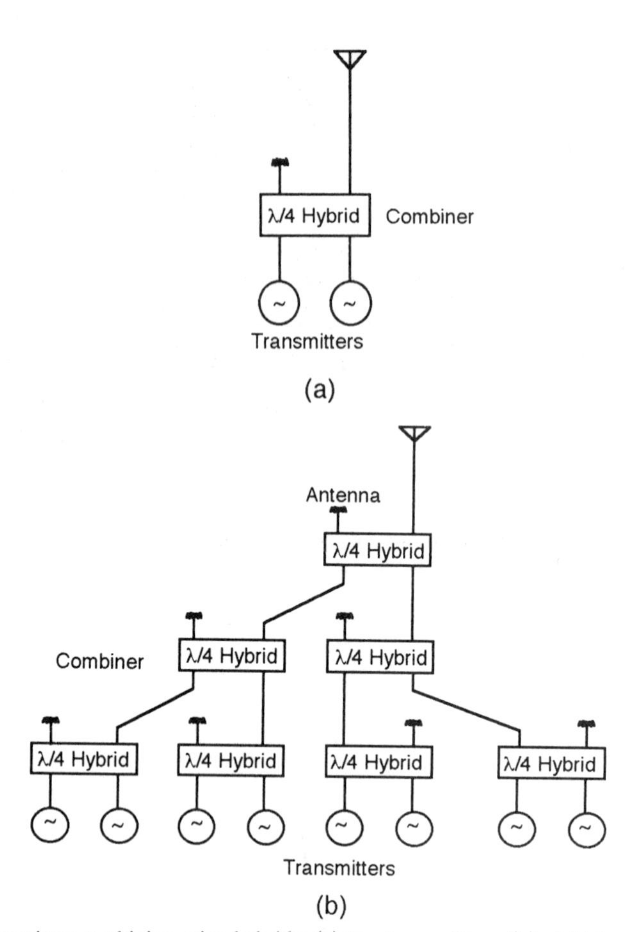

Figure 5.23 Transmitter combining using hybrids: (a) two transmitters, (b) eight transmitters.

3 dB each time the number of transmitters is doubled. However, hybrids have several advantages not the least of which is their small size.

Often, the designer has to do a tradeoff between the nearness of two frequencies that can be combined in the hybrid combiner versus the high loss that the signal has to withstand while passing through the hybrid combiner. If modifying the basic hybrid approach results in an increased radiated power instead of wasting the power in the dummy load, it provides both the flexibility of combining closely spaced frequencies and a reduced insertion loss. Figure 5.24 shows the implementation of the modified hybrid approach where the loss is limited to about 1 dB. The transmission lines feeding the hybrids have been crossed to produce a quadrature phase progression at the output. In the configuration, transmission lines are of the same length and the radiators are added so that power is not wasted in dummy loads. If properly aligned, the vector sum of the entire field produces an

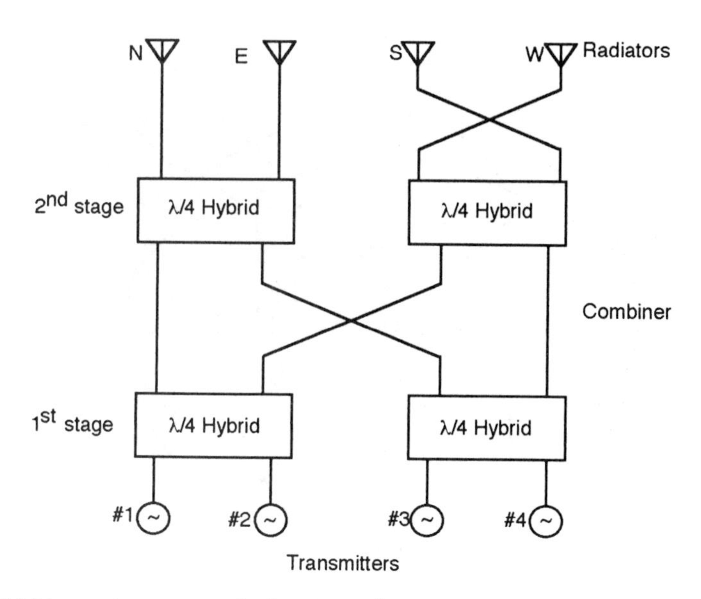

Figure 5.24 Multitransmitter antenna for four transmitters.

omnidirectional pattern, as shown in Figure 5.25. Table 5.3 shows the phase progression for the signal for each transmitter at the individual antenna outputs. Four quarter-wave hybrids are configured to achieve a quadrature phase relationship at their outputs by crossing the transmission lines that feed them. Thus, transmitter 1 produces no phase shifts in the first and the second stage of hybrids while feeding to the north antenna. The same signal to the east antenna is shifted by 90 deg while passing through the second stage of the hybrid. A signal applied to the south antenna gets shifted by 180 deg, the first 90 deg due to the first stage of the hybrid and the second 90 deg to the second stage of the hybrid. Similarly, transmitter 1 feeding to the west antenna appears at 90 deg. We conclude our discussion of combiners by providing an example of a 16-channel combiner system.

Example 5.2

For a 16-channel transmitter combiner, the input of each transmitter is −40 dBm and provides 40 dB attenuation to other ports. See Figure 5.26. Assuming the permissible noise at the receiver multicoupler input to be −130 dBm, find the required isolation between the transmit and the receive antenna. The cable loss on both the transmit and receive side can be assumed to be 3 dB. The T_x combiner noise output = $-40.0 - 40.0 + 10\log_{10}(16) = -68$ dBm. Assuming −130 dBm

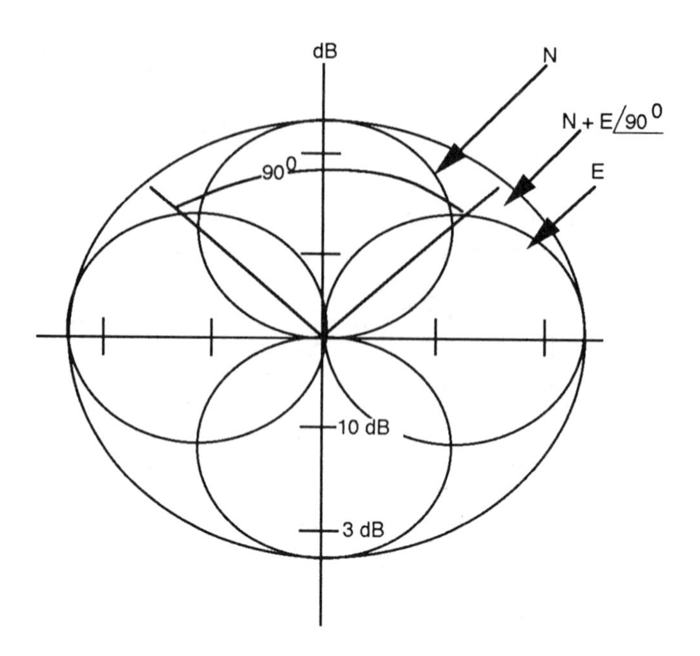

Figure 5.25 Pattern addition of separate radiators in a multitransmitter array of four.

to be the maximum allowed out-of-band noise at the multicoupler input and a total of 6-dB cable loss, we get $-68 - 3 - 3 - L_s \leq -130$ dBm or L_s $-74 + 130.0 =$ 56 dB. Thus, the minimum required isolation between the transmit and the receive antenna is 56 dB.

5.5.2. Receiver Multicoupler

The receiver multicoupler provides a means of routing the radio frequency signal from the antenna to the receiver for diversity combining. The components of the

Table 5.3
Transmitter Phase at the Radiator Outputs

Transmitter Number	*Phase at Radiator (Degrees)*			
	N	*E*	*S*	*W*
1	0	90	180	90
2	90	180	90	0
3	90	0	90	180
4	180	90	0	90

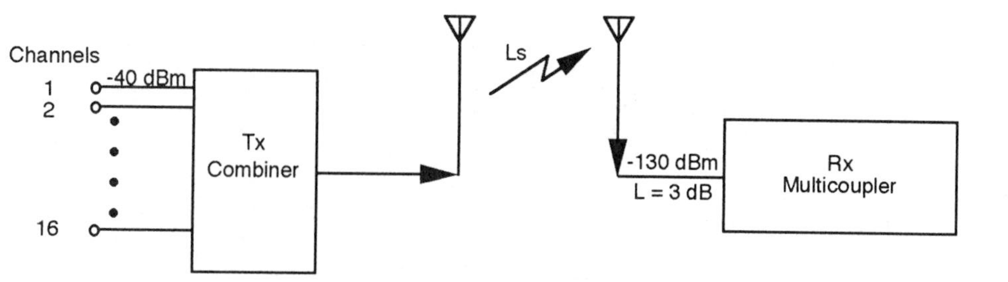

Figure 5.26 Typical multicoupler configuration.

multicoupler consist of a bandpass filter, a preamplifier, and the power divider. See Figure 5.27. The bandpass filter is often called a receiver preselector, which passes the signals of the desired band while rejecting the out-of-band signals. The preamplifier is usually mounted at a tower top close to the antenna in order to compensate for the loss in the long transmission line. However, any overcompensation of the transmission line loss may cause instability in the system. The power divider splits the signal into a number of channels where each channel has a fixed loss based on the number of splits of the signals. A commonly used configuration for a power splitter is 16 in cellular applications. A typical multicoupler configuration is shown in Figure 5.27.

For cell-site usage, the band of interest where the signal must be received without much distortion and attenuation is 825–845 MHz (AMP). Simultaneously, the proper rejection of the unwanted cosite-located transmit band of 869–894 MHz must be obtained. Typical gain characteristics of the cellular multicoupler are shown in Figure 5.28. Typical values of A and B are 10 dB and 60 dB, respectively. Here, A is the overall gain of the coupler, while B is the required stopband attenuation. The overall gain is A = amplifier gain − BPF loss − power splitter loss. The other important characteristics of the multicoupler are high third-order intercept points and the noise figure of the multicoupler system. Typical values of these parameters from different manufactures of multicouplers are shown in Table 5.4.

The concept of noise figures are discussed in Section 5.7. For third-order intercept points, we only want to say that the higher the value, the better it is. Here, we note that it is difficult to get a low value for a noise figure with a high value for a third-order intercept point and often a compromise is made. We provide an example to illustrate the concept of overall gain of the receiver multicoupler system.

Example 5.3

For the configuration shown in Figure 5.29, find the net gain with and without the power splitter. The net gain of the amplifier filter and splitter combination without

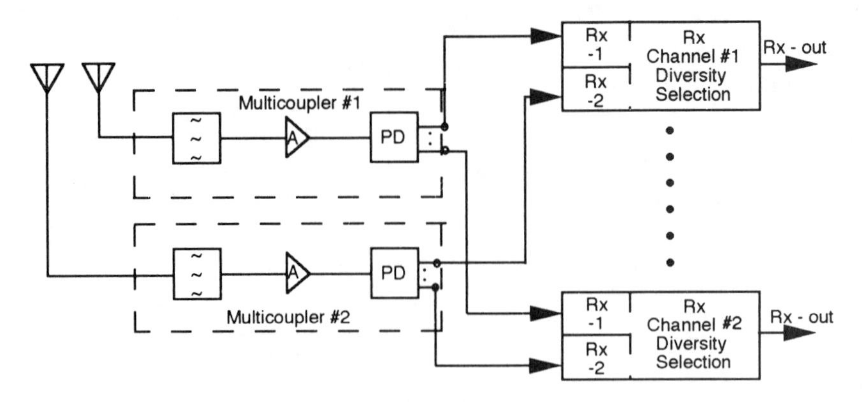

Figure 5.27 Cell-site receiver multicoupler configuration.

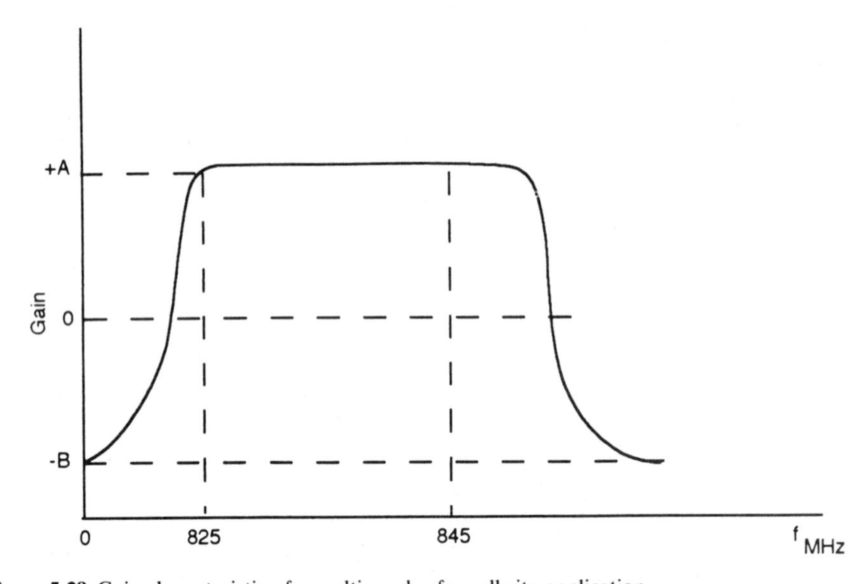

Figure 5.28 Gain characteristic of a multicoupler for cell-site application.

pad loss is $G_1 + G_2 - L_1 - L_2 = 15 + 20 - 1.5 - 17.5 = 16.0$ dB. The net gain with pad $= G_1 + G_2 - L_1 - L_2 - L_3 = 10$ dB. We shall carry out this example further in Section 5.7 for computing the overall noise figure.

5.5.3. Duplexer

For simultaneous transmit and receiver operation from a single antenna, a duplexer is used at the cell site. Sometimes due to space limitations on the tower and in the equipment frame, it may be necessary to use a single antenna with the duplexer arrangement. See Figure 5.30. The basic design of the duplexer requires a bandwidth of at least 70 MHz to cover both the transmit and the receive bands in the U.S. Some common manufacturer's parameters of the 16-channel duplexer are shown in Table 5.5.

As shown in Table 5.5, noise suppression at the receive frequency and receiver isolation at the transmit frequency are required. The suppressions can be achieved

Table 5.4
Technical Specifications for the Receiver Multicoupler

Manufacturer and Model Number	*Number of Ports*	*Typical Gain*	*Noise Figure*	*Third-Order Intercept Point*
Microwave Associate				
RC-800-8	8	3	8	+32
Phelps Dodge				
RMC800-8	8	2*	4	+30
RMC800-12	12	2	4	+30
RMC800-20	20	2	3.5	+34
Decibel				
DB-8508-101	8	6	3.5	+36
DB-8516-101	16	3	3.5	+34

*Minimum.

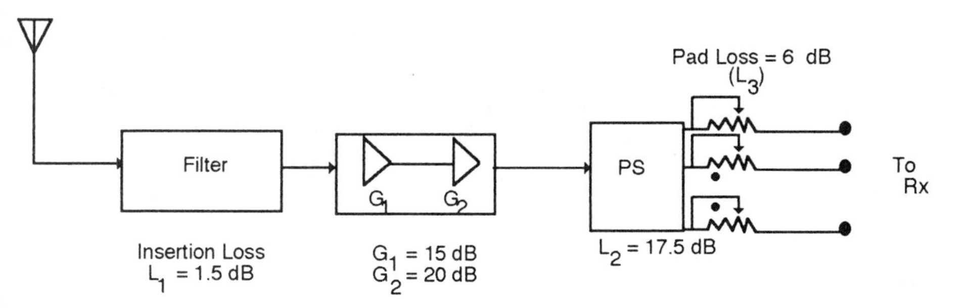

Figure 5.29 Typical multicoupler set-up.

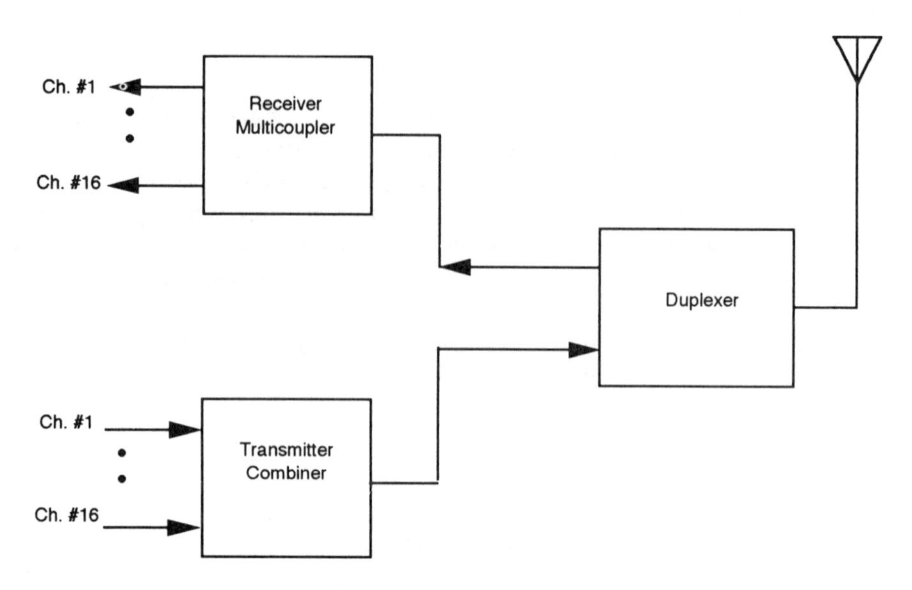

Figure 5.30 Duplexer arrangement in typical cell site.

by two different filter configurations, namely the stopband and the passband. The stop-band filter configuration inserts a bandstop filter at the transmit frequency in the receiver input path to prevent transmit energy from entering into receiver input. Similarly, the stop-band filter provided at the transmit path eliminates the receive frequency from the transmit path. The same performance can also be achieved by

Table 5.5
Duplexer Characteristics

Manufacturer and Model Number	*Antenna Loss*		*Transmitter Noise at Receiver Frequency (isolation dB)*	*Receiver Isolation at Transmitter Frequency (dB)*	*Power (W)*
	Transmitter	*Receiver*			
Microwave Associates					
DU-800-50	0.5	0.5	50	50	200
DU-800-70H	0.4	0.4	70	70	1000
Phelps Dodge					
PD-896	1.0	1.0	80	80	125
PD-898	1.0	1.0	65	65	125
Decibel					
DB-4090	0.7	0.7	65	65	150
DB-4185	0.9	2.0	60	90	1000

inserting bandpass filters of the transmit band in the transmit path and a receive band filter in the receive path. We illustrate below an example of a duplexer.

Example 5.4

A ten-channel combiner system (DB-4458-2) has an insertion loss of 3.3 dB. Assuming the individual transmitter power input to the combiner is 90W, find the minimum power rating of the duplexer. Per channel power output of the combiner = $10\ \mathrm{Log}_{10}(90) - 3.3 = 15.24$ dBw. Therefore, the combiner power outputs = 25.24 dBW = 420.7W. Thus both the Decibel DB – 4185 and the Microwave Associates DU – 800 – 70H will meet the requirements.

5.5.4 Transmission Line

The proper choice of a transmission line is important for the overall design of a cellular system. For applications in the 900-MHz range, a low-loss coaxial cable with 50 ohm characteristic impedance is used. Due to the exposed nature of its usage, the cable should be able to withstand water migration. The cable may be pressurized with foam as a dielectric, or air can be used as dielectric material. Due to its outside usage, flexibility in the cable is more desirable than rigidity. The cable should be attached to the tower in as many points as possible, allowing the cable to move with the tower during windy periods. Coaxial-type transmission lines generally fall into two categories, flexible braid coaxial hardline.

A braided coaxial transmission line should not be used if possible due to its high attenuation. Power attenuation and the power rating for the braided cable is shown in Table 5.6 [2–6].

Among the hardline cables, aluminum and copper foams are the main choices. As seen from the table, the same diameter copper with foam cable as the dielectric material is about 1 dB better in its attenuation characteristics than aluminum with foam as dielectric. For few hundred feet of feeder cable, the choice of copper with foam as dielectric material is usually a better choice. For very large runs where it is important to save even one-tenth of one dB, it is desired to choose copper with air as a dielectric material. Jumper cables are also required for connecting the main feeder cable to the antenna and at the antenna base for connection to the combiner. For cases where hardline is used for the main feeder cable, the jumper wire, if required, should also be hardline. Cabling between the combiner, multicoupler, transmitter, and receivers should also be hardline. The same cable should be used on the receiver side in order to minimize the different spare parts. Table 5.6 provides the attenuation and power rating of the different types of cables according to their decreasing value of attenuation. We conclude the discussion of this section by taking up an example

Example 5.5

The input for each of the five channels to the combiner is 35W. Assuming the combiner loss to be 2 dB, compute the power at the antenna input terminal if 50 ft of RG-58 cable is used in the layout with the stated attenuation value of 20 dB/100 ft. Figure 5.31 shows an example of the proper choice of transmission line. The combiner output per channel = 10 log 35 − 2 dB = 13.4 dBW. The total combiner output = $5\ 10^{1.34}$ = 109.4W ≈ 110W = 20.4 dBw. With 50 ft of RG-58, the total available power at the antenna input terminal is 20.4 − 10.0 = 10.4 dBw or 10.9W. Obviously, most of the transmitter power is lost in the cable system.

Table 5.6
Transmission Line Characteristics at 850 MHz

Description	*Type*	*Attenuation (dB/100 ft)*	*Power (W)*
Braided shield	RG-58	17–20	47
	RG-8	8	200
	RG-213	8	200
	RG-303	12	750
Hardline aluminum (foam)			
1/2″		3.1	550
7/8″		2.2	1.1 K
Copper (foam)			
1/4″	FHJ1-50	5.8	125
1/2″	LDF4-50	2.2	570
7/8″	FLC78-50J	1.05	1.5 K
1-5/8″	LDF7-50	0.83	3.7 K
Copper (air)			
1/2″	HJ4-50	2.3	680
7/8″	HJ5-50	1.1	2.0 K

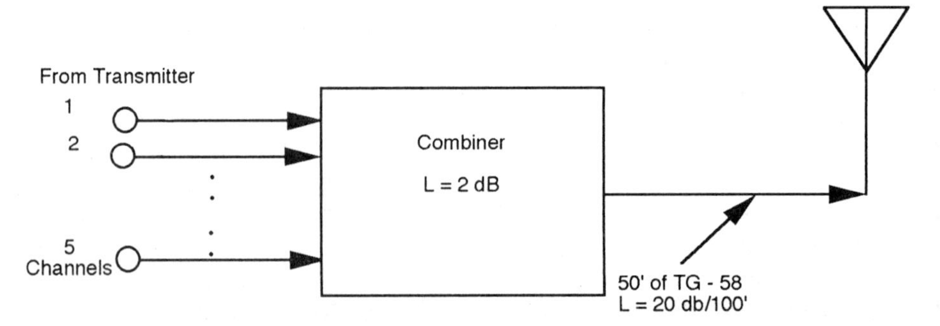

Figure 5.31 Proper choice of transmission line.

Alternately, if the 7/8 in HJ5-50 solid copper with air dielectric is used, with an attenuation value of 1.1 dB per 100 ft, the total available power at the antenna input is increased to (20.4 − 0.55) dBw or 95.9W.

5.6 PROBLEMS WITH CELLULAR ANTENNAS [24]

In this section we describe the various problems one encounters in the design and installation of a cellular antenna. The antenna being an important element in cellular system, the associated problems must be understood and properly tackled. Three problems discussed are: the dead spot problem with the cell-site antenna, isolation between the transmit and receive antennas, and the cell-site collinear antenna mounting. The high-gain antenna has to be mounted carefully; that is, a slight unintentional tilt in the antenna causes a large loss of gain in the desired direction. A large reduction in antenna gain may cause a coverage problem and create holes or dead spots. In order to meet the sensitivity requirement of the receiver, the input noise of the receiver has to be controlled. One source of noise is due to transmitter generated out-of-band noise coupling the receiver through the receiving antenna system. Obviously, the more spacing between transmitter and receiver antennas, less the coupling. Thus, high antenna isolation has to be maintained. A higher antenna gain can be achieved when the antenna is mounted at the topmost point on the tower. Since there is only one tower top, several antennas cannot be mounted at the same point. Therefore, the choice is to mount one antenna at the top and other antennas below it or mounted on the sides. Obviously, side mounting antennas perform well. However, there are several problems that must be understood.

5.6.1 Dead Spot Problem [2–4]

Higher antenna gain is achieved by compressing the beamwidth in the elevation plane. Unfortunately, with compression, more minor lobes appear with the pattern. Close-in dead spots may be encountered due to minor lobe nulls even though the distant coverage is good because of a high major lobe. Due to the presence of minor lobes, dead spots may occur in an area where the coverage is desired. Figure 5.32 indicates the presence of null at point X. Point X can easily be computed by

$$D_x = h/\tan\theta \tag{5.31}$$

Here, θ represents the angle of the first null with respect to the antenna bore sight. With knowing the 3-dB point on ground, the equation can be modified such that 2α represents the vertical beamwidth of the antenna. Some computed values of

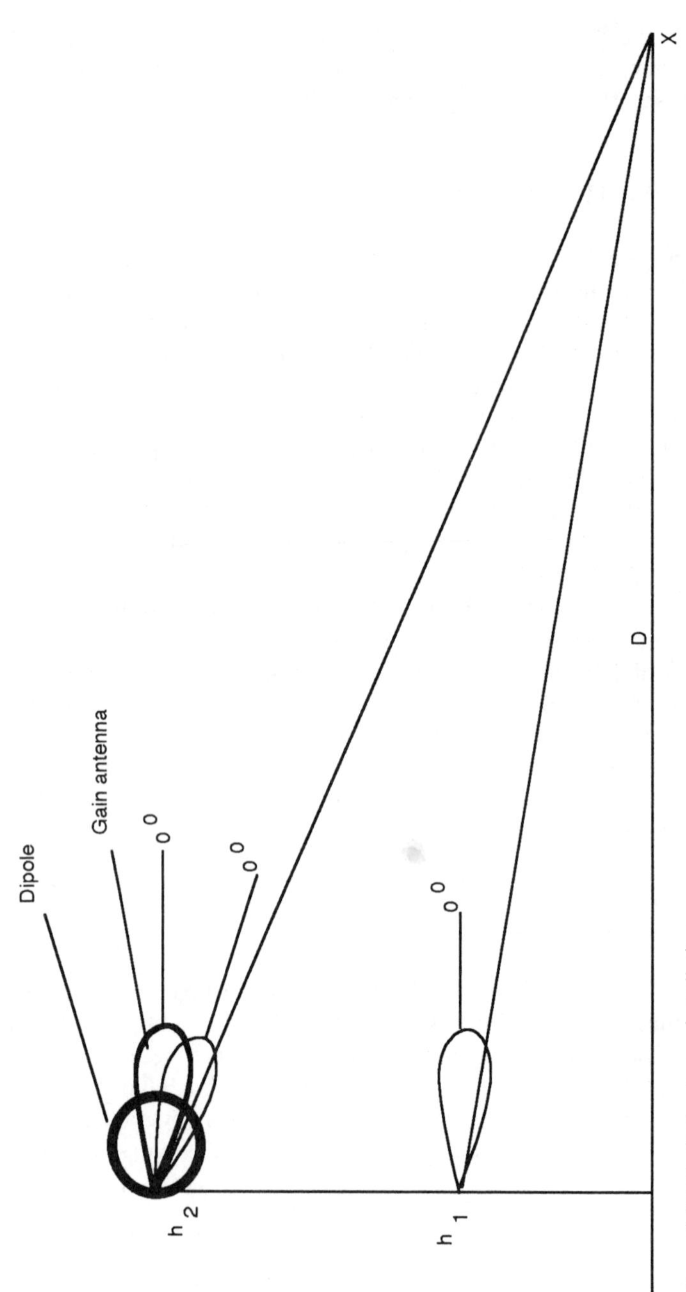

Figure 5.32 Typical antenna installation.

the 3-dB intercept distance X as a function of vertical beamwidth is shown in Table 5.7. Note that the antenna bore sight in this case is parallel to ground.

As discussed above, gain in the vertically polarized omnidirectional antenna can be achieved by compressing the beam in the vertical direction. Unfortunately, more compression results in more minor lobes with pattern nulls. For example, a typical collinear antenna array with ten times the power gain would have its first null approximately six degrees below the horizon. The depth of null in this case is of the order of 30 dB. As shown in Figure 5.33, for the same value of null angle the dead spot will move closer to the cell site as the antenna height is reduced. Assuming a 10-dBd antenna gain located at h_2, which is 2,000-ft above the ground with a null angle of -6 deg will put the null at about three miles from the cell site. Thus, the mobile will experience about 30-dB attenuation due to site geometry. Performance will improve as the mobile moves away from cell site. At a distance of about seven miles, the mobile will have 3-dB attenuation with respect to maximum performance, which may only be attainable 20–30 miles away from cell site.

The problems of improving the signal level at point x without increasing the transmitter power are: a reduction of antenna height, use of a lower gain antenna, and electrically tilting the antenna beam. As shown in Figures 5.32, and 5.33, reducing the antenna height improves the coverage as the horizontal line h_1X intercepts the antenna gain curve at a higher value of gain. The same concept holds if the gain of the antenna is reduced, thereby making the intercept h_2X at a higher value of gain. The null can also be avoided by electrically tilting the major lobe of the antenna gain pattern at height h_2 so that the point X is no longer in null. This is shown in Figure 5.32. However, care should be taken when considering the use of down-tilt antennas because: normal antenna gain and bandwidth is reduced; once the down-tilt is designed, it cannot be removed; and shadowing of a close-in area may cause serious problems.

5.6.2 Isolation Between Transmit and Receiver Antennas [10]

Adequate transmitter and receiver isolation is required because of the receiver desensitization, which is defined as the reduction in receiver sensitivity caused by:

- Receiver inband noise increase caused by the cosite transmitter, (out-of-band signal);
- Gain reduction of the receiver low-noise amplifier caused by a strong off-channel signal.

The amount of isolation required is based on the out-of-band noise generated at the transmitter and the receiver desensitization characteristics. Techniques used for minimizing the effects of this are an adequate choice of antenna spacing and filtering the out-of-band transmitter channel noise. Filtering is provided by different

Table 5.7
Distance Where 3-dB Point of Vertical Pattern Intercepts the Ground

Vertical	*Antenna Height (ft)*								
BW	*100*	*400*	*1,000*	*1,500*	*2,000*	*2,500*	*3,000*	*4,000*	*5,000*
5.5	2,082 ft	1.6 mi	3.9 mi	5.9 mi	7.9 mi	9.9 mi	11.8 mi	15.8 mi	19.7 mi
5.0	1,908 ft	1.4 mi	3.6 mi	5.4 mi	7.2 mi	9.0 mi	10.8 mi	14.4 mi	18.1 mi
8.0	1,430 ft	1.1 mi	2.7 mi	4.1 mi	5.4 mi	5.8 mi	8.2 mi	10.8 mi	13.5 mi
20.0	576 ft	2,268 ft	1.1 mi	1.6 ft	2.1 ft	2.7 mi	3.2 mi	4.3 mi	5.4 mi
40.0	275 ft	1,099 ft	2,747 ft	4,121 ft	1.0 mi	1.3 ft	1.6 ft	2.1 mi	2.6 mi
60.0	173 ft	693 ft	1,732 ft	2,598 ft	3,464 ft	4,330 ft	1.0 mi	1.3 mi	1.6 mi

Note: BW = bandwidth.

devices, such as receiver multicouplers, duplexers, and isolators. Here, we shall discuss the isolation produced due to the vertical and the horizontal antenna spacing, which is necessary for noise reduction in the receiver band. Horizontal and vertical separation between two antennas are shown in Figure 5.34(a).

Isolation

We will now discuss the approximate expression required for isolation to minimize out-of-band transmitter noise. The expression is experimental and includes the effects of transmission line loss, transmitter combiner loss, output, and permissible noise level.

Horizontal spacing

Isolation L_h, in decibels, is given by

$$L_h = 22.0 + 20 \log_{10}(d/\lambda) - (A_t + A_r) \tag{5.32a}$$

The factor 22.0 represents a propagation constant $(4\pi)^2$ in decibels, d is the antenna spacing in feet λ is the wavelength in feet A_t and A_r are the transmit and receive antenna gains, respectively. Thus, at 850 MHz for 20-ft spacing and with $A_t = A_r = 10$ dB, the isolation value is

$$L_h = 22.0 + 20 \log_{10}(20.0/1.15) - 20.0 = 25.8 \text{ dB} \tag{5.32b}$$

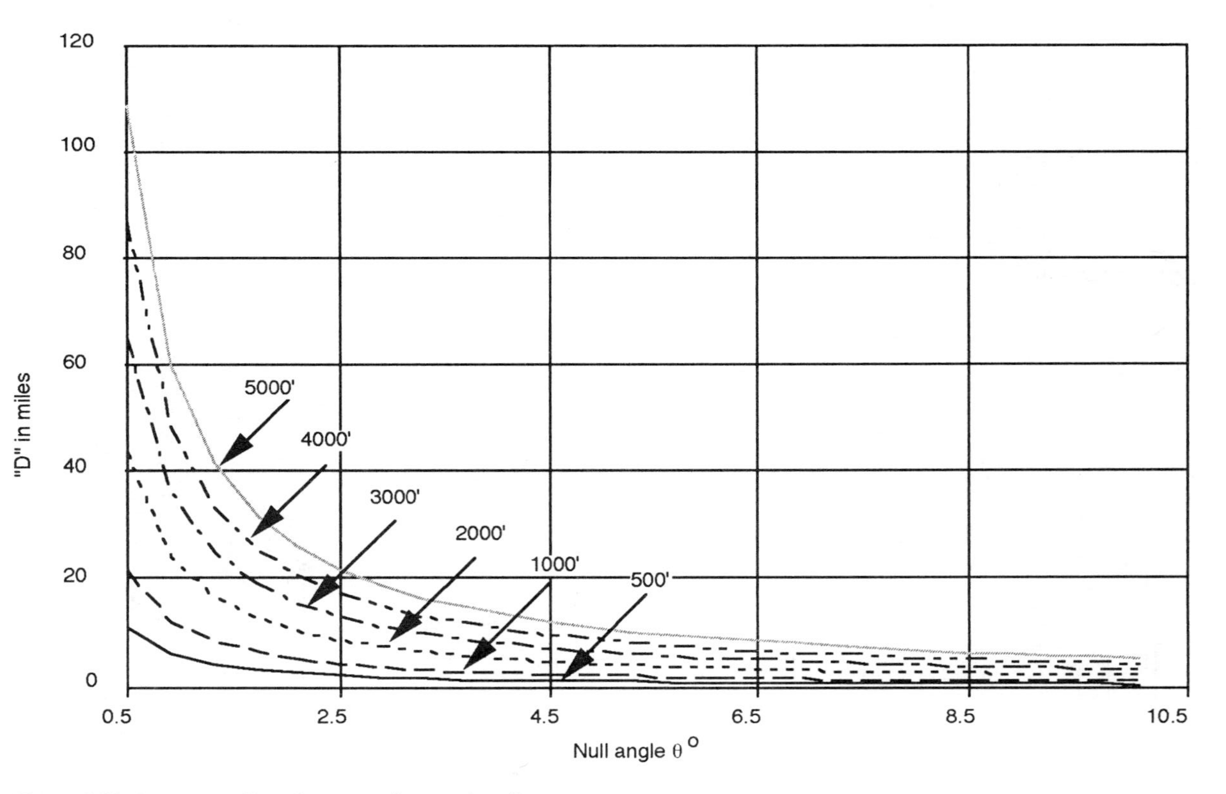

Figure 5.33 Antenna null angle versus distance in miles.

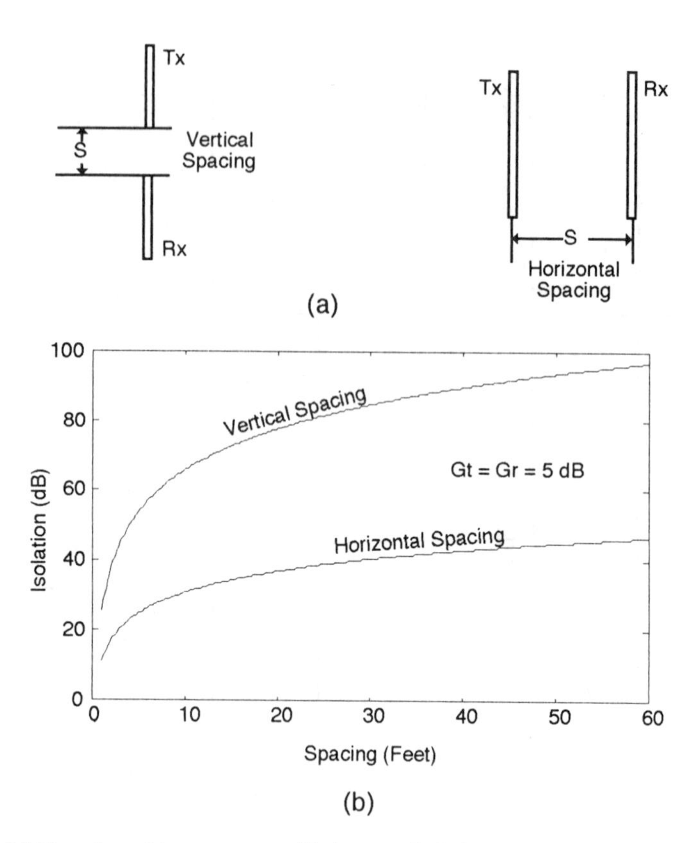

Figure 5.34 (a) Mounting of two antennas. (b) Antenna isolation versus spacing at 850 MHz.

Vertical Spacing

Isolation L_v, in decibels, is given by

$$L_v = 28.0 + 40 \log_{10}(d/\lambda) \tag{5.33}$$

where $d = 10.0$ ft, $\lambda = 1.15$ ft, and the computed value of $L_v = 65.6$ dB. Figure 5.34(b) is the plot of antenna isolation versus vertical and horizontal spacing in feet at 850 MHz when the transmitter and receiver antenna gains are 5 dB each.

5.6.3 Antenna Mounting

As stated above, cellular systems with a large number of channels and state-of-the-art transmitter combiners use more than one transmitter antenna at the cell sites. Since each tower has only one top and since there is always a need to place several antennas, some must be mounted at the sides at the same level. In this section we shall discuss the effect of mounting a number antennas at UHF band with different spacings. The resultant antenna patterns in all cases are plotted when only one antenna is fed while the other antennas are terminated in 50 ohms impedance.

Antenna patterns with two antennas at different UHF frequencies and top-mounted side by side with different spacing are shown in Figure 5.35. For the spacing of 23-3/8 in, which is slightly more than 1.5λ, the number of maximum and minimum is six. At 866 MHz, with 54 in spacing, it is observed that a +1 dB to −2 dB gain variation occurs. For the same spacing between elements, as the frequency is increased the ratio of maximum to minimum is increased. Also, for the same frequency, as the spacing between two antenna elements is increased, a smoothness in the antenna pattern is seen.

Isolation between antennas with different spacings is shown in Figure 5.36. For a 9-dB omnidirectional gain, minimum isolations of 20 dB, 25 dB, and 27 dB are seen with 23-3/8 in, 54 in, and 102 in of spacing between the elements.

5.7 RECEIVER NOISE CONSIDERATIONS

Filtering and amplification are the two most important signal processing operations performed in any typical receiver. In a cellular system, these operations are performed by two port devices consisting of active and passive components. Since the input to these devices is always accompanied by noise, the noise will also be amplified along with the signal. Furthermore, components internal to the two-port network also generate noise, which further degrades the signal quality at the receiver output. In this section we will develop an appropriate measure for describing the extent to which the input signal to the network is degraded while passing through a two-port network. Signal quality at the input is described by the input signal-to-noise ratio (SNR). Because of the internal noise within the two-port, the SNR at the output of the network is further reduced. The degradation in signal quality is characterized in terms of noise equivalent bandwidth, effective noise temperature, and noise figure, which is the topic of our discussion here. Noise figure values of the cellular radio at 900 MHz are around 6 dB, and are technology dependent. In the future, as the performance of components improve, the noise figure will also be reduced.

5.7.1 Noise Equivalent Bandwidth

In the absence of internally generated noise, noise equivalent bandwidth is the measure of total input noise, which appears at the output and provides an estimate

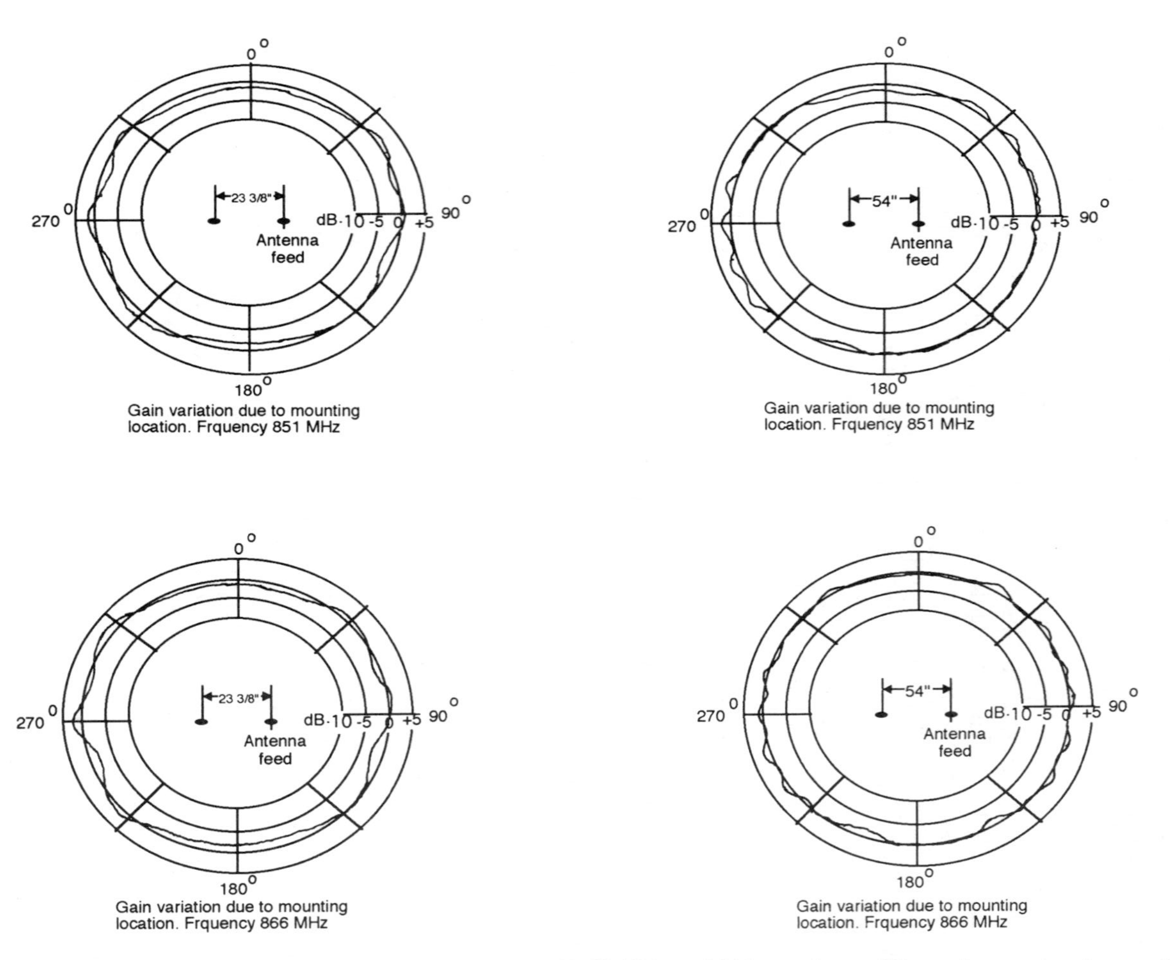

Figure 5.35 Gain variation on side-by-side mounted antennas with 23-3/8 in and 54 in spacing at different frequencies. Source: [24].

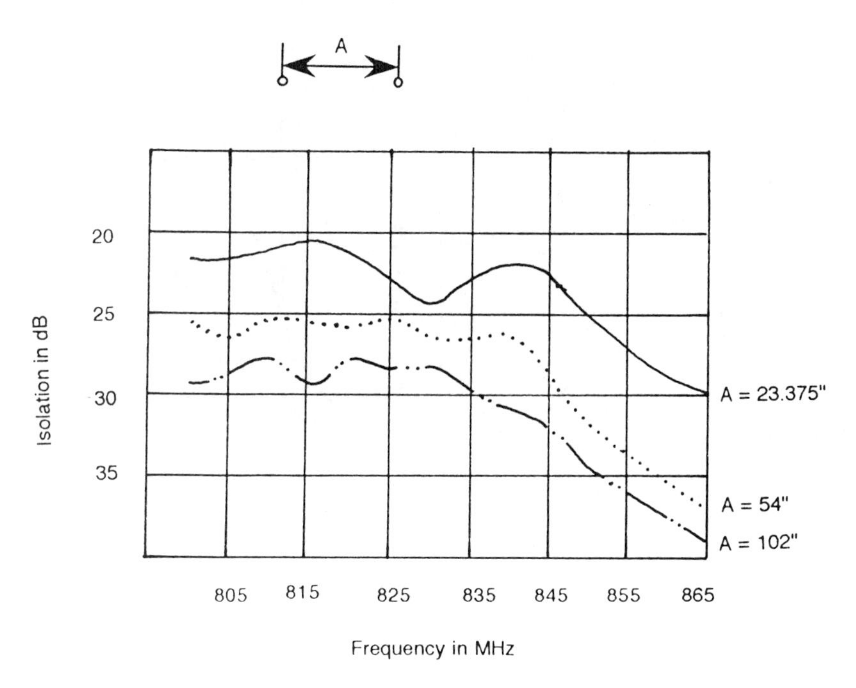

Figure 5.36 Isolation versus frequency for 9-dB omnidirectional antennas mounted as shown. Source: [24].

of the output SNR. If the input noise is white with a two-sided power spectral density equal to $\eta/2$, then the output noise power N_o is given by

$$N_o = \int_{-\infty}^{\infty} \frac{\eta}{2} |H(f)|^2 df = \eta \int_0^{\infty} |H(f)|^2 df$$

Let

$$B_N = \frac{1}{g_a} \int_0^{\infty} |H(f)|^2 df \tag{5.34}$$

where $g_a = |H(f)|^2_{\max}$, defined as the maximum available power gain, often the midband gain of the network. Thus,

$$N_o = \eta g_a B_N \tag{5.35}$$

where B_N is known as the noise equivalent bandwidth, and is shown in Figure 5.37.

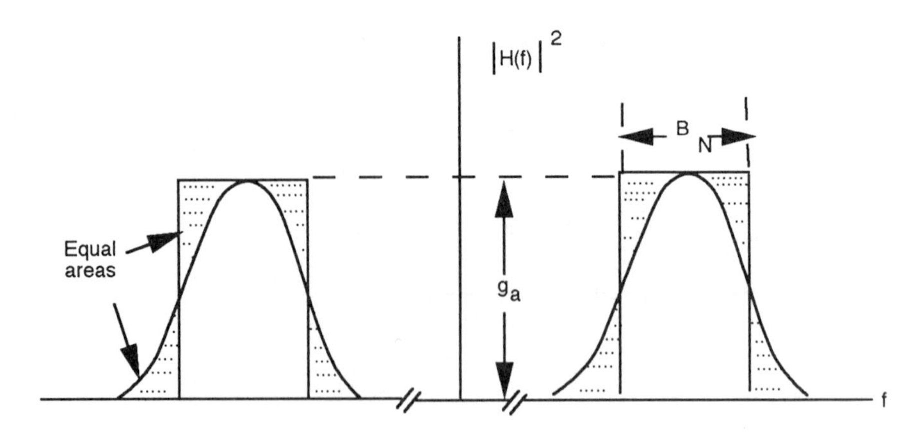

Figure 5.37 Noise equivalent bandwidth of a bandpass filter.

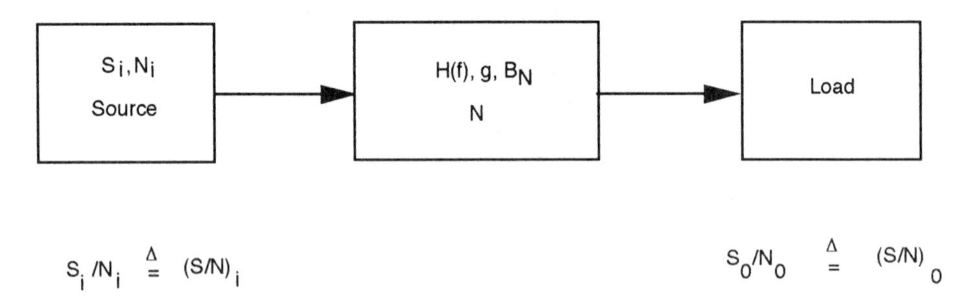

Figure 5.38 Filter network with matched source and matched load.

Equation (5.35) states that the noise power at the output of a filter is characterized by the parameters g_a and B_N of the filter and the input noise *PSD*. Furthermore, B_N is equal to the bandwidth of an ideal rectangular filter with a power gain g_a that would pass as much white noise power as the actual filter will pass. In other words, the noise bandwidth of a rectangular filter is the actual filter bandwidth. However, for practical and well behaved filters, B_N will be somewhat greater than the half-power (3 dB) bandwidth.

5.7.2 Noise Figure (Effective Noise Temperature)

Let us assume that the two-port network is driven by a matched source and also that the output of the filter drives a matched load. Thus, the network accepts the maximum power at the input and delivers maximum power to the output. Then, (from Figure 5.38)

$$S_o = S_i g \qquad N_o = N + N_i g \tag{5.36}$$

where $H(f)$ is the transfer function of the network, g is the power gain of the network, B_N is the noise bandwidth as defined above, and N is the additional noise generated inside the two-port network, which adds to the output noise. If the input noise is white and represented by an equivalent noise temperature T_i, then $N_i = kT_iB_N$. And then the output SNR can now be computed as

$$\begin{aligned}\frac{S_o}{N_o} &= \frac{S_i g}{N + N_i g} = \frac{S_i g}{N + kT_iB_N g} \\ &= \frac{S_i g}{kgB_N\left(T_i + \dfrac{N}{kgB_N}\right)} = \frac{S_i}{kB_N(T_i + T_e)}\end{aligned} \tag{5.37}$$

where

$$T_e = \frac{N}{kgB_N} \tag{5.38}$$

where T_e is defined as the equivalent noise temperature of the two-port network.

In the above equation, output noise power appears to originate from a thermal noise source at a temperature of $T_i + T_e$. Hence, the two-port device itself can be modeled as noise free and accounts for the noise generated within the two-port device by assigning the input noise source a new temperature higher than T_i by the amount T_e. The output SNR from the above equation is

$$\frac{S_o}{N_o} \triangleq \frac{S_i}{kB_NT_i(1 + T_e/T_i)} = \left(\frac{S}{N}\right)_i \frac{1}{1 + T_e/T_i} \tag{5.39}$$

As $T_e/T_i \rightarrow 0$, the output signal quality is the same as the input signal quality as measured by the SNR at the input. Thus $(1 + T_e/T_i)$ is a factor by which the output SNR is reduced due to internal noise.

Let $1 + T_e/T_i = F$. Then the output SNR can be expressed as

$$\left(\frac{S}{N}\right)_o = \frac{S_i}{N_i}\frac{1}{F} = \left(\frac{S}{N}\right)_i \frac{1}{F}$$
$$F = \frac{\frac{S_i}{N_i}}{\left(\frac{S}{N}\right)_o} = \frac{\left(\frac{S}{N}\right)_i}{\left(\frac{S}{N}\right)_o} \tag{5.40}$$

Thus, the noise figure of the networks in cascade provides an estimate of input SNR reduction due to the two-port network. If F is 0 dB, then $(S/N)_i = (S/N)_o$ is an ideal case to achieve.

In addition, F can be considered as the actual noise temperature divided by the output noise temperature if the two-port network were noiseless. The concept of noise figure also applies to lossy elements such as attenuators, transmission lines, and so forth. These devices are represented by power loss rather than gain, and from (5.37) we see $N_o = kgB_N(T_i + T_e)$, if $g = 1/L$ (for lossy network). Then $N_o = (k/L)B_N(T_i + T_o)$. If the input temperature of the network T_e is assumed to be 270° k $= T_o$, then

$$N_o = \frac{k}{L} B_N(T_i + T_o) \tag{5.41}$$

5.7.3 Network in Cascade

Consider the two-port networks in cascade, each having the same noise bandwidth, B, but with different available power gains and noise figures. Let the effective noise temperatures of the two networks be T_{e1} and T_{e2}, and the driver source temperature be T_o, as shown in Figure 5.39. Thus,

- Noise power at the output due to driver temperature $T_o = BkT_o g_1 g_2$;
- Noise power at the output due to $T_{e1} = BkT_{e1} g_1 g_2$;
- Noise power at the output due to $T_{e2} = BkT_{e2} g_2$;

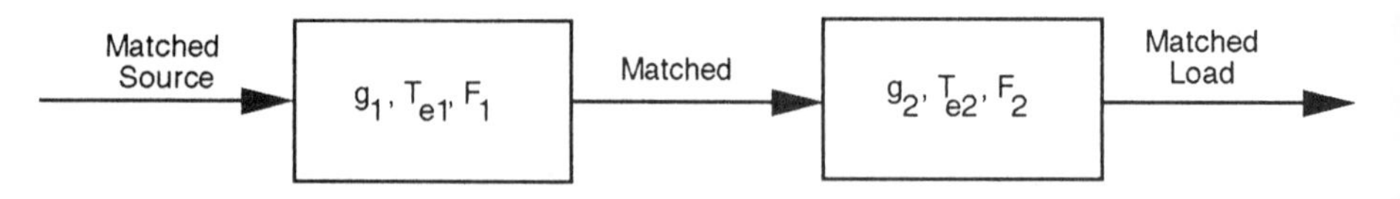

Figure 5.39 Network in cascade.

- The total noise power at the output is $N_o = BkT_o g_1 g_2 + BkT_{e1} g_1 g_2 + BkT_{e2} g_2$;
- Noise power due to two network $= Bkg_2\,(T_{e2} + g_1 T_{e1})$.

Let the effective noise temperature at the input be T_{e12}. Then

$$T_{e12} = \frac{Bkg_2(T_{e2} + g_1 T_{e1})}{g_1 g_2 kB} = T_{e1} + \frac{T_{e2}}{g_1} \tag{5.42}$$

Equation (5.42) can be extended to n networks in cascade where the equivalent noise temperature can be given by

$$T_{e12\ldots n} = T_{e1} + \frac{T_{e2}}{g_1} + \frac{T_{e2}}{g_1 g_2} + \cdots + \frac{T_{en}}{g_1 g_2 \cdots g_{n-1}}$$

Since $F \triangleq 1 + T_e/T_o$, we obtain

$$T_o(F_{12\ldots n} - 1) = T_o(F_1 - 1) + \frac{T_o}{g_1}(F_2 - 1) + \cdots + \frac{T_o(F_n - 1)}{g_1 g_2 \cdots g_{n-1}}$$

or

$$F_{12\ldots n} = F_1 + \frac{F_2 - 1}{g_1} + \cdots + \frac{F_n - 1}{g_1 g_2 g_3 \cdots g_{n-1}} \tag{5.43}$$

This equation is known as the Friis formula and it implies that the overall noise figure of n stages is highly dependent on the first stage noise figure F_1. Therefore, the front-end of the receiver should contain a low noise figure amplifier (low F) and should be of high gain (g_1). In this situation, the succeeding stages will not contribute much noise to the receiver input. We now illustrate this theory with several examples.

Example 5.6

For the mobile receiver shown in Figure 5.40, find the overall noise figure of the system looking into the bandpass filter. Assume the filter inband loss is L_1, low-noise amplifier has a gain and noise figure of G_1, and F_1. Power is divided by a

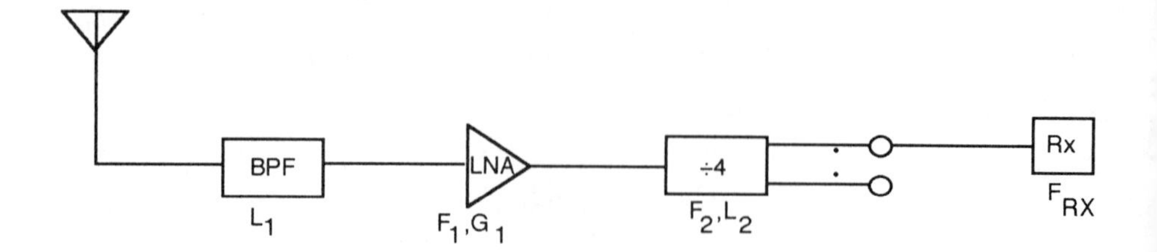

Figure 5.40 Front-end setup of a typical cell-site receiver.

factor of four and has the loss and noise figure of L_2 and F_2. The noise figure of the mobile receiver is F_{RX}. The overall noise figure of the mobile receiver is

$$
\begin{aligned}
F &= L_1 + (F_1 - 1)L_1 + \frac{F_2 - 1}{G_1} L_1 + \frac{(F_{RX} - 1)}{G_1} L_1 L_2 \\
&= F_1 L_1 + \frac{(F_2 - 1)}{G_1} L_1 + \frac{(F_{RX} - 1)L_1 L_2}{G_1}
\end{aligned}
\tag{5.44}
$$

Taking some typical values for UHF: $F_1 = 3.5$ dB, $L_1 = L_2 = 1$ dB, $G_1 = 16$ dB, $F_2 = 5.2$ dB, $F_{RX} = 10$ dB, the equivalent noise figure $F = 5.12$ dB.

Example 5.7

The configuration shown in Figure 5.41 is for cellular rural statistical area receiver channel. Find the overall noise figure of the system. The total noise figure, NF_{TOT} is

$$
10 \log_{10} \left[\begin{aligned} & 10^{0.1} + (10^{0.06} - 1)10^{0.1} + \frac{(10^{0.35} - 1)10^{(0.1+0.608)}}{10^{1.9}} \\ & + 10^{0.608} \frac{10^{0.1}}{10^{1.9}} + (10^{1.41} - 1) \frac{10(0.1 + 0.608 + 1.46)}{10^{(2.3+1.9)}} + \frac{10^{(1.46+0.1+0.608)}}{10^{(2.3+1.9)}} \end{aligned} \right]
$$

$$
= 10\log_{10}[10^{0.1} + (10^{0.06} - 1)10^{0.01} + (10^{0.35} \times 10^{0.708})/(10^{1.9}) + (10^{(1.41+2.168)}/(10^{(2.3+1.9)}]
$$

$$
= 10 \log_{10}[1.82818] = 2.6 \text{ dB}
$$

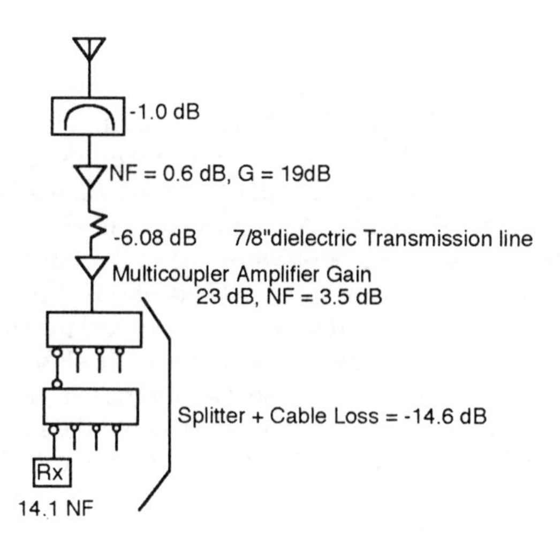

Figure 5.41 Tower-mounted amplifier receiver system.

5.8 INTERMODULATION [11]

RF intermodulation (IM) is an area that must be addressed when discussing the antenna system. IM products caused by transmitters can be avoided with proper combiner design. However, some nonlinear circuits, such as a loose contact of different metals at the receiver antenna system, may generate IM interference falling in the receive band due to transmitter power impinging on the loose contact. The basic source of intermodulation is impinging RF energy on severe corrosion of the metallic joints. Therefore, it is important to consider proper protection of the metal surfaces, including connectors and mounting hardware.

One method of representing a nonlinear device is by means of polynomial transfer characteristics:

$$Y = C_o + C_1x + C_2x^2 + \cdots + C_nx^n \tag{5.45}$$

where the input or excitation parameter is x, the response or the output is Y, and $C_o, C_1, \ldots, C_n$ are constant coefficients. For a linear system, every term of the series is zero except C_1. In general, for a practical system the dc term $C_o = O$ and $C_1 > C_2 > \cdots > C_n$. In this case, (5.45) can be written in the closed form as

$$Y = \sum_{i=0}^{n} C_i x^i \tag{5.46}$$

where n is the degree of the polynomial

Intermodulation in Cellular Systems

Spectral distribution of two carrier frequencies that are close to each other and passing through a nonlinear element is shown in Figure 5.42. Rather than using actual valves, amplitudes of the higher order terms are shown to decrease in value and act as an aid to understanding. As represented in the diagram, even-order products are far away from the carriers, while the odd-order products of the form $2f_1 - f_2$ or $2f_2 - f_1$ appear close to f_1 and f_2. This observation can be generalized to show that for a narrowband system where excitation frequencies are close to each other, odd-order intermodulation product terms lie close to the original band (inband). These inband products are of special interest to us in cellular systems where the cell-site transmitter products can affect either the mobile receiver or the return path of the cell-site receiver itself. Therefore, the cell-site transmitter frequency assignment should be such that the resulting intermodulation frequency does not fall in the receive band.

A typical cellular system with transmit and receive bandwidth of F_B, channel frequency separation of F_C, and guardband of F_G is shown in Figure 5.43. A higher frequency band is assigned for transmitting, while the lower frequency band is for receiving the incoming signal.

The intermodulation interference range is shown in Figure 5.44. Here, the guardband is 20-MHz wide. Generally, the odd order intermodulation products appear close to the carrier. Thus, third, fifth, seventh, and ninth-order products will appear around the transmit band with increasing width. Due to guardband third-order intermodulation, products probably will not appear in the receiver band. However, higher order intermodulation products, such as fifth and seventh, may lie in the receiver band. These intermodulation products will increase the noise floor of the receiver, causing a degradation of the desired signal performance.

We will consider below a typical cellular system transmitter assignment scheme and arrive at the general equation for the odd-order intermodulation products. Then we will impose constraints such that the intermodulation products do not coincide with the cell-site receive carrier frequencies. Assuming m channels to be assigned to a typical cell site, then the transmitting and receiving channel frequencies can be written as

$$T_K = T_1 + (k - 1)F_C \tag{5.47}$$

where T_1 is the center frequency of the first channel assigned to a specific cell site, F_C is the channel separation (30 kHz in the U.S.), and $(k - 1)$ is the channel separation factor. Thus, if the second assigned channel is separated by ten channels, then the factor $(k - 1)$ assumes the value 9.

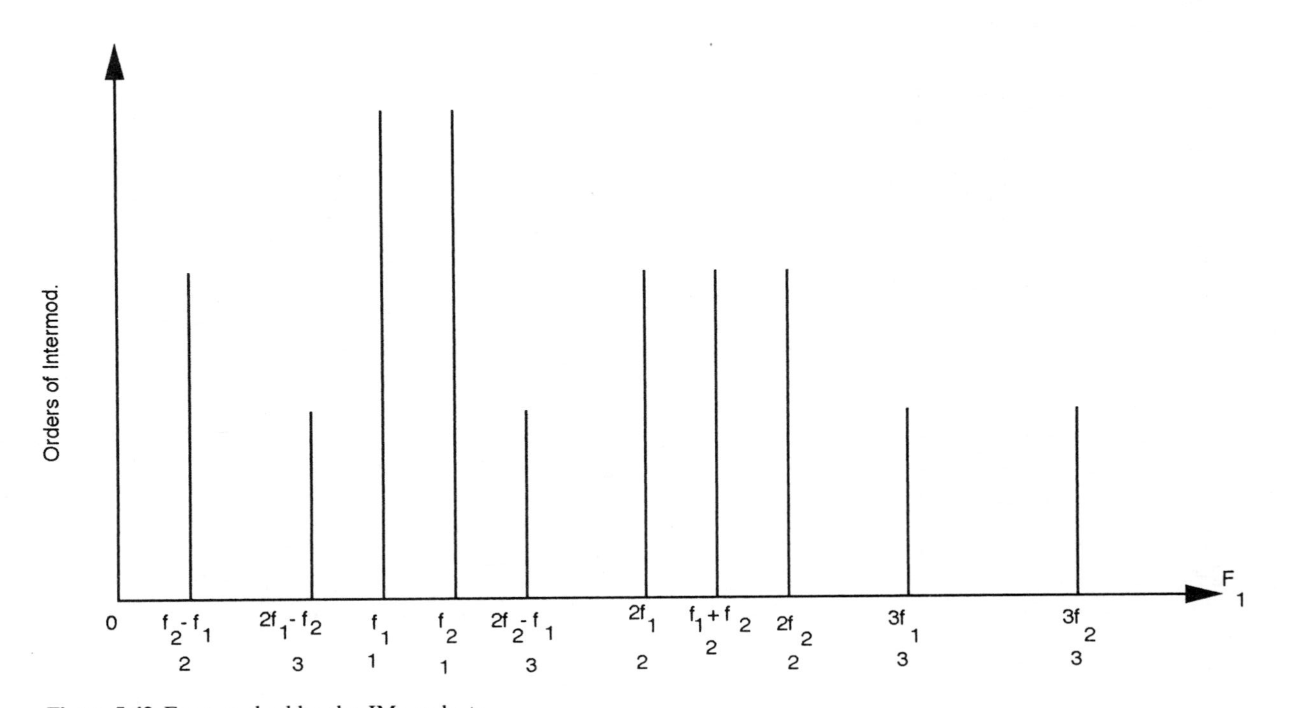

Figure 5.42 Even- and odd-order IM products.

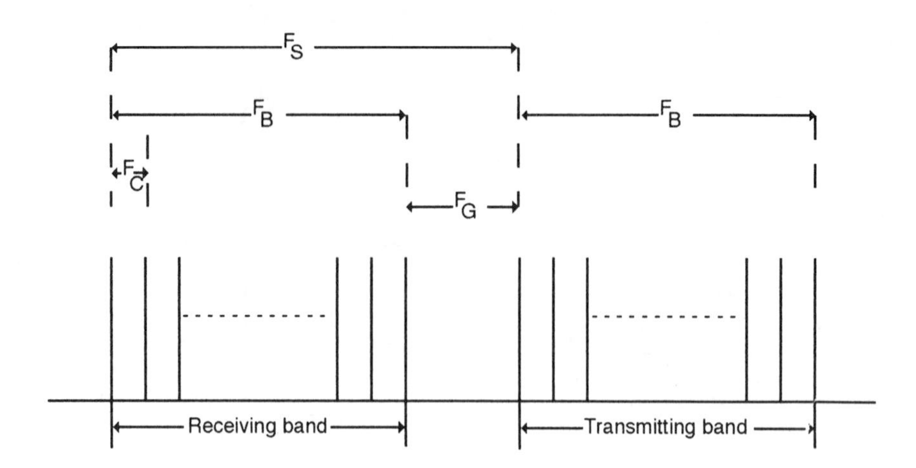

Figure 5.43 Frequency layout of cellular radio.

The corresponding receiving frequency R_x is

$$R_x = T_K - F_S = T_1 + (k - 1)F_C - F_S \tag{5.48a}$$

The intermodulation product's frequencies for third-order intermodulation products at the transmitter output can be computed as

$$\begin{aligned} 2T_m - T_n &= 2[T_1 + (m - 1)F_C] - [T_1 + (n - 1)F_C] \quad m \neq n \\ &= T_1 + (2m - n - 1)F_C \end{aligned} \tag{5.48b}$$

The fifth-order intermodulation products can be expressed as

$$3T_m - 2T_n = T_1 + (3m - 2n - 1)F_C \tag{5.48c}$$

Similarly, the seventh-order intermodulation products can be expressed as

$$4T_m - 3T_n = T_1 + (4m - 3n - 1)F_C \tag{5.48d}$$

Thus, the general expression of intermodulation products can be written as

$$T_1 + lF_C \tag{5.48e}$$

where l is a function of m and n and is an integer. The condition that the frequency of intermodulation does not fall at the carrier of some other channel in the receiver

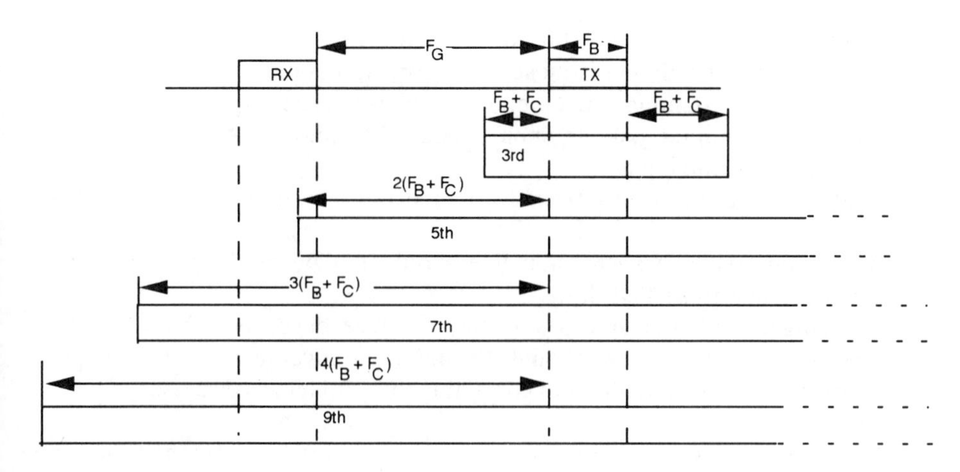

Figure 5.44 Odd-order intermodulation products.

band can now be easily arrived at by equating (5.48e) to (5.48a): $T_1 + lF_C = T_1 + (k - 1)F_C - F_S$ or $l = (k - 1) - F_S/F_C$. The condition that l can only be an integer is satisfied if the ratio F_S/F_C is an integer. Therefore, so long as F_S/F_C is not an integer, there is no possibility that the intermodulation product's frequency will exactly coincide to the receive channel carrier frequency. We can illustrate the above concept with an example.

Example 5.8

For a 16-channels/cell system, find the value of F_S/F_C assuming the channel separation to be 30 kHz and the total number of channels to be 333. Maximum spacing for a 16-channel per cell system with 30-kHz channel spacing is determined as follows. The number of cells = 333/16 ≅ 21. Therefore, the frequency separation = 21 × 30 = 630 kHz. Thus, F_S/F_C = 45,000/630 = 71.43. Therefore, this ratio is not an integer and the IM products cannot fall on the carrier frequencies of receiver channels.

5.9 CONCLUSIONS

The following conclusions can be made.

- The use of corner reflectors at the cell site reduces the system interference.

- A unity gain antenna is most useful for densely populated urban areas, while a 5-dB collinear antenna is most useful for rural applications.
- The best place to mount the mobile antenna is the roof top of an automobile.
- For a well-mounted glass antenna, a negligible amount of RF energy is lost as it passes through glass.
- Multicouplers that have power dividers at the output are used for diversity combining.
- A braided coaxial transmission line should not be used unless absolutely necessary due to its high loss.
- A high-gain antenna at cell site may produce dead spots because of the presence of minor lobes, though the far field coverage may be adequate. There are several techniques of correcting this problem as discussed in Section 5.6.1.
- Due to the presence of several antennas at the cell site, antenna mounting is important. Composite gain variation of the antenna system is a function of frequency of operation and the antenna spacing.
- Often a compromise is sought in the design of low noise figure amplifiers and the high intercept point.

PROBLEMS

5.1 Find the antenna gain having half-power beamwidth of 45 deg in both azimuth and elevation planes.

5.2 Provide reasons in favor or against the following statements:
a. 0-dB mobile antenna gain can be used in urban areas.
b. 3-dB mobile antenna gain can be used in suburban areas.
c. 5-dB mobile antenna gain can be used in rural areas.

5.3 For the configuration shown in Figure 5.45, find the values of D_{min} and D_{max} when the tower heights are 1,000 ft and 2,000 ft. Downtilt angles are 6 deg and 8 deg, respectively. Assume the antenna 3-dB beamwidth to be 10 deg.

5.4 For the configuration shown in Figure 5.46, find the SNR at the receiver input where L_o and L_1 are the front-end filter and the power divider loss. Power division is by a factor n and the receiver is connected to port 1 of the PD, which has a noise figure F_2. You can assume the antenna to be at a temperature of $T_o = 290°K$

5.5 The figure in problem 5.4 is modified by inserting an amplifier and a resistive network in each arm of the receiver, as shown in Figure 5.47. Find the SNR at the receiver input when L_o and L_1 are the filter and power divider attenuation. F_1 and F_2 are the amplifier and the receiver noise figures. You can assume the antenna to be at a temperature of $T_o = 290°K$.

5.6 For a 32-channel combined system, find the value of F_s/F_c and conclude whether the IM products can fall at any of the receive channel's carrier frequencies or not.

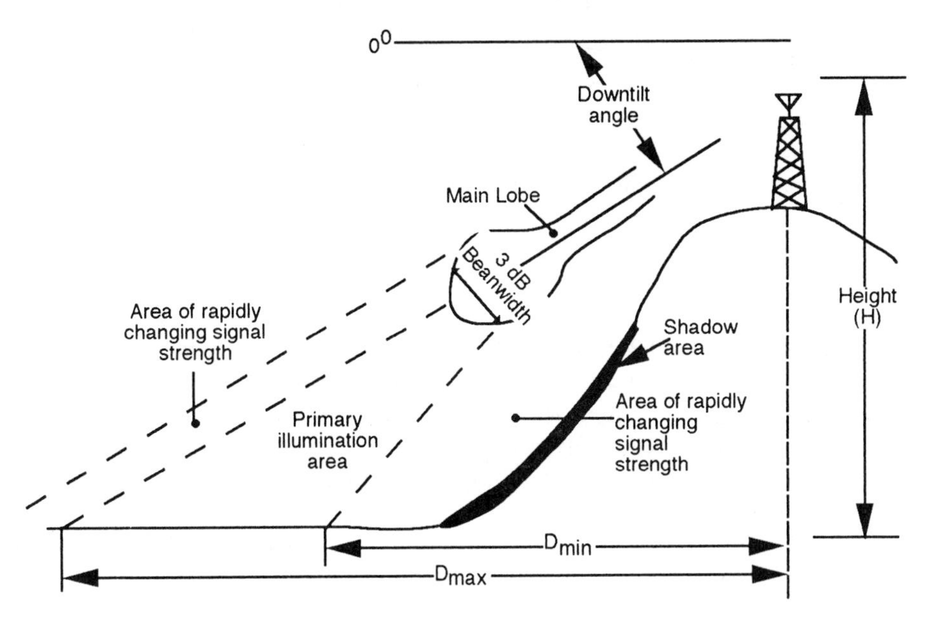

Figure 5.45 Downtilting the antenna to cover some shadow areas.

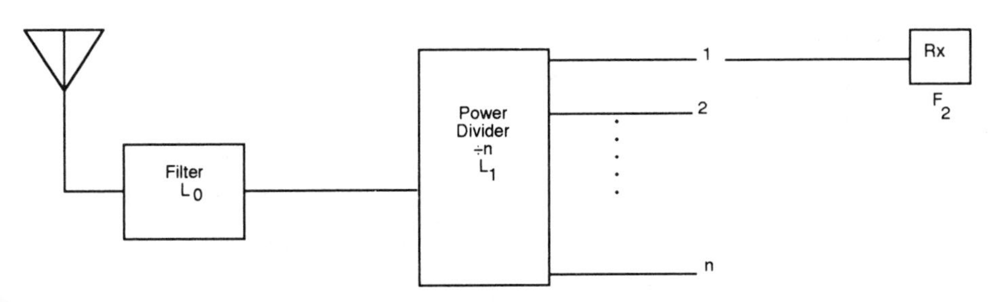

Figure 5.46 Receiver front end.

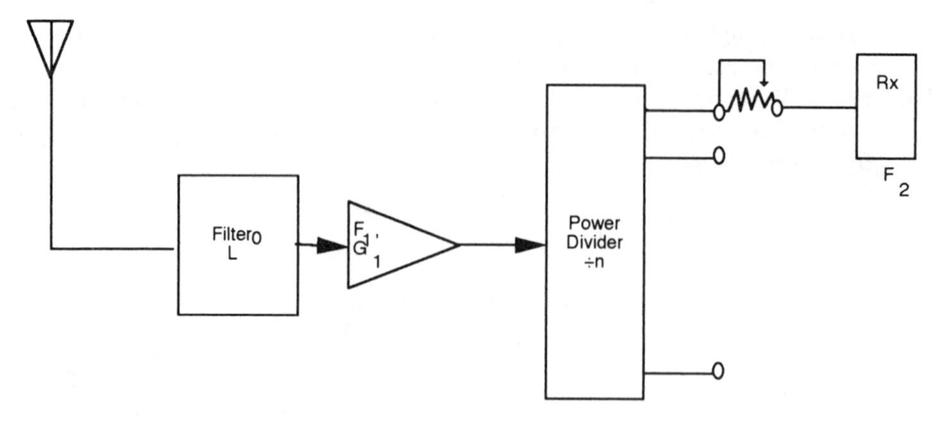

Figure 5.47 Receiver front end with amplifier and power divider.

5.7 Find 17, 19, 21, and 23rd-order IM product frequencies if two cell-site active transmitters are operating at 861.4625 MHz and 865.4625 MHz. Decide whether these IM frequencies fall at the receiver channel carrier frequency or not. Assume 16 channels per system.

5.8 For the configuration shown in Figure 5.48, find the SNR at the second amplifier output.

5.9 The RSA transmitter antenna system is shown in Figure 5.49. Limiting the ERP to 500W, compute the actual transmitter output power.

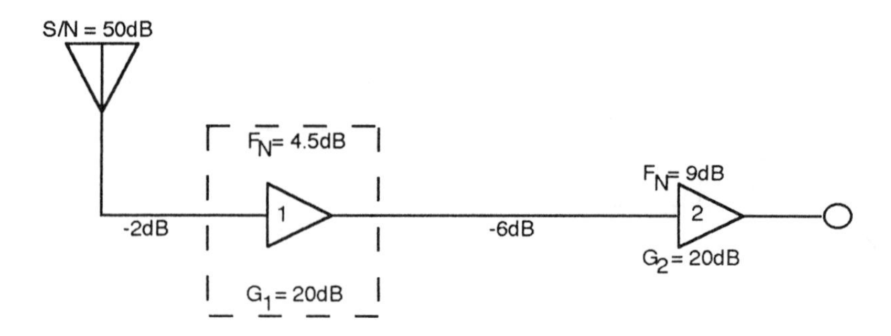

Figure 5.48 Front end of a receiver.

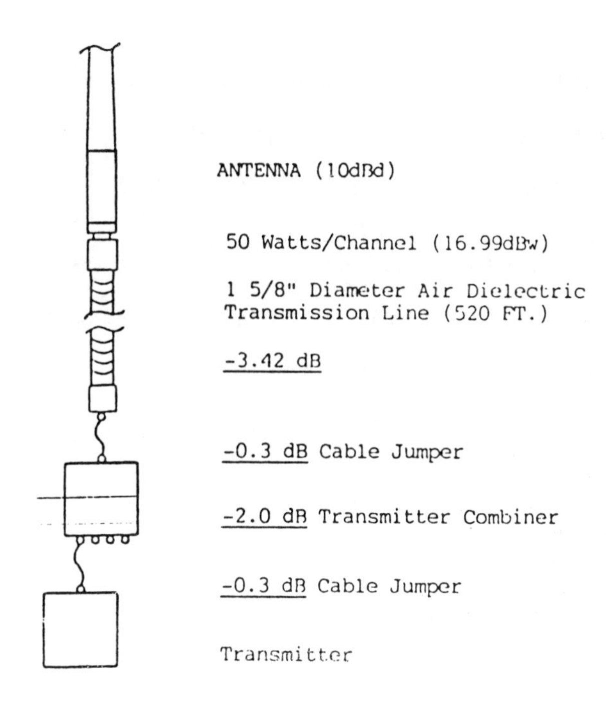

Figure 5.49 Typical RSA setup.

REFERENCES

[1] Tilston, W. V. "On Evaluating the Performance of Communications Antennas," *IEEE Communications Society magazine*, Sept. 1981.

[2] Andrew Corporation. Antenna Systems Catalog 32, 1983.

[3] Antenna Specialists Co. "Base Station, Mobile and Portable Antennas and Accessories," 1984-1985.

[4] Telewave, Inc. Antenna Systems 25 MHz-1,600 MHz.

[5] Harris, O., and P. Blerins. "Making Right 800-MHz Antenna Choice Requires Analysis for Top Performance," *Communication News,* June 1983.

[6] Wong, J. L. and H. E. King. "Wide-Band Corner-Reflector Antenna for 240 to 400 MHz," *IEEE Trans. on Antennas and Propagation*, Vol. AP-33, No. 8., Aug. 1985.

[7] Horn, D. W. "Vehicle-Caused Pattern Distortion at 800 MHz."

[8] Tesch, R. L. "Measured Vertical Antenna Performance," *IEEE Trans. on Vehicular Technology*, Vol. VT-34, No. 2, May 1985.

[9] EN830508, "Study Report for Transmitter Combiner Design," Matsushita Communication Industrial Co., Ltd.: Yokohama, Japan, May 1983.

[10] Lee, W.C.Y. "Antenna Spacing Requirements for a Mobile Radio Base-Station Diversity," *The Bell System Technical Journal*, Vol. 50, No. 6, July–Aug. 1971.

[11] EN830503, "Study Report for Intermodulation Interference in the Same Radio System" Matsushita Communication Industrial Co., Ltd.: Yokohama, Japan, May, 1983.

[12] Jakubowski, R. J. "Determining Antenna Gain," *Cellular Business*, Aug. 1988.
[13] Chenoweth, J. "Expand Your Coverage Area With 5 dB Gain Antennas," *Mobile Radio Technology*, Sept. 1991.
[14] Blaese, H. "Cellular Antenna Theory With the Installer in Mind," *Mobile Radio Technology*, Jan. 1991.
[15] Johnson, A. K., and R. Myer. "Linear Amplifier Combiner," *IEEE Vehicular Technology Conference*, 1987.
[16] Singer, A. " Combine Transmitters With Little or No Loss," *Mobil Radio Technology*, Sept. 1991.
[17] Singer, A. " Use Downtilt Selectively to Improve System Coverage," *Mobile Radio Technology*, Nov. 1990.
[18] Clark, J. "Cellular Rural Statistical Area (RSA) Transmit and Receive Combining," *IEEE Vehicular Technology Conference*, 1989.
[19] Sagers, R. C. "Intercept Point and Undesired Responses," *IEEE Vehicular Technology Conference*, 1982.
[20] Kraus, J. D. *Antennas*, McGraw-Hill Book Co.: New York, NY 1988.
[21] Dipizza, G. C., A. Plitkins, and G. I. Zysman. "The Cellular Test Bed," *The Bell System Technical Journal*, Jan. 1979.
[22] Wickline, W. A., "Computer Simulation Provides Accurate Propagation Prediction." *IEEE Vehicular Technology Conference*, Washington, D.C., 1981, pp. 183.
[23] Davidson, A. L., "Omnidirectional Transmitter Combining Antenna," *IEEE Vehicular Technology Conference*, San Diego, 1982, p. 295.
[24] Bryson, W. B., "Antenna Systems for 800 MHz," *IEEE Conference on Vehicular Technology*, San Diego, May, 1982, p. 287.

Chapter 6
Cochannel and Adjacent-Channel Interference

6.1 INTRODUCTION

As discussed in earlier chapters, the utility of cellular radio lies in the frequency reuse concept. Channels used at one cell site are repeated at other cell sites, thereby increasing the capacity of the system. However, these channels can only be repeated if the separation between cochannel cells is kept high, such that the cochannel interference is not objectionable. In order to obtain a tolerable value of cochannel interference, the system designer has to maintain a minimum separation distance to the cochannel site. The performance of the system is judged by a factor known as the D/R ratio. Other than the frequency reuse factor D/R, the cochannel interference also depends upon the number of cells N in the cluster and the number of cochannel cells. For a normal cellular system, cochannel frequencies are spaced so far apart geographically that when the desired signal is off, the cochannel station is not expected to introduce a sufficient signal to open the squelch of the receiver. Since the analog cellular system is FM modulated, one can take advantage of the capture effect property of FM receivers, whereby the desired signal, if stronger by a given factor (known as capture ratio), takes control of the receiver and suppresses the weak signals. In order to make the FM receiver work in the presence of cochannel interference, the designer must understand and manage this interference rather than avoid it completely. Complete avoidance of interference simply means no repetition of frequencies, which destroys the basic concept of a cellular system. Therefore, the system has to operate, perhaps in the best possible way, in the presence of cochannel interference.

Similar to the cochannel cells that are geographically separated by a distance D, the adjacent-channel interference is the function of a specific channel assignment scheme. In cellular systems, cochannel interference is the prime source of noise, and in that sense the system is interference limited rather than noise limited. Unlike cochannel interference, which falls in the receiver inband, the adjacent-channel interference results from splattering of the modulated RF signal into neighboring channels. The degree to which this form of secondary interference constrains the system performance is dependent on the spacing between adjacent channels, the selectivity of the receiver, and the statistically dependent desired carrier level to the adjacent-channel carrier level. It should be noted that the adjacent channel here is not the adjacent channel in a strict communication sense, but rather it is the nearest assigned channel in the same cell and can be several channels apart. The true adjacent-channel interference can be considered a secondary source of interference. The controlling factor for both cochannel and adjacent-channel interference is the specific channel assignment scheme used by the designer. In view of the above discussion, the objective of this chapter is to develop the basic theory of cochannel and adjacent-channel interference along with the possible techniques for their reduction. In Section 6.2, we develop the dependence of the carrier-to-interference ratio (CIR) on the D/R ratio and the groundwave propagation exponent n. We also derive the receiver probability of cochannel interference based on the receiver capture ratio. Lastly, different methods of reducing cochannel interference are discussed. In Section 6.3, we develop the model of adjacent-channel interference along with techniques of controlling this, which are mainly based on frequency management and receiver selectivity. In Section 6.4, we discuss the interference produced within a specific cell due to the closeness of the interfering mobile with respect to the desired mobile from the cell site. We summarize the findings of this chapter in Section 6.5.

6.2 COCHANNEL INTERFERENCE

The objective of cellular systems is to provide an RF cochannel CIR of 18 dB or greater for over 90% of the coverage area. The number 18 dB is arbitrary and is based on the fact that at this value of carrier-to-noise ratio (CNR), 75% of all users consider the voice quality to be excellent or good on a 30-kHz wide FM channel in a multipath environment. It is assumed that the following factors are accounted for in our present discussion:

- Signal fade is associated with Rayleigh fading.
- Distance of each cochannel interfering station with respect to the base station is known.
- Propagation laws follow either Okamura, Longley-Rice, Carey, Bullington, or other similar models.
- A cell site's radiated power must be adjusted for terrain profile and thus for the antenna height above the average terrain.
- The activity factor of the interfering transmitters that are operated for less than 100% of the time during the busy hour has to be considered.
- Cochannel interference is also a function of traffic. There is a higher probability of cochannel interference during the peak traffic hours of the day than during the off-peak hours.

The ratio of cochannel cell site distance to the cell radius, R, is given by

$$D/R = \sqrt{3N} \tag{6.1}$$

where $N = i_2 + j_2 + ij$ for hexagonal cells, i and j are nonnegative integers, and N is the number of cells per cluster. Figure 6.1 shows the value of this ratio in terms of R for the first ring of surrounding cochannel cells. Note that for $N = 4$,

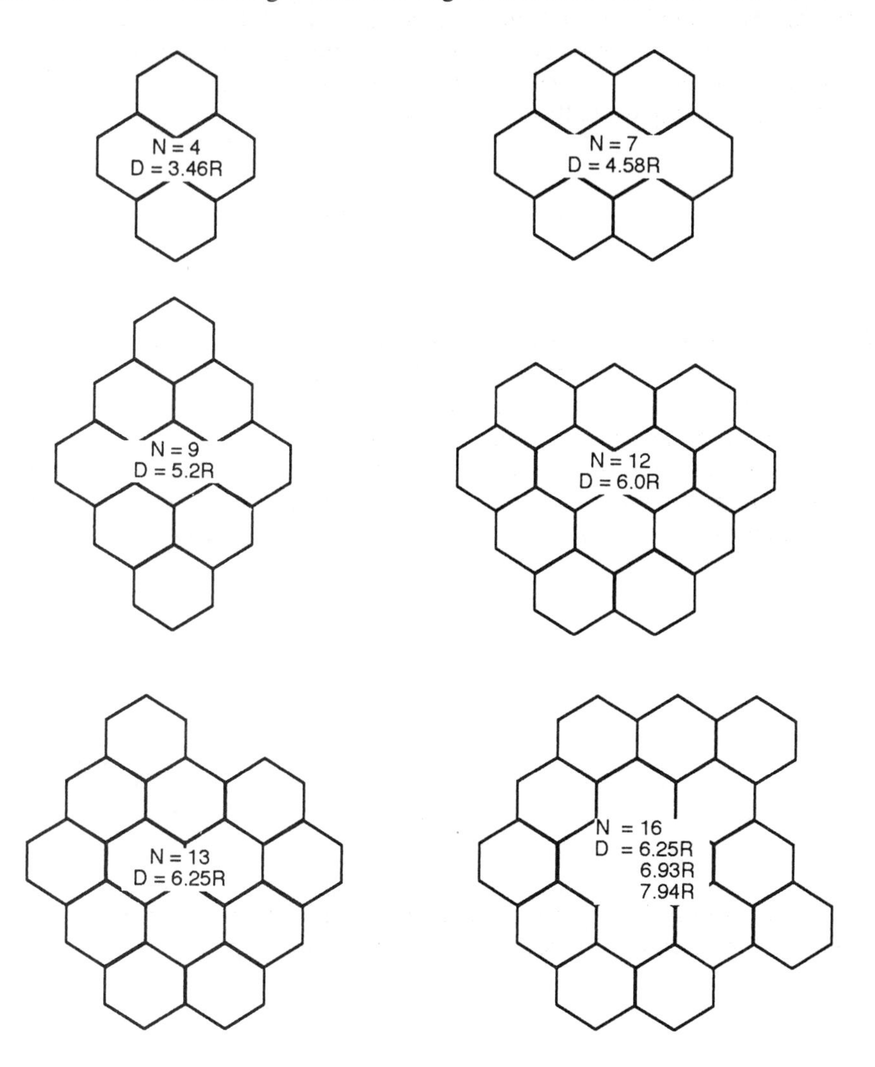

Figure 6.1 Cell structure for $N = 4, 7, 9, 12, 13$, and 16 along with their D/R ratios.

7, 9, 12, and 13, this ratio is unique. However, due to the asymmetry for $N = 16$, this ratio is different along three axes [1]. Thus, once N is known, the distance to the cochannel cells can be uniquely determined.

The seven-cell configuration ($N = 7$) for $i = 2$ and $j = 1$, and with anti-clockwise location of the cochannel cells is shown in Figure 6.2. If the same channel is simultaneously active in all six first tier of cells, the cell site not only receives the desired-channel signal from the mobile but also receives the interfering signals from undesired mobiles located in the cochannel cells [2]. Let us assume that the second set of six cochannel cells in the second tier of cells introduces negligible interference and the omniantennas are used in all cells. Then the desired *CIR* at the cell site, from Figure 6.3(a), is given by

$$\frac{C}{I} = \frac{S_m}{N_c + \sum_{i=1}^{6} I_i} \geq 18 \text{ dB} \tag{6.2}$$

where S_m is the desired mobile signal power received at the cell site, N_c is the noise power at the receiving cell site, and I_i is the cochannel interfering power received at the cell site due to ith interfering cell. 18 dB is the desired value of the CIR, which has been established by subjective tests.

Assume that the distances to the desired center cell site are approximately the same for the mobiles located at different cochannel cells and the transmitted powers of all mobiles are the same. Then (6.2) can be approximated as

$$\frac{S_m}{N_c + 6I} \geq 18 \tag{6.3}$$

where

$$\sum_{i=1}^{i=6} I_i \approx 6I$$

If the noise power N_c is far lower than the interference power $6I$; that is, the system is cochannel interference limited, then (6.3) is approximated as

$$\frac{S_m}{6I} \geq 18 \tag{6.4}$$

The situation that the interfering mobile units are on the nearest fringe of the

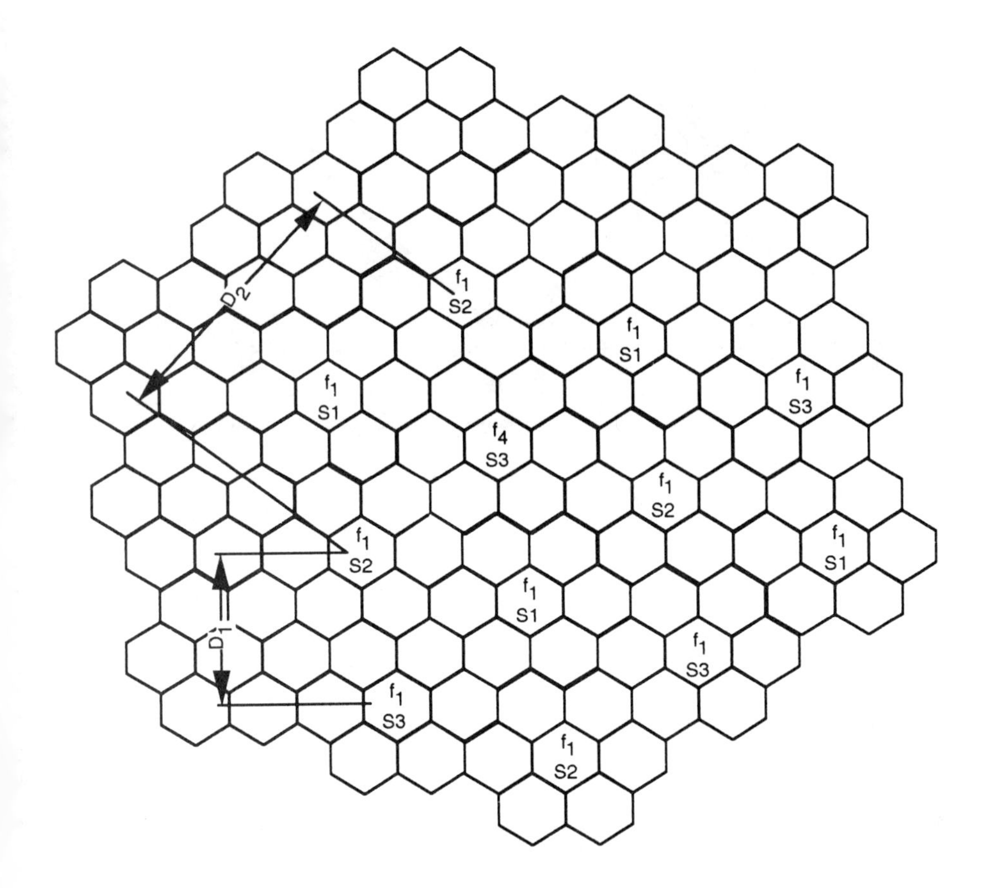

Figure 6.2 Arrangement of seven-cell blocks to cover a large area.

neighboring six cochannel zones transmitting simultaneously can be regarded as the worst case. In this situation, the relation between S_m and I is given by

$$\frac{S_m}{6I} = \frac{1}{6} \frac{R^{-n}}{(D - R)^{-n}} \geq 18 \text{ dB} \tag{6.5}$$

It is assumed here that the local mean of the received signal at the cell site is proportional to the inverse exponent, the nth power of the distance between the fixed cell site and the moving vehicle.

Since S_m and I are the desired RF and the undesired interfering power at the cell site, the received CNR at the cell site is given by

$$\frac{C}{I'} = \frac{S_m}{6I} = \frac{1}{6}\left(\frac{D - R}{R}\right)^n \geq 18 \text{ dB} \tag{6.6}$$

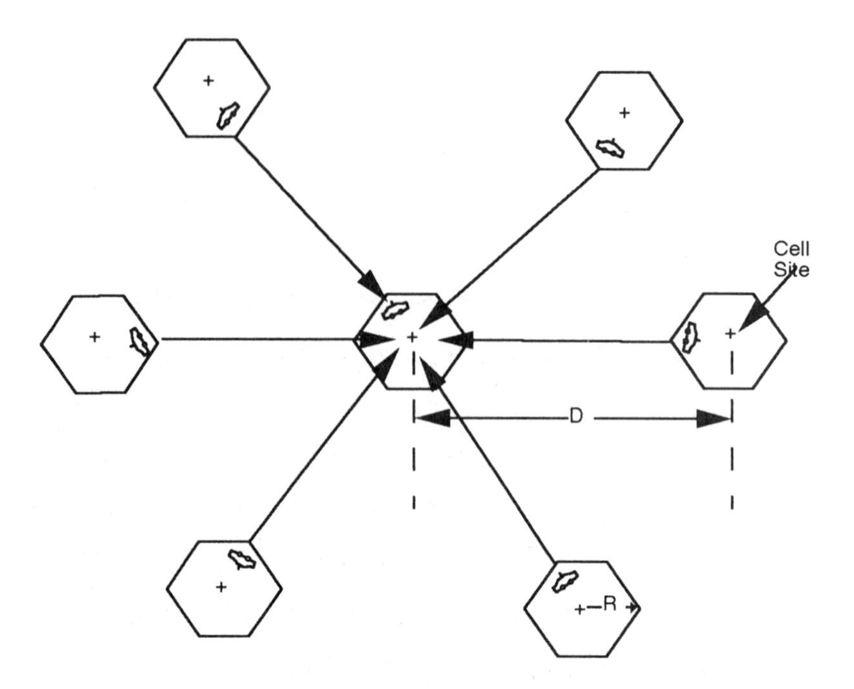

Figure 6.3(a) Interference at the cell site due to first tier of six cochannel mobiles.

where $C = S_m$ = desired carrier power and $I' = 6I$ = undesired interference power. Considering the worst case of equality, the above equation can be rewritten as

$$\left(\frac{D}{R} - 1\right) = \left(6\frac{C}{I'}\right)^{1/n} = (6.10^{1.8})^{1/n} \tag{6.7}$$

Assuming the terrain to be such that the value of the exponent n is 4, we have $D/R = 5.4$. It should be noted that this is the worst case D/R ratio considering the worst location of all the six interfering mobiles at the boundary of their respective cells closest to the cell site under consideration. In general $(D - R)$ in the numerator of (6.6) can be approximated by D and the D/R ratio in the normal case comes out to be 4.6. This is shown in Figure 6.1(b).

Similar to the above case, if we compute the desired carrier and the interference power at the mobile due to six cochannel cells, as shown in Figure 6.3(b), then (6.2) can be modified as

$$C/I = S_c/\left(N_m + \sum_{i=1}^{6} I_i\right) \geq 18 \text{ dB} \tag{6.8}$$

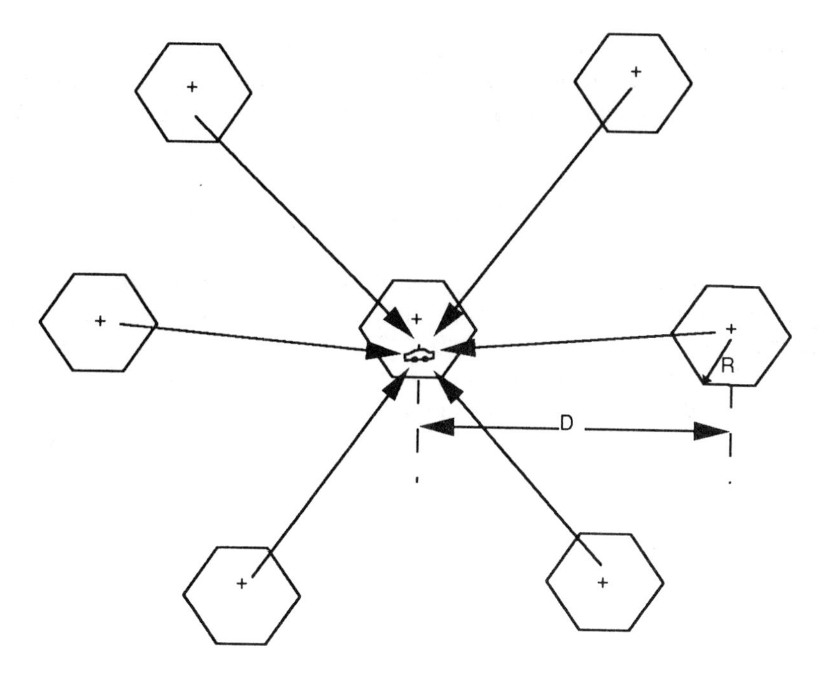

Figure 6.3(b) Interference at the mobile due to first tier of six cochannel cells.

where N_m is the noise power at the mobile receiver, S_c is the desired cell site power received at the mobile, and I_i is the interference power at the mobile due to the ith cochannel cell. The above equation is only approximately true as the interfering power from the second and higher tiers of cochannel cells are neglected. Assuming the noise power to be considerably lower than the cochannel interference power at the mobile, the C/I power ratio can be approximated as

$$C/I' = S_c/6I \geq 18 \text{ dB} \tag{6.9}$$

where

$$I = \sum_{i=1}^{6} I_i$$

Since S_c and I are the desired RF carrier and the undesired interference power at the mobile, the received CNR at the mobile is given by

$$C/I' = \frac{S_c}{6I} = 1/6[R/(D - R)]^{-n} \geq 18 \text{ dB} \tag{6.10}$$

This is the worst case CIR and assumes that the mobile is at the boundary of the desired cell and at a distance of $(D - R)$ from the undesired cells. Simultaneously having $(D - R)$ distance from all the cochannel cells is a physical impossibility. Thus,

$$(D/R) - 1 = (6C/I')^{1/n} = (6 \times 10^{1.8})^{1/n} \tag{6.11}$$

Thus, $D \neq R$ equals 5.4 for $n = 4$ or based on a 40-dB per decade value of radio propagation exponent n. Again, for the large value of D/R, the left-hand side of the equation can be approximated as D/R. The path loss variation model is shown in Figure 6.4. Once we know the ratio, the actual distance D to the cochannel cell can be computed for a given cell of radius R. In other words

$$D = R[(6C/I')^{1/n} + 1] \tag{6.12}$$

Making use of (6.1), (6.12) becomes

$$N = 1/3[(6C/I')^{1/n} + 1]^2 \tag{6.13}$$

From (6.13), it can be seen that for low values of the CIR, the value of N, the

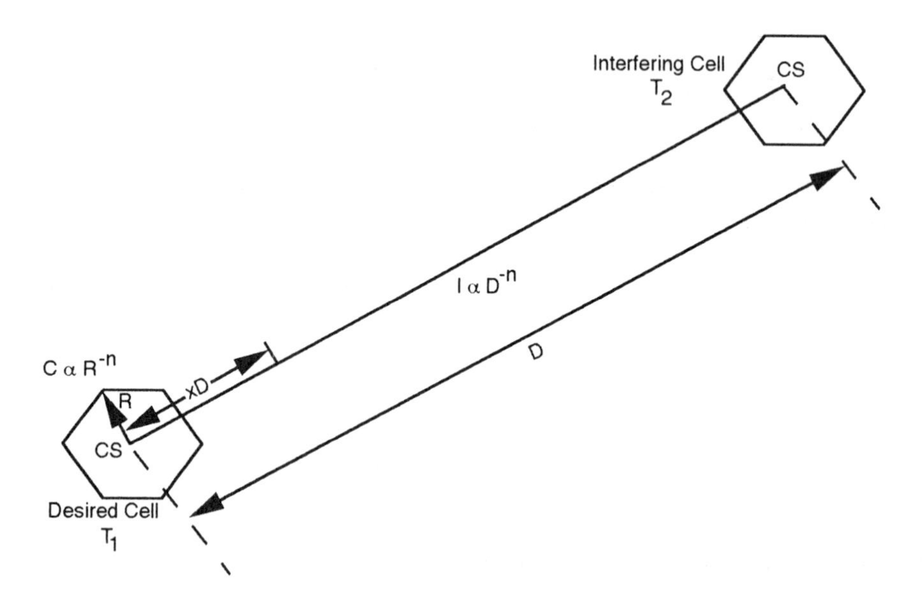

Figure 6.4 Path loss variation model for a cellular system in a straight-line path between two cells.

number of cells per cluster, is reduced. This implies a larger number of channels per cell, and thus a large traffic capacity.

Assuming $n = 4$, (6.13) is reduced to

$$N = 1/3[(6C/I')^{1/4} + 1]^2 \tag{6.14}$$

Thus, for a CIR of 18 dB, the computed value of N is approximately 10. Neglecting the factor 1, for a higher CIR and for $n = 4$, the above expression can be approximated as

$$N = \sqrt{2/3(C/I')} \tag{6.15}$$

As the value of C/I' is increased (less cochannel interference), a higher value of N is obtained, implying less channel capacity per cell. Defining n to be the number of channels per cell, one can write

$$n = \frac{B_T}{NB_c} = \frac{B_T}{B_c\sqrt{2/3(C/I')}} \tag{6.16}$$

where B_T = total available bandwidth (12.5 MHz, AMPS) in transmit and (12.5 MHz, AMPS) in receive direction, and B_c = channel bandwidth (30 kHz, AMPS). Thus for an 18-dB CNR, the number of channels n on a per cell basis equals 64. We provide below an example illustrating the use of the above formulas.

Example 6.1

For corner excited 120-deg sectored configuration, as shown in Figure 6.5, compute the CIR at A, B, C, and D due to a cochannel cell at A'. The computed values of the distances are:

$$A'A = 4R,\ A'C = 5R$$

$$A'B = \sqrt{\left(\frac{4\sqrt{3}}{2}R\right)^2 + (3R)^2} = \sqrt{21}\,R$$

$$A'D = \sqrt{\left(5\frac{\sqrt{3}}{2}R\right)^2 + \frac{R^2}{4}} = \sqrt{19}\,R$$

Assuming $n = 4.0$, the CIR at A, B, C, and D due to single cochannel cell A' are:

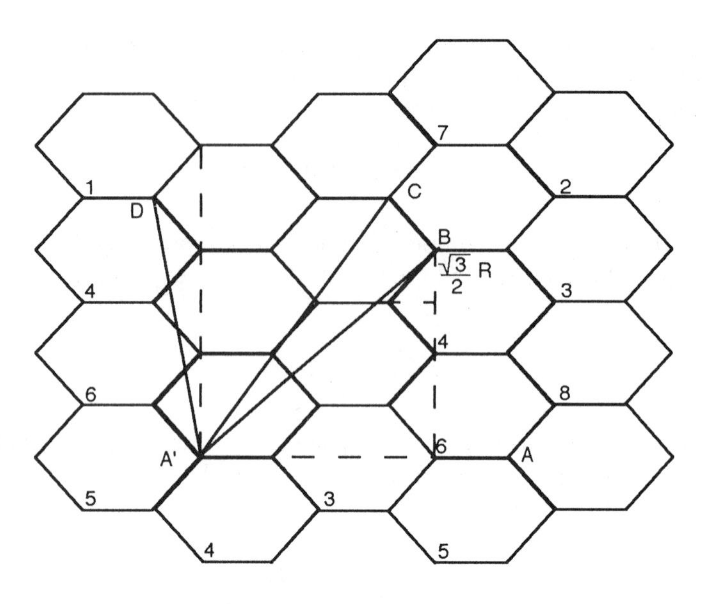

Figure 6.5 Cell configuration for sectored antenna.

- Point A: $(C/I) = (4R/R)^4$ or $C/I = 24.0$ dB;
- Point B: $(C/I) = (\sqrt{21}\ R/R)^4$ or $C/I = 26.4$ dB;
- Point C: $(C/I) = (5\ R/R)^4$ or $C/I = 27.9$ dB; and
- Point D: $(C/I) = (\sqrt{19}\ R/R)^4$ or $C/I = 25.6$ dB.

One can easily see that the CIR is higher for points located further from the cochannel cell A'.

6.2.1 Probabilistic Model

As stated in the introduction of this chapter, the cellular geographic statistical area (CGSA) is divided into a number of smaller cells. Each cell site is equipped with a number of channels and the low-power transceivers, which are capable of communicating with the mobile located within the cell [1, 3–6]. Adjacent cells are assigned from different channel sets to avoid interference. However, each channel set can be assigned to cells within one system that are geographically separated so that they limit the cochannel interference. In this way, each channel may be used simultaneously by several mobiles in the system, resulting in an effective multiplication of the total number of voice circuits. In such a system, the cochannel interference depends upon the D/R ratio and the capture ratio of the receivers. If the distance between the cochannel cells is reduced (low D/R), the cochannel interference increases. On the other hand, if the cochannel distance (high D/R) is increased, the number of channels per cell site is reduced, thereby reducing the capacity of the system. The effect of the cochannel interference can also be reduced by more stringent and hence, more costly, receiver designs. In order to reuse the channels effectively, the receiver should exhibit a capture effect as seen in an FM receiver. As long as the desired received signal is a times greater than the undesired signal, the interference can be subdued. For avoiding the cochannel interference, the capture ratio should be made as small as possible, offering strong protection against cochannel interference. However, this will also require higher linearity of the receiver over the wide dynamic range and the large bandwidth of the receiver. A higher linearity requirement for the FM receiver makes the receiver design more complex and therefore, in the interest of cost and the complexity, a compromise with a higher value of the capture ratio is made. If the desired signal is less than a times the interference signal, some cochannel interference results. Assuming that the FM receiver is properly designed with a capture ratio of a, then the cochannel transmitter needs only be spaced far enough so that the interfering signal rarely exceeds the local desired signal, as shown in the Figure 6.6. Here, the receiver will be captured by signal y_1 as this signal is greater then signal y_2 for most of the time,

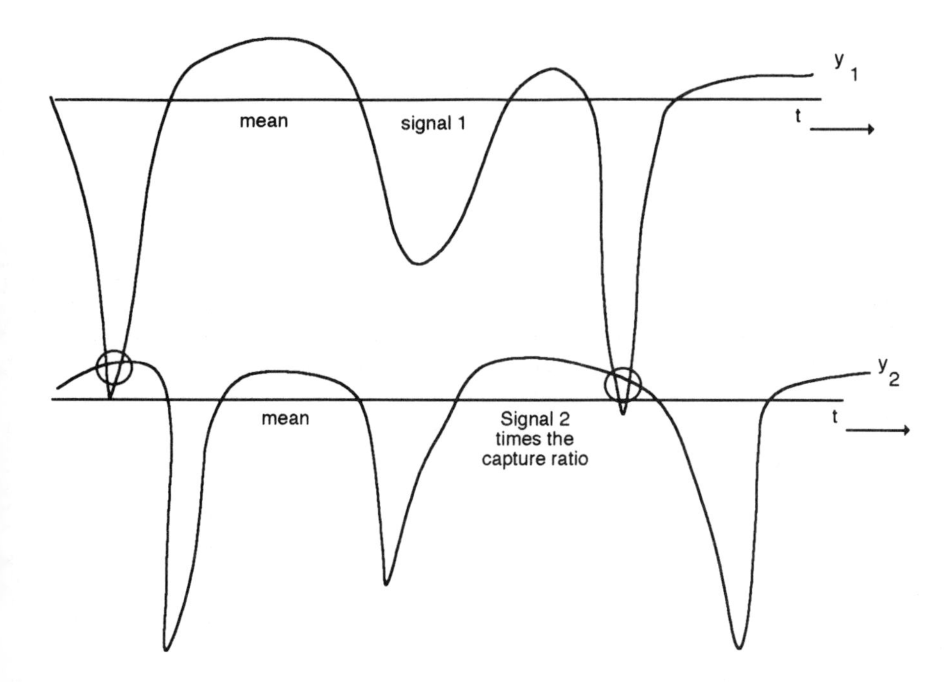

Figure 6.6 Amplitudes of signals from two cochannel cells.

after accounting for the capture effect. As shown in the figure, the circled points are the only cases where the capture ratio times the undesired signal y_2 equals the desired signal y_1. We derive below the probability of interference on the axis joining two cochannel cells T_1 and T_2 [1–4]. The configuration is shown in Figure 6.4.

In the configuration, the mobile is located at xD $(0 < x < 1)$ from the desired cell T_1. If y_1 and y_2 are the desired and undesired signal at the mobile, then for minimizing the cochannel interference the signal $y_1 > ay_2$, where a is the capture ratio of the FM receiver. Since the mobile operates in a Rayleigh fading, we can write

$$P(y_1 > ay_2 | y_2) = \int_{ay_2}^{\infty} \frac{y_1}{\sigma_1^2} e^{-y_1^2/2\sigma_1^2} dy_1 = e^{-a^2 y_2^2} \neq 2\sigma_1^2 = P_1 \tag{6.17}$$

Since the inequality $y_1 > ay_2$ must be satisfied for all values of y_2, we can integrate out the variable y_2 from the above equation. Thus

$$P_2 = \int_0^{\infty} P_1 p(y_2) dy_2 = \int_0^{\infty} P(y_1 > ay_2 | y_2) P(y_2) dy_2 \tag{6.18}$$

or

$$\begin{aligned} P_2 &= \int_0^{\infty} \left(e^{-a^2 y_2^2/2\sigma_1^2} \right) \left(\frac{y_2}{\sigma_2^2} e^{-y_2^2/2\sigma_2^2} \right) dy_2 \\ &= \int_0^{\infty} \frac{y_2}{\sigma_2^2} \exp\left[\frac{-y_2^2}{2} \left(\frac{a^2\sigma_2^2 + \sigma_1^2}{\sigma_1^2 \sigma_2^2} \right) \right] dy_2 = \frac{\sigma_1^2}{a^2\sigma_2^2 + \sigma_1^2} \end{aligned} \tag{6.19}$$

Equation (6.19) provides the probability of having no cochannel interference; that is., the case where the FM receiver is captured by the desired signal. In other words, the FM signal has overcome the cochannel interference. Assuming the propagation to follow the nth-power law, from Figure 6.4, we get

$$\begin{aligned} \frac{\sigma_1^2}{\sigma_2^2} &= \frac{W_1}{W_2} \left(\frac{xD}{D - xD} \right)^{-n} \\ &= \frac{W_1}{W_2} \left(\frac{1 - x}{x} \right)^{n} \end{aligned} \tag{6.20}$$

where W_1 and W_2 are the transmitted powers from the desired cell site T_1 and the interference cell site T_2.

It should be noted that in the above analysis the clutter factor (local mean) associated with signals, which follows a log-normal distribution, has been neglected.

This is rightly so as we are only interested in the ratios of the two signals that are basically in the same geographical areas, and therefore the mean level of these signals cancel out. Substituting (6.20) in (6.19), the probability that the desired signal y_1 will be received at a level above ay_2 is given by

$$P_2 = \frac{W_1(1 - x)^n}{a^2 W_2 x^n + W_1(1 - x)^n} \tag{6.21a}$$

Assuming, the power levels of the desired and undesired signals W_1 and W_2 are the same, then the above equation is reduced to

$$P_2' = \frac{(1 - x)^n}{a^2 x^n + (1 - x)^n} \tag{6.21b}$$

In the above equation $a > 1$, and the receiver is captured by the desired signal. The above establishes a relationship between the desired signal y_1 and the undesired signal y_2 received by the mobile. On the other hand, if $y_2 > ay_1$ or $y_1 < y_2 \neq a = by_2$ where $b > 1$, then the receiver is captured by the undesired interfering signal and the desired signal is lost. In this case, cochannel interference probability can be found as follows:

$$P(y_1 < by_2 | y_2) = 1 - e^{-b^2 y_2^2 / 2\sigma_1^2} \tag{6.22}$$

Since the above inequality has to be satisfied for all values of y_2 (similar to the last section), we integrate the value of y_2 from the above equation. Following this procedure, the interference probability P_3 is given by [1]

$$P(y_1 < by_2 | y_2) P(y_2) = P(y_1 < by_2) = \frac{b^2 x^n}{b^2 x^n + (1 - x)^n} = P_3 \tag{6.23}$$

The equation provides the total probability of having cochannel interference at a distance x from the desired cell site T_1. It should be noted that we are assuming $W_1 = W_2$ in the above equation, which means that both the desired and the undesired interferer powers are the same as in (6.21b).

It should also be noted that in the above analysis the exponent has been given a general value of n. For car-mounted antennas (receiver antenna), the data by Okumura shows an attenuation constant close to 3 for a cell transmitting antenna height up to 600 feet. The value of n increases as the transmitting antenna height is reduced, approaching 4 for a 20-ft high transmitting antenna.

Figure 6.7 is the plot of (6.23) for attenuation constants of 3.5. For normal cell site antenna heights, these values are appropriate. The values of b are varied from 1 to 5. At a normalized distance x from the desired cell site, the interference probability P_3 is higher for higher values of b or a lower protection ratio a. Alternately, for a given system and for a fixed exponent n, the coverage distance x on the line of travel increases as the value of b is reduced or the protection ratio a is increased.

From (6.21b), the normalized value of distance x is

$$\hat{x} = \frac{1}{1 + \sqrt{a}[P_2'/(1 - P_2')]^{1/4}}$$

A plot of the normalized maximum distance $\hat{x}$ for two values of P_2' (6.21), for both speech and data, is shown in Figure 6.8 for different values of the protection ratio a, in decibels. Here, P_2' is the probability that the signal y_1 is received at a level above y_2 by the desired protection ratio a. The lowest quality speech can have an error rate of one in ten, while one error in a hundred can be regarded as a moderate-to-good speech quality. For data, the minimum acceptable value of error can be

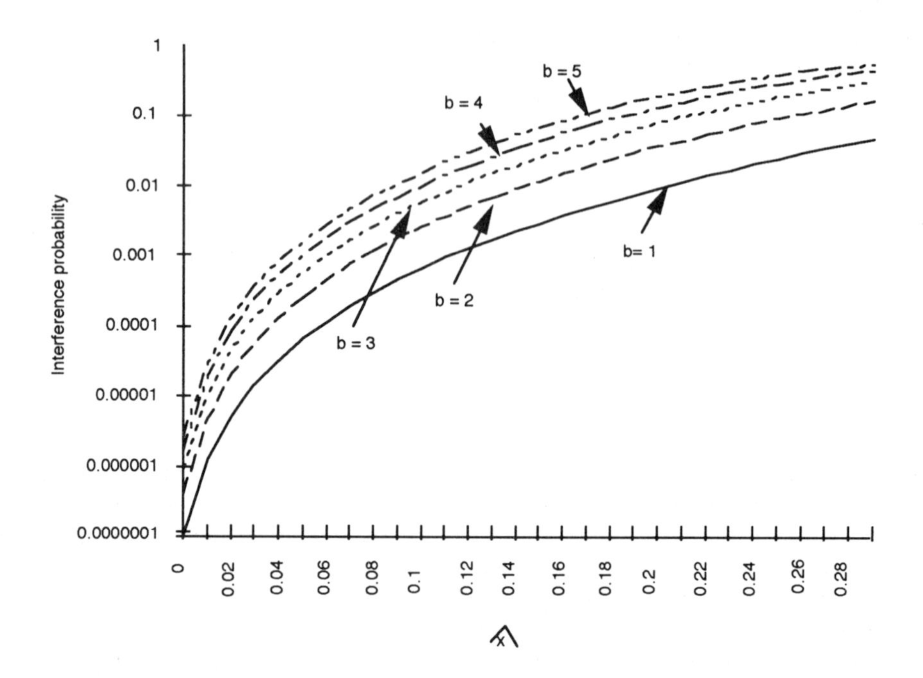

Figure 6.7 Probability of interference from one cochannel station on the line of travel (attenuation constant $n = 3.5$).

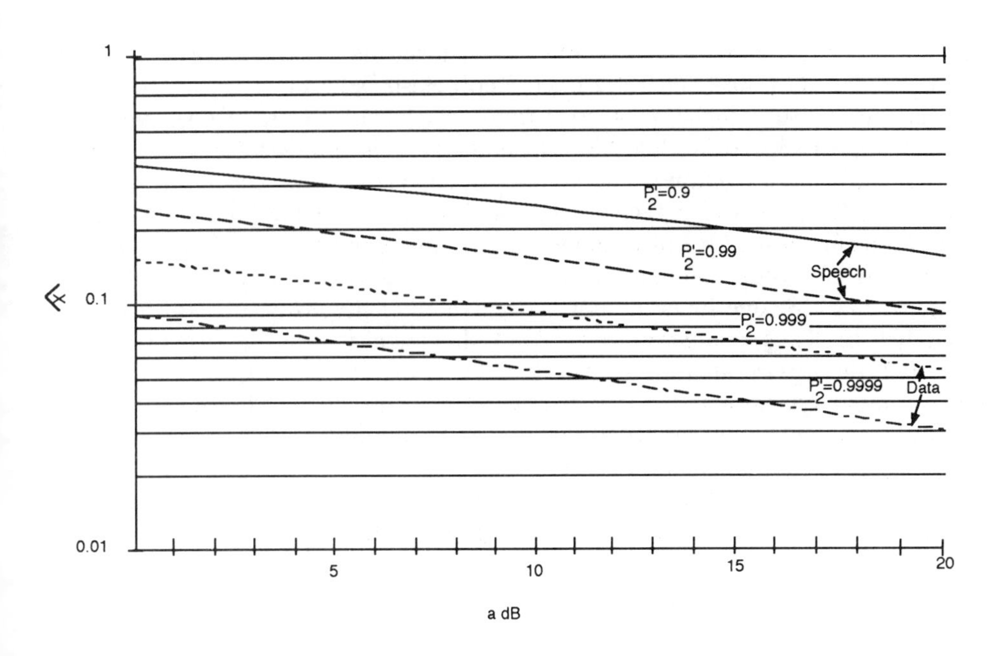

Figure 6.8 Coverage distance $\hat{x}$ versus protection ratio for different values of interference probability.

one in one thousand, while one error in 10^4 may be desired in most cases. There is no hard and fast rule in choosing the protection ratio *a*, but the CCIR recommends determining its values as follows. "A degradation of the initial *SNR* of 20 dB to a signal-to-noise plus interference ratio of 14 dB. This criterion corresponds to the minimum acceptable grade of service." For a nonlinear system like FM, it is difficult to arrive at a value of protection ratio *a* by applying the above definition. In this case, the parameter can only be arrived at empirically. For a linear modulation like SSB, it is however straight forward to compute the value of *a*. By applying the above CCIR definition, we obtain $P_N = 10^{-2}\ P_s = 0.01\ P_s$ and $P_N + P_I = 10^{-1.4}\ P_s = 0.04\ P_s$, where P_s, P_N, and P_I are the signal, noise, and interference powers. Thus, the allowed value of $P_I = 0.03\ P_s$, or the ratio $P_s/P_I = 15.2$ dB. Therefore the value of the protection ratio is allowed to drop below 15.2 dB with a very low probability. However, this value applies to only linear modulation. In an FM system, a tradeoff exists between lower bandwidth and the higher protection ratio versus higher bandwidth and the lower protection ratio. Assuming an 8-dB protection ratio for a 12.5-kHz channel and 4 dB for 25 kHz, we obtain (as in Figure 6.8) the normalized coverage distances $\hat{x}$ to be approximately 0.17 and 0.2 for good quality speech having the value of P_2' to be 0.99. Thus, the normalized distance with a 25-kHz bandwidth is increased to 1.17 and the corresponding area by a ratio of 1.4. However, as pointed out before, the number of channels in this case is reduced by 50%. Therefore, a tradeoff exists between an increased number of channels versus a higher protection ratio, which leads to a reduced service area. It may appear superficial that the wideband system is at a disadvantage when one accounts for the reduction of the number of channels. However, as shown below this is not the case. CCIR recommends an 8-dB value for the protection ratio for a FM system irrespective of whether it is a narrowband or wideband FM system. Extending the above argument further to a 5-kHz SSB system with the protection ratio of 15.2 dB, one finds the normalized value of the distance $\hat{x}$ to be approximately 0.12. Thus, compared to 25-kHz FM channel that has a normalized value $\hat{x}$ of 0.2, the service area in this case is reduced by a factor of 0.36. Therefore an SSB cellular system with a 5-kHz bandwidth may not be an advantage when one considers the reduction of service area, though the number of channels are increased by a factor of five or six based upon 25 kHz or 30 kHz as the standard FM system. In other words, an SSB system with a 5-kHz channel bandwidth may not be a good choice for a cellular system as its tolerance to cochannel interference is greatly reduced.

6.2.2 Cochannel Interference Avoidance

The reduction of cochannel interference in the cellular system can be achieved by:

- Use of a higher values of N, the number of cells per cluster;
- The judicious choice of channel assignment;
- Use of sector antennas;
- Use of SAT in conjunction with cochannel assignment;
- The diversity reception.

From (6.1), the use of a higher values of N increases the cochannel separation distance D with respect to cell radius R. This reduces the interference level to the cell site under consideration. However, the negative aspect of this reduction is the lowering of the traffic carrying capacity of the system.

Another technique that can effectively be used in reducing the cochannel interference is the judicious choice of channel assignment. Channel assignment can properly be monitored so that the cochannel assignment can be minimized as far as practical. This will help in reducing the cochannel interference, at least during the off-peak period. The probability of interference for a seven-cell cluster with a protection ratio of 2 and the propagation exponent of 3.5, is shown in Figure 6.9 [1]. As shown in the figure, for $\hat{x} \cong 0.2$, the probability of interference is reduced roughly from 10 to 5% when all six cells use the same channel (peak traffic indicator) versus a single cell acting (off-peak indicator) as a cochannel interferer. Use of a directive antenna helps in improving the cochannel interference and is shown in

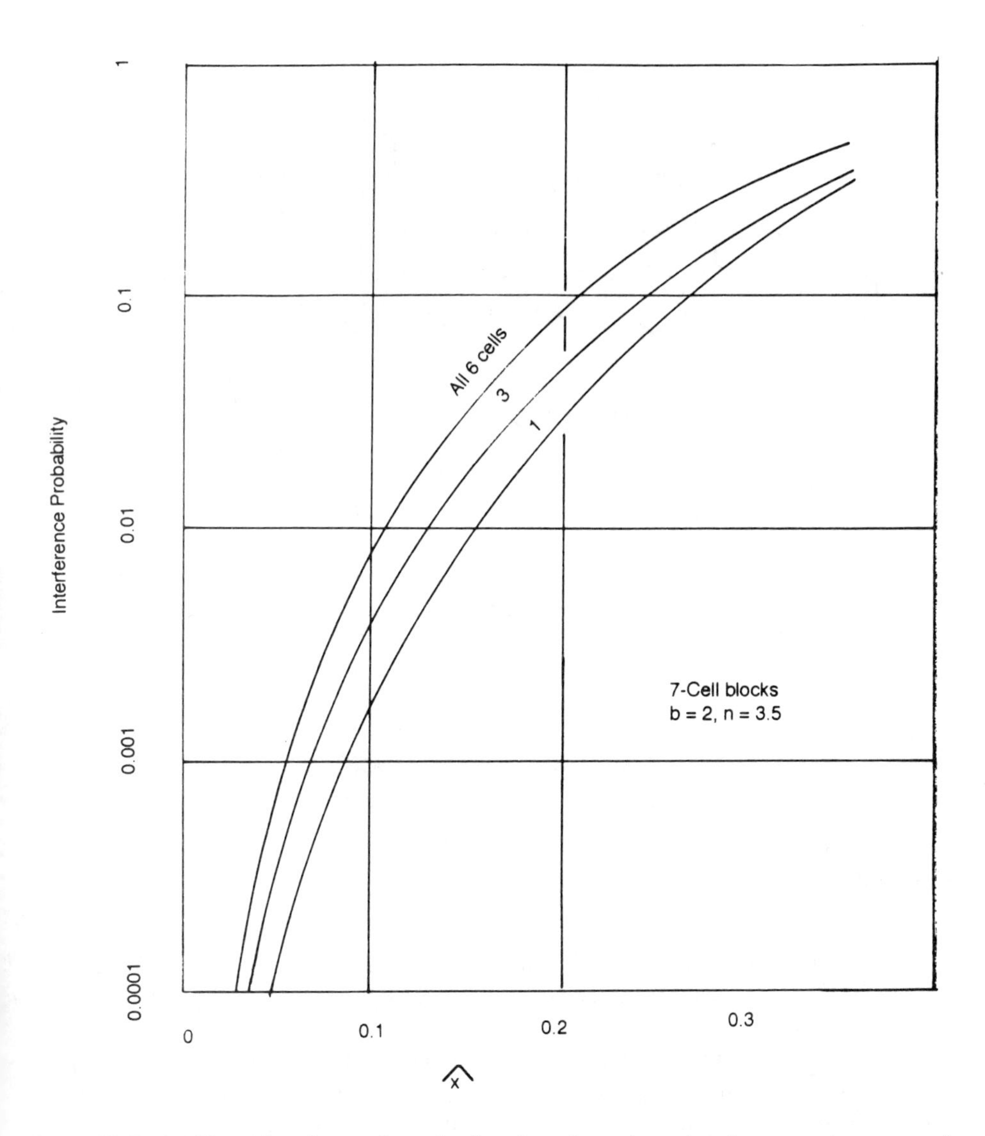

Figure 6.9 Probability of interference from the first ring of a cochannel station, assuming seven-cell blocks. Source: [1].

Figure 6.10. Normally when all six cochannel cells use the same channel, the magnitude of the interference for an omnidirectional antenna can be as high as six times the interference of a single cochannel cell. For a sector antenna with a seven-cell cluster, the interference received from site C to the cell under consideration, Z, is of the same magnitude as that of the omnicell, however, the interference

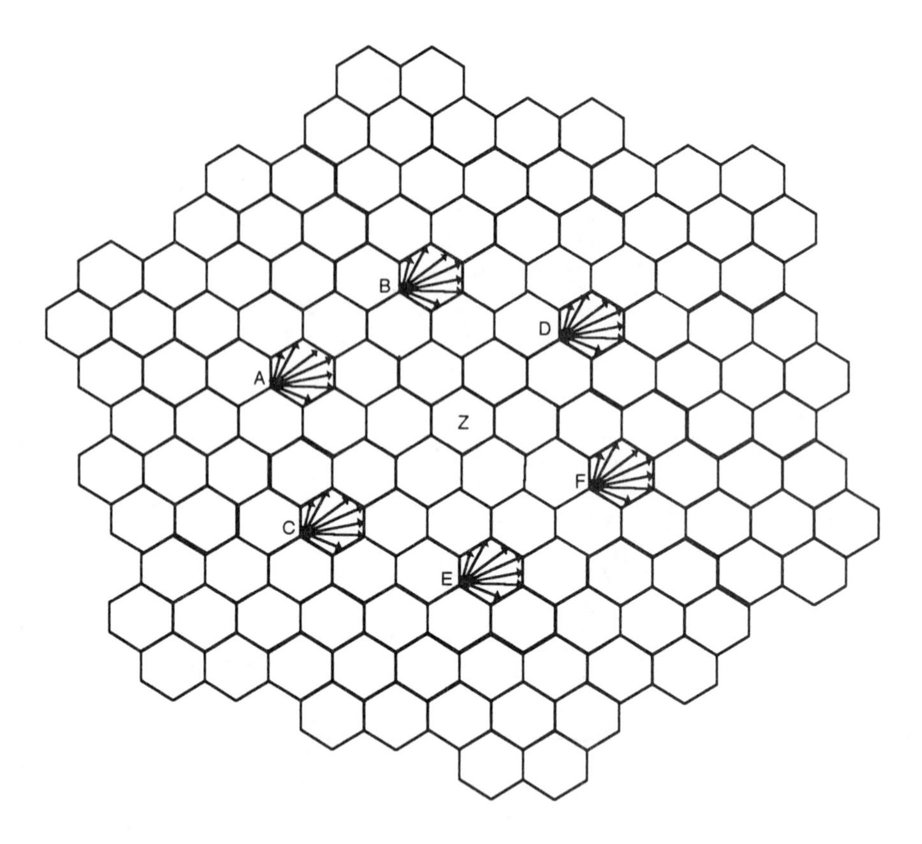

Figure 6.10 Use of directive antennas.

received from cells B and A are reduced and is almost eliminated from cells D, E, and F. Thus, the use of the sectored antenna helps to minimize the cochannel interference.

Achieving reduced cochannel interference is one of the advantages of the four-cell sectored antenna system used in the Dyna TAC cellular portable radio telephone system. Although the main advantage of a sectored antenna system is the high directive gain so that the low-powered portable can easily be serviced, they were also proved to be advantageous in reducing the cochannel interference [6]. Figure 6.11 depicts the joint omni and sector cell scheme. For the omni scheme, the cochannel interference is at its greatest when the mobile is located at the boundary of the desired cell C nearest to the interferer, since the desired power it receives is at its lowest. When the mobile is located at the boundary of the cell of radius R and the cochannel separation is D, the CIR in the worst case will be proportional to $(D - R)$ to R, which limits the performance of the matured

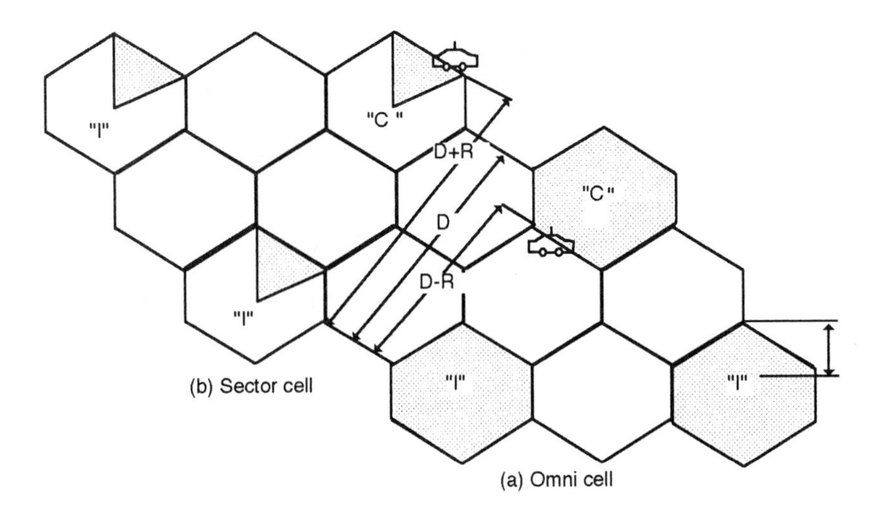

Figure 6.11 Sector transmit cells.

omnidirectional cell pattern. The improvements that the sectored antenna system provide over omni is shown in Figure 6.12. For this case, the cell radius and the channel reuse distance have been kept the same as R and D for an omnidirectional antenna. If the reuse pattern employed is such that the frequency reuse is limited to sectors that have the same orientation in the cell, this results in an improvement of the cochannel interference ratio. One such sectored frequency reuse plan is shown in Figure 6.13. As stated above, the worst case of cochannel interference results when the mobile is at the boundary of the cell with radius R, where the desired signal strength is at its lowest. However, as shown in Figure 6.11, this condition can now occur at the far side of the sectored cell where the distance from the cochannel cell is $(D + R)$. Consequently, the CIR in the sectored cell case is proportional to $(D + R)$ to R. Assuming the groundwave loss exponent to be 4, the SIR for sectored and omniantennas can be expressed as

$$\left(\frac{C}{I}\right)_{\text{sec}} = \left(\frac{D + R}{R}\right)^{n} = \left(\frac{D}{R} + 1\right)^{4} \tag{6.24}$$

$$\left(\frac{C}{I}\right)_{\text{omni}} = \left(\frac{D - R}{R}\right)^{n} = \left(\frac{D}{R} - 1\right)^{4} \tag{6.25}$$

Thus, for the same CIR and cell size R, and the exponent n the cochannel separation distance, D, can be reduced in a sectored cell layout. Since the D/R ratio determines

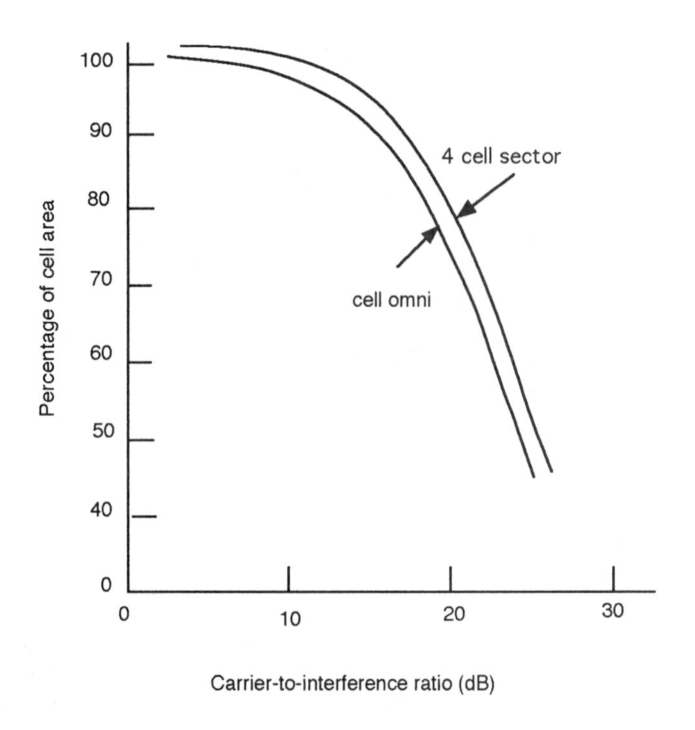

Figure 6.12 Distribution of CIR (dB) over cell area. Source: [11].

the number of cells in a reuse pattern, sectored cells provide the same level of interference probability with a larger number of cells than their omni counterparts. Assuming a CIR of 18 dB with an exponent value of 4 and with the same cell radius R, the D/R ratio for a sectored cell is 1.82 versus a D/R ratio of 3.82 for an omnidirectional cell. As a result, the channel capacity for the sectored cell is increased by a factor of 4.41 with respect to omnidirectional antennas.

If the presence of a proper SAT (same SAT) in the assigned channel is made essential before switching on to conversation, the cochannel interference can be reduced further. This is shown in Figure 6.2 (same S_s), where the geographical separation between the cochannel cells are increased, thereby reducing the CIR. From this point of view, the second ring of cochannel cells are more important than the first ring of cells, as the same SAT only appears at the second ring of cells.

Lastly, various forms of diversity reception can be used to improve the performance of cellular systems under multipath surroundings. One form of antenna-switched diversity is by Rustako et al., as discussed in Chapter 8 [10]. For this form of diversity, the author has shown that when the received signal is below threshold,

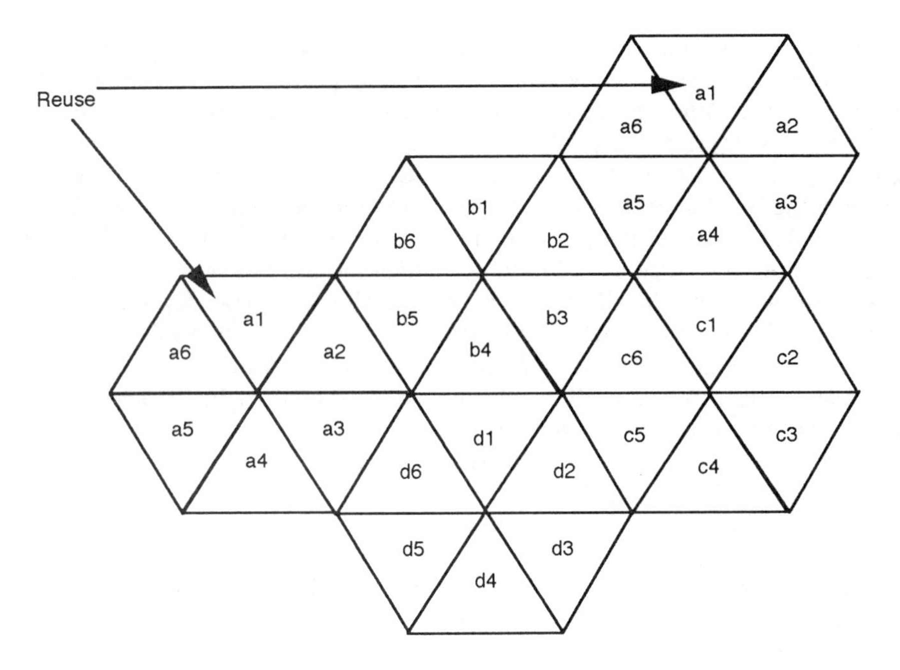

Figure 6.13 Four-cell pattern.

the probability of interference is below its single antenna value (no diversity) by the ratio of the threshold level to the average signal level.

6.3 ADJACENT-CHANNEL INTERFERENCE

In cellular systems, channels used in one cell are reused in other cells that are sufficiently spaced so that the cochannel interference is minimized. Similarly, adjacent channels are not assigned in the same cell to avoid excessive adjacent-channel interference. Thus, the controlling factor for adjacent-channel interference is the specific channel assignment in the cell. The adjacent channel under discussion here is the nearest assigned channel frequency in the cell and is not the true adjacent channel. However, due to the splatter of modulation, the adjacent-channel interference results even though the interfering channel is not the true adjacent channel. It should be noted that, due to practical reasons (increased mobile complexity), RF channels cannot be band limited prior to transmission. The degree to which these types of impairments constrain system performance is a function of channel spacing between adjacent channels, receiver predetection filter response, selectivity of the receiver, and the statistical distribution of the ratio of desired-channel carrier level to adjacent-channel carrier level. In our discussion here, we shall assume that the adjacent channels are transmitted at the same power level and show that the adjacent-channel interference is usually of less importance when compared with the cochannel interference [8–9]. The design objective for the cellular radio is assumed to be 18 dB or greater for an RF CIR over 90% of the coverage area. Studies at the Bell Laboratory have shown that a voice-modulated adjacent-channel carrier level at 10-dB greater power than the desired-channel carrier level produces baseband interference comparable to cochannel interference when the cochannel power level is adjusted 17-dB below the desired-channel carrier level under fading conditions [8]. This result includes the effects of a maximum allowable transmitter (1.5 parts per million (ppm)) and receiver frequency drifts (2.5 ppm). In this experiment, the receiver 6-dB bandwidth was assumed to be 30 kHz, and the adjacent-channel carrier was attenuated by more than 48 dB when located 27-kHz away from the center of the desired band. The effects of adjacent-channel interference on the control channel were also studied at the Bell Laboratory. The result, as reported in [8], states that the adjacent channel, when modulated by a carrier level about 5-dB above the desired control channel carrier, causes the bit error rate, which is about the same as cochannel interference when the cochannel data is modulated at a carrier level 17-dB below the desired control channel carrier under fading conditions. The effect of adjacent-channel interference can be controlled by proper designing of the receiver predetection filter. In this section we will develop techniques for computing the adjacent-channel interference when the receiver front-end filter characteristics are known. Once again the reader is reminded that the adjacent channel under discussion here is the system adjacent channel, which is a function of the channel assignment in a specific cell and is not the true adjacent channel. We shall discuss this below.

6.3.1 Adjacent-Channel Interference Control

In this section we shall consider the interference produced in the channel of interest due to system adjacent channel. One can assign channels to cells, which provides the maximum frequency separation between channels; that is, to the kth cell assign; $k, k + N, k + 2N, \ldots k + nN$ channel frequencies where n is an integer such that $(395 - nN)/N < 1$. If we consider the case for $N = 12$, we obtain the voice assignment as shown in Table 6.1. Here, we assume a total of 395 available voice channels, as in AMPS system.

The channel pattern for the group of 12 cells per cluster is shown in Figure 6.14. This pattern can be used in the startup cellular system, where mostly the omnidirectional antenna is used. It should be noted that sets with an adjacent set number (including 12 and 1) contain an adjacent channel. Therefore, they cannot be the closest neighbor. As seen in Figure 6.14, the nearest neighbor of the cell 1

Table 6.1
Channel Assignment for $N = 12$

Cell/Group Number	*Channel Assignment*
1	1, 13, 25, 37, 49, 61, 73, 65, 97, 109, 121, 133, 145, 157, 169, 161, 193, 205, 217, 229, 241, 253, 265, 277, 269, 301, 365
2	2, 14, 26, 36, 50, 62, 74, 64, 96, 110, 122, 134, 146, 156, 170, 162, 194, 206, 216, 230, 242, 254, 266, 276, 290, 302, 366
3	3, 15, 27, 39, 51, 63, 75, . . . , 291, 303, 367
4	4, 16, 26, . . . , 292, 304, 366
5	5, 17, 29, . . . , 293, 305, 369
6	6, 16, 30, . . . , 294, 306, 390
7	7, 19, 31, . . . , 295, 307, 391
6	6, 20, 32, . . . , 296, 306, 392
9	9, 21, 33, . . . , 297, 309, 393
10	10, 22, 34, . . . , 296, 310, 394
11	11, 23, 35, . . . , 299, 311, 395
12	12, 24, 36, . . . , 300, 312, 364

is cell 4. These cells are assigned frequencies from group 1 and group 4. As seen in Table 6.1 the closest two frequencies in these two cells are two groups apart, but for our discussion here, we are not concerned with this case. On the other hand, the subject of interest to us is the nearest assigned channel frequency difference in the same cell. In the same cell, channels are 360-kHz apart (30 × 12 kHz). As an example, in cell 1 for channel 1, the nearest channel is number 13. Frequency separation between these two channels is shown in Figure 6.15.

For avoiding interference, the adjacent-channel signal should be sufficiently attenuated at the band edge of the desired channel so that an adequate CIR can be preserved. This is achieved by having the proper selectivity of the receiver predetection filter. The receiver predetection selectivity mask, as used in the design of AMPS cellular channel receiver, is shown in Figure 6.16.

The selectivity plot has a response of 6 dB at a frequency not greater than 18 kHz from the channel center frequency. The selectivity attenuates the unmodulated RF carrier signals by at least 48 dB at an offset value of 29 kHz. The plot provides attenuation relative to an unmodulated RF carrier at zero frequency. The selectivity requirements must be met with the worst environmental changes, alignment tolerances, and the frequency instabilities associated with the receiver.

Assuming that channels 1 and 13 are simultaneously active in cell 1, then an attenuation value of 65 dB or more is seen by the adjacent channel, which is 330-kHz apart from the band edge of the desired channel. For this analysis, we are assuming that the mobiles using channels 1 and 13 are at equal distances from the cell site. Adjustment to the interference power relative to the desired-channel power at the cell site has to be made according to the relative interference power

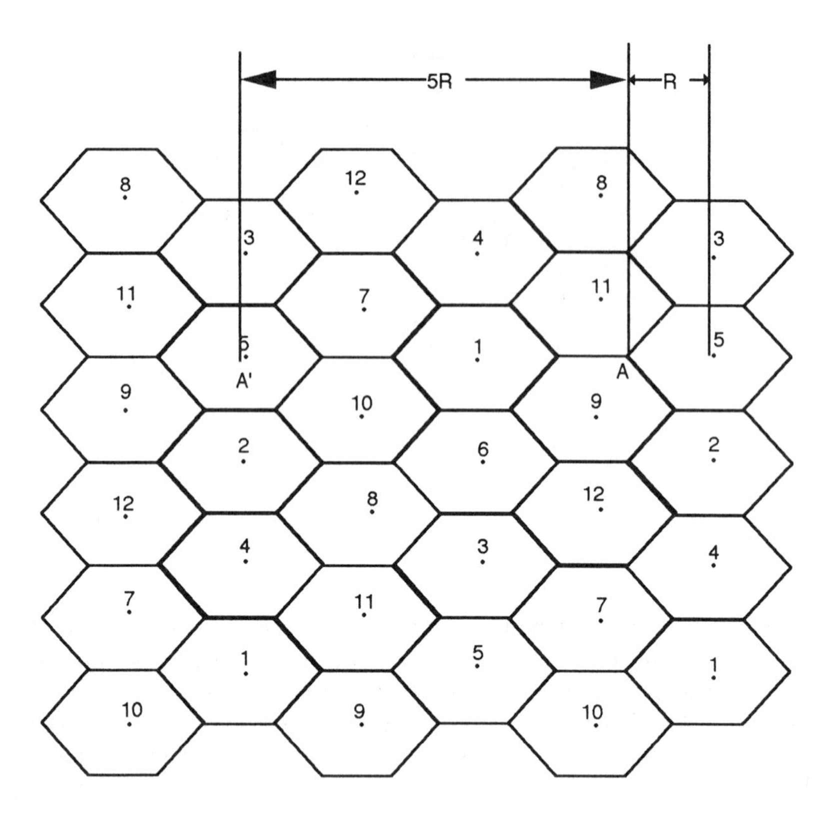

Figure 6.14 Channel-set deployment for 12 cells per cluster.

being less than $-65 - 10 \log_{10}(d_2/d_1)^4$. Thus, the adjacent-channel interference power received by the desired receiver will be below 65 dB if the interfering mobile distance d_2 from the cell site is more than the desired mobile distance d_1. It should be noted that we are assuming here the exponent value of 4. In general, the value of the exponent lies between 2 and 4. On the other hand, if the interfering mobile distance d_2 is less than the desired mobile distance d_1, the interference received at the cell site will be above 65 dB. This difference in received power due to different geographical separation of two mobiles is discussed under near-end to far-end ratio in the next section. Here, we assume that either the distance d_2 is greater than d_1 or at best they are equal. In these cases, the relative interference power is always below 65 dB. We shall now describe the process of designing the receiver predetection filter, whose characteristics are shown in Figure 6.16, such that the adjacent-channel interference is reduced below the channel (cell 1, channel 1) on the left as referenced at dc. and draw an equivalent low-pass model of Figure 6.17.

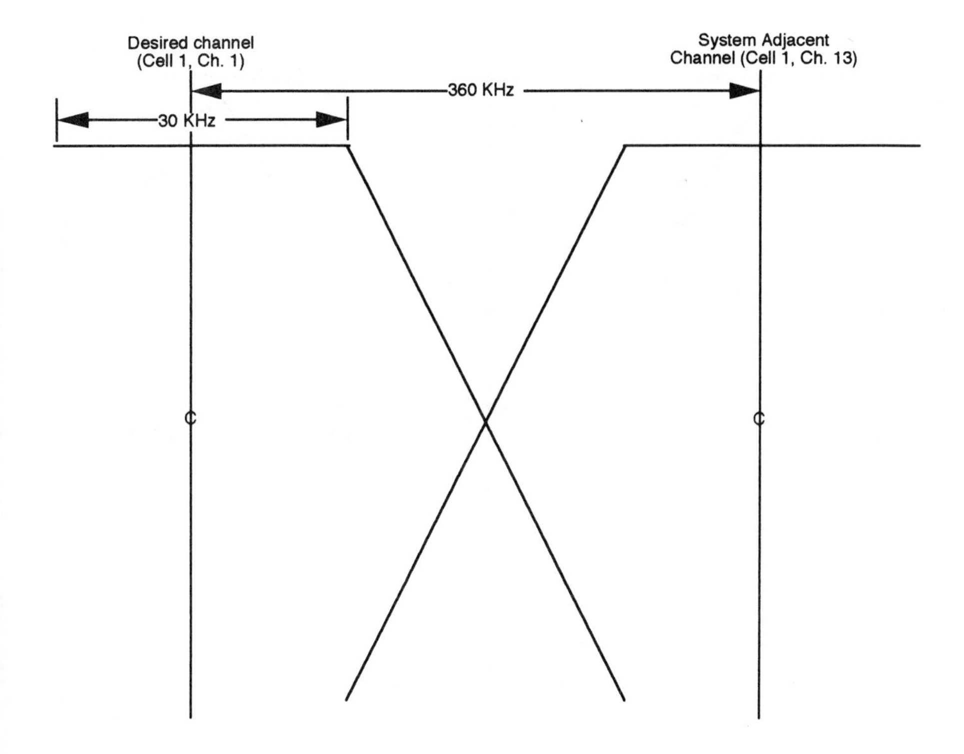

Figure 6.15 Frequency separation between desired and system-adjacent channel.

From the model, the desired passband radian frequency is w_p, below which the attenuation has to be less than equal to a specified maximum value of α_{max}. Beyond a stopband radian frequency of w_s, the attenuation has to exceed the value of α_{min}. Assuming that we desire the Butterworth low-pass configuration, the equations for attenuation α, in decibels, is given by $10 \log_{10}[1 + (w/w_0)^{2n}]$, where w_0 is an arbitrary reference frequency that is normally taken as 1 radian/sec and n is the order of the filter. Thus,

$$(w/w_0)^{2n} = (10^{\alpha/10} - 1) \tag{6.26}$$

or

$$w_0 = \frac{w}{(10^{\alpha/10} - 1)^{1/2n}} \tag{6.27}$$

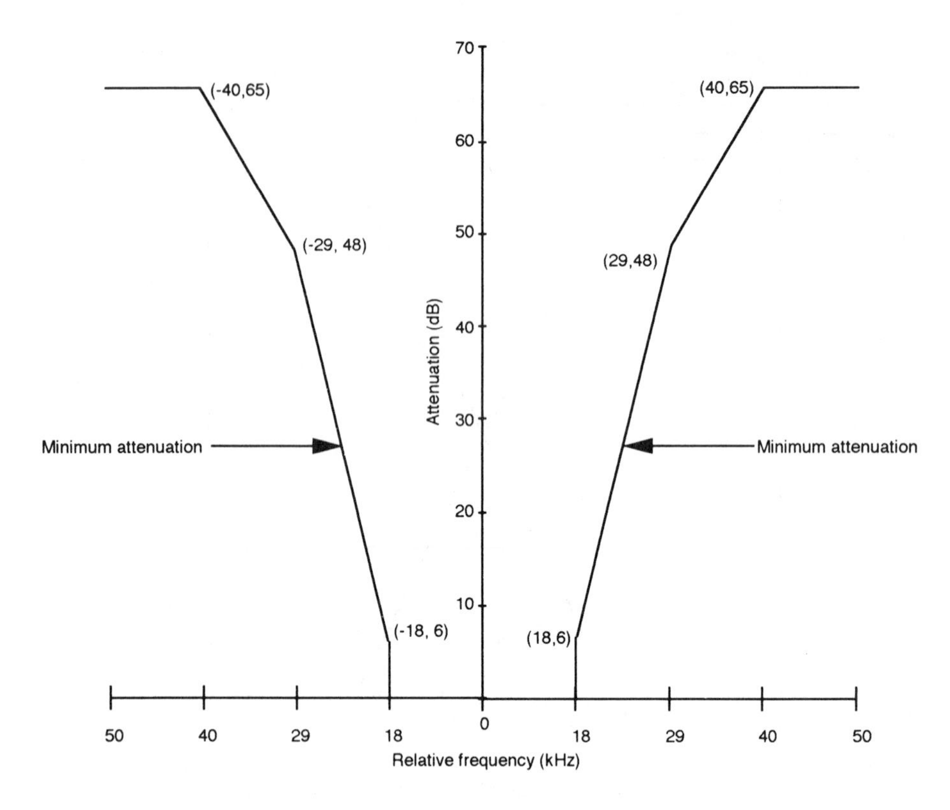

Figure 6.16 Receiver predectection selectivity mask in AMPS.

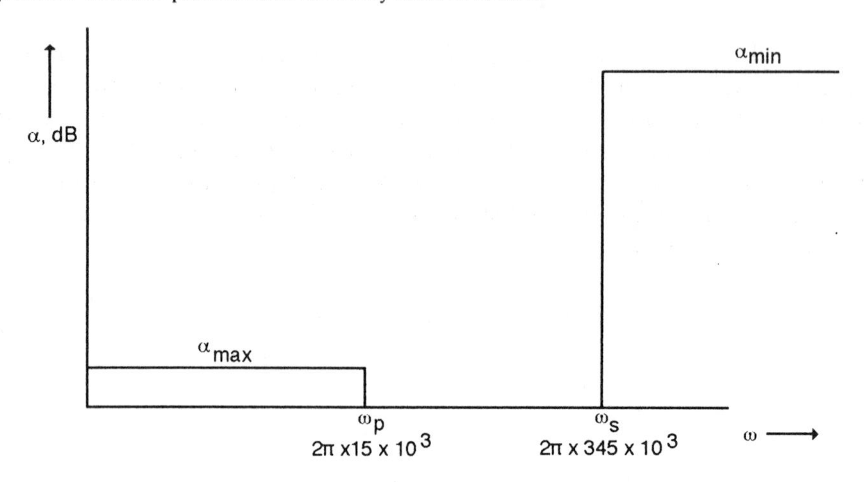

Figure 6.17 Low-pass equivalent model for adjacent channels 1 and 13 in cell 1.

Applying this equation to the passband, we have

$$w_0 = \frac{w_p}{(10^{\alpha_{max}}/10 - 1)^{1/2n}} \tag{6.28}$$

Applying the same equation for the stopband

$$w_0 = \frac{w_s}{(10^{\alpha_{min}/10} - 1)^{1/2n}} \tag{6.29}$$

Taking the ratio of (6.28) and (6.29), we have

$$\left(\frac{w_s}{w_p}\right)^{2n} = \frac{10^{\alpha_{min}/10} - 1}{10^{\alpha_{max}/10} - 1} \tag{6.30}$$

Taking the logarithm of this equation and solving for n gives

$$n = \frac{\log_{10}[(10^{\alpha_{min}/10} - 1)/(10^{\alpha_{max}/10} - 1)]}{2 \log_{10}(w_s/w_p)} \tag{6.31}$$

Normalized plots (with $w_0 = 1$) of passband and stopband attenuation for different values of n are shown in Figure 6.18. The curve provides the passband and stopband attenuation versus normalized frequency W for different values of n.

Assuming $\alpha_{max} = 0.5$ dB for frequencies less than 15 kHz and $\alpha_{min} = 60$ dB for frequencies greater than 345 kHz, the value of n from (6.26) is 3. Thus, using the third-order Butterworth filter one can satisfy the equivalent lowpass requirement of the cellular receiver predetection filter. With this filter, an adjacent-channel attenuation requirement of 60 dB is met. Since the center frequency of channel 1 is not at dc., the usual lowpass to bandpass transformations will be required. The reader is referred to the excellent handbook by A. I. Zvereu for transforming a lowpass model to a bandpass equivalent model. The above design clearly shows that a moderate complexity of filter easily meets the adjacent channel requirement.

6.4 NEAR-END TO FAR-END RATIO

As shown in Table 6.1, channels assigned to the same cell are separated by a minimum number of channels so that the adjacent channel interference can be minimized. Assuming that the mobile transmits the same power, the signal received at the cell site is proportional to the geographical distance between the cell site and the mobile under consideration. When the separation from the cell site for

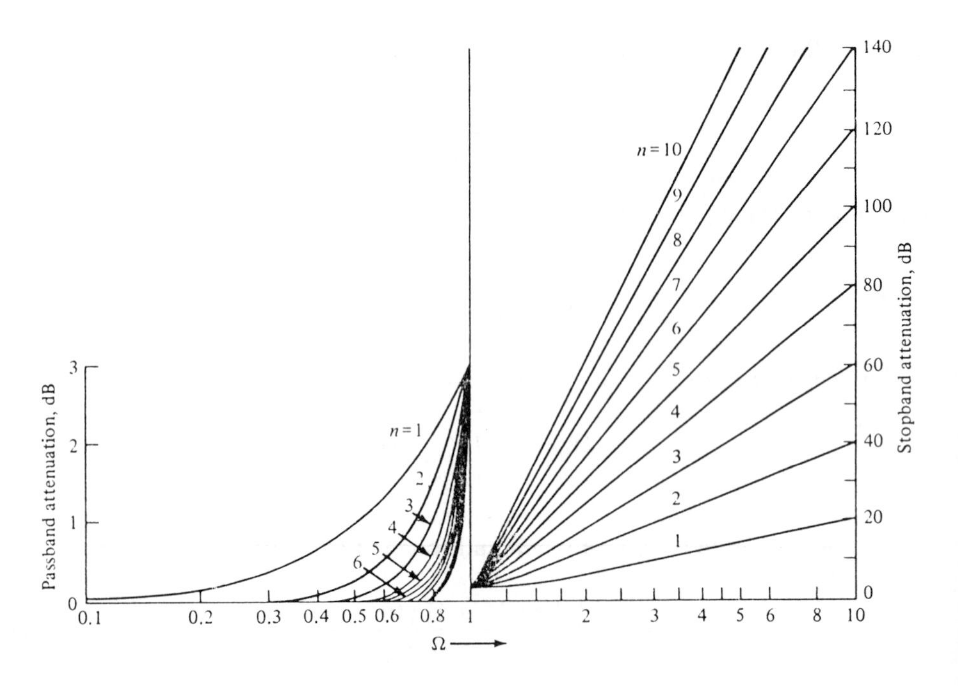

Figure 6.18 Passband and stopband attenuation for Butterworth filter. Source: [12].

two mobiles operating on adjacent channels are widely different, there can be a situation where the power received at the cell from a nearby mobile is far higher than the power received from the mobile farther away. This unbalanced received power is against the selectivity of the receiver predetection filter. The power difference at the cell site due to different path losses is known as near-end to far-end ratio interference.

$$\text{near-end to far-end ratio} = \frac{\text{propagation loss due to path } d_1}{\text{propagation loss due to path } d_2} \tag{6.32}$$

where the interfering mobile is at a distance d_1 (nearby interferer) from the cell site while the desired mobile is at a larger distance d_2 from the cell site. See Figure 6.19.

Assuming that the path loss exponent is four, the near-end to far-end ratio is reduced to

$$\text{near-end to far-end ratio} = (d_1/d_2)^{-4} \tag{6.33}$$

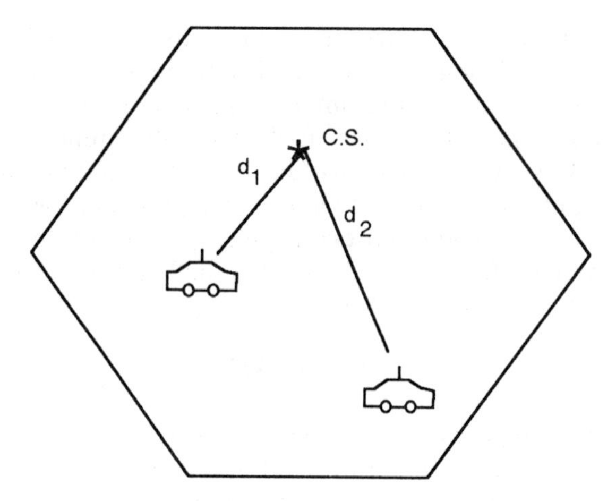

Figure 6.19 Mobile location within a cell.

Assuming $d_1 = 1$ mi and $d_2 = 6$ mi, this ratio is = 36.1 dB; that is, the closer mobile comes at a level 36.1-dB higher than the mobile farther away. From the predetection selective curve (shown in Fig. 6.16), at a frequency separation of 29 kHz we get the attenuation value of 48 dB. Thus even if the adjacent-channel assignment to the same cell is truly adjacent, the attenuation encountered by the interferer far exceeds his or her higher level superiority due to its closeness to the cell site. If the channel assignment is as shown in the Table 6.1, the normalized channel separation is 12 or 360 kHz. Therefore, with this assignment of channel, a higher power of the nearby interferer is always compensated due to the receiver filter selectivity characteristic, and thus this is not of much concern here.

6.5 CONCLUSIONS

This chapter discussed cochannel interference, adjacent-channel interference, and near-end to far-end ratio. We began the discussion with cochannel interference as this is the most prominent source of interference in a cellular system. There, we discussed the basic model of interference for a seven-cell cellular configuration and arrived at the approximate expression of the CIR. Due to the simultaneous activation of the same frequency in all of the six first-tier of cochannel cells, this case is considered to be the worst. A probabilistic model outlining the FM receiver designing techniques based on protection ratio was discussed. The probability of interference as a function of normalized distance $\hat{x}$, the attenuation constant n, and the protection ratio were arrived at in (6.23). We have critically examined the behavior of interference as a function of protection ratio and the propagation exponent n. Different techniques of cochannel interference avoidance were outlined in Section 6.2.2. The use of directive antenna appears very effective in minimizing the cochannel interference. The adjacent-channel interference problem was outlined in Section 6.3, where it was shown that this does not play such a big role in cellular radio design as the cochannel interference problem. Lastly, in Section 6.4 we have outlined the near-end to far-end problem, where we have shown that this is not a serious problem. In the next chapter we shall take up the study of different modulation schemes for cellular radio.

PROBLEMS

6.1 For the four-cell repeat pattern shown in Figure 6.20, with a 60-deg sectored antenna, compute the value of the CIR at points P, Q, R, and E. The value of the propagation exponent n can be assumed to be 4.

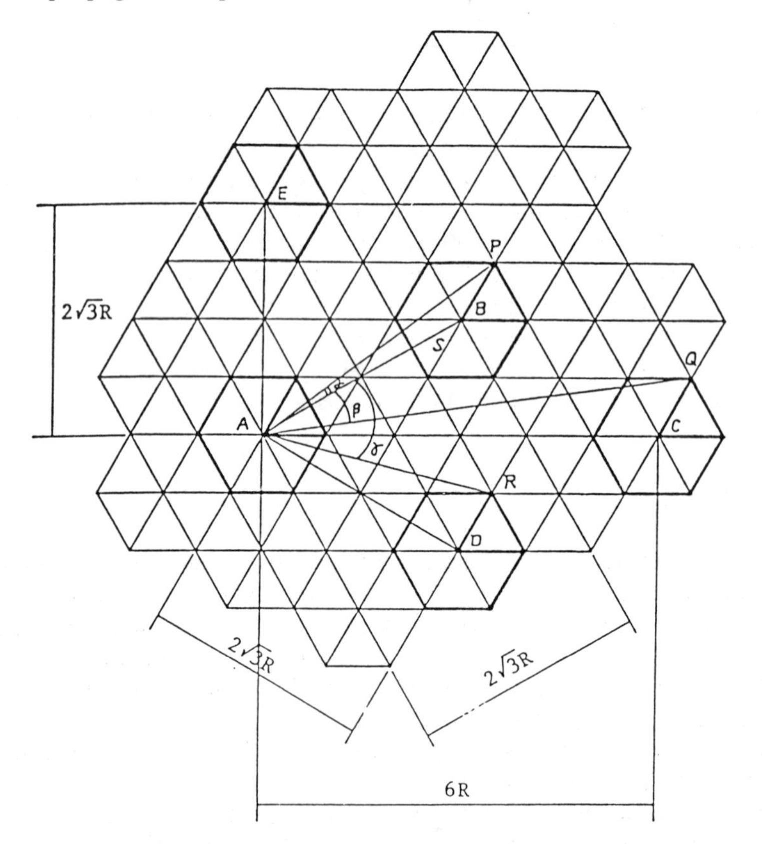

Figure 6.20 Four-cell repeat pattern.

6.2 For $N = 7, 12, 19, 21$, and 33, compute the value of the D/R ratio assuming omnidirectional antenna and 36.5-dB per decade as the propagation path loss factor. Compute the value of CIR when all of the six cochannels cells are excited and when a single cochannel interfering cell is excited.

6.3 Repeat problem 3 with the same data but assuming a 120-deg sectored antenna. Comment on the improvement obtained by using the sectored over the omnidirectional system.

6.4 Find the frequency assignment table for $N = 4$ and 7 assuming omnidirectional antennas at cell sites. For both these cases, find the order of the low-pass Butterworth filter n, assuming $\alpha_{min} = 60$ dB and $\alpha_{max} = 0.1$ dB.

6.5 Find the number of channels per cell for a CIR of 18 dB for $N = 7, 12$, and 19. Comment on the cell traffic carrying capacity for these cases. Assume the total available bandwidth to be 15 MHz and the channel bandwidth to be 25 kHz.

6.6 Justify why SSB transmission with a 5-kHz bandwith may not be of advantage in the cellular system when compared to FM with a 25-kHz bandwith.

6.7 Explain conceptually the role of protection ratio in the design of a cellular receiver. Will the protection ratio be higher or lower when the channel bandwidth is 25 kHz instead of 12.5 kHz?

REFERENCES

[1] Stocker, C. A. "Co-channel Interference and its Avoidance in Close-Spaced systems," *IEEE Trans. on Vehicular Technology*, Vol. VT-31, No. 3, Aug. 1982.

[2] Lee, W.C.Y. "Fundamentals of Mobile Cellular Systems," *IEEE Vehicular Technology Newsletter*, Aug. 1985.

[3] Gosling, W. "A Simple Model of Co-channel and Adjacent Channel Interference in Land Mobile Radio," *IEEE Trans. on Vehicular Technology*, Vol. VT-29. No. 4, Nov. 1980.

[4] Engle, J. S. "The Effects of Co-Channel Interference on the Parameters of a Small-Cell Mobile Telephone System," *IEEE Trans. on Vehicular Technology*, Vol. VT-16, No. 3 Nov. 1989.

[5] Cox, D. C. "Co-Channel Interference Consideration in Frequency Reuse Small-Coverage Area Radio Systems," *IEEE Trans. on Communications*, Vol. COM-30, No. 1, Jan. 1982.

[6] Schrenk, G. L. "Interference Management in Cellular System Design," *IEEE Vehicular Technology Conference*, 1984.

[7] "Study Report of Cell Pattern and Co-channel Interfial," Co., Ltd., Yokohama, Japan May 1983.

[8] Halper, Feggeler, MacDonald, and Whitehead. "Adjacent and Co-Channel Interference in Large-Cell Cellular Systems," *Telecommunications*, March 1984.

[9] Lee, W.C.Y. *Mobile Communication Engineering*, McGraw-Hill Book Co: New York, NY, 1983.

[10] Rustako, A. J., Y. S. Yeh, and R. R. Murray. "Performance of Feedback and Switch Space Diversity 900 MHz FM Mobile Radio Systems With Rayleigh Fading," *IEEE Trans. on Communication*. Vol. Com-21, No. 11, Nov. 1973, pp. 1257–1288.

[11] Milkulski, J. J., "DynaTAC Cellular Portable Radiotelephone System Experience in the U.S. and the UK," *IEEE Communications Magazine*, Vol. 24, No. 2, February 1986, p. 42.

[12] Ziemer R. E. and R. L. Peterson, *Digital Communications and Spread Spectrum Systems*, Macmillan Publishing Co.: New York, 1985, p. 88.

Chapter 7
Analog and Digital Modulation

7.1 INTRODUCTION

Transmission of voice and digital control information by microwave radio channel in the 800–900 MHz frequency range presents significantly different problems than those encountered in conventional land communication systems. Unlike wireline systems, the channel characteristics are never fixed, but vary with movement of the vehicle and changes in its surroundings. These dynamics give rise to a formidable set of design challenges since the character of the radio channel can change drastically during a single call as the vehicle moves through the service area and is "handed off" to successive cell sites. Although these channel variations will occur, the radio transmission parameters, and consequently the voice and data transmission functions, have to be designed to prevent as much as possible the user from experiencing changes in voice quality and in control and signaling reliability.

Considering the dynamic behavior of channels, most analog cellular systems of the world have adapted FM modulation for voice and FSK for data transmission. For low-capacity requirements, analog systems with these modulations serve well. However, with dynamic growth in cellular demand, analog systems have to be replaced by new generation, high-capacity digital systems.

The primary requirements of any digital communication system are the transmitted power and the channel bandwidth. A general system design objective is to use these two resources as efficiently as possible. In particular communication channels, one of these two resources may be more precious than the other and thus most channels can either be classified as power limited or band limited. In mobile communication, both power and bandwidth efficiencies are desired. Based on bandwidth and power, the requirements imposed on digital modulation are:

- The compact output power spectrum;
- An efficient power utilization;
- The variable frequency oscillator (synthesizer);

- Immunity to noise and interference;
- The low hardware and software complexity.

The cellular radio requires that the adjacent-channel interference should be below -60 dB, which imposes that the channel spectrum density be compact (several dB below -60 dBc). In mobile radio a deep fades of the order of 40 dB is common, which will still leave 20 dB carrier-to-interference ratio (CIR) to the desired signal provided the adjacent-channel interference requirement is met. Due to restricted spectrum allocation, bandwidth-efficient modulation techniques can only be chosen. Also, with ever increasing demand for service, the problem of restricted bandwidth allocation can cause significant degradation of the system. Therefore, there is a need to provide subscribers with a quality of service which is demand independent. Digital transmission achieves high immunity to noise and interference, enabling more frequent use of the channels. Demand-assigned multiple access techniques can be applied to digital systems. Digital radio is also far more compatible with future developments in telecommunication networks and is much more adaptable to future changes in speech and channel coding rates. Due to low power availability at the mobile, use of class-C amplification is also desired. Unfortunately the use of a class-C amplifier does generate nonlinearities, resulting into AM/PM conversion. This nonlinear conversion of the amplitude modulation on the carrier into the phase modulation results in phase jitter in the carrier and symbol recovery circuits, reducing their efficiency and consequently the performance of the system.

One way to reduce the nonlinear effects of the class-C amplifier is to make use of a constant-envelope modulation format. This constraint of a constant-envelope feature for the modulation scheme narrows the search to the frequency shift keying (FSK) family of modulations. Binary phase shift keying (BPSK) and higher order modulations, belonging to the PSK family, display a varying amount of nonconstant-envelope characteristics and thus require more careful consideration. Of course, other advantages of the PSK family have to be considered over the FSK family. In general, it has been demonstrated that minimum shift keying (MSK) modulation, belonging to the FSK family, provides better performance for wide channel spacings and worse for narrow channel separations, relative to four-phase modulations. Among these two families of modulations one has to select schemes having high noise immunity and thus requiring a low carrier-to-noise ratio (CNR). This in turn will permit small cell size and thus low geographical channel separation between cochannel cells, resulting in a high-capacity system. Thus, a careful selection of digital modulation has a direct impact on capacity of the system, which is one of the main reasons for the changeover from analog to digital, both in Europe and in the United States.

In view of the above discussion, the content of this chapter is solely devoted to modulation schemes presently adapted in analog and forthcoming digital cellular radios, present and future cordless telephone systems, and proposed personal communications networks, as shown in Table 7.1. The objective is to develop a unified picture of these modulation schemes, and provide reasons for choosing them. To this respect, the reader should have basic fundamentals of analog and digital modulation as background.

We begin this chapter by discussing FM modulation presently adopted in almost all analog cellular radios through out the world. The digital section starts by discussing the FSK and PSK family from the point of view of power and bandwidth efficiency. Important parameters affecting the bandwidth efficiency are the baseband pulse shape, phase transition characteristics, and envelope fluctuations. The background modulation schemes binary frequency shift keying (BFSK), BPSK, quadrature phase shift keying (QPSK), and offset quadrature phase shift keying (OQPSK) are discussed in Section 7.4. Reasons for choosing Gaussian minimum shift keying (GMSK) and $\pi/4$ DQPSK, belonging to the FSK and PSK families, for mobile applications are discussed. Further improvement on spectral efficiency

Table 7.1
Modulation Schemes in Different Mobile Systems

	Modulation Scheme		
Systems	*Voice*	*Data*	*Comments*
Analog cellular			
AMPS (U.S.), TACS (U.K.)	FM	FSK	AMPS: R_d = 10 Kbps, spectral efficiency = 0.33 bits/sec/Hz
NTT (Japan)	FM	FSK	
MATS-E (German)	PM	FFSK	
Nordic 450/900	PM	FSK (MSK)	
C-450 (German)	PM	FSK	
Digital cellular			
GSM	GMSK	GMSK	R_d = 270.8 Kbps, spectral efficiency = 1.35 bits/sec/Hz
NADC	$\pi/4$-DQPSK	$\pi/4$-DQPSK	R_d = 48.6 Kbps, spectral efficiency = 1.62 bits/sec/Hz
JDC	$\pi/4$-DQPSK	$\pi/4$-DQPSK	R_d = 42 Kbps, spectral efficiency = 1.6 bits/sec/Hz, B_bT = 0.3
Cordless Telephone			
CT1	Analog, FM	—	
CT2	Digital, MSK	MSK	R_d = 72 Kbps, spectral efficiency = 0.72 bits/sec/Hz
CT3, DECT	GMSK	GMSK	R_d = 1.152 Mbps, spectral efficiency = 0.67 bits/sec/Hz, B_bT = 0.5
PCN (USA)	QPSK*	QPSK	
PCN (Japan)	$\pi/4$-DQPSK†	$\pi/4$-DQPSK	

*Proposed CDMA. †Proposed.

of GMSK is possible by using correlative coding, which is the basis of tamed frequency modulation (TFM), and is discussed in section 7.6. In order to complete the whole area of mobile communication, we have discussed the FM modulation briefly in Section 7.2. We should note that FM is once again a constant-envelope modulation.

7.2 ANALOG ANGLE (FM AND PM) MODULATION

FM-modulated speech with channel spacing of 25 kHz or more can meet the toll-quality voice standard demanded by the telephone industry in multipath surroundings. Thus, for an acceptable grade of service, linear modulation, such as AM and SSB, is not used very much for mobile communication. As shown in Table 7.1, all analog cellular systems of the world have adapted either FM or PM, the two different forms of angle modulation. With AM and SSB modulations, reproduced speech may contain distortions even at good received-signal levels. With FM, the distortion drops to a very low level, provided the frequency deviation due to modulation is sufficiently high, say a 5-kHz peak or more. Also, signal levels below a certain threshold value where noise increases rapidly can be removed by proper use of squelch circuits in the receiver. Angle modulation, being nonlinear, provides different gains to the unequal signals at the receiver input. The overall effect is to suppress the weaker interfering signals, thus enhancing the desired signal quality at the output.

The phenomenon is known as the capturing effect, and for narrowband (NB) FM systems it provides an interference rejection of about 6 dB compared with linear receivers. It has been shown that FM in the single-channel-per-carrier mode is a very effective modulation in interfering surroundings. Studies have also shown that there is an optimum bandwidth of about 40 kHz to obtain a maximum number of conversations per km^2 (for a 30-dB signal-to-noise ratio (SNR) at the baseband output) [1–6].

The optimum bandwidth is caused by the fact that with FM modulation the SNR and signal-to-impairment ratio (SIR) after the demodulation can be increased by increasing the modulation index, thus wider channel spacing makes frequency reuse possible at a smaller distance from the cell. This results in a better reuse of the frequencies per km^2, which initially overcompensates the loss in channels per cell. In view of the above, channel bandwidths of 25 kHz and 30 kHz adapted by most cellular systems of the world are nearly optimum.

7.2.1 Characteristics of Angle Modulation

Unlike linear modulation such as AM and SSB where operations performed on the signal are linear and the spectrum of the modulated signal is basically the translated message spectrum, angle modulation is a nonlinear process and the spectral components of the modulated waveform are not related in any simple form to the message spectrum. In linear modulation, the modulated signal bandwidth is never more than twice the baseband bandwidth. In the case of angle modulation, the modulated signal is usually much greater than twice the message bandwidth. The increase in bandwidth is amply compensated for by an improved SIR at the receiver output. As a matter of fact, improvements of output signal to noise is directly related to an increase in signal bandwidth and is independent of a signal power increase. An angle modulated signal can be represented as

$$v_c(t) = A\cos[\omega_c t + \Phi(t)] = \mathrm{Re}\{Ae^{j[\omega_c t + \Phi(t)]}\} \tag{7.1}$$

The instantaneous phase is represented as

$$\Theta_i(t) = \omega_c t + \Phi(t) \tag{7.2}$$

Thus, the instantaneous frequency of the modulated signal is defined as

$$\omega_i(t) = d\Theta_i/dt = \omega_c + d\Phi/dt \tag{7.3}$$

The functions $\phi(t)$ and $d\phi/dt$ are referred to as the (instantaneous) phase and frequency deviations, respectively. The phase deviation of the carrier $\phi(t)$ is related to the baseband message signal $v(t)$. Depending on the nature of the relationship between $\phi(t)$ and $v(t)$, we have different forms of angle modulation

In phase modulation, the instantaneous phase deviation of the carrier is proportional to the message signal; that is,

$$\Phi(t) = k_p v(t) \tag{7.4}$$

where k_p is the phase deviation constant (expressed in radians/volt). For frequency modulated signals, the frequency deviation of the carrier is proportional to the message signal, that is,

$$\frac{d\Phi}{dt} = k_f v(t) \tag{7.5}$$

$$\Phi(t) = k_f \int_{t_0}^{t} v(\lambda)d\lambda + \Phi(t_0) \tag{7.6}$$

where k_f is the frequency deviation constant (radians/sec/volt). Assuming $t_0 = -\infty$ and $\phi(t_0 = -\infty) = 0$, we can express the angle-modulated waveform equation (7.1) in two different forms for PM and FM, respectively:

$$v_c(t) = A\cos[\omega_c t + k_p v(t)] \tag{7.7a}$$

$$= A\cos\left[\omega_c t + k_f \int_{-\infty}^{t} v(\tau)d\tau\right] \tag{7.7b}$$

Comparing these equations, they are of the same general form except that PM is an angle modulation with ϕ proportion to unmodulated voice $v(t)$, while FM is an angle modulation with ϕ proportional to

$$\int_{-\infty}^{t} v(\tau)d\tau \tag{7.8}$$

A block diagram representation of FM (7.7a) and PM (7.7b) modulations is shown in Figure 7.1.

7.2.2 FM Spectra

Since angle modulation is a nonlinear process, an exact description of the spectrum of the output signal for an arbitrary $v(t)$ is difficult to arrive at. However, if $v(t)$ is sinusoidal, then the instantaneous phase deviation of the angle-modulated signal (for both PM and FM) is also sinusoidal and the spectrum can easily be obtained as shown below. Let

$$v(t) = A\cos\omega_m t \tag{7.9}$$

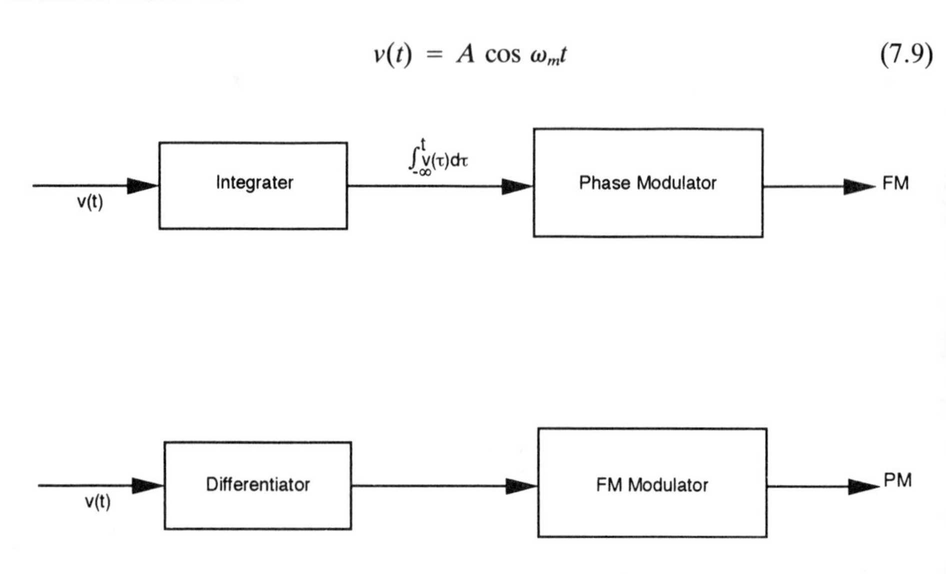

Figure 7.1 Relationship between FM and PM modulators.

Then the instantaneous phase and frequency deviations are

$$k_p A \cos \omega_m t \tag{7.10a}$$

$$\frac{k_f A}{\omega_m} \sin \omega_m t \tag{7.10b}$$

for PM and FM, respectively. The modulated signal for an FM case can be written as

$$\begin{aligned} v(t) &= A \cos\left[\omega_c t + \frac{k_f A}{\omega_m} \sin \omega_m t\right] \\ &= A \cos[\omega_c t + \beta \sin \omega_m t] \end{aligned} \tag{7.11}$$

where the parameter β is only defined for tone modulation and represents the maximum phase deviation produced by a single tone. Equation (7.11) can be written as

$$v(t) = A \cos(\beta \sin \omega_m t)\cos \omega_c t - A \sin(\beta \sin \omega_m t)\sin \omega_c t \tag{7.12}$$

using the identity

$$\cos(\beta \sin \omega_m t) = J_0(\beta) + 2 \sum_{n=1}^{\infty} J_{2n}(\beta)\cos(2n\omega_m)t \tag{7.13}$$

and

$$\sin(\beta \sin \omega_m t) = 2 \sum_{n=0}^{\infty} J_{2n+1}(\beta)\sin(2n + 1)\omega_m t \tag{7.14}$$

where $J_n(\beta)$ is the Bessel function of the first kind and order n with argument β. Thus, the modulated signal in (7.12) can be expressed as

$$\begin{aligned} v(t) = {} & AJ_0(\beta)\cos \omega_c t - 2AJ_1(\beta)\sin \omega_m t \sin \omega_c t \\ & + 2AJ_2(\beta)\cos 2\omega_m t \cos \omega_c t - 2AJ_3(\beta)\sin 3\omega_m t \sin \omega_c t + \cdots \end{aligned} \tag{7.15}$$

The plots of $|V(\omega)|$ are shown in Figure 7.2. From the above graph, the following should be noted. Spectra of $v(t)$ are a set of discrete impulses of strength $AJ_n(\beta)$ spaced ω_m apart. For sufficiently large n, $J_n(\beta)$ becomes very small (less than 0.1) and we therefore say $v(t)$ is band-limited to $2n\omega_m$ where n is the largest n such that

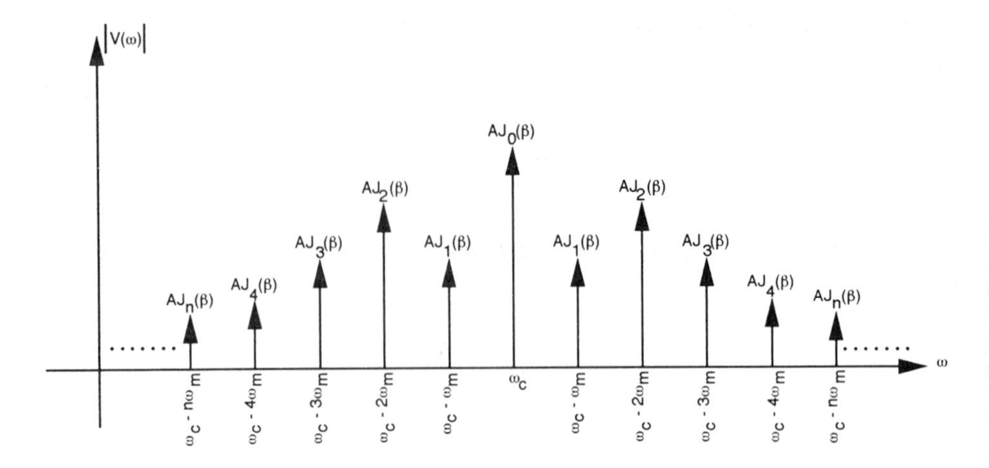

Figure 7.2 Absolute value of FM spectrum.

$J_n(\beta) \leq 0.01$. Using the above definition of bandwidth (BW) we obtain the following universal BW curve for a single-tone sinusoidally modulated FM signal. Figure 7.3 shows a universal bandwidth for a single-tone FM signal.

The following observations can be made from the above curve.

- For large β ($\beta > 20$), $BW \rightarrow 2\Delta\omega$ (asymptotic value): wideband (WB) FM.
- For small β ($\beta < 0.3$), $BW \rightarrow 2\omega_m$: narrowband FM.
- For intermediate values of β, Carson suggested a convenient rule given by

$$BW = 2\omega_m(1 + \beta) = 2(\Delta\omega + \omega_m) \tag{7.16}$$

One should note that the Carson's rule estimates the BW on the low side. The above results are only good when $v(t)$ is a tone. These results also hold for a PM signal where $\beta = Ak_p$.

For an arbitrary modulated signal $m(t)$, the following general guidelines can be used. An FM waveform is represented as

$$v_c(t) = A\cos\left(\omega_c t + k_f \int_{-\infty}^{t} v(\tau)d\tau\right) = A\cos[\omega_c t + m(t)] \tag{7.17a}$$

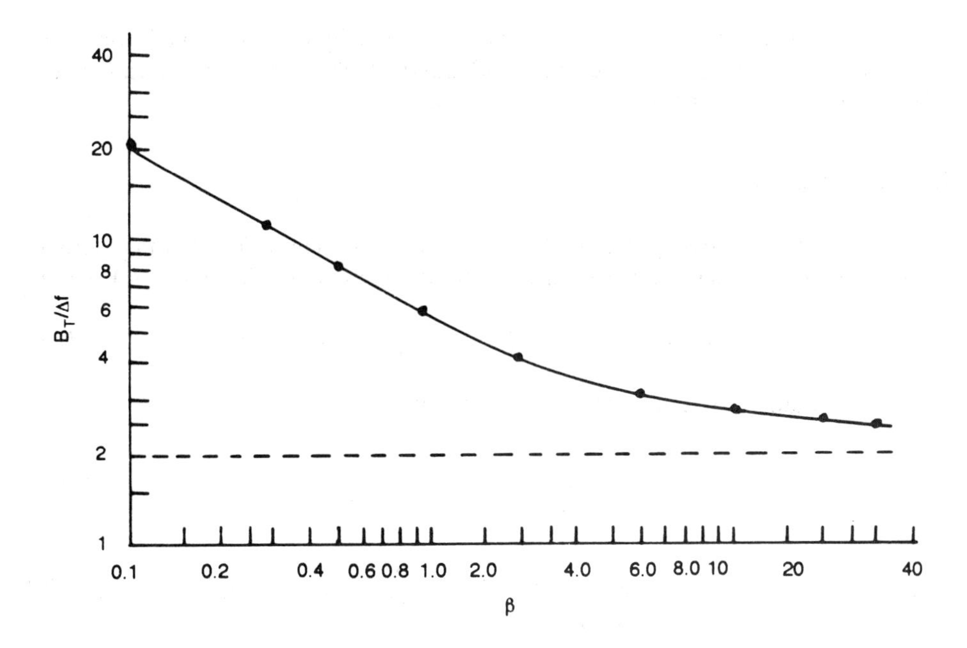

Figure 7.3 Universal bandwidth curve for a single tone FM signal. After [35].

where

$$m(t) = k_f \int_{-\infty}^{t} v(\tau)d\tau \tag{7.17b}$$

In the narrowband FM case, (7.17a) is approximated as

$$v_c(t) = A \cos \omega_c t - A[m(t)]\sin \omega_c t; \ |m(t)| \ll \pi/2 \tag{7.18}$$

and the modulated signal BW is nearly equal to $2\omega_m$. For WBFM, the spectrum of $v_c(t)$ follows the shape of PDF of $v(t)$ displaced around the carrier frequency.

7.2.3 Generation of FM Signals

There are basically two techniques of generating FM signals, direct and indirect. The direct method makes use of a device known as a voltage-controlled oscillator (VCO) whose frequency depends linearly upon the applied voltage. In the indirect method, a narrowband FM signal is first generated by an AM modulator. Then the narrowband FM is converted to a wideband FM by frequency multiplication. We shall discuss both of these methods below.

7.2.3.1 Direct FM

Here we allow the message signal $v(t)$ to control a frequency determining element in an oscillator. Consider a tuned circuit oscillator, as shown in Figure 7.4, where

$$C = C_0 + \Delta C(t) = C_0 + kv(t) \qquad |\Delta C| \ll C_0 \tag{7.19}$$

The instantaneous frequency of the oscillator is

$$\omega_i = \frac{1}{\sqrt{LC}} = \frac{1}{\sqrt{L(C_0 + \Delta C)}} \tag{7.20a}$$

$$= \frac{1}{\sqrt{LC_0\left(1 + \Delta C/C_0\right)}} \approx \omega_0\left(1 - \frac{\Delta C}{2C_0}\right) = \omega_0\left(1 - \frac{kv(t)}{2C_0}\right) \tag{7.20b}$$

where

$$\omega_0 = \frac{1}{\sqrt{LC_0}} \qquad \left|\frac{\Delta C}{C_0}\right| \ll 1$$

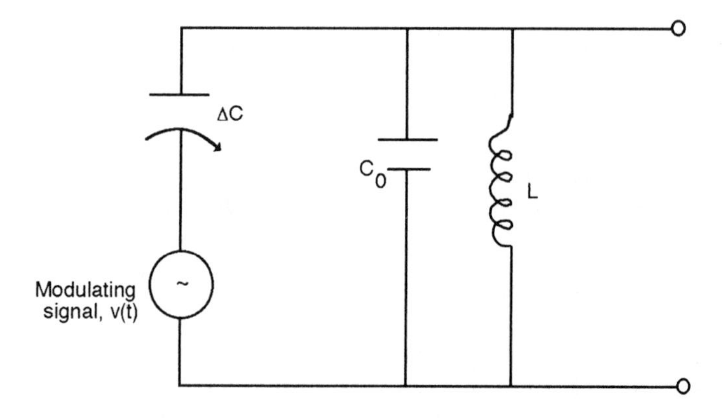

Figure 7.4 Direct FM generation.

Thus,

$$\omega_i = \omega_0 - [(k\omega_0)/(2C_0)]v(t) \tag{7.20c}$$
$$= \omega_0 + \Delta\omega v(t)$$

Here, $\Delta\omega = -(k\omega_0)/(2C_0)$ = peak frequency deviation. The advantage of this approach is that no frequency conversion is required, which always produces some phase distortion. The disadvantage is the poor frequency stability that will require an automatic frequency compensation (AFC) circuit. See Figure 7.5.

The variable capacitance required in the generation of an FM signal can be implemented by using different devices, such as reversed biased PIN diodes, reactance tube, reflex klystron, and a saturable reactor.

7.2.3.2 Indirect FM (Armstrong Modulator)

This method generates a narrowband FM signal and when passed through a frequency multiplier creates a WBFM signal. Equation (7.18) for a narrowband FM is

$$v_c(t) = A\cos\omega_c t - A\left[k_f\int_{-\infty}^{t} v(\tau)d\tau\right]\sin\omega_c t \tag{7.21}$$

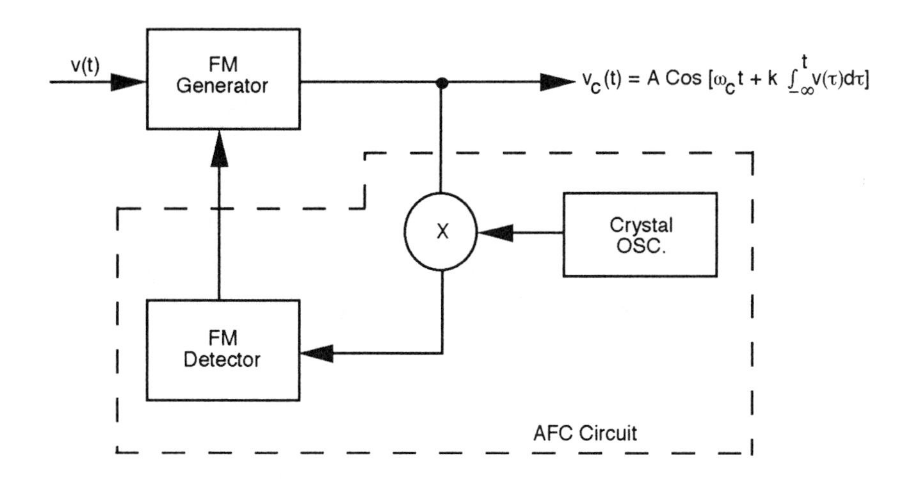

Figure 7.5 Direct FM generation with automatic frequency compensation.

The implementation of this equation is shown in Figure 7.6(a). A WBFM signal can be obtained from the NBFM signal by a frequency multiplier. The frequency multiplier is a nonlinear device designed to multiply the incoming frequencies by a factor n, as shown in Figure 7.6(b). Equation (7.21) is represented as

$$v_c(t) = A\cos\omega_c t - B\sin\omega_c t = \sqrt{A^2 + B^2}\cos[(\omega_c t + \tan^{-1}(B/A)] \quad (7.22)$$

where

$$B = Ak_f\int_{-\infty}^{t} v(\tau)d\tau$$

Since the practical frequency multiplier acts as a limiter, AM will be stripped off and will not cause any problem. On the other hand, it is difficult to design a multiplier without introducing phase distortion. Also approximating $\tan^{-1}(B/A)$ by B/A does introduce harmonic distortion.

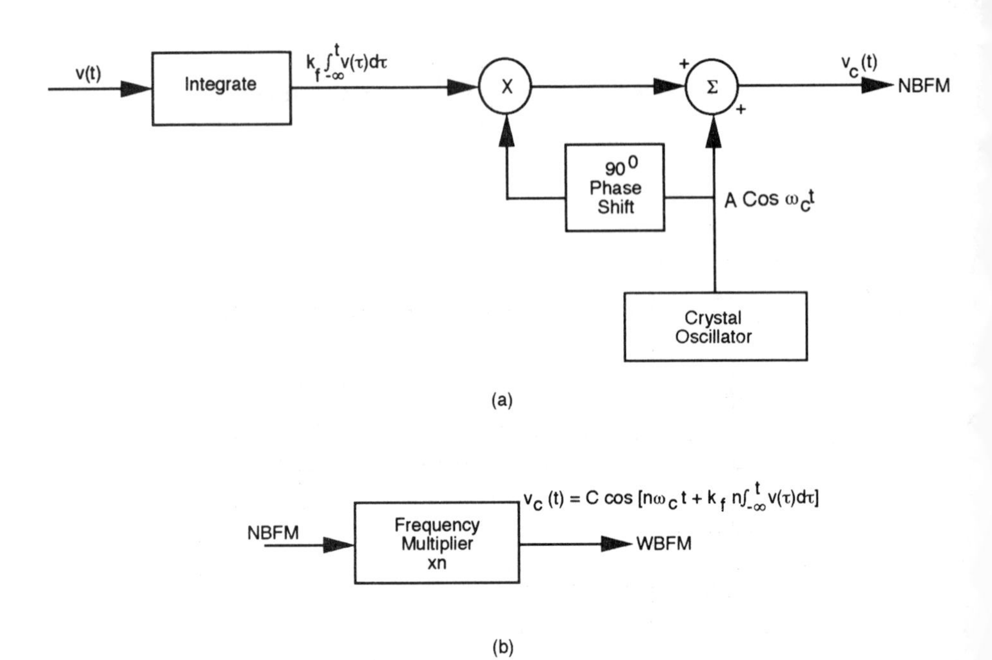

Figure 7.6 (a) NBFM transmitter realization; (b) frequency multiplier for WBFM generation.

7.2.4 FM Demodulators

FM demodulation is required to produce an output voltage that is linearly proportional to the incoming frequency. One such circuit that provides this type of response is known as a discriminator circuit. Figure 7.7 provides a block diagram of a limiter discriminator for FM detection. Input to the discriminator in this case is represented as

$$v_2(t) = A \cos\left[\omega_c t + k_f \int_{-\infty}^{t} v(\tau)d\tau\right] \tag{7.23}$$

The characteristics of an ideal discriminator are obtained through the use of a differentiator followed by an envelope detector. Thus, differentiating (7.23), we have

$$v_3(t) = A[\omega_c + k_f v(t)] \sin\left[\omega_c t + k_f \int_{-\infty}^{t} v(\tau)d\tau\right] \tag{7.24}$$

Since the output of the differentiator is an AM signal, an envelope detector can be used to recover the message signal $v(t)$. In order to recover the signal without distortion max $|k_f v(t)| \ll \omega_c$. Since the discriminator is sensitive to any AM signal, the limiter is used in the front end to remove amplitude variation from the incoming signal. This will ensure that only the frequency-modulated signal is present at the discriminator input. Components outside the frequency band are eliminated by including the bandpass filter, $H_{BP}(f)$. The characteristics of the intermediate frequency (IF) filter, discriminator, and baseband filter are shown in Figure 7.7. From (7.24), it is clear that the output voltage of the discriminator is a function of the deviation constant k_f. Thus, the output signal power can be increased by increasing k_f. The output SNR is a function of the frequency deviation at the discriminator input and is not a function of the input signal power.

One of the problems faced in the realization of an FM discriminator is the linear operating range. The use of double discriminators, known as a balanced discriminator, is usually adapted for this purpose [5–6]. PLL demodulators are also widely in use these days due to their superior performance, ease of implementation when using integrated circuits, and low cost. Since this is covered in almost all books on modulation, we will not cover it here.

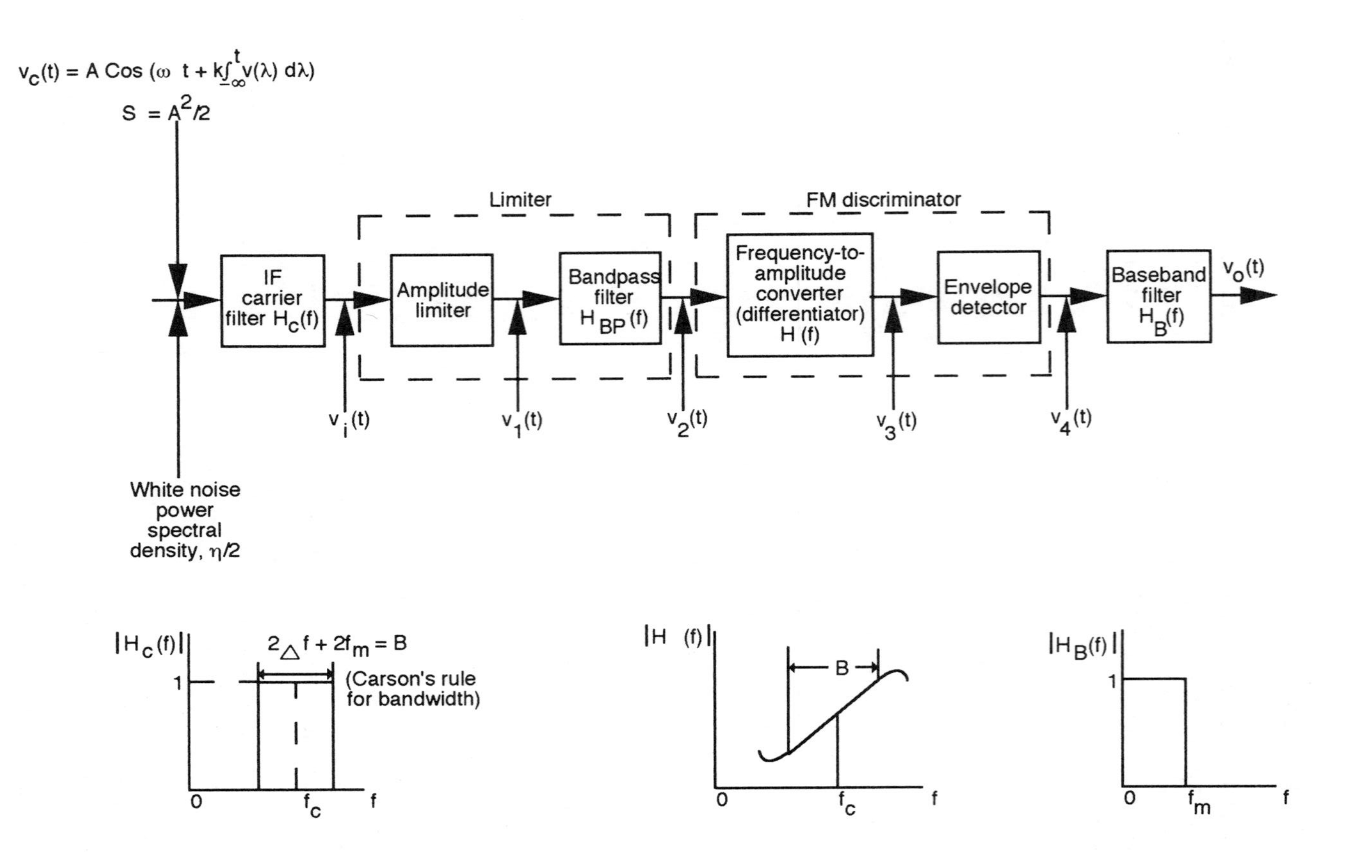

Figure 7.7 A limiter discriminator to demodulate an FM signal.

7.2.4.1 The Effects of Noise on the FM System

The effect of noise on FM communication systems is slightly more complicated than with AM systems. From the receiver block diagram shown in Figure 7.8, we can define the output SNR as

$$\mathrm{SNR}_0 = \frac{\text{mean signal power}}{\text{mean noise power with unmodulated carrier}} \tag{7.25a}$$

Because of the inclusion of a limiter, we are now concerned only with the effect of noise on the phase of the signal.

The predetection signal-to-noise ratio SNR_i is often called the CNR and, in the simple case of single-tone modulation, is given by

$$\mathrm{SNR}_0 = 3(\beta + 1)\beta^2\mathrm{SNR}_i \approx 3\beta^3\mathrm{SNR}_i \tag{7.25b}$$

where $3\beta^3$ is the SNR improvement factor because, for $\beta > 1$, the output SNR is increased. Experimentally it has been determined that the FM exhibits a threshold. For a given value of β, as the SNR_i decreases, a point is reached where SNR_0 falls off more rapidly than SNR_i. The threshold value of SNR_i is arbitrarily taken to be the value at which SNR_0 falls 1-dB below the dashed extension shown in Figure 7.9.

FM characteristics, both below and above the threshold, are given by Shimbo's equation [7]:

$$(\mathrm{SNR}_0)^{-1} = \frac{1}{3(2F_e)A^2(C/N)} + \frac{2F_e e^{-C/N}}{\sqrt{3\pi}A^2(1 - e^{-C/N})^2}\sqrt{N/C}\left[1 + 24\left(\frac{A}{2F_e}\right)^2 C/N\right] \tag{7.26}$$

where $C/N = \mathrm{SNR}_i$ is the SNR at the input to the discriminator. The parameters are defined as

$$F_e = \frac{B_{IF}}{2f_m} \qquad A = \frac{\Delta f_{\mathrm{rms}}}{f_m} \tag{7.27}$$

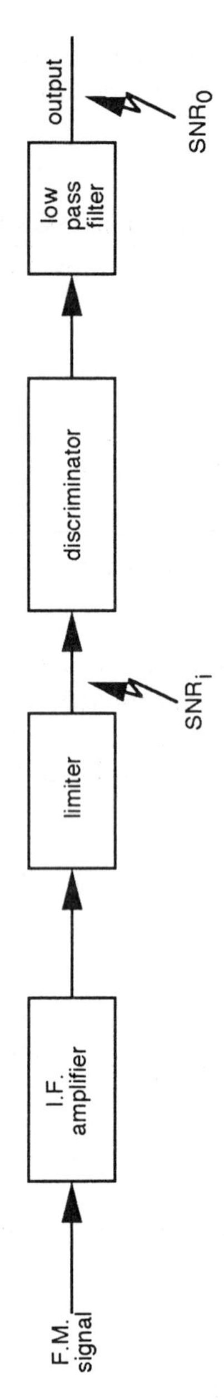

Figure 7.8 FM discriminator demodulation circuit.

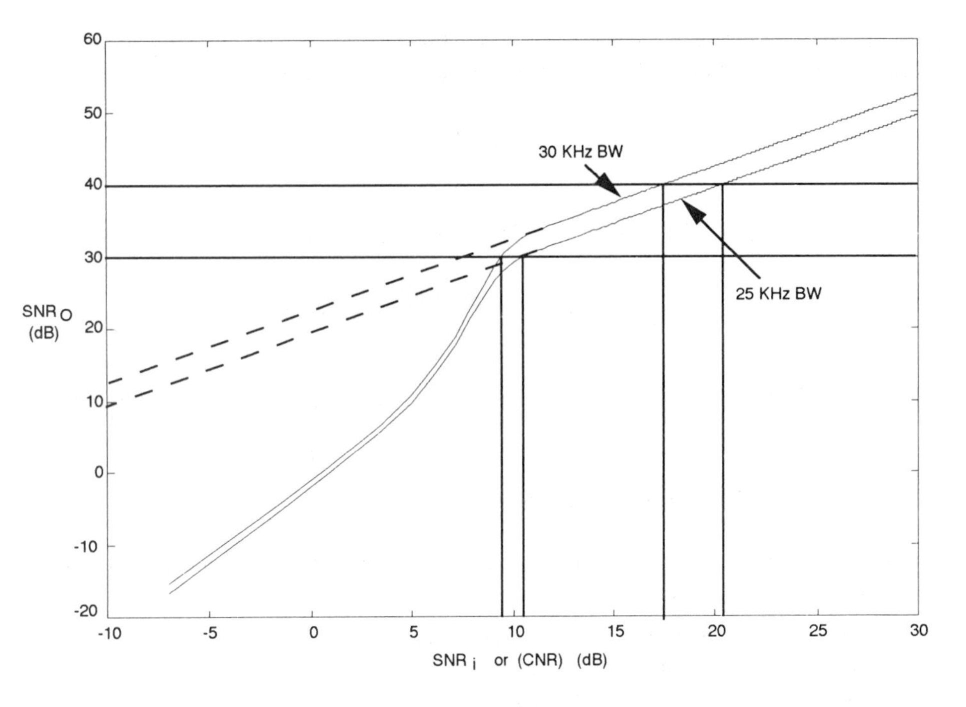

Figure 7.9 SNR_0 versus SNR_i for 30 and 25-kHz channel bandwidths.

With these substitutions, (7.26) is reduced to

$$(SNR_0)^{-1} = \frac{1}{3(B_{IF}/f_m)\,(\Delta f_{rms}/f_m)^2\; C/N} + \frac{(B_{IF}/f_m)\; e^{-C/N}}{\sqrt{3}\pi(\Delta f_{rms}/f_m)^2(1 - e^{-C/N})^2}\sqrt{N/C}\left[1 + 24\left(\frac{\Delta f_{rms}}{B_{IF}}\right)^2 C/N\right] \quad (7.28)$$

Plots of (7.28) for 25 kHz and 30-kHz channels, with Δf_{rms} = 6.65 and 8.485 and f_m = 3.3 kHz, are shown in Figure 7.9. As seen from these curves, the CNR at the threshold is slightly higher for 25 kHz than for a 30-kHz channel. Thus, for the same quality of voice, the input SNR is lower in wider bandwidth systems. In FM modulation, one can always trade off BW with CNR at the receiver input. A CNR value specified for a 30-kHz channel FM system, in the United States is 18 dB. For this CNR, the baseband output SNR is 40 dB. For the same output SNR for a 25-kHz channel, the required value of CNR is about 3-dB higher. In addition, preemphasis/deemphasis gain has to be accounted for before arriving at the final number for output SNR. Preemphasis/deemphasis gain for $f_2 \gg f_1$ is given by [8]:

$$\rho_{FM} = \frac{(f_2/f_1)^2}{3} \tag{7.29}$$

Thus, for $f_2 = 3{,}200$ Hz and $f_1 = 200$ Hz, the improvement factor is

$$\rho_{FM} = 10 \log_{10} \frac{(3{,}200/200)^2}{3} = 19 \text{ dB}$$

Thus, the effective SNR at the baseband output will be approximately 59 dB.

7.3 DETAILED DIGITAL TECHNICAL REQUIREMENTS

There are several criteria that must be considered in choosing a modulation technique for a digital cellular system:

- Spectrally efficient modulation technique;
- Low adjacent-channel interference;
- Good bit error rate (BER) performance;
- Efficient use of mobile dc power;
- Applicability to the cellular environment; and
- Implementation ease.

There are many different modulation techniques that can be considered for digital cellular systems. Although the basic criteria for choosing a modulation format for a cellular system are the same as for any other system, the severe fading conditions present in the cellular channel impose additional constraints on the modulation choice. In addition, the limited spectrum available for cellular use dictates that high spectral efficiency is very important.

Two broad classes of digital modulation are constant envelope modulation and linear modulation. Constant-envelope modulation includes such schemes as MSK, GMSK, and tamed frequency modulation (TFM), which have received much attention in the past few years. MSK is a subset of FSK where the frequency shift is set to one-fourth of the bit rate. Both FSK and MSK can use discriminator for the detection. In order to reduce the transmitted bandwidth, premodulation filters can be added to both these systems and still retain the constant-envelope property. A Gaussian filter is typically used as a premodulation filter in data systems for their linear phase characteristics. The filtered version of MSK is known as GMSK. To detect these filtered versions of FSK and MSK using discriminator detectors with minimum degradation in sensitivity, the premodulation filter cannot be set lower than one-half of the bit rate. By coherently detecting MSK and using a premodulation filter as low as 0.2 times the bit rate, a minimum bandwidth system is produced [9–11]. Because of their constant-envelope property, these schemes allow the use of class-C power amplifiers at the expense of spectral efficiency. Constant-envelope schemes are limited to a spectral efficiency of approximately 1 bit/sec/Hz, regardless of the number of modulation levels used. However, linear modulation schemes such as shaped PSK and QAM, which use linear power amplifiers, can achieve efficiencies of greater than 1 bit/sec/Hz. Greater efficiency can be attained with a higher number of modulation levels used. However, an increase in the number of modulation levels must be traded off with a higher bit error rate due to reduced distance between constellation points. Because of the use of linear amplifiers, linear modulation techniques can provide better out-of-band radiation performance, increasing the spectral efficiency of the system further. Linear modulation techniques have not received as much attention as constant-envelope schemes, partly because of the need for linear RF amplifiers. Advances in technology and devices have made the use of linear amplifiers feasible and cost effective.

7.3.1 Spectral Efficient Modulation Techniques [9]

Channel separation, which is commonly used as a measure of modulation efficiency, should be made as narrow as possible. Assuming that the RF spectra in the adjacent channels cannot be overlapped with each other, the required channel separation f_s is given by

$$f_s = B + 2\Delta f \tag{7.30}$$

where B is the transmission bandwidth occupied by the RF signal power spectrum and Δf is the carrier frequency drift in each transmitter. The transmission bandwidth is given by $B = R_d/n$, where R_d denotes the channel data rate and n is the transmission efficiency that determines the digital modulation method. Equation (7.30) can be written as

$$f_s = (R_d/n) + 2\Delta f \tag{7.31}$$

Thus, to make the channel separation narrower in a digital system for voice transmission systems, it is necessary to:

1. Reduce the encoded speech channel data rate;
2. Narrow band-efficient digital modulation with a high value of n;
3. Lower oscillator drift (i.e., stabilized carrier frequency).

The carrier frequency drift is given as the product of the RF carrier frequency and the frequency stability of the local oscillators. Considering that the mobile radio unit has to be necessarily simplified, miniaturized, and economized, it will not be easy to realize the local oscillator having a frequency stability less than 10^{-6} per year unless any special frequency stabilization technique is adopted. Thus, assuming the frequency stability of the two adjacent channel oscillators are limited to 2×10^{-6}/year, the value of $\Delta f \leq 2$ kHz at 900 MHz.

For cellular radio application where the transmission is over the multipath surroundings, the value of m lies between 1 and 2 bits/Hz. Choosing m above 2 bits/Hz degrades the BER performance of the system as the constellation comes closer. Using the above value of Δf and transmission efficiency m of 1 and 2, the required channel separation f_s versus channel data rate R_d is plotted in Figure 7.10.

Thus, for a 25-kHz channel bandwidth, the maximum data rate can be 22–45 Kbps for transmission efficiencies lying between 1 and 2 bits/sec/Hz. For a 30-kHz channel bandwidth, the corresponding data rates are 28–55 Kbps/Hz. Of course, there are other factors one has to consider before arriving at the final choice of data rate for a given channel bandwidth. Let us simply note here that the data rates for GSM is nearly equal to 34 Kbps per 25-kHz channel bandwidth and for NADCS it is 48.6 Kbps in a channel bandwidth of 30 kHz. Both these numbers fall in this range (but toward the high end).

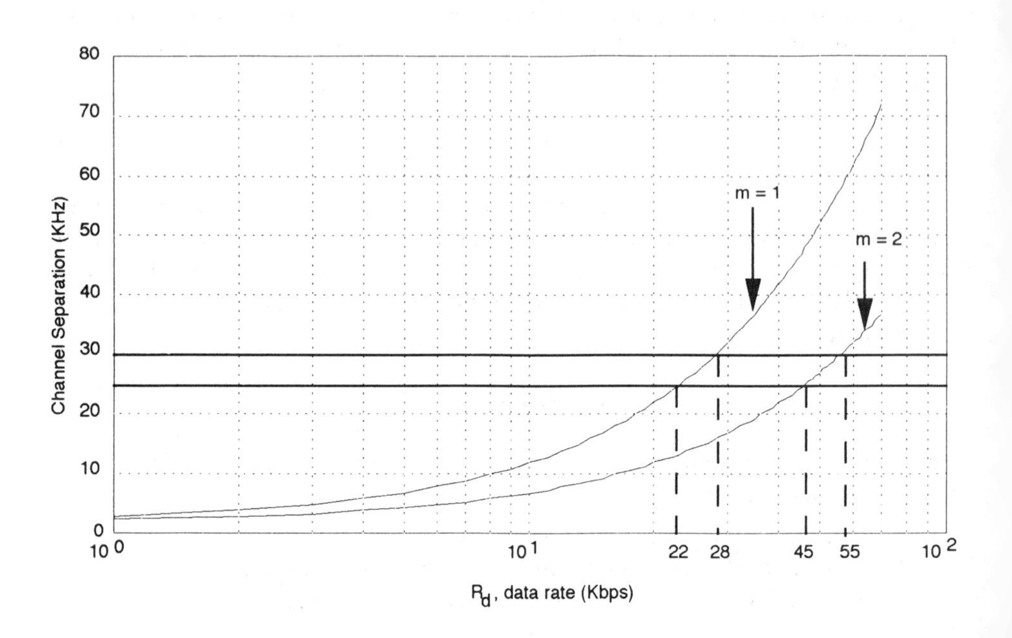

Figure 7.10 Channel separation versus data rate.

7.3.2 Adjacent-Channel Interference [12]

In order to meet the minimum performance standard in digital cellular radio, power in either adjacent channel must be less than 60-dB below the desired-channel power. For 25-kHz channel bandwidth, transmitting 16-Kbps speech, adjacent-channel power is defined as a part of the total power output of the transmitter which falls within the adjacent channel, centered 25-kHz from the desired-carrier frequency. From the above definition, the adjacent-channel interfering power must satisfy

$$10 \log \int_a^b 2s(f)|H(f)|^2 df = -60 \text{ dB} \tag{7.32a}$$

The factor 2 in (7.23a) is due to two adjacent channels interfering with the desired channel. If the receiver IF filtering is assumed to have a gain of unity at the center of the band, as shown in Figure 7.11, then the power received in the IF band must satisfy

$$3 + 10 \log \int_a^b s(f) df = -60 \text{ dB} \tag{7.32b}$$

where a and b are the edges of the rectangular adjacent-channel filter nearest to and furthest from the center frequency of the interfering channel, $s(f)$ is the desired spectrum density of the modulated signal, $H(f)$ is the transfer function of the adjacent channel assumed to have magnitude of unity in (7.32a), the values of a and b are 12.5 kHz and 37.5 kHz, respectively.

As stated in Section 7.1, if the adjacent channel interference power of -60 dB is satisfied, then even with deep fades of the order of of 40 dB, adjacent-channel interference will allow a 20-dB CIR in the channel of interest. Obviously, this requirement can be met if we choose a modulation scheme with narrow output power spectra of the mainlobe and low spectral trails. It should be noted that the true adjacent channel is never allocated in the same cell. However, for a seven-cell configuration where the adjacent channel will be allocated to an adjacent cell, the interference becomes important for a mobile at the boundary of a cell.

7.3.3 Good BER Performance

Linear modulation schemes have the property that increasing the modulation levels increases the spectral efficiency. Thus the modulation level $m = 4$ should be better than $m = 2$. However, a higher level of modulation increases the BER as the signal constellation points come closer for the same power level. Thus, higher level schemes must be traded off with increased BER. On the other hand, the modulation

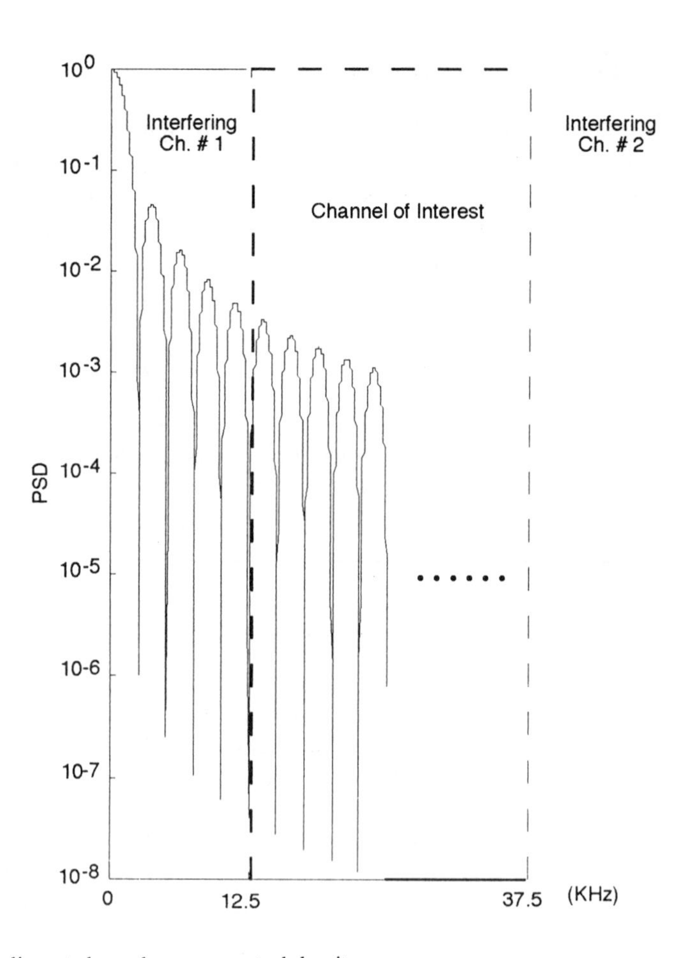

Figure 7.11 Adjacent-channel power spectral density.

scheme should be such that the BER requirements are met at a lower value of CIR or in the presence of a high interfering power. The low required CIR will permit cochannel cells with lower geographical separation, and hence, a higher traffic carrying capacity.

7.3.4 Efficient Use of Mobile dc Power

In cellular radio, the mobile is always power limited, especially those with handheld radios. Constant-envelope modulation will allow efficient utilization of available dc power using class-C power amplifiers. In view of the above, MSK and GMSK are good choices. Linear modulation such as QPSK can provide higher spectral density but requires using a linear power amplifier and is power inefficient.

7.3.5 Applicability to the Cellular Environment

No matter which modulation we choose, it must not only be able to withstand the severe multipath fading but also perform satisfactorily; that is, the BER should at least be less than 10^{-2} or better. This is the minimum acceptable BER for speech communication.

7.3.6 Implementation Ease

It is absolutely essential that the chosen modulation scheme be easily implemented, preferably using large-scale integrated circuits. Use of a digital signal processor may also be desirable, especially in the United States where a dual-mode mobile unit is required due to compatibility requirements with the present analog FM system.

7.4 DIGITAL MODULATION

In this section we shall discuss basic two and four-level modulations and build reasoning for further discussions on actual modulations used for mobile communication. The binary modulations chosen are FSK and PSK. The four-level modulations chosen for discussion belong to basic BPSK and are QPSK and its modified form OQPSK.

7.4.1 BFSK Modulation

The constraint of a constant-envelope feature for a modulation scheme narrows the search to two major signaling techniques, namely FSK and PSK. Consider binary communication—transmitting a pulse every T seconds (at the signaling rate of $1/T$ baud) to denote one of two equally likely information symbols, $+1$ or -1. FSK denotes the two states by transmitting a sinusoidal carrier at one of two possible frequencies. See Figure 7.12(a). Figure 7.12(b,c) shows the signal constellation and tone spacing for FSK.

These signals are generated by switching between two oscillations or by frequency modulating an oscillator, as shown in Figure 7.13(a, b).

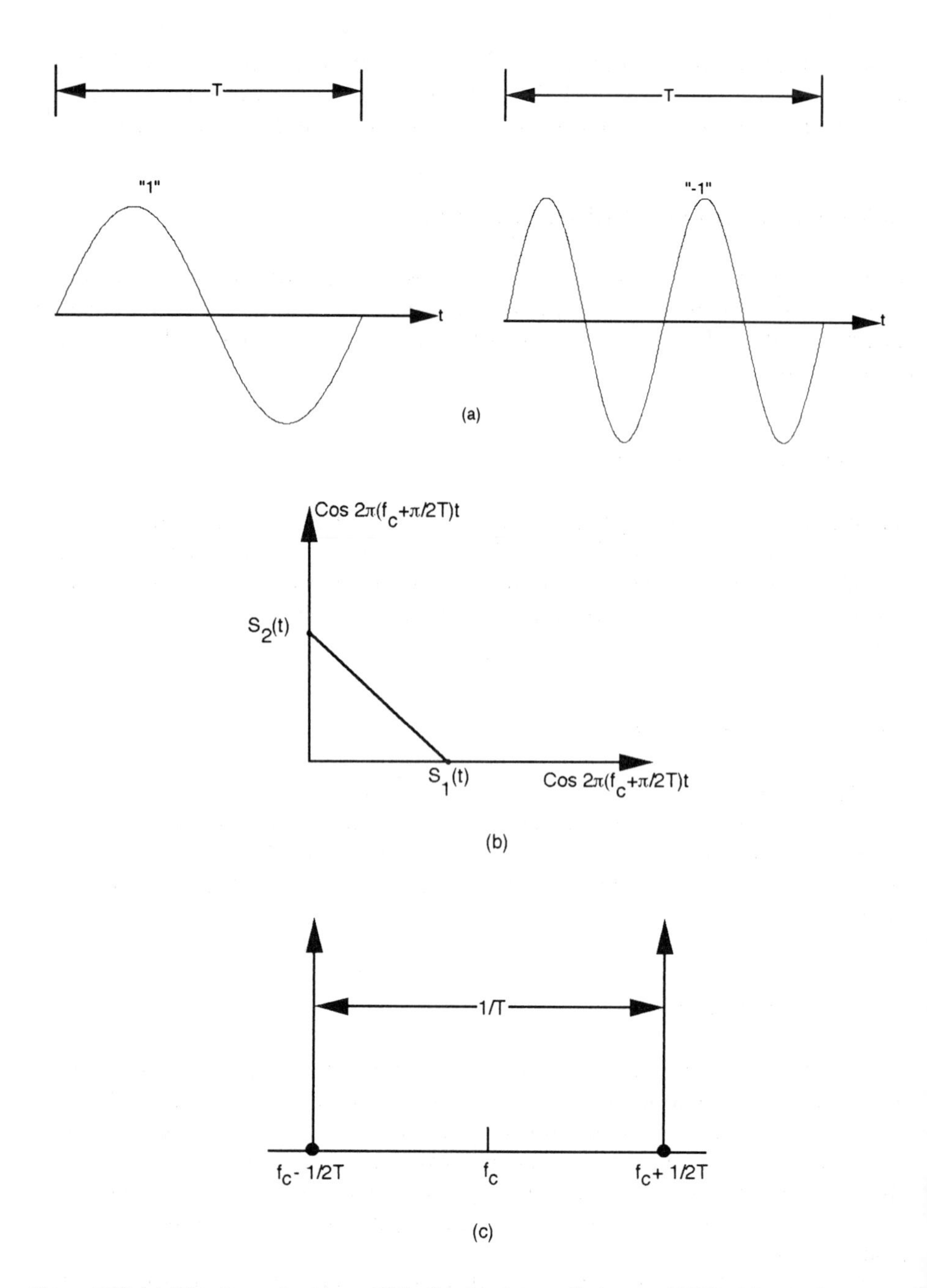

Figure 7.12 (a) Waveforms for binary FSK; (b) signal constellation for BFSK; (c) tone spacing for FSK.

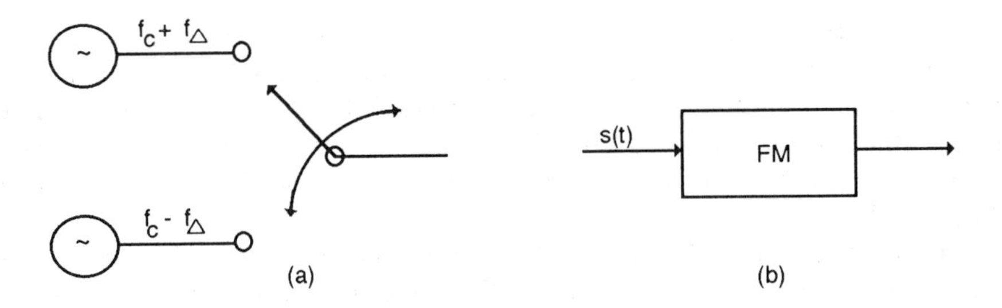

Figure 7.13 (a) BFSK modulator; (b) FM modulator as FSK generator.

Assuming that the modulator constant is set to a maximum frequency deviation of Δf, the modulator output signals are represented as

$$s(t) = A\cos\left[2\pi f_c t + 2\pi\Delta f\int s_b(t)dt\right] \tag{7.33a}$$

$$= A\cos[2\pi(f_c \pm \Delta f)t] \tag{7.33b}$$

In the above equation, $s_b(t)$ is the baseband signal that assumes the value of ± 1, which leads to frequency change of $\pm\Delta f$. Here, we are dealing with a piecewise-constant baseband signal, which leads to FSK signals where the frequency is constant over a bit period T. This abrupt transitions in frequency leads to wide bandwidth. In order to reduce the bandwidth, the baseband pulses can be shaped. This also reduces the intersymbol interference. The instantaneous frequency in this case can be written as

$$f(t) = f_c + \Delta f\cos(2\pi f_c T) \tag{7.34}$$

where f_c is the carrier frequency. Here we are assuming sinusoidal frequency shaping. The frequency deviation is given by

$$\Delta f = (f_2 - f_1)/2 \tag{7.35}$$

where $f_2 = f_c + \Delta f$ and $f_1 = f_c - \Delta f$. Both noncoherent and coherent detection can be used for data recovery at the receiver.

From the viewpoint of simpler receiver implementation, noncoherent detection schemes, which do not make use of the carrier phase reference, can be used. FSK-modulated signals can be detected by two bandpass filters tuned to two different frequencies, followed by an envelope detector, a bit-rate sampler, and a decision circuitry based on comparing which of the two signals is larger. Signals at the output of the two bandpass filters are regarded as noncoherently orthogonal when there is no crosstalk, that is, when an f_1 tone is transmitted, the sampled envelope of the output of the receiving filter tuned to f_2 is zero. For noncoherent orthogonality, the minimum required separation between two tones equals the signaling rate $1/T$. See Figure 7.12(c). The block diagram of this system is shown in Figure 7.14.

Alternatively, an FM discriminator can be used for noncoherent detection of a signal. In this case, the demodulator first changes the FM to AM using time differentiation. The time derivative of the used FM waveform is given by

$$\frac{d}{dt} A \cos\left[2\pi f_c t + K \int s_b(t)dt\right] = A[2\pi f_c + Ks_b(t)] \sin\left[2\pi f_c t + K \int s_b(t)dt\right] \tag{7.36}$$

As shown in the above equation and in Figure 7.15, the FM-modulated signal is first converted to an AM signal, which can be demodulated by envelope detector. The output of the demodulator is sampled and compared to the threshold to decide the presence of +1 or −1. This is the scheme used for detection of FSK signals in most analog cellular systems of the world.

The error rate equation for noncoherent detection with Gaussian noise is given by

$$P_e = 1/2\, e^{-E_b/2N_0} \tag{7.37}$$

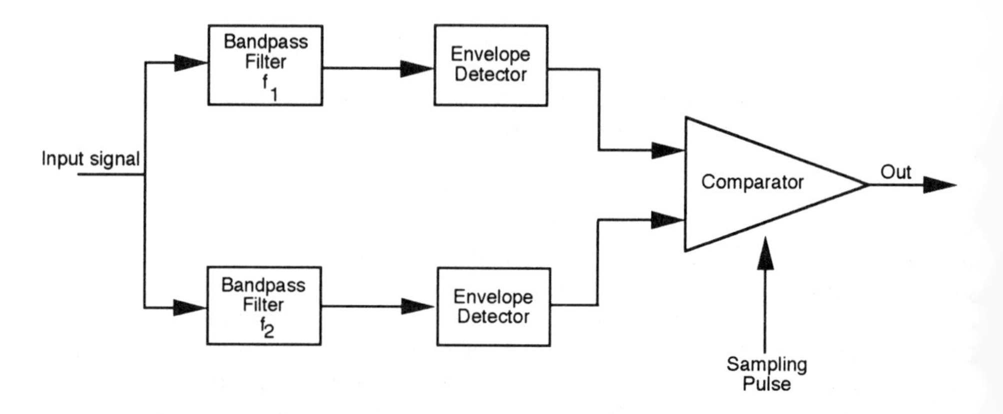

Figure 7.14 Noncoherent FSK detector.

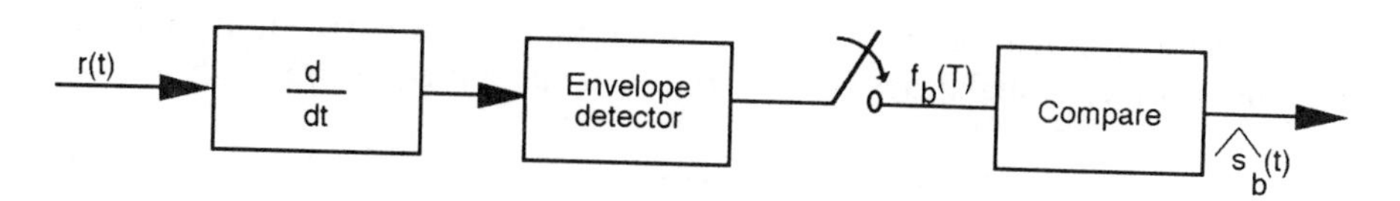

Figure 7.15 FM discriminator.

where the noise variance $\sigma^2 = N_0$. Lower bit error rates can be obtained with phase-coherent detectors. A block diagram is shown in Figure 7.16. Coherent detectors require the availability of a reference signal that is phase synchronous with the desired signal. The detector is sensitive to the phase of the input signal. It can thus reject noise components that are in phase quadrature with the desired signal, resulting in an improved output SNR. The output of the first bandpass filter is multiplied by a phase-synchronous signal of frequency f_1 and the output of the second bandpass filter is multiplied by a phase-synchronous signal at frequency f_2. The outputs of the two multipliers are filtered to eliminate harmonic components and the amplitudes are compared at the sampling instant to decide which frequency was actually transmitted.

The performance of the correlation detector for any binary communication is given by [13–15]:

$$P_e = \frac{1}{2} \operatorname{erf} \sqrt{\frac{(1 - \rho)E_b}{2N_o}} \tag{7.38}$$

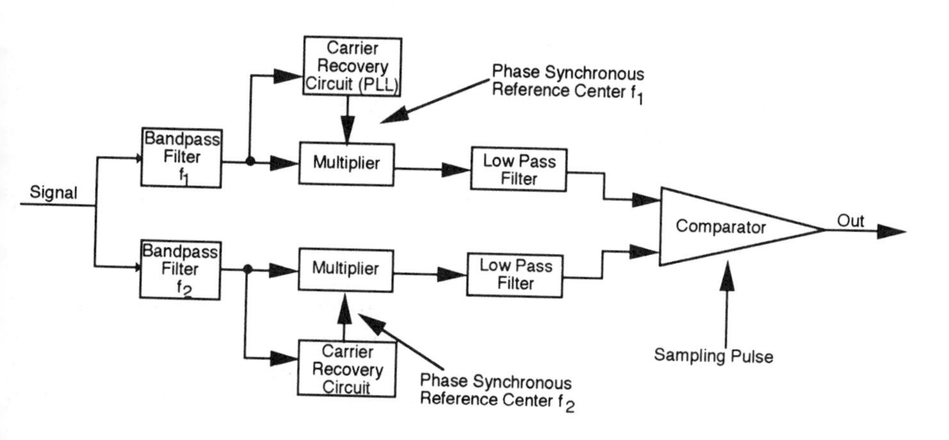

Figure 7.16 Coherent FSK detector.

where E_b is the average energy of the two signals and ρ is the correlation coefficient. Here, the a priori probabilities of ± 1 are the same and equal 1/2, with

$$s_1(t) = A \cos 2\pi f_1 t \qquad 0 \le t \le T \tag{7.39a}$$

$$s_2(t) = A \cos 2\pi f_2 t \qquad 0 \le t \le T \tag{7.39b}$$

Thus,

$$E_b = \int_0^T A^2 \cos^2 2\pi f_1 t dt = \frac{A^2 T}{2} \tag{7.40}$$

if $T \gg 1/f_0$ and

$$\rho = \frac{A^2 \int_0^T \cos 2\pi f_1 t \cos 2\pi f_2 t \, dt}{A^2 T/2} \tag{7.41a}$$

$$\rho = \frac{\sin 2\pi(f_2 + f_1)T}{2\pi(f_2 + f_1)T} + \frac{\sin 2\pi(f_2 - f_1)T}{2\pi(f_2 - f_1)T} \tag{7.41b}$$

where ρ is the correlation coefficient, $f_c = (f_1 + f_2)/2$ under the assumption that $2\pi f_c T \gg 1$ or $2\pi f_c T = k\pi$. The correlation coefficient is reduced to

$$\rho = \frac{\sin 2\pi(f_2 - f_1)T}{2\pi(f_2 - f_1)T} \tag{7.42}$$

The plot of ρ versus $2\pi(f_2 - f_1)T$ is shown in Figure 7.17. If we use $\rho = 0$, the error probability P_e becomes

$$P_e = \frac{1}{2} \operatorname{erf}_c \sqrt{\frac{E}{2N_0}} \tag{7.43}$$

From (7.38), it should be clear that a $\rho < 0$ is desirable as it reduces the error rate for the same value (E_b/N_0). For $\rho \le 0$, signals are uncorrelated. The plots of the coherent and noncoherent error probability curves are shown in Figure 7.18. It is seen that the coherent system always performs better than the noncoherent system under Gaussian noise.

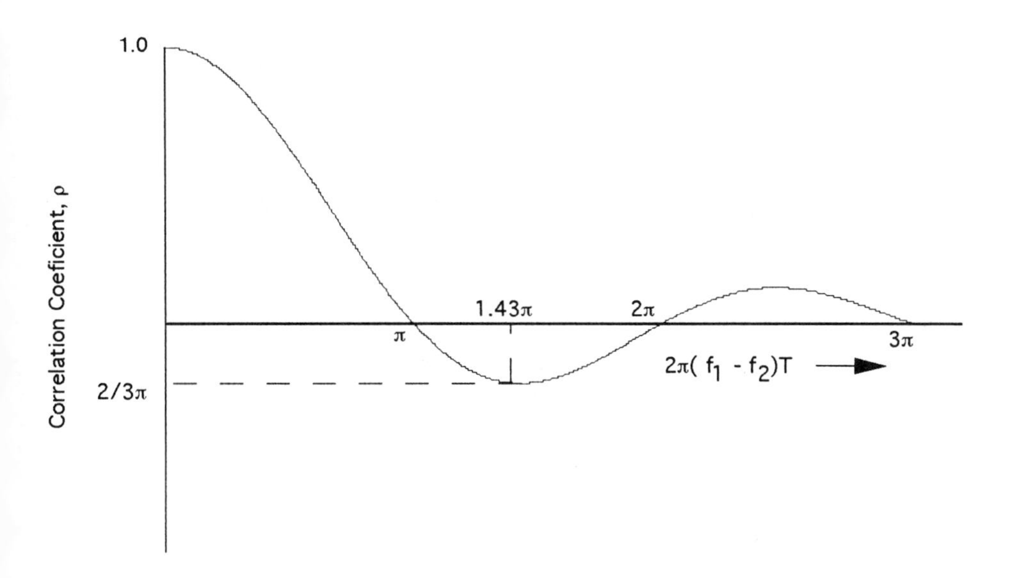

Figure 7.17 FSK correlation coefficient.

7.4.2 BPSK Modulation

The second simplest type of modulation is binary phase shift keying. In BPSK, opposite phases of the carrier (i.e., 0 deg and 180 deg) are transmitted every T seconds based on the information symbols $+1$ or -1. BPSK can also be regarded as an amplitude modulation scheme where the carrier is multiplied by $+1$ or -1 based on the bit polarity, as shown in Figure 7.19. As can be seen from the plot, the carrier flips its phase 180 deg when transition from one to zero occurs. Displayed on the I/Q plane, BPSK appears as two points on the I axis which are 180-deg apart, as shown in Figure 7.20.
DATA $+1$ output modulation is

$$\cos(\omega t + 0 \text{ deg}) = +\cos \omega_c t \tag{7.44a}$$

DATA -1 output modulation is

$$\cos(\omega t + 180 \text{ deg}) = -\cos \omega_c t \tag{7.44b}$$

The modulator structure is shown in Figure 7.21. For BPSK, it can be shown that the best performance calls for an optimum receiver that can be implemented by

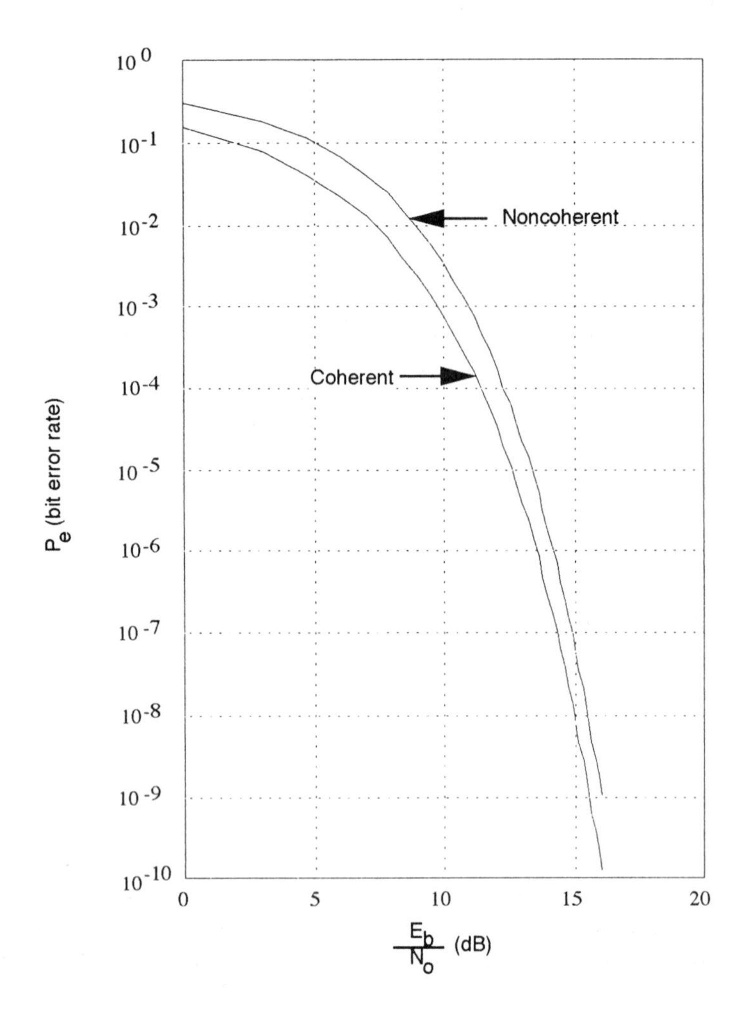

Figure 7.18 Performance comparison of noncoherent and coherent FSK detectors.

matched filters and is otherwise known as a correlation detector. The implementation of an optimum receiver requires a perfect carrier reference phase at the receiver. For coherent optimum detection of an "antipodal" BPSK, the required value of E_b/N_0 is minimum for a specified BER. A BPSK-modulated signal is regarded as antipodal if the two signals denoting the two possible information symbols have exactly the same shape but are opposite in polarity. The correlation receiver structure is shown in Figure 7.22. Here, the first stage is a coherent demodulator followed by an integrator over each signaling interval and a sampling

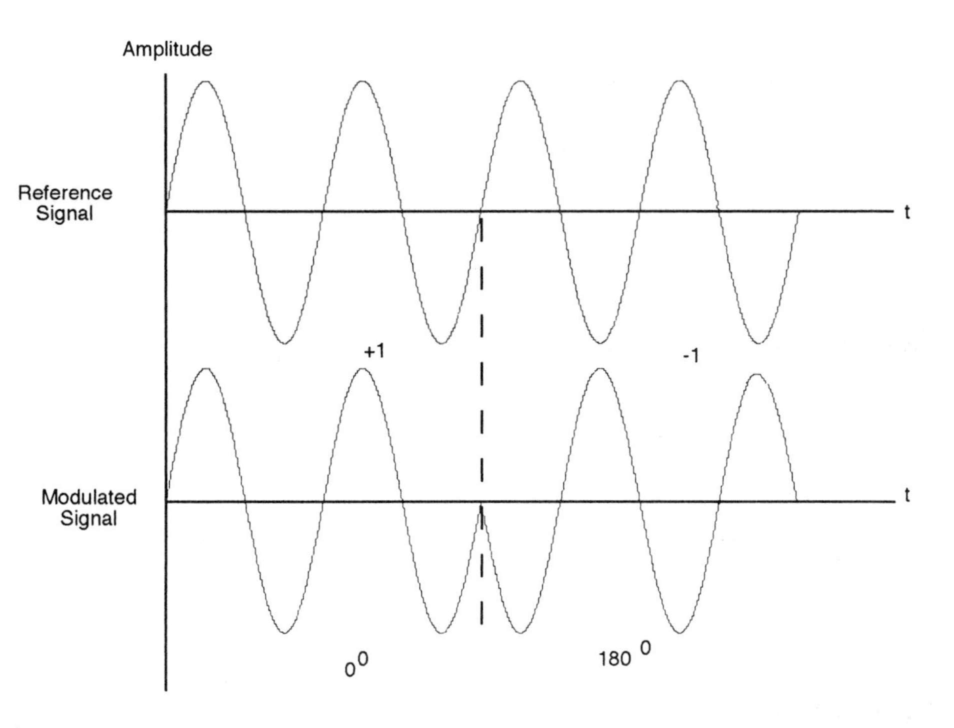

Figure 7.19 Time waveform of BPSK signal.

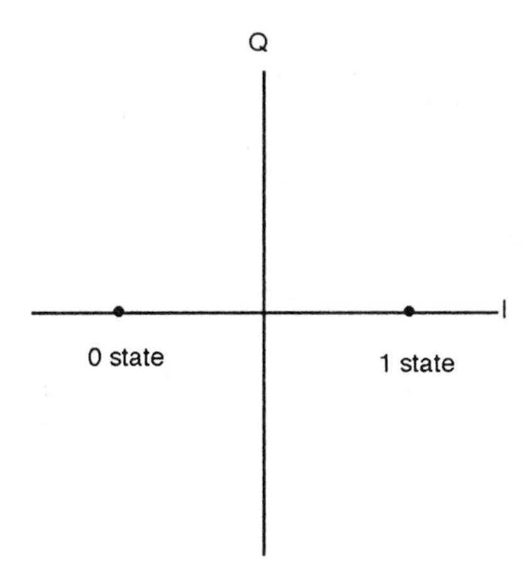

Figure 7.20 BPSK signal constellation.

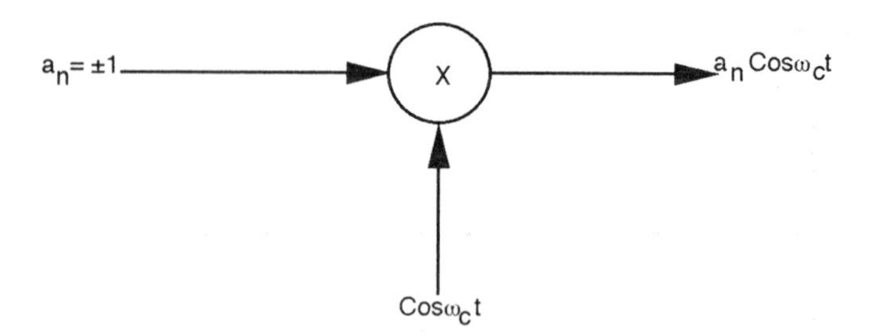

Figure 7.21 BPSK modulator structure.

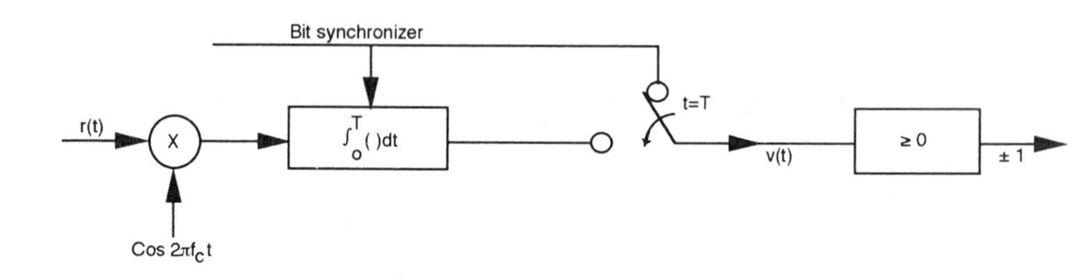

Figure 7.22 BPSK demodulator.

of the integrator output at the end of the signaling intervals. Thus, two stages of synchronization are required. The first is carrier synchronization, wherein the coherent carrier reference is established to perform coherent demodulation. The second, often referred to as bit synchronization, involves the establishment of a clock signal in order to perform the integrate-and-dump operation in synchronism with the incoming stream of contiguous signaling elements. The BER equation can be found by evaluating the correlation factor ρ and substituting in the general expression (7.38) with

$$s_1(t) = \sqrt{2P} \cos 2\pi f_c t \qquad s_2(t) = -s_1(t) \tag{7.45}$$

where the power in the continuous waveform $P = E_b/T$ and the value of correlation coefficient $\rho = -1$. Thus, from (7.38) the BER is represented as

$$P_e = 1/2 \ \mathrm{erf}_c \sqrt{E_b/N_0} \tag{7.46}$$

The plot of P_e versus E_b/N_0 along with the coherent and noncoherent FSK is shown in Figure 7.23.

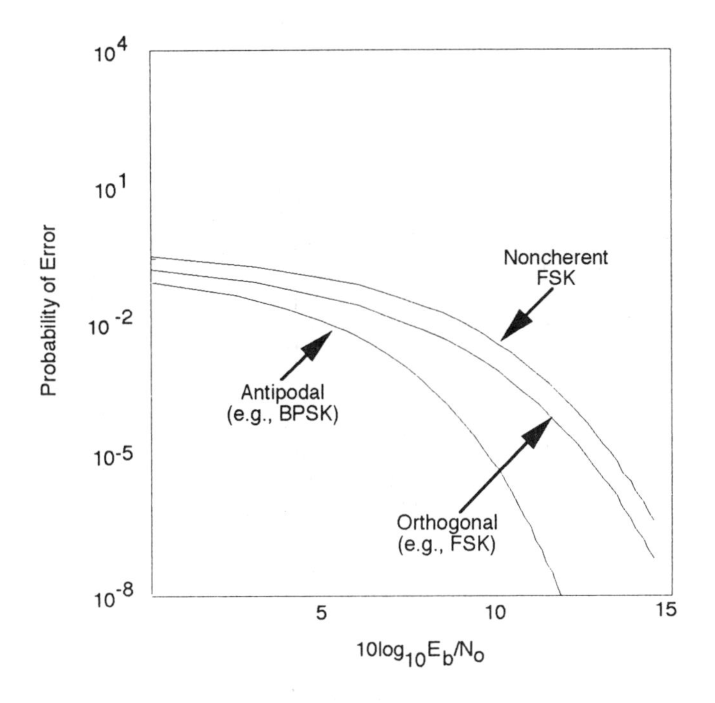

Figure 7.23 BER plots of BPSK and coherent and noncoherent FSK.

Though the realization of BPSK is simple, the bandwidth requirement for a given data rate R_d is large. This makes it unsuitable for mobile application. As a general rule, 99% of the power in a BPSK-modulated carrier will require nearly 8 R_d. Different definitions of bandwidth are shown in Figure 7.24. Among the most commonly used bandwidth definitions are:

1. Half-power bandwidth—the interval between frequencies at which $S(f)$ has dropped to half power, or 3-dB below the peak value.
2. Noise-equivalent bandwidth—the bandwidth that satisfies the relationship $P = W_N S(f_c)$, where P is the total signal power over all frequencies, W_N is the noise-equivalent bandwidth, and $S(f_c)$ is the value of $S(f)$ at the band center assumed to have a maximum value over all frequencies.
3. Null-to-null bandwidth—the most popular measure of bandwidth is the width of the main spectral lobe, where most of the signal power is contained.
4. 99% of power—adopted by the FCC (*FCC Rules and Regulations*, Section 2.202) and states that the occupied bandwidth is the band which leaves exactly 0.5% of the signal power above the upper band limit, and exactly 0.5% of the signal power below the lower band limit. Thus, 99% of the signal power

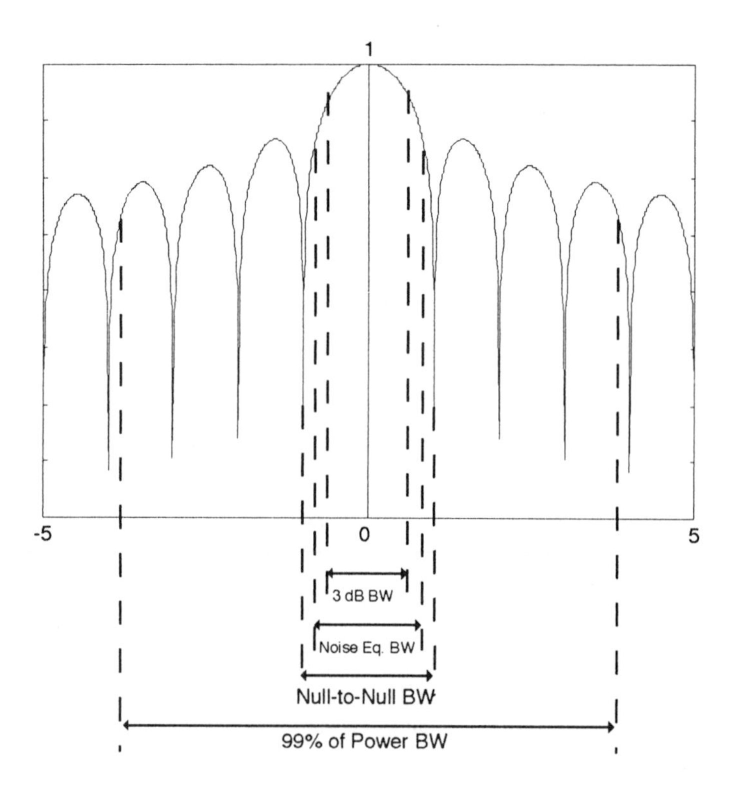

Figure 7.24 Various bandwidth criteria.

is contained inside the occupied band. Obviously, other definitions of bandwidth are less constraining than 99% of power bandwidth.

Using the 99% of bandwidth criterion for a channel data rate of nearly 48 Kbps for the American digital system, BPSK will require 400 kHz of bandwidth, which makes it unsuitable. This poor bandwidth efficiency can be improved by using QPSK, where quadrature carrier is employed.

Comparison of error rate curves for BPSK and FSK clearly shows that both coherent and noncoherent FSK performs poorer than BPSK. Even if coherent orthogonal FSK is used and detected by coherent methods in an effort to improve the performance, it is still poorer by 3 dB (additional E_b/N_0) than BPSK. For this reason, BFSK and a higher order of FSK are generally considered only for low data rate and in low efficiency applications such as the various analog cellular systems of the world, as shown in Table 7.1. For high data rate applications, BPSK or higher order PSK modulations are generally a good choice.

7.4.3 QPSK and OQPSK

As seen in BPSK modulation, for a data rate of 48 Kbps, the required channel bandwidth is nearly equal to 400 kHz. The bandwidth efficiency can be improved by allowing a second orthogonal channel over a basic BPSK-modulated signal:

$$\alpha(t)\cos 2\pi f_c t \tag{7.47}$$

where f_c is the carrier frequency and $\alpha(t)$ is the modulating data stream. Since $\sin 2\pi f_c t$ and $\cos 2\pi f_c t$ are coherently orthogonal signals, we have

$$\int_0^{n/f_c} \sin(2\pi f_c t)\cos(2\pi f_c t)dt = 0 \tag{7.48a}$$

where n is an integer. Thus, a second BPSK signal can be placed in quadrature with the first without causing interference to either of the two signals. Such a modulating scheme, which increases the bandwidth efficiency of a BPSK by two is known as QPSK and is shown in Figure 7.25.

The input binary bit stream $\{a_k\}$, $a_k = \pm 1$; $k = 0, 1, 2, \ldots$, arrives at the modulator input at a rate of $1/T$ bits/sec and is separated into two data streams, $a_I(t)$ and $a_Q(t)$, consisting of even and odd bits, respectively. The resulting modulated QPSK signal $s(t)$ is represented as

$$\begin{aligned} s(t) &= 1/\sqrt{2}\, a_I(t)\cos(2\pi f_c t + \pi/4) + 1/\sqrt{2}\, a_Q(t)\sin(2\pi f_c t + \pi/4) \\ &= A\cos[2\pi f_c t + \theta(t)] \\ A &= \sqrt{1/2(a_I^2 + a_Q^2)} = 1 \\ \theta(t) &= -\tan^{-1} a_Q(t)/a_I(t) \end{aligned} \tag{7.48b}$$

where $\theta(t) \in= \{0 \text{ deg}, \pm\pi/2, \pi\}$ corresponding to the four combinations of $a_I(t)$, and $a_Q(t)$. On the I/Q plane, QPSK appears as four equally spaced points separated by 90 deg, as shown in Figure 7.26(a).

Each of the four possible phase states of carriers represents two bits of data. This leads to a new term called symbol rate. By definition, symbol rate is the bit rate divided by the number of bits per symbol. Thus for QPSK, there are two bits per symbol. Since the symbol rate for QPSK is only half the bit rate, twice the information can be carried in the same amount of channel bandwidth (compared to BPSK). Assuming that the required channel bandwidth is nearly equal to $8R_s$ for containing 99% of the power, the total channel bandwidth equals 200 kHz for a data rate of 48 kHz. Thus, QPSK modulation provides a substantial savings (factor by 2) in channel bandwidth over BPSK.

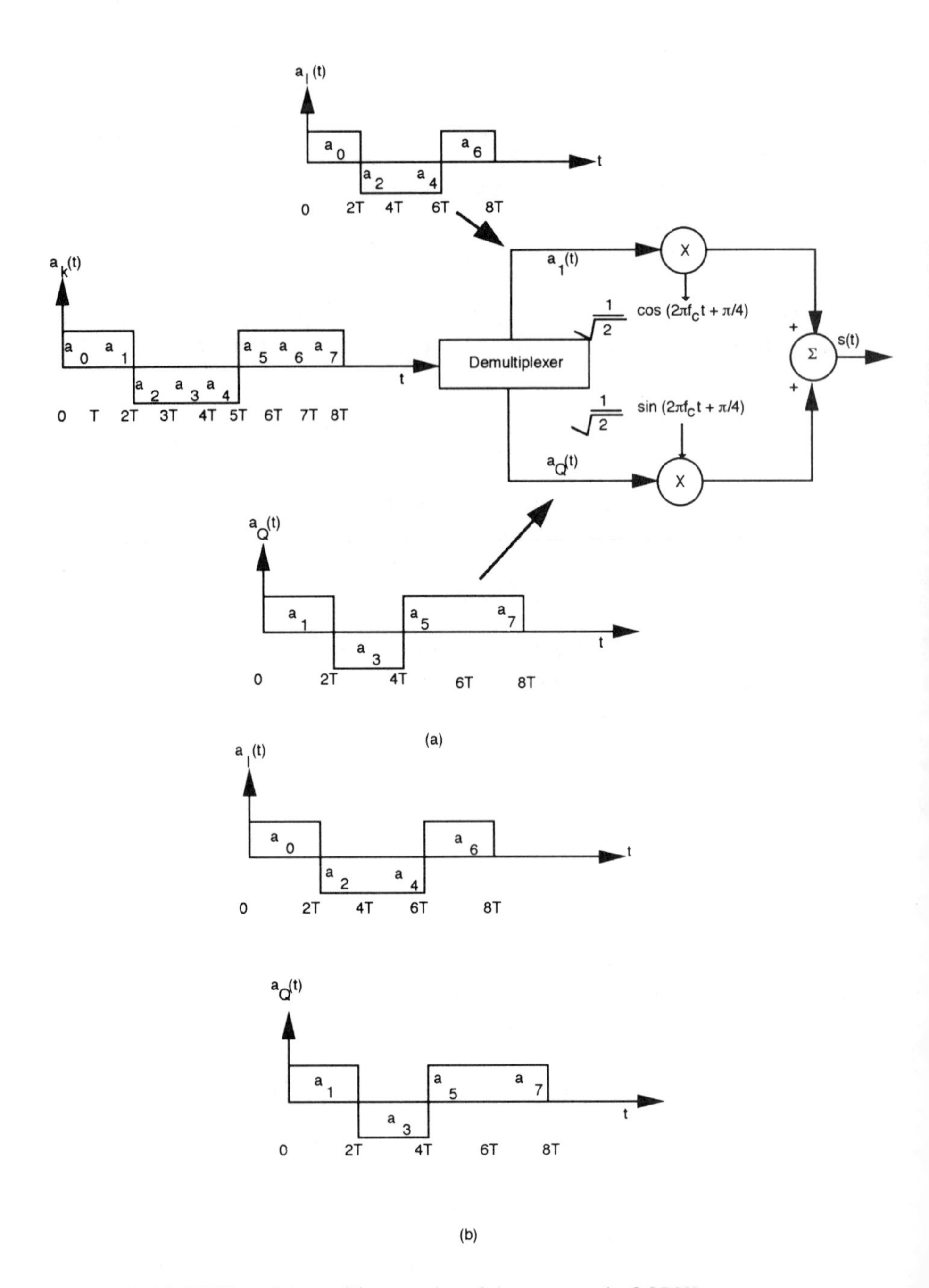

Figure 7.25 (a) QPSK modulator; (b) staggering of data streams in OQPSK.

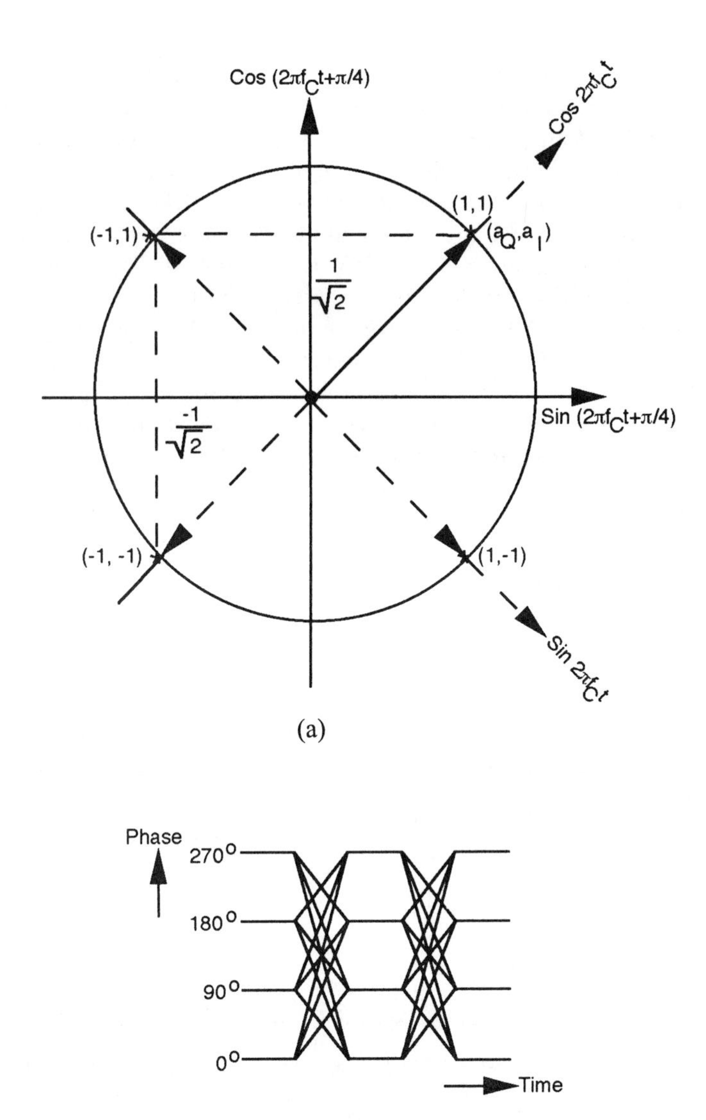

Figure 7.26 Signal space diagram (a) and phase function (b) for QPSK.

In QPSK, due to the coincident alignment of $a_I(t)$ and $a_Q(t)$ streams, the carrier phase can change only once every $2T$. The carrier phase over any $2T$ interval is any one of the four phases shown in Fig. 7.26(b), depending on the values of $a_Q(t)$ and $a_I(t)$. In the next $2T$ interval, if neither bit stream changes sign, the carrier phase remains the same. If one component $a_I(t)$ or $a_Q(t)$ changes sign, a phase shift of 90 deg occurs. A change in both components results in a phase shift of 180 deg. This 180-deg phase transition leads to an envelope collapse of a band-limited transmission (transition through origin), as shown in Figure 7.27(a).

In order to limit the envelope fluctuation, bit streams are staggered, as shown in Figure 7.27(b). An OQPSK is obtained from the conventional QPSK by delaying the odd-bit stream by a half-bit interval with respect to the even one, as shown in Figure 7.27(c). This reduces the range of phase transition to 0 deg and 90 deg in OQPSK, as shown in Figure 7.27(d). Thus, the phase transitions in OQPSK occur twice as often as in the QPSK carrier, but with only half the intensity. This in turn reduces the envelope fluctuation at the modulator output. In order to eliminate such envelope fluctuations, a phase-locked loop (PLL) can be inserted at the modulator output. Since the voltage-controlled oscillator (VCO) output has a constant envelope, the envelope fluctuations at the modulator output completely van-

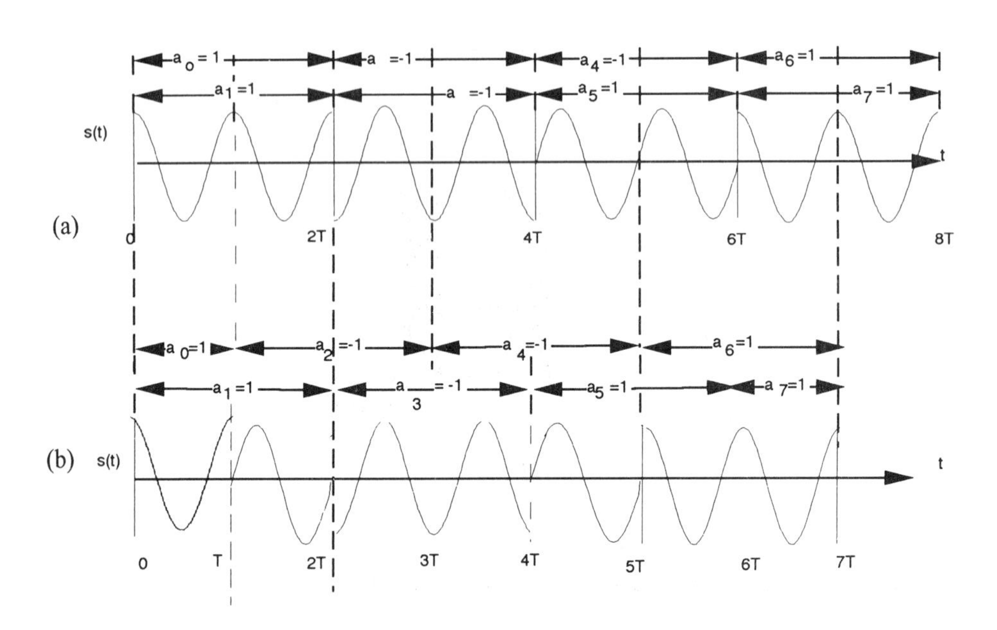

Figure 7.27 (a) QPSK waveform (Source: [34]); (b) OQPSK waveform.

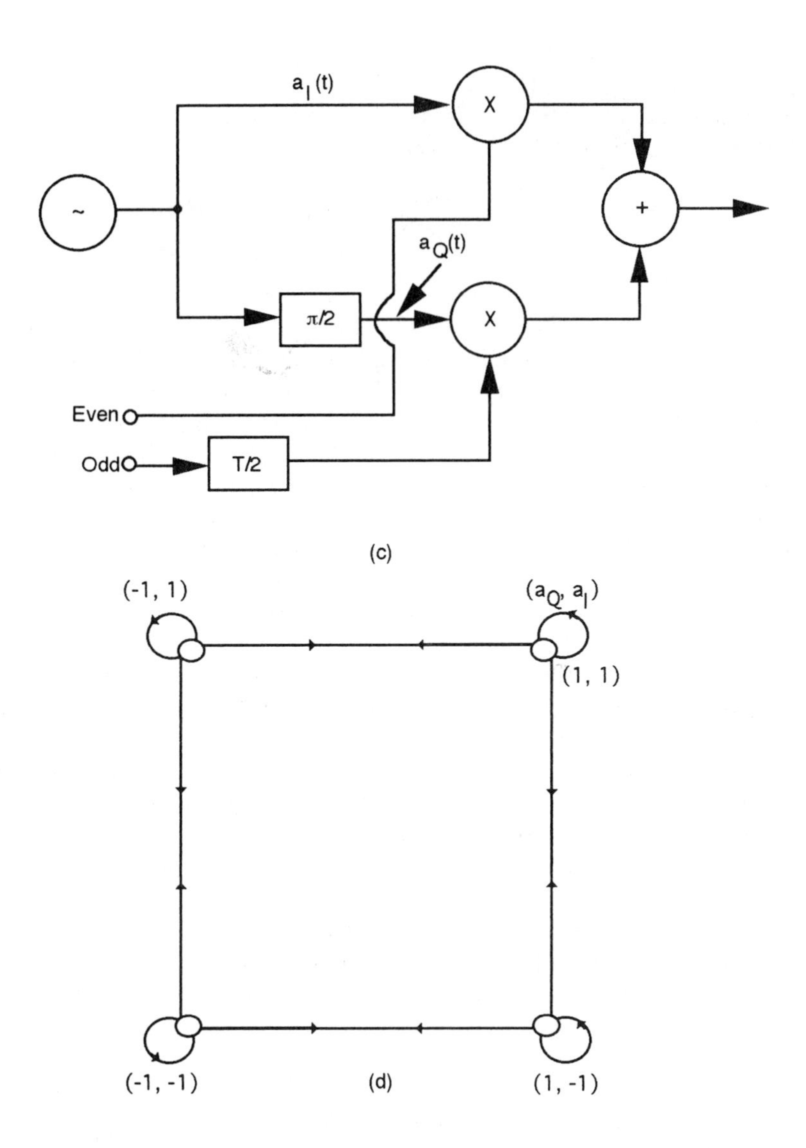

Figure 7.27 (cont.) (c) OQPSK modulator; (d) signal space diagram for OQPSK.

ish. The difference in time alignment in the bit streams has no impact on the output power spectral density of OQPSK over QPSK.

For rectangular baseband pulses, output power spectrum density (*PSD*) for both QPSK and OQPSK is of the form

$$S(f) = \left(\frac{\sin 2\pi f T}{2\pi f T}\right)^2 \tag{7.49}$$

However, these two modulations respond differently when a band-limited QPSK and OQPSK pass through a nonlinear amplifier. In cellular radio, the spectrum is shared among users. Thus, an individual channel is filtered out to reduce adjacent-channel interference. However, due to fast envelope fluctuation in QPSK, passing this signal through a nonlinear amplifier regenerates the high-frequency components originally removed by filtering, as shown in Figure 7.28(a).

On the other hand, due to the absence of fast-phase transition (less envelope fluctuation), filtered OQPSK when passing through a nonlinear amplifier will not fully regenerate the high-frequency components originally removed by a band-limiting filter. The filtered OQPSK signal suffers some sidelobe regeneration, but not anywhere close to QPSK, as shown in Figure 7.28 [16]. Here the normalized filtered bandwidth, BT product, is 1. The other advantage of OQPSK is the ease of bit synchronization at the receiver compared to QPSK due to more frequent bit transition.

Figure 7.29 depicts a generalized quadrature demodulator. It consists of a coherent demodulation section where the incoming signal is split into two parallel paths and each is amplified by locally generated quadrature carriers. We assume that I and Q baseband signals are produced by coherent demodulation and are individually processed through ideally matched filters. Matched filters for QPSK/OQPSK signals are integrator and dump circuits.

The BER probability for QPSK/OQPSK is the same for BPSK. Symbol rate error can be obtained by combining the BER probability for I and Q channels, which is given by

$$\begin{aligned} P_s &= \frac{1}{2}\,\mathrm{erf}_c(\sqrt{E_s/2N_0})\,[1 - \frac{1}{2}\,\mathrm{erf}_c(\sqrt{E_s/2N_0})] \times 2 + \frac{1}{4}\,\mathrm{erf}_c^2\left(\sqrt{\frac{E_s}{2N_0}}\right) \\ &= \mathrm{erf}_c\left(\sqrt{\frac{E_s}{2N_0}}\right) - \frac{1}{4}\,\mathrm{erf}_c^2\left(\sqrt{\frac{E_s}{2N_0}}\right) \\ &\approx \mathrm{erf}_c\left(\frac{\sqrt{E_s}}{2N_0}\right) \end{aligned} \tag{7.50}$$

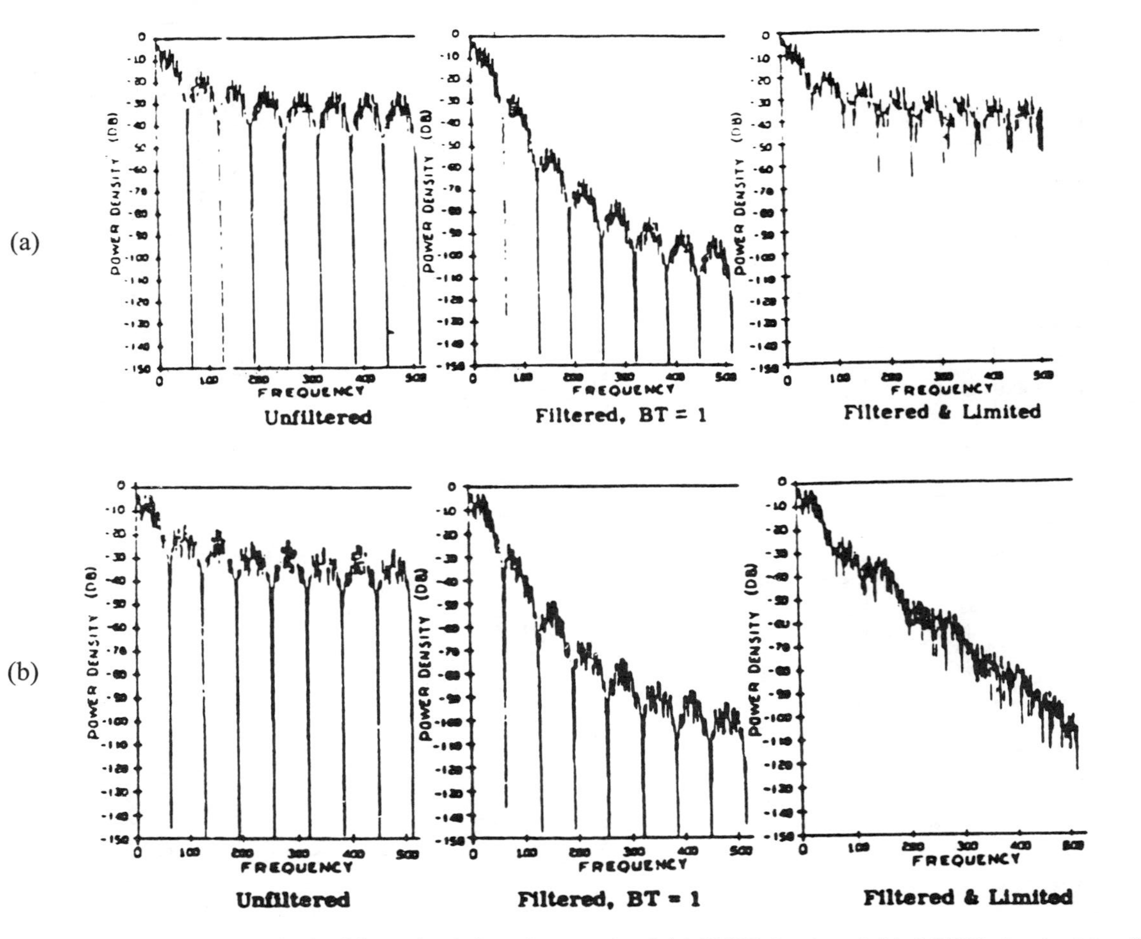

Figure 7.28 Power spectra obtained from simulation of conventional (a) QPSK signals and (b) OQPSK signals. After [36].

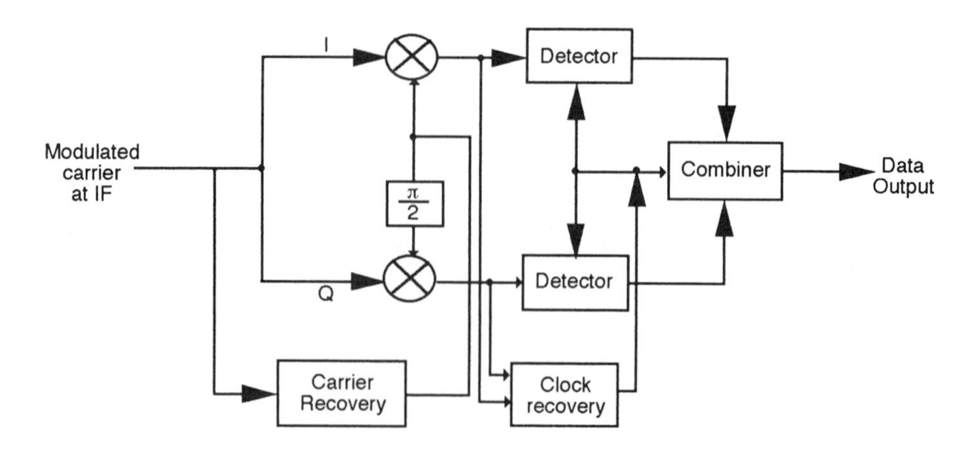

Figure 7.29 Generalized quadrature demodulator.

Here the first term is due to bit error in I or Q channel, but not in both channels simultaneously. The second term is when errors are present in both I and Q channels. The symbol energy is E_s and equals $2E_b$.

As stated above, the spectral efficiency of QPSK/OQPSK is doubled when compared to BPSK. Also, the spectral containment of QPSK is better than BPSK. Both these facts favorably recommend for QPSK/OQPSK. Unfortunately, the bandwidth requirement is still high, as discussed above, and the channel spectrum leakage of < -60 dBc cannot be easily met without placing a complex filter at the output of the QPSK/OQPSK modulator. One can think of two approaches to solving these problems:

1. Use 8 − PSK to increase the spectral efficiency further;
2. Reduce the sidelobe levels by minimizing phase discontinuities.

Use of 8 − PSK is not recommended due to poor *BER* performance. Since the total mobile signal power is the same, and as the size of the signal set is doubled, the spacing between signal constellation points is reduced to half when compared to QPSK/OQPSK. Thus, the signals are now more susceptible to noise. The second approach to reduce the sidelobe levels by minimizing phase discontinuities is the preferred technique, and is discussed below under minimum shift keying (MSK).

Minimum shift keying (MSK) is considered as a form of OQPSK in which the symbol pulse shape is a half-cycle sinusoid, rather than the usual rectangular form.

7.5 ACTUAL MODULATIONS USED IN MOBILE COMMUNICATIONS

In this section we shall discuss those modulations that are used in digital cellular (GSM and NADCS), various cordless telephones (CT1, CT2, and DECT), and proposed schemes for PCN.

7.5.1 MSK Modulation [17–20]

As discussed above, due to the limited availability of power at the mobile, a higher order modulation such as 8 − PSK cannot be adopted for cellular radio application. At the same time we cannot live with QPSK or OQPSK due to the difficult filtering requirements to limit the adjacent total channel leakage to −60 dB. This leads us to choose a modified OQPSK modulation known as MSK. MSK is an adaptation of OQPSK in which the modulating pulses are sinusoidal instead of rectangular. Thus, MSK can be generated and coherently detected as OQPSK with sinusoidal pulses. Having a constant-envelope property allows the power amplifier to work at saturation without significant distortion. MSK is also a special case of coherent FSK modulation, known as fast frequency shift keying (FFSK), where the minimum spacing between high and low tones is 0.5 times the data rate. This is the minimum required tone spacing that allows the two frequency states to be orthogonal to each other. MSK derives its name from the fact that this is the minimum frequency spacing for a coherently orthogonal sinusoid.

Amplitude versus time graph of MSK results in one complete cycle of the lower shift frequency, and one and one-half cycles of the higher shift frequency. The I/Q diagram, time waveform, and phase function for MSK are shown in Figure 7.30.

Since MSK is a special case of OQPSK with sinusoidal pulse weighing, the signal can be defined as

$$s(t) = a_I(t)\cos\left(\frac{\pi t}{2T}\right)\cos 2\pi f_c t + a_Q(t)\sin\left(\frac{\pi t}{2T}\right)\sin 2\pi f_c t \qquad (7.51a)$$

This equation can be expressed alternatively as

$$s(t) = \cos\left[2\pi f_c t + b_k(t)\frac{\pi t}{2T} + \Phi_k\right] \qquad (7.51b)$$

where $b_k = +1$ for $a_I a_Q = -1$; $b_K = -1$ for $a_I a_Q = +1$; and $\Phi_k = 0$ for $a_I = 1$; $\Phi_k = \pi$ for $a_I = -1$. The various component waveforms of (7.51a) are shown

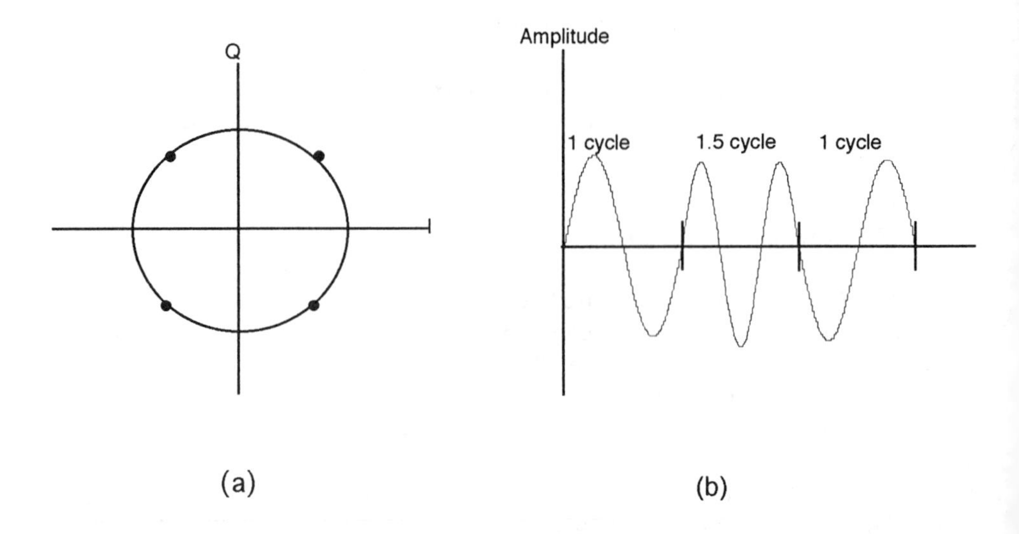

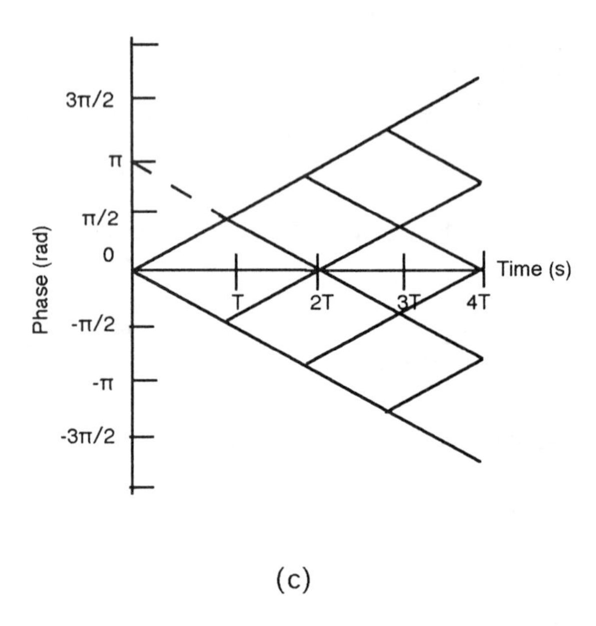

Figure 7.30 (a) I/Q diagram; (b) signal time waveform; (c) phase function of FFSK.

in Figure 7.31. From the figure and (7.51a) and (7.51b), the following conclusions can be drawn for MSK modulation

1. It has a constant envelope.
2. During each bit interval, a phase of MSK carrier is shifted linearly with time by $\pm\pi/2$.
3. The value $b_k = +1$ corresponds to a higher frequency f_H and $b_k = -1$ corresponds to a lower frequency f_L.
4. The high tone f_H contains one more half-cycle of carrier than does the low tone f_L.
5. The carrier frequency f_c is never transmitted, and only f_L or f_H are present in the modulated signal.
6. For a modulation bit rate of R_b, $f_H = f_c + 0.25R_b$, when $b_k = +1$, and $f_L = f_c - 0.25R_b$, when $b_k = -1$.

Coherent detection allows orthogonal detection of an MSK tone with a minimum tone spacing $\Delta = f_H - f_L = 0.5R_b$. MSK is a two-tone (binary) continuous-phase form of frequency shift keying (CP-FSK), which is also referred to as binary FM (FFSK).

The index of $m_f = 0.5$ and is given by

$$m_f = \Delta/R_b = (f_H - f_L)/R_b = 0.5 \tag{7.52}$$

One method of MSK realization (serial realization, shown in Figure 7.32) is from (7.51b), which can be rewritten as

$$s(t) = \cos[2\pi f_c t + \Phi_k + b_k(t)\omega_T t] \tag{7.53}$$

where $\omega_T = 2\pi(0.25/T) = 0.5\pi R_b$ or $f_T = R_b/4$, where, $f_H = f_c + f_T$ and $f_L = f_c - f_T$. The corresponding matched receiver structure is shown in Figure 7.33. The matched filter output is passed through an A/D converter before the I and Q channel mixer and decision circuit. The power spectra of QPSK, OQPSK, and MSK can be found by taking the Fourier transform of the symbol shaping function. The $p(t)$ for QPSK, OQPSK, and MSK are

$$p(t) = 1/\sqrt{2} \qquad |t| \leq T \tag{7.54a}$$

for QPSK and OQPSK and zero elsewhere and

$$p(t) = \cos \pi t/2T \qquad |t| \leq T \tag{7.54b}$$

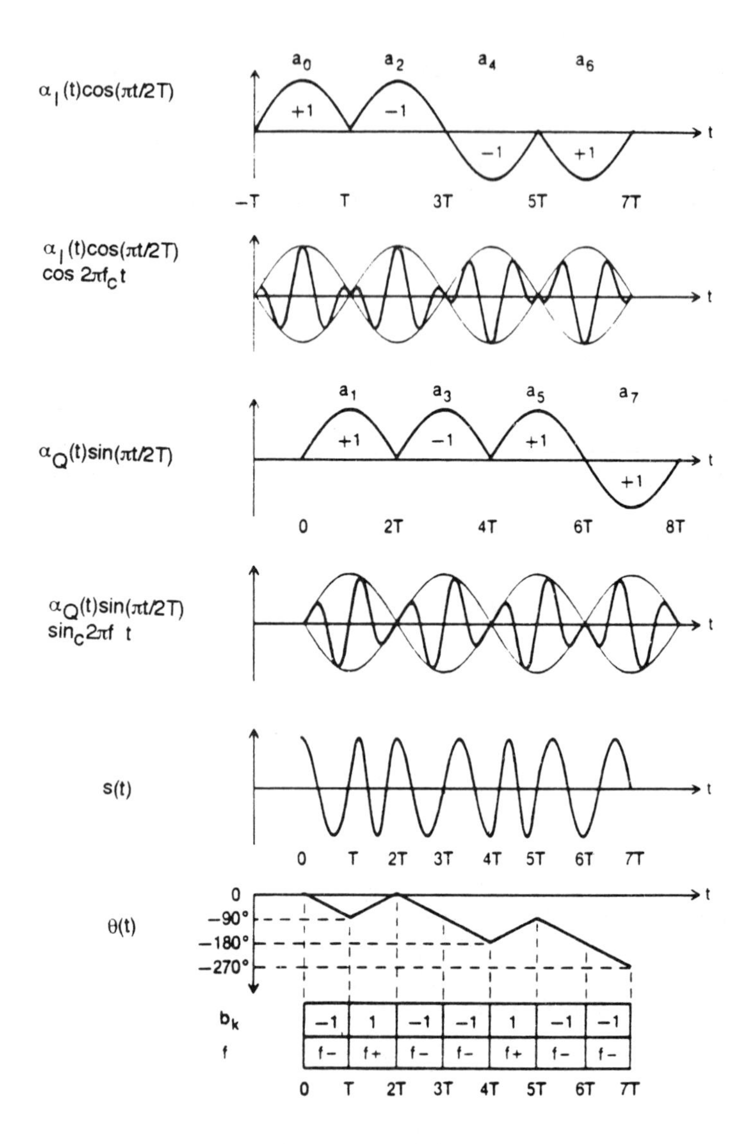

Figure 7.31 Component waveforms of (7.51a). After [34].

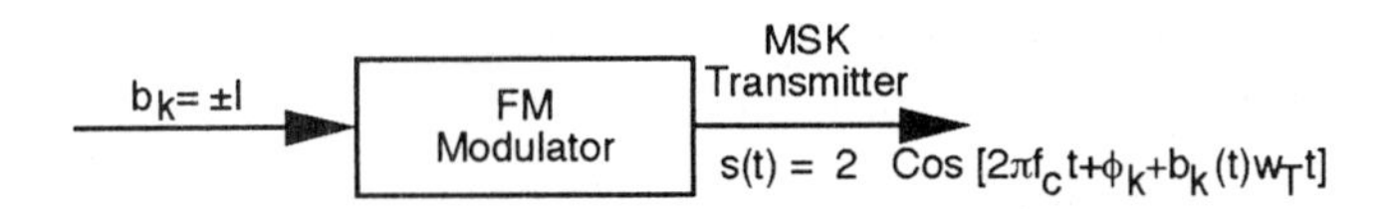

Figure 7.32 One form of MSK generation.

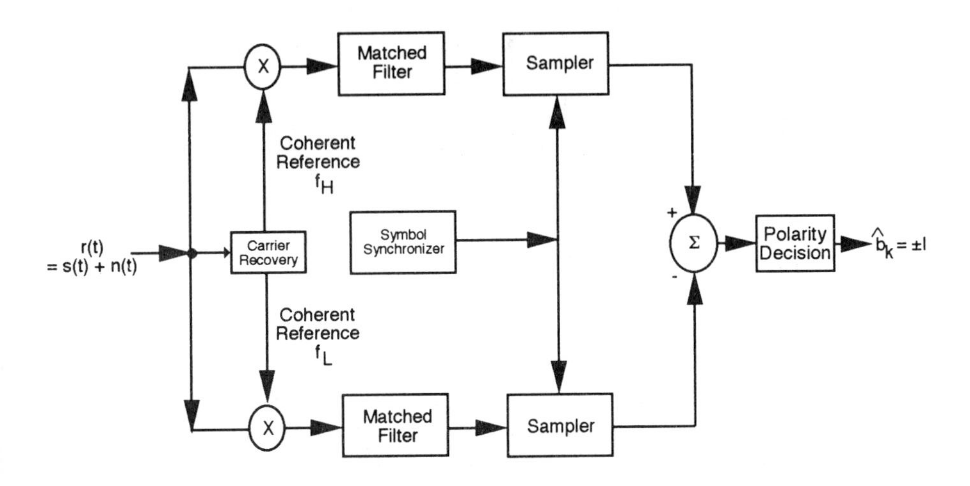

Figure 7.33 MSK receiver (serial form).

for MSK and zero elsewhere. The normalized spectral density $G(f)$ for QPSK and OQPSK is

$$\frac{G(f)}{T} = 2\left(\frac{\sin 2\pi fT}{2\pi fT}\right)^2 \tag{7.55a}$$

The normalized spectral density for MSK is

$$\frac{G(f)}{T} = \frac{16}{\pi^2}\left(\frac{\cos 2\pi fT}{1 - 16f^2T^2}\right)^2 \tag{7.55b}$$

The spectra are sketched in Figure 7.34. The difference in rates of falloff of these spectra can be explained on the basis of the smoothness of the pulse shape $p(t)$.

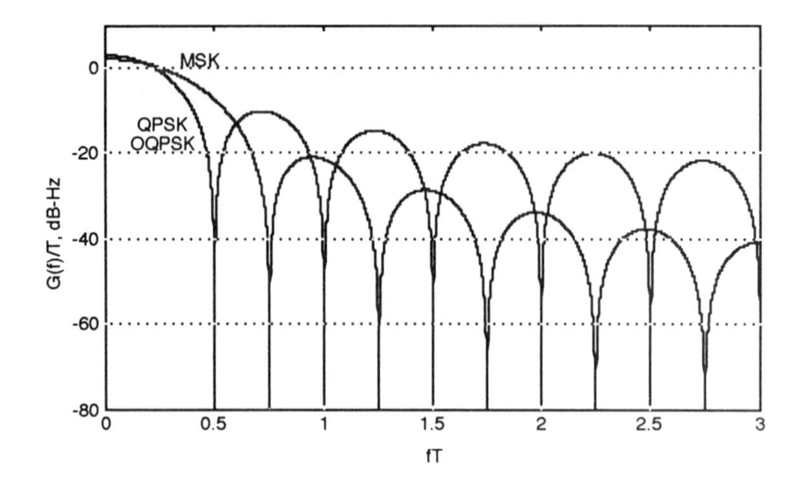

Figure 7.34 Spectral density of QPSK, OPQSK, and MSK.

The smoother the pulse shape, the faster the drop of spectral tails to zero. Thus MSK, having a smoother pulse, has lower sidelobes than QPSK and OQPSK. MSK is phase continuous and QPSK is discontinuous, which corresponds to a slope of 12-dB per octave and 6-dB per octave, respectively, at further sidelobes. Unfortunately, this performance improvement is somewhat offset by an increase in the mainlobe width with the first zero crossing at a normalized bandwidth of 1.5 (two-sided). Thus, MSK is not suitable for narrowband applications. As discussed earlier, a measure of the compactness of a modulation spectrum is the bandwidth B, which contains 99% of the total power. Figure 7.35 compares out-of-band power versus normalized bandwidth. For MSK, B is nearly equal to $(1.2/T_b)$ (99% of the total power contained, two-sided), while for QPSK and OQPSK, B is nearly equal to $(8/T_b)$. Thus, for North American digital cellular radio where the data rate is 48.6 Kbps, the required value of channel bandwidth using MSK modulation is $1.2 \times 48.6 \approx 58$ kHz. Therefore, a 30-kHz channel bandwidth is not adequate. If we accept the null-to-null bandwidth, we need $0.75 \times 48.6 \approx 36$ kHz, which is also not adequate for NADCS. Therefore, a spectrally more efficient modulation scheme is required for NADCS.

As a rule of the thumb, if the BT product is greater than 2.4 (two-sided), MSK will outperform QPSK and OQPSK in spectral containment tests, as seen from Figure 7.35. Therefore, MSK modulation can be a good choice for present analog FSK data channels in the U.S. where the product BT equals 3 (B = 30 kHz and $T = 1/(10 \times 10^3)$).

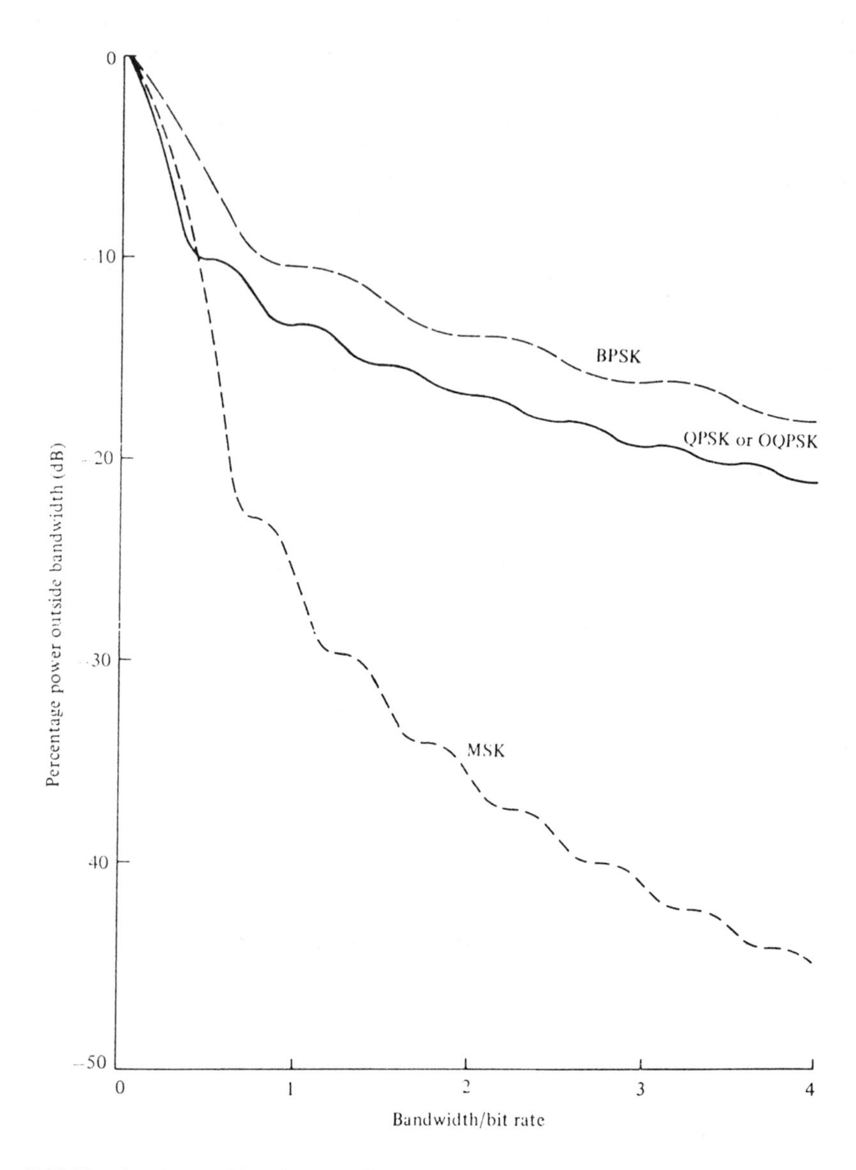

Figure 7.35 Fractional out-of-band power for BPSK, QPSK or OQPSK, and MSK. Source: [17, p. 128].

The second form of MSK modulator, known as the parallel form, is a direct implementation of (7.51a), and is shown in Figure 7.36. The front-end multiplier provides f_H and f_L, which are separated out by two bandpass filters. Components $x(t)$ and $y(t)$ are the two quadrature components of (7.51a). Components $x(t)$ and $y(t)$ are multiplied with odd and even bit streams $a_I(t)$ and $a_Q(t)$ to provide the modulated waveform $s(t)$. Component waveforms of the top and bottom arms of the modulator are shown in Figure 7.31(b,d).

An MSK receiver block diagram is shown in Figure 7.37(a). Here the received signal $s(t)$ (from (7.55a) in the absence of noise and interference) when multiplied by $x(t)$ and $y(t)$ and passed through an integrate-and-dump circuit (integration interval $2T$), and when properly sampled and detected will provide the polarity of baseband pulses $a_I(t)$ and $a_Q(t)$. This is shown from the following equations.

The low-pass filter output of the I channel is

$$s(t) \cos(\pi t/2T) \cos 2\pi f_c t\big|_{LP} = Ka_I(t)\ 1/4(1 + \cos \pi t/T) \tag{7.56a}$$

The low-pass filter output of the Q channel is

$$s(t) \sin(\pi t/2T) \sin 2\pi f_c t\big|_{LP} = Ka_Q(t)\ 1/4(1 - \cos \pi t/T) \tag{7.56b}$$

where K is the gain associated with the low-pass filter. Sampling this every $2T$ intervals and passing it through threshold detectors provides an estimate of $a_I(t)$ and $a_Q(t)$. The reference waveform $x(t)$ and $y(t)$ and the clock at half the bit rate $[\cos(2\pi t)/(2T)]$ are needed for the receiver. These can be derived from Figure 7.37(b).

Although signal $s(t)$ has no discrete components that can be used for synchronization, it produces strong discrete spectral components at $2f_H$ and $2f_L$ when passed through a squarer. These components are extracted by bandpass filters and then frequency division circuits produce the signals

$$s_1(t) = 1/2 \cos[(2\pi f_c t + \pi t/2T)] \tag{7.57a}$$

$$s_2(t) = 1/2 \cos[2\pi f_c t - (\pi t/2T)] \tag{7.57b}$$

respectively. The sum and differences $s_1 + s_2$ and $s_2 - s_1$ produce the reference carriers $x(t)$ and $y(t)$, respectively. The components $s_1(t)$ and $s_2(t)$, when multiplied and low-pass filtered, provide the required timing signal at half the bit rate.

Error Rate Performance

Coherent detection of MSK tones as FSK utilizes signal energy of only one bit interval as shown in Figure 7.33. An MSK signal detected as OQPSK (parallel

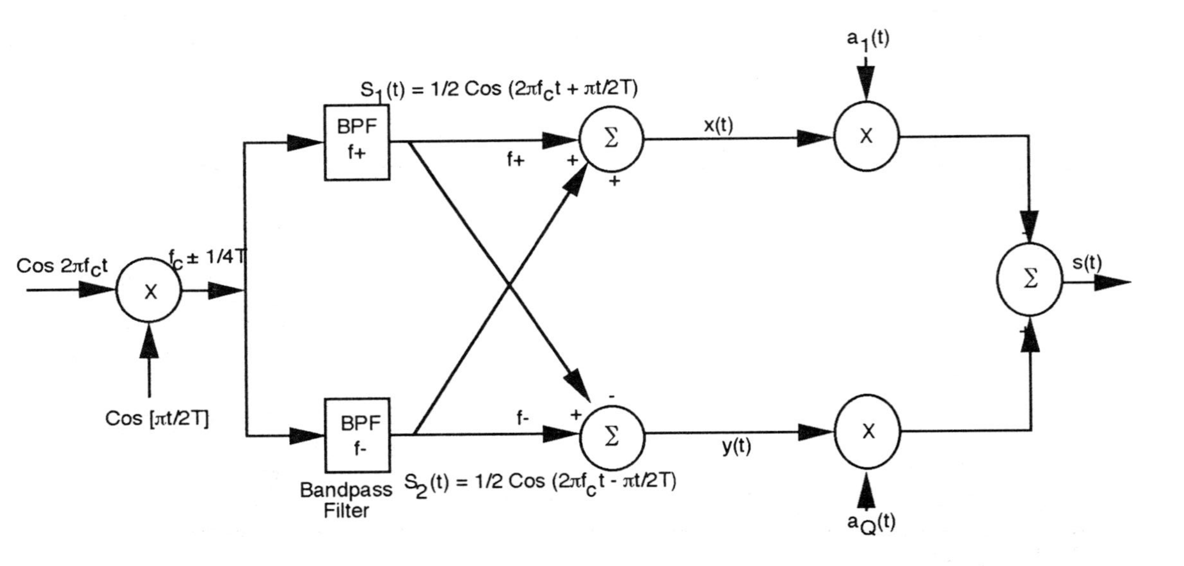

Figure 7.36 MSK modulator: $x(t) = \cos(\pi t/2T)\cos 2\pi f_c t$; $y(t) = \sin(\pi t/2T)\sin 2\pi f_c t$

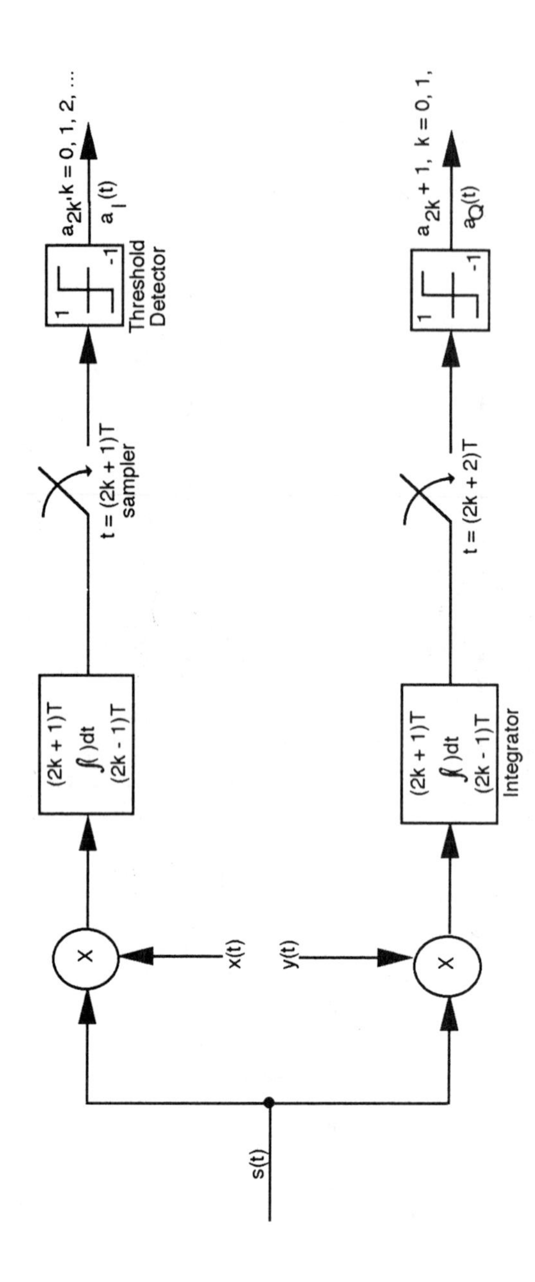

Figure 7.37(a) MSK receiver: $x(t) = \cos(\pi t/2T)\cos \pi f_c t$; $y(t) = \sin(\pi t/2T)\sin 2\pi f_c t$.

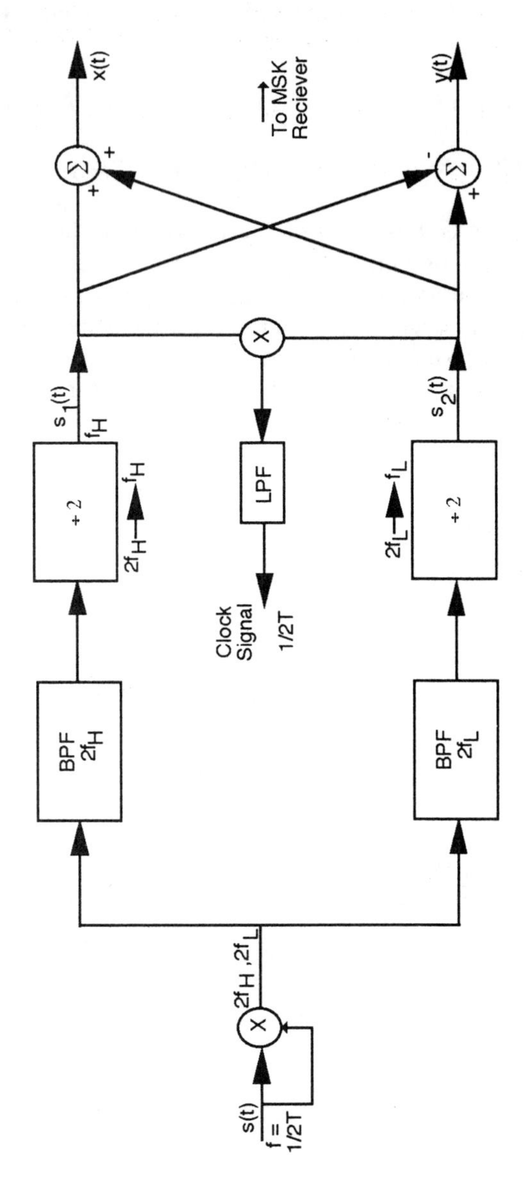

Figure 7.37(b) Carrier and bit recovery circuits for MSK.

form, Figure 7.37(a)) integrates signal energy over $2T$ intervals, and thus performs 3-dB better than as serial FSK realization. In other words, MSK tone detection (as FFSK, serial form) is 3-dB inferior to coherently detected OQPSK (parallel form). Thus, the BER is

$$P_e(\text{MSK as OQPSK}) = P_e(\text{QPSK}) = \text{erf}_c(\sqrt{E_b/N_0}) \tag{7.58a}$$

$$P_e(\text{MSK as FFSK}) = 1/2\ \text{erf}_c(\sqrt{E_b/2N_0}) \tag{7.58b}$$

Plots of BER for MSK as OQPSK and FSK are shown in Figure 7.38.

Another explanation of improved performance of MSK when realized as OQPSK is through time interval of integration. When T is the duration of a one-bit interval, then only one decision per transmitted bit is needed. Better performance is obtained by observing the received waveform over a period of $2T$, which provides more knowledge about the underlying signal and/or noise process.

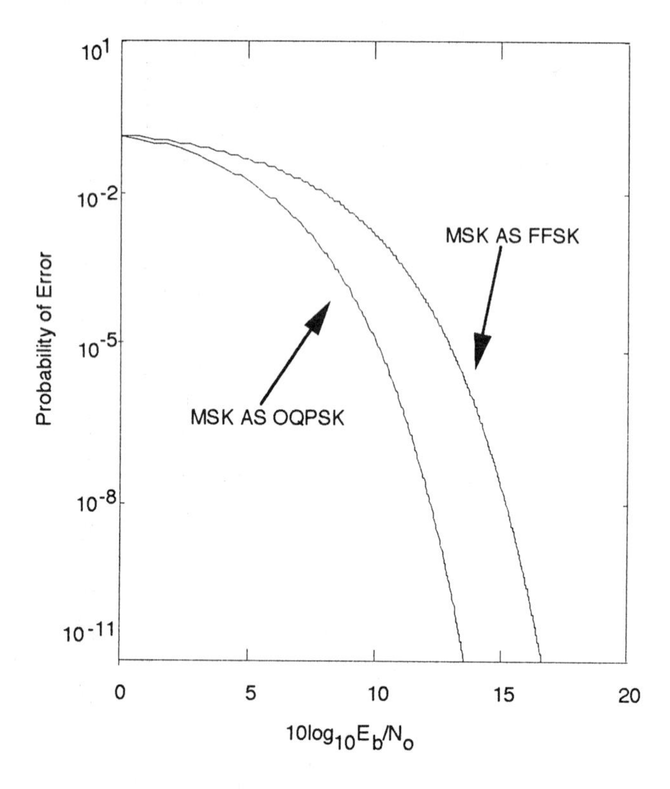

Figure 7.38 BER curves for MSK.

7.5.2 GMSK Modulation [18–23]

As discussed above, MSK is an improvement over QPSK (in terms of out-of-band power) as it makes the phase change linear and limits it to $\pm\pi/2$ over a bit interval. Low sidelobe in output power spectral density represents the effect of this linear phase change.

Unfortunately, the mainlobe is even wider than in QPSK/OQPSK cases (Figure 7.34). A low sidelobe definitely helps to control adjacent-channel interference. However, it is still difficult to satisfy the maximum permissible adjacent-channel interference limit of -60 dB. Both the sidelobe power level and the width of the mainlobe can be reduced by introducing a baseband Gaussian-shaped filtering of the rectangular pulses before modulation. This controlled amount of inter-symbol interface (ISI) further reduces the phase discontinuities in the carrier. As a result, roll-off of the mainlobe of the transmitter spectral density is increased and the sidelobe level is reduced.

The power spectrum density plot as a function of normalized frequency difference from the carrier center frequency $(f - f_c)T$, with the normalized bandwidth of the baseband Gaussian filter B_bT as a parameter is shown in Figure 7.39 [9]. Here B_bT is the 3-dB bandwidth of the Gaussian filter and T is the data pulse width. A normalized bandwidth value of ∞ represents a normal MSK power spectrum density. For a normalized frequency difference of 1.5 (representing $(f - f_c)$ $= 200$ kHz, $R_d = 270.8$ Kbps, $B_{IF}T_s = 200 \times 2/270.8 \approx 1.5$) and with $B_bT =$ 0.36), the power spectrum density shows to have a value below -60 dBc. It is also observed that the PSD of GMSK approximates the PSD of tamed frequency modulation with $B_bT = 0.2$.

A plot of the spurious radiated power in the adjacent channel to the desired-channel power, with normalized frequency separation f_sT as abscissa and B_bT as a parameter is shown in Figure 7.40. With $f_s = 200$ kHz, and $T = 1/(270.8 \times 10^3)$ sec, which provides normalized frequency separation of 1.5, the power in the adjacent channel is below 60 dB for $B_bT = 0.24$. Thus, GMSK with $B_bT = 0.24$ (maximum value) can be adapted as a digital modulation scheme that will satisfy the adjacent-channel interference requirement; that is, the ratio of spurious radiated power in the adjacent channel to that of the desired channel will be about -60 dB.

GMSK has been adapted by European GSM system as well as by DECT (digital European cordless telephone system). For the GSM system, a normalized bandwidth of B_bT is 0.3 with a total channel data rate of 270.8 Kbps (IF bandwidth $= 0.3 \times 271$ kHz), and DECT has adapted $B_bT = 0.5$ with a data rate of 1.152 Mbps. Therefore, the CCIR recommended value of -60 dB may not be met in GSM system. A choice of $B_bT = 0.3$ appears to be a compromise between BER and out-of-band interference as $B_bT < 0.3$ increases the BER exponentially due to a drastic reduction in signal power. For DECT adjacent, an adjacent-channel

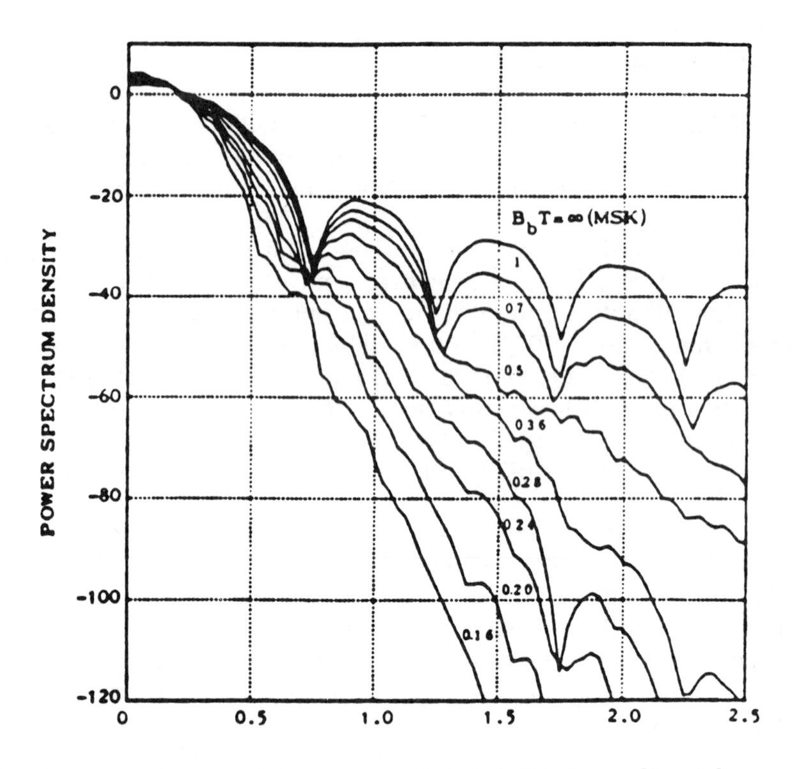

Figure 7.39 Normalized frequency difference versus PSD of GMSK. Source: [9, p. 15].

interference requirement of nearly −40 dB only can be met, which may be sufficient in view of the fact that the system reassigns channels as the interference goes up.

7.5.2.1 GMSK Representation

GMSK is a continuous phase and a constant amplitude modulation represented by

$$s(t) = A\cos[2\pi f_0 t + \Phi(t, \alpha)] \qquad nT \leq t \leq (n+1)T \tag{7.59}$$

The information is contained in the phase term $\Phi(t, \alpha)$. For n bits, the total accumulated phase is given by

$$\Phi(t, \alpha) = 2\pi h \sum_{i=0}^{n} \alpha_i q(t - iT) \tag{7.60}$$

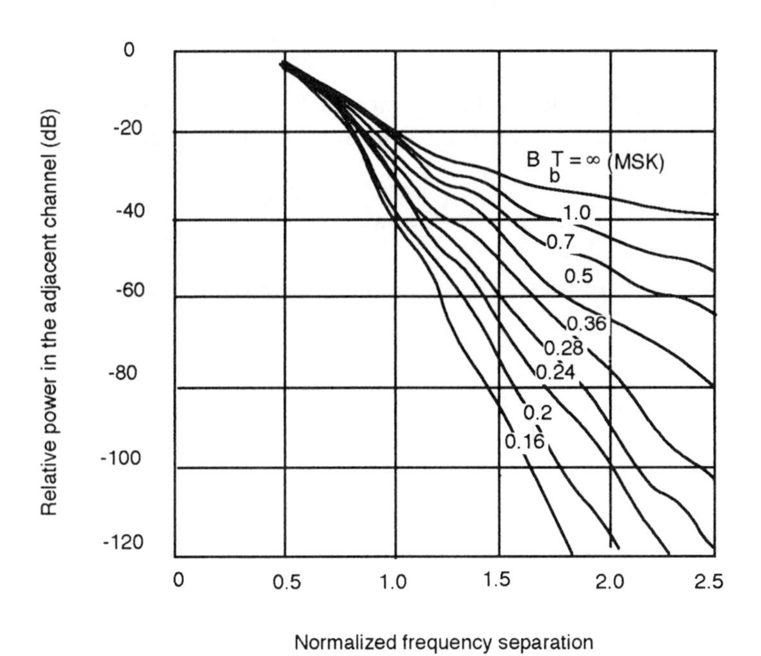

Figure 7.40 Relative power radiated in the adjacent channel. Source: [9, p. 16].

where α_i represents data symbols (± 1) and h is the modulation index defined as Δf per bit rate. Thus,

$$s(t) = A \cos\left[2\pi f_0 t + 2\pi h \sum_{i=0}^{N} \alpha_i q(t - iT)\right] \tag{7.61}$$

The baseband pulse response $g(t)$ is related to the baseband phase response $q(t)$ by an integral given by

$$q(t) = \int_{-\infty}^{t} g(\tau) d\tau \tag{7.62}$$

The pulse response $g(t)$ can be found by knowing the impulse response of the Gaussian filter. The premodulation Gaussian filter has a transfer function of the form

$$H(f) = Ke^{[(1-f/B)^2(\ln^2/2)]} \tag{7.63}$$

Here, B is the filter 3-dB bandwidth point and K is a constant.

Taking the inverse Fourier transform, the impulse response is given by

$$h(t) = K\sqrt{2\pi/\ln 2}\; Be^{-2\pi^2B^2t^2/\ln 2} \tag{7.64}$$

Considering the filter response $g(t)$ to a unit rectangular pulse of width T centered at the origin, we write

$$g(t) = K\sqrt{2\pi/\ln 2}\; B\int_{t-T/2}^{t+T/2} e^{-2\pi^2B^2x^2/\ln 2}\,dx \tag{7.65}$$

$$= K/2\{\mathrm{erf}[\sqrt{2/\ln 2}\;\pi B(t - T/2)] + \mathrm{erf}[\sqrt{2/\ln 2}\;\pi B(t + T/2)]\} \qquad t > 0 \tag{7.66}$$

where

$$\mathrm{erf}(y) = \frac{2}{\sqrt{\pi}}\int_0^y e^{-u^2}du \tag{7.67}$$

and $\mathrm{erf}(y) = \mathrm{erf}(-y)$. From the above, $g(t) = g(-t)$. Thus, by knowing the pulse response $g(t)$, the phase response $q(t)$ can be found by applying (7.62).

7.5.2.2 GMSK Modulator

The most straightforward way of implementing a GMSK modulator is to transmit the data stream through a Gaussian low-pass filter and apply the resultant waveform to a voltage-controlled oscillator, as shown in Figure 7.41. The output of the VCO is then a frequency-modulated signal with a Gaussian response. The premodulation filter should have a narrow bandwidth and a sharp cutoff as well as a low overshoot impulse response while preserving the filter output area to assure a $\pi/2$ phase shift at the modulator output at the end of every bit interval.

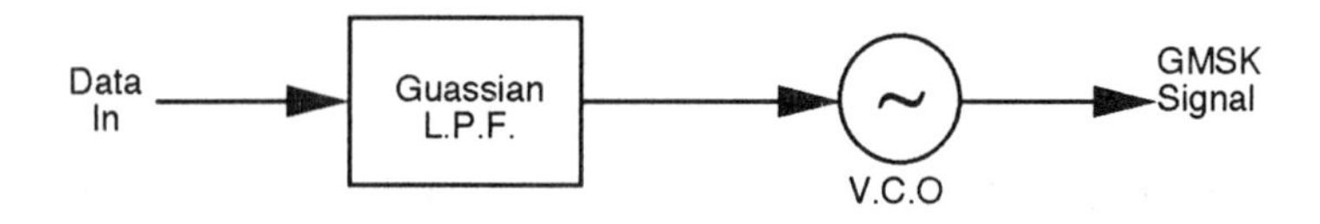

Figure 7.41 Simple GMSK modulator.

In order to solve the phase ambiguity problem in acquisition of a coherent local reference, differential encoding is included in the transmitter with a corresponding differential decoding in the receiver. The modified GSM transmitter block diagram is shown in Figure 7.42.

The main disadvantage of the modulator is the instability of VCO; as a result, a signal produced by this is not suitable for coherent demodulation. It is difficult to keep the center frequency within the allowable value under the restriction of maintaining the linearity and the sensitivity for the required FM modulation.

An alternate realization of a GMSK modulator by Murota and Hirade is shown in Figure 7.43(a) [18], where the data input first phase shifts the BPSK modulator by $\pi/2$ before passing through a phase-locked loop. Unfortunately, very large-scale integration (VSLI) implementation of the above modulator is difficult to realize.

7.5.2.3 Demodulation

Three different techniques can be used for the demodulation of GMSK signals: differential detection, coherent detection, and FM discriminator detection. Since differential detection does not require an absolute phase reference in the receiver,

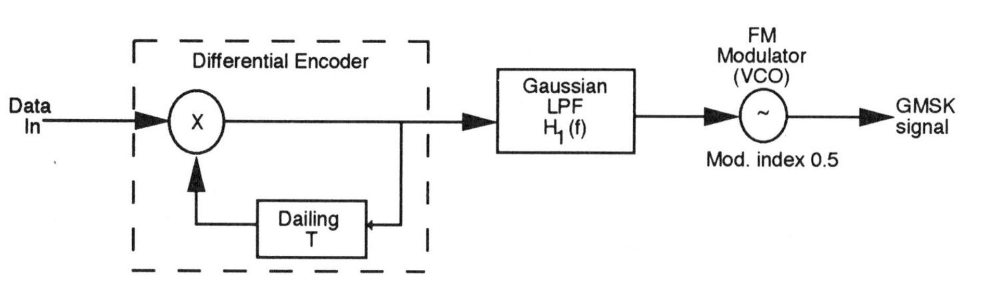

Figure 7.42 Modified GMSK modulator with differential encoder.

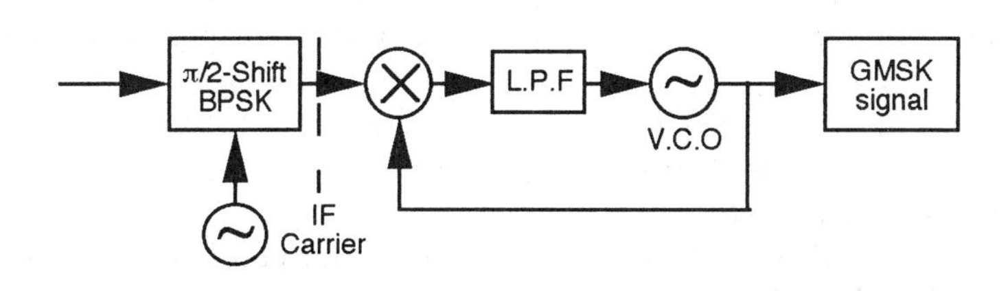

Figure 7.43(a) PLL GMSK modulator.

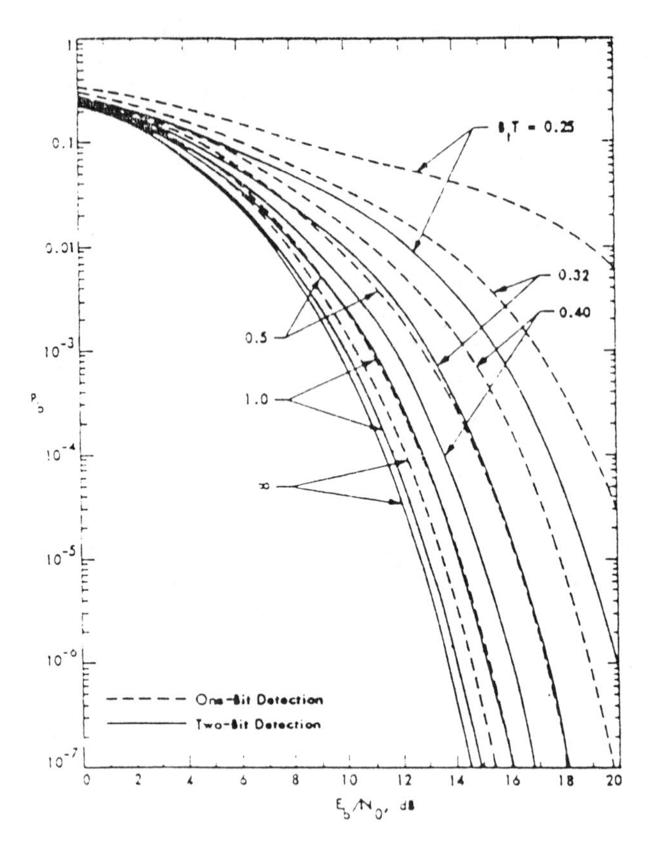

Figure 7.43(b) Performance of one and two-bit differential detection of GMSK; receiving B_tT is optimized for each B_tT (for a two-bit detection case, the detection threshold is optimized for each E_b/N_o). Source: [19, p. 315].

this is a preferred choice for signal recovery in multipath surroundings. Both one-bit and two-bit delay versions have been proposed in the literature. Receiver performance in the two-bit delayed version is superior to that of the one-bit differential detection because the collected energy over a two-bit interval is larger. Coherent detection is similar to that of detecting straight MSK signals. The FM discriminator (noncoherent demodulation) does not account for phase, and thus may not be very desirable.

The analysis of one-bit and two-bit differential detector has been done by Simon and Wang [19]. BER curves for one-bit and two-bit detection with different values of B_bT are shown in Figure 7.43(b). As seen from this curve, the BER

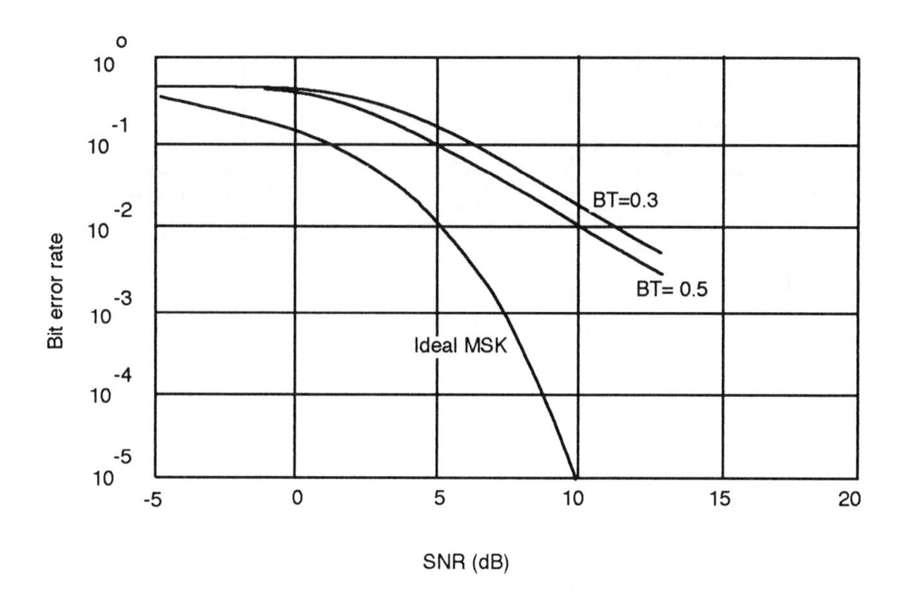

Figure 7.43(c) BER performance of Costas loop receiver in AWGN. Source: [37].

degrades substantially below B_bT of 0.32. On the other hand, performance improvement is marginal above this value. This once again establishes the fact that the choice of a B_bT of 0.3 for GSM is influenced by BER performance at the expense of not exactly meeting the CCIR recommended value of -60 dB for adjacent-channel interference.

A coherent receiver structure is shown in Figure 7.44. The input bandpass filtered signal is multiplied by the in-phase and the quadrature phase recovered carrier. The carrier recovery is by Costas loop. The data and clock are recovered after low-pass filtering of both I and Q channels. BER performance in the presence of white Gaussian noise is shown in Figure 7.43(c) [20–21]. As shown in this figure, BER with $BT = 0.3$ (LPF $BW = 81$ kHz, $R_d = 270.8$ Kbps) requires additional signal power by nearly 0.8 dB compared to $BT = 0.5$ ($R_d = 1.152$ Mbps).

7.5.3 π/4-DQPSK Modulation [24–30]

As discussed above, a constant-envelope GMSK modulation does not cause spectral spreading when the signal passes through a nonlinear amplifier, thus satisfying the adjacent-interference requirement. Nonlinear amplification is also essential in mobile communication due to the limited power availability of the mobile, as in

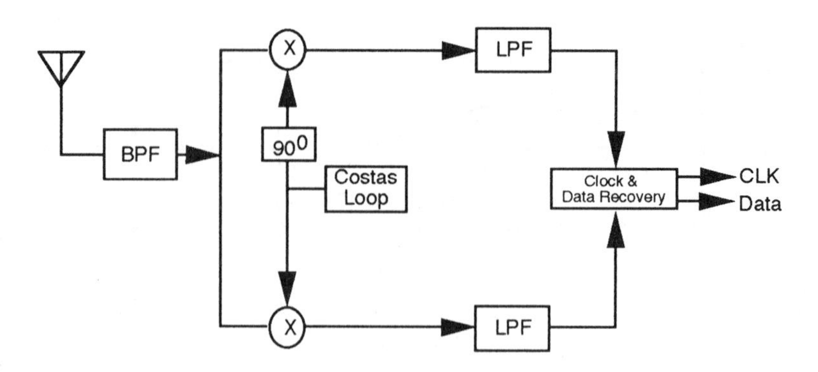

Figure 7.44 Coherent GMSK receiver.

GMSK. However, as the demand of mobile communication increases, GMSK may not be able to support increased traffic. To improve the spectral efficiency of the mobile communication, a higher order power-efficient linear modulation, such as QPSK/OQPSK, can be a better choice. As described in Section 7.4.3, OQPSK is a better choice over QPSK. In multipath fading, channel surroundings, the use of coherent detection is difficult and often leads to poorer performance over noncoherently based systems. In differential detection, OQPSK is inferior to QPSK. Differential detection of OQPSK is subject to ISI with resulting poor system performance.

$\pi/4$-DQPSK is a compromise, as its phase fluctuation is restricted to $\pm\pi/4$ and $\pm 3\pi/4$, which is worse than $\pm\pi/2$ for OQPSK but allows noncoherent detection, which is a must for NADCS due to compatibility requirements with the existing analog FM cellular radio. Also, the spectral efficiency of $\pi/4$-DQPSK is about 20% more than that of GMSK.

The U.S. digital cellular system TIA/EIA IS-54 standard recommends using $\pi/4$-DQPSK. $\pi/4$-DQPSK is essentially a $\pi/4$-shifted QPSK with differential encoding of symbol phases. The differential encoding mitigates loss of data due to phase slips. However, differential encoding results in the loss of a pair of symbols when channel errors occur, which can be translated to approximately a 3-dB loss in E_b/N_0 relative to coherent $\pi/4$ QPSK.

7.5.3.1 Modulation

A $\pi/4$-shifted QPSK signal constellation comprises symbols corresponding to eight phases. These eight phase points can be considered to be formed by superimposing two QPSK signal constellations, offset by 45-deg relative to each other. During each symbol period a phase angle from only one of the two QPSK constellations is transmitted. The two constellations are used alternately to transmit every pair of bits (di-bits), and as a result, successive symbols have a relative phase difference that is one of the four phases shown in Table 7.2. Figure 7.45 shows the $\pi/4$-shifted QPSK signal constellation. When the phase angles of $\pi/4$-shifted QPSK symbols are differentially encoded, the resulting modulation is $\pi/4$-shifted DQPSK. This can be done by either differential encoding of the source bits and mapping them onto absolute phase angles or, alternately, by directly mapping the pairs of input

Table 7.2
Comparison Between DQPSK and
$\pi/4$ DQPSK Modulation Formats

Symbol	*DQPSK Phase Transition*	*Transition*
00	0 deg	45 deg
01	90 deg	135 deg
10	−90 deg	−45 deg
11	180 deg	−135 deg

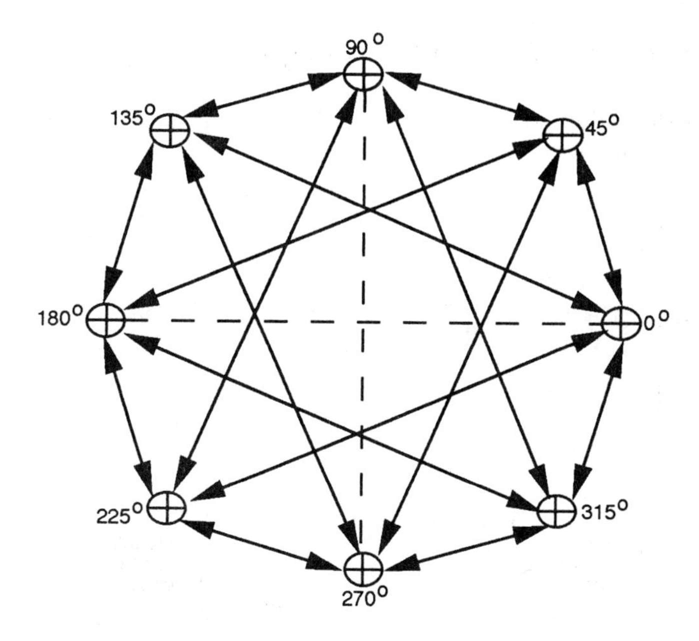

Figure 7.45 $\pi/4$ DQPSK modulation.

bits onto relative phases ($\pm\pi/4$, $\pm3\pi/4$), as shown in Figure 7.46. Here, the binary data stream entering the modulator b_m is converted by a serial to parallel converter into two binary streams x_k and y_k before these bits are differentially encoded.

7.5.3.2 Transmitter Model

Differentially encoded bits I_k and Q_k represented as the kth symbol (I_k, Q_k) are generated by using the information in Table 7.2. The (I_k, Q_k) symbol can be represented as

$$I_k = I_{k-1}\cos(\Delta\Phi_k) - Q_{k-1}\sin(\Delta\Phi_k) \tag{7.68a}$$

$$Q_k = I_{k-1}\sin(\Delta\Phi_k) + Q_{k-1}\cos(\Delta\Phi_k) \tag{7.68b}$$

where I_k and Q_k are the in-phase and quadrature components of the $\pi/4$-shifted DQPSK signal corresponding to the kth symbol. The amplitudes of I_k and Q_k are ±1, 0, $\pm1/\sqrt{2}$. Since the absolute phase of $(k-1)$th symbol is Φ_{k-1}, the above in-phase and quadrature components can alternately be expressed as

$$I_k = \cos\Phi_{k-1}\cos(\Delta\Phi_k) - \sin\Phi_{k-1}\sin(\Delta\Phi_k) = \cos(\Phi_{k-1} + \Delta\Phi_k) \tag{7.69a}$$

$$Q_k = \cos\Phi_{k-1}\sin(\Delta\Phi_k) + \sin\Phi_{k-1}\cos(\Delta\Phi_k) = \sin(\Phi_{k-1} + \Delta\Phi_k) \tag{7.69b}$$

These component signals (I_k, Q_k) are then passed through baseband filters having a raised cosine frequency response as

$$|H(f)| = \begin{cases} T & 0 \le |f| \le \dfrac{1-\alpha}{2T} \\ T\sqrt{\dfrac{1}{2}\left\{1 - \sin\left[\dfrac{\pi T}{\alpha}\left(|f| - \dfrac{1}{2T}\right)\right]\right\}} & \dfrac{1-\alpha}{2T} \le |f| \le \dfrac{1+\alpha}{2T} \\ 0 & |f| > \dfrac{1+\alpha}{2T} \end{cases} \tag{7.70}$$

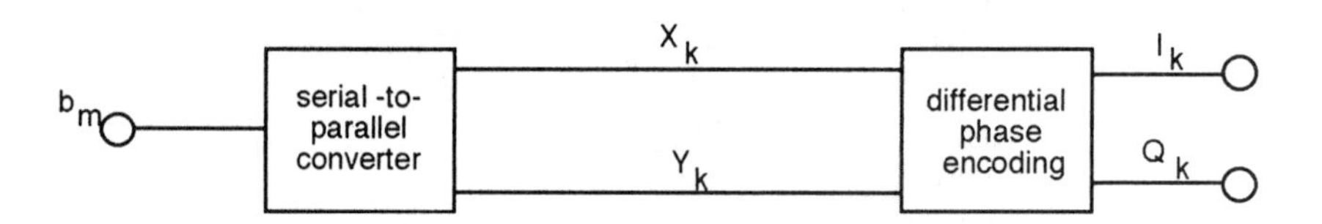

Figure 7.46 Differential encoding of $\mu/4$ DQPSK.

where α is the roll-off factor and T is the symbol duration. If $g(t)$ is the response to pulses I_k and Q_k at the square root raised cosine input, then the resultant transmitted signal is written as

$$
\begin{aligned}
s(t) &= \sum_k g(t - kT) \cos \Phi_k \cos \omega_c t - \sum_k g(t - kT) \sin \Phi_k \sin \omega_c t \\
&= \sum_k g(t - kT) \cos(\omega_c t + \Phi_k)
\end{aligned}
\tag{7.71}
$$

where ω_c is the carrier frequency of transmission. The component Φ_k results from differential encoding (i.e., $\Phi_k = \Phi_{k-1} + \Delta\Phi_k$).

From (7.71) and the block diagram in Figure 7.47 of the transmitter, it is clear that the signal $s(t)$ can be demodulated by detecting Φ_k the carrier phase at the sampling instant or by knowing the phase difference of the carrier between the sampling instants. Thus, $\pi/4$-shifted DQPSK can be demodulated using coherent detection, differential detection, and discriminator detection. As mentioned at the beginning of this section, discriminator detection is of more use in NADCS due to the required coexistence of analog cellular with digital. It should be noted that any detection scheme can be chosen so long as it meets the desired performance.

7.5.3.3 $\pi/4$ DQPSK Detection

As stated above, the main advantage of $\pi/4$-shifted-QPSK is that it can be detected either by coherent demodulation, a differential detector, or a discriminator followed by an integrate-and-dump filter. Differential detection and discriminator detection

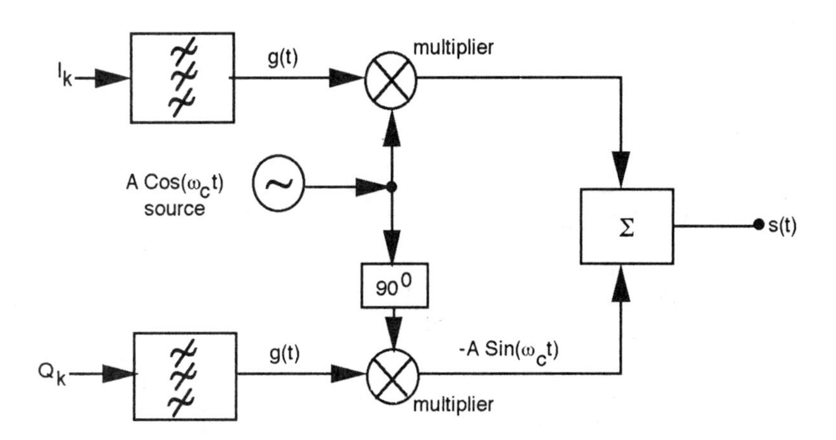

Figure 7.47 Block diagram of $\rho/4$ DQPSK transmitter.

allow simpler receiver realizations. Coherent detection is more complex than either differential or discriminator detection due to the carrier recovery process. Also, fast-fading coherent detection results in a higher irreducible BER than differential and discriminator detection. Thus, in this section we shall discuss the noncoherent scheme only. The reader is referred to reference [29] for further reading on coherent detection schemes.

7.5.3.4 Noncoherent detection

There are three types of noncoherent detection realizations which have been proposed in the literature [26–30]: baseband differential detector, IF band differential detector, and FM discriminator type. With optimal filtering, performance of the three types is about the same. We shall only discuss the IF band differential detector and FM discriminator type schemes, as they are more useful in cellular surroundings. For baseband differential detection the reader is referred to [26, 27].

Delay Differential Detector [26, 27]

A $\pi/4$-shifted DQPSK detector with time delay of one symbol interval T_s is shown in Figure 7.48. Detection is achieved by multiplying the desired signal by its delayed version (one symbol delay). Use of the raised cosine roll-off BPF to match the transmitted signals preserves the transmitted carrier phase under noise and ISI free conditions. Measured BER performance under a white Gaussian noise (AWGN) is shown in Figure 7.49 [27]. The plot is the input CNR in BPF versus probability of error. The theoretical curve is the plot of (7.72) except E_b/N_0 is adjusted by 3 dB to account for dibit in a symbol [26–30].

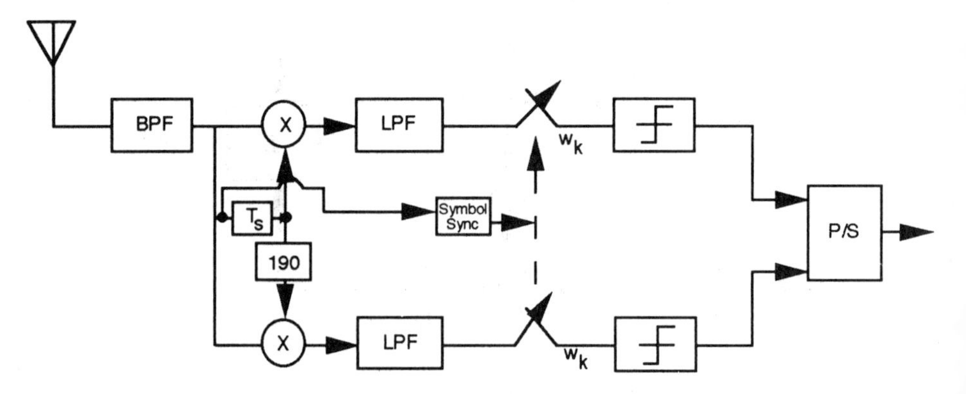

Figure 7.48 ρ/4 DQPSK delay detector.

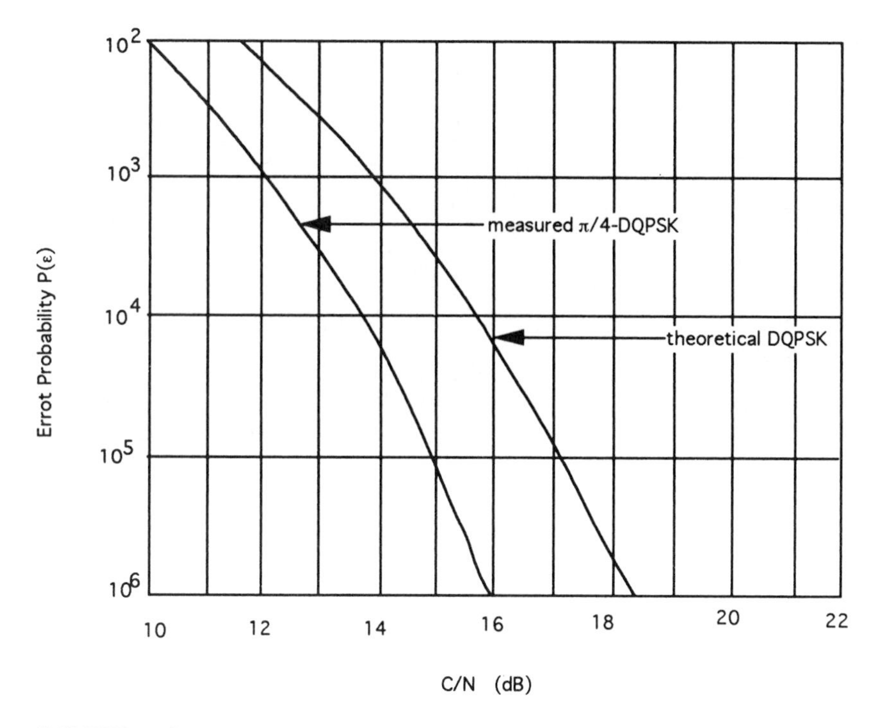

Figure 7.49 BER performance of $\pi/4$-DQPSK in AWGN channel. Source: [27, p. 241].

$$p(\epsilon) = e^{-2E_b/N_0} \sum_{k=0}^{\infty} (\sqrt{2} - 1)^k I_k\left(\sqrt{2}\,\frac{E_b}{N_0}\right) - \frac{1}{2}\, I_0\left(\sqrt{2}\,\frac{E_b}{N_0}\right) e^{-2E_b/N_0} \tag{7.72}$$

where I_k is the kth order Bessel function of the first kind.

FM Discriminator Detector

Due to compatibility requirements in NADCS, $\pi/4$ OQPSK detection based on an FM discriminator is most desirable. Considering a common subscriber unit for both analog and digital systems in the initial stage of introduction of the digital system, this feature is of great advantage since $\pi/4$-shifted DQPSK can be detected by a limiter-discriminator as the signal always varies by $\pm\pi/4$ or $\pm 3\pi/4$. The implementation is shown in Figure 7.50. The eye diagram represents the four levels produced at the output of the filter, as shown in Figure 7.51.

One symbol integrator integrates the frequency deviation between $(k - 1)T$ to KT, which is the phase difference between two sampling instants. The decoder

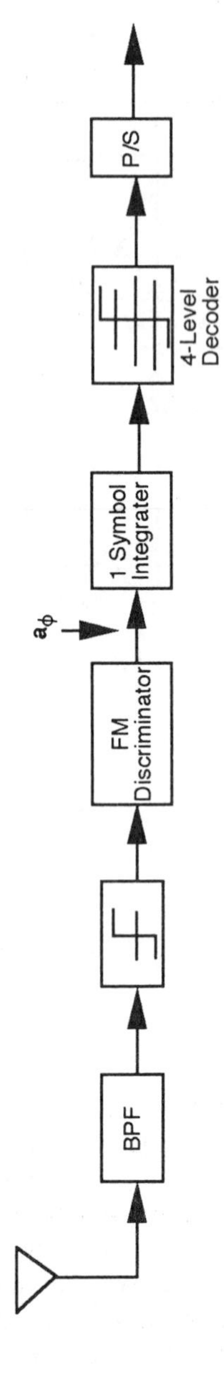

Figure 7.50 FM discriminator detector.

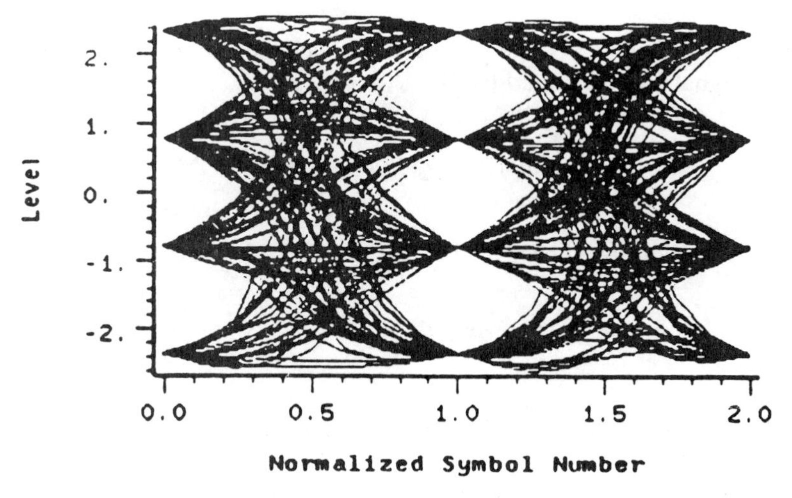

Figure 7.51 Eye diagram for discrimination detector. Source: [29].

has thresholds at 0 and $\pm\pi/2$. Integrated outputs are compared against these thresholds before a symbol decision is made.

The second type of detector that can be used for $\pi/4$ DQPSK is the coherent detector. A coherent detector is more complex to implement than non-coherent detectors due to the required additional circuitry for carrier recovery. However, the performance of a coherent detector is 2–3 dB better than a non-coherent detector in Gaussian and slowly fading channels. In Rician fading channels with compensation, coherent detection may be preferable.

7.6 TAMED FREQUENCY MODULATION [31–34]

As seen in Section 7.5, MSK is an improvement over QPSK as it makes the phase change linear and limits it to $\pm\pi/2$ over a bit interval. The low sidelobe in output PSD was obtained as a result of this linear phase change. Unfortunately, the mainlobe was even wider than QPSK/OQPSK modulation. Further reduction in the sidelobe *PSD* was attained by passing the baseband rectangular pulses through a Gaussian filter before modulation. Here we found that the output *PSD* became a function of the normalized filter bandwidth B_bT, as shown in Figure 7.39. The baseband filtering resulted in introducing a controlled amount of ISI, which in turn smoothes out the output phase by reducing discontinuities in the slope of the phase function. Thus, proper injection of ISI smoothes the phase function during bit intervals. TFM takes this process a step further. In TFM, pulses are wider than a single symbol period. Signal phase shifts are correlated so that the phase shift over a bit interval is a function of the present bit as well as several previous bits. Thus, if the data signal $a(t)$ is defined by

$$a(t) = \sum_{n=-\infty}^{n=+\infty} a_n \delta(t - nT) \tag{7.73}$$

with $a_n = \pm 1$; then, for TFM,

$$\Phi(mT + T) - \Phi(mT) = \frac{\pi}{2}\left(\frac{a_{m-1}}{4} + \frac{a_m}{2} + \frac{a_{m+1}}{4}\right) \tag{7.74}$$

Thus, when all three bits are of the same polarity, the maximum change is limited to $\pm\pi/2$. When three bits are alternating in polarity, the maximum phase change is zero. Phase shift is limited to $\pm\pi/4$ when two adjacent bits are of the same polarity and the third is of the opposite polarity. This includes the cases where $a_{m-1}\ a_m\ a_{m+1} = (++-,\ +--,\ -++,$ and $--+)$. The phase functions for straight MSK, MSK with sinusoidal smoothing, and TFM are shown in Figure 7.52.

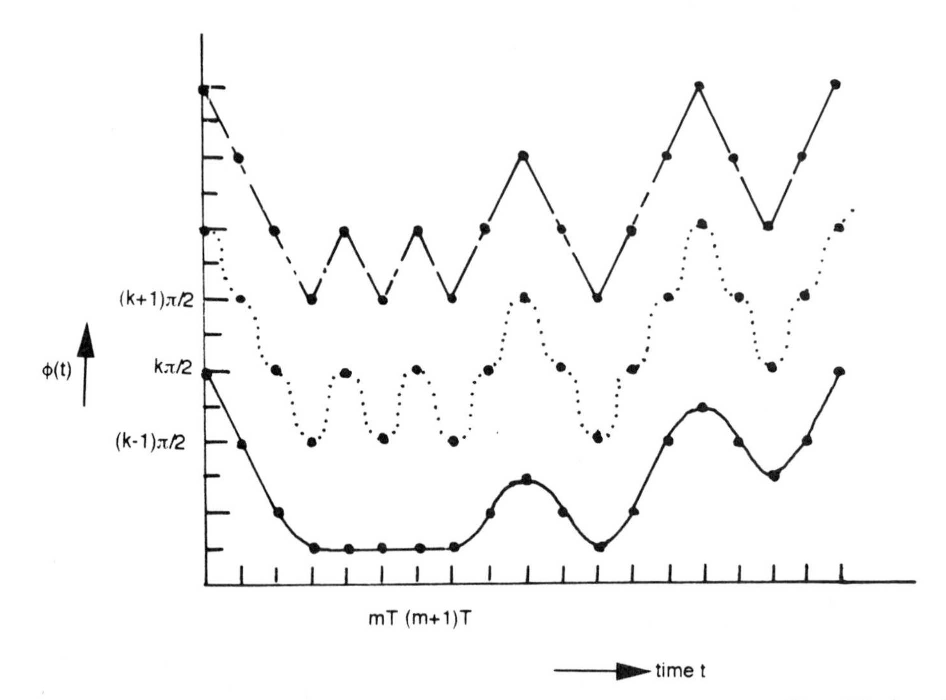

Figure 7.52 Phase behavior of MSK (dashed line), MSK with sinusoidal smoothing (dotted line), and TFM (solid line). Source: [29, p. 535].

A phase plot for MSK shows a phase change of $\pm\pi/2$ during two adjacent bits. MSK with sinusoidal smoothing shows a smoother version of MSK phase. Here, the sharp edges have disappeared but the maximum slope of the phase path has considerably increased. In TFM, phase function is smooth and the slope of the phase function is not as steep as MSK with sinusoidal smoothing. Figure 7.53 shows the PSD plots for all three cases. The PSD of TFM is straight drooping and does not bounce back. This results in a further reduction of adjacent channel interference at a moderate increase in complexity.

BER plots of 4 PSK and TFM are shown in Figure 7.54.

$$P_{\mathrm{QPSK}}(\epsilon) = \frac{1}{2}\,\mathrm{erf}_c\left(\sqrt{\frac{E_b}{N_0}}\right) \tag{7.75}$$

$$P_{\mathrm{TFM}}(\epsilon) = \frac{1}{4}\,\mathrm{erf}\left(\frac{\sqrt{E_b}}{N_0}\right) + \frac{1}{4}\,\mathrm{erf}_c\left(\sqrt{0.69\,\frac{E_b}{N_0}}\right) \tag{7.76}$$

For the details of premodulation filters and modulator and demodulator structures, the reader is referred to [32].

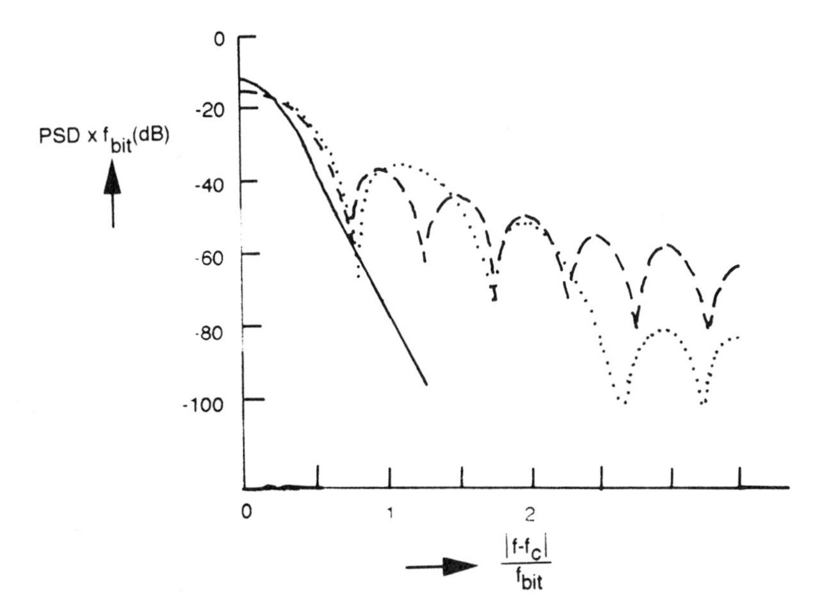

Figure 7.53 Power spectral density functions of MSK (dashed line), MSK with sinusoidal smoothing (dotted line), and TFM (solid line). Source: [29, p. 535].

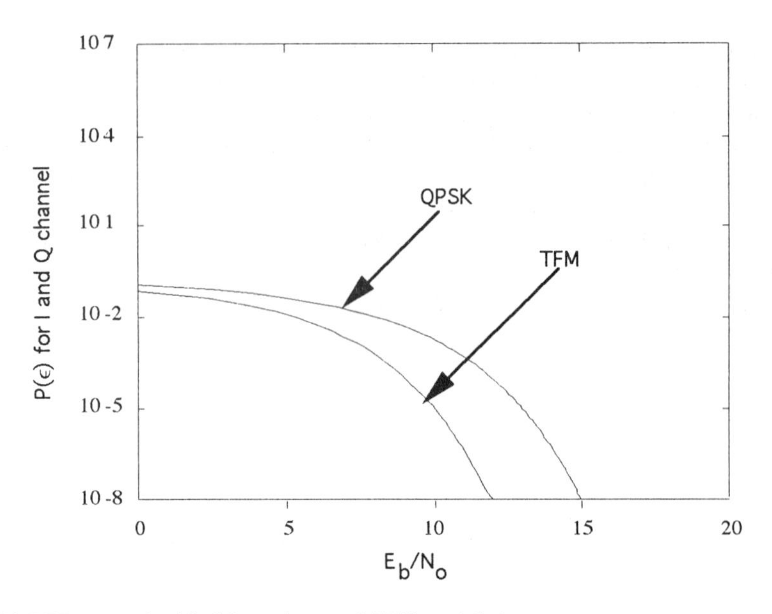

Figure 7.54 BER curves for ideal four-phase and TFM modulations.

7.7 CONCLUSIONS

In this chapter we have consolidated all the modulations used in different mobile communications. Key to the success of a nonlinear modulation lies in continuous and smooth phase transitions at the modulator output. This is achieved by providing pre-modulation gentle filter, such as Gaussian, having linear phase characteristics. This reduces the output bandwidth requirement along with a reduction of adjacent-channel interference. An example of belonging to the non-linear class is MSK and GMSK, extensively used in various cordless telephones and the European cellular system. Unfortunately, the spectral efficiency of these modulations is not high. Higher values of bits/sec/Hz can be achieved by choosing linear modulation such as $\pi/4$ DQPSK. Due to extraordinary traffic demand, this scheme has been adopted in the U.S. cellular system.

PROBLEMS

7.1 Prove that the Carson's rule BW is satisfied for both U.S. and European analog cellular systems. Assume Δ_{pk} for U.S. and European systems to be 12 kHz and 8 kHz, respectively. What percentage of power will the channels have? What is the modulation index in these cases? Will these systems be regarded as NBFM or WBFM?

7.2 With the values of the modulation index computed in problem 7.1, how many terms in the expansion of Bessel's series are important, assuming the maximum modulating tone frequency to be 3.3 kHz.

7.3 Consider an angle modulated signal

$$V_c(t) = 10\cos(2\pi f_c t + 3\cos 2\pi 1{,}500t)$$

a. Assuming FM modulation, determine the modulation index and find the required transmission bandwidth when f_m is increased by a factor of 2 and when f_m is decreased by a factor of 2.

b. Repeat the part a assuming modulation to be PM.

7.4 An FM signal $v_c = A\cos(w_c t + \beta\sin\omega_m t)$ is applied to RC highpass filter. Assume that $\omega RC \ll 1$ in the frequency band occupied by $v_c(t)$ and show that the output voltage across the resistor is an amplitude modulator signal. Find the AM modulation index.

7.5 The plot of (7.28) in Figure 7.9 shows that 30-dB output SNR point for 30 kHz channel bandwidth lies on the left of a 25-kHz channel curve. If *I* increases the channel bandwidth further, the 30-dB point should shift more to the left. Can this process continue indefinitely? Give your reasons in favor or against this argument.

7.6 Given a data rate of 270.8 Kbps (GSM) and a channel BW of 200 kHz divided among eight users, find the spectral density of this system. For this computed value, plot (7.32) and draw conclusions with respect to the desired channel separation versus the actual channel separation of GSM.

7.7 Spectrum given in Figure 7.55 is flat up to 0.25 dB and rolls off at approximately 75 dB/B. Therefore, the GMSK spectrum can be approximated by $G_{dB}(f) = -K$ for $0 \le f \le 0.25$ and $G_{dB} = -75(f/B - 0.25) - K$ for $f > 0.25$B, where $G(f)$ is the transmitted power, f is the frequency offset from center, B is the transmitted bit rate, and K is the power in the mainlobe from 0 to $0.25B$. Find the value of K and the transmitted power falling into the adjacent channel. You can assume the system to be GMSK.

7.8 Show that (7.51b) can be expressed as

$$s(t) = a_k\cos 2\pi f_c t\left(\cos\Phi_k\cos\frac{\pi t}{2T}\right) - b_k\sin 2\pi f_c t\left(\cos\Phi_k\sin\frac{\pi t}{2T}\right)$$

with $\phi_k = 0, \pi$ modulo 2π. Show that $\cos\phi_k$ can only change at the zero crossings of $\sin(\pi t/2T)$. What is the symbol weighting on each of the quadrature carriers, and what is the symbol duration in terms of T?

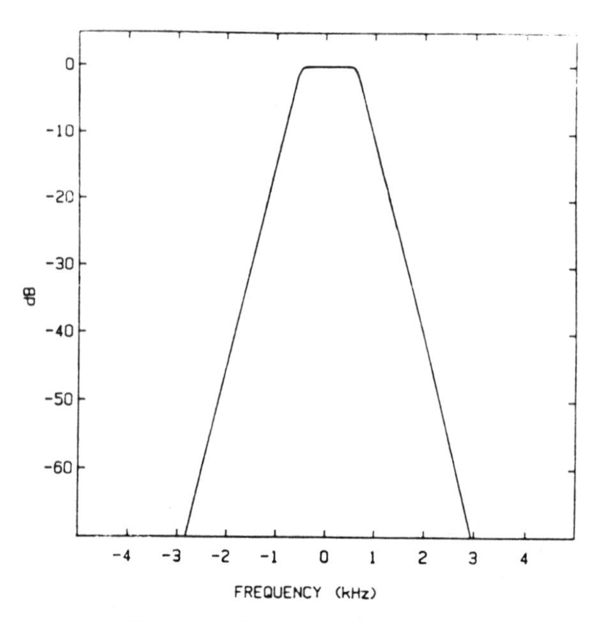

Figure 7.55 Spectrum for problem 7.7.

7.9 Assume that the sequence of transmitted FFSK (MSK with $\Delta = 0.5R_b$) frequencies are f_H, f_L, f_L, f_H, f_H, f_L, f_H, where $f_H = f_c - 0.25R_b$ and $f_L = f_c + 0.25R_b$. Assuming further that the initial carrier phase is 0 deg, draw the baseband quadrature waveforms.

7.10 If the channel bit error rate is given by P_e, show that the error rate at the output of the differential decoder for both BPSK and QPSK is given by $P_b = 2P_e(1 - P_e)$. For small P_e, what is the effect of differential decoding? What is the (E_b/N_0) degradation if $P_e = 10^{-3}$ for BPSK and QPSK?

7.11 With reference to Figure 7.28, explain clearly why a band-limited (filtered) QPSK spectrum regenerates its sidelobes when passed through a nonlinear amplifier. Why does the sidelobe remain low for OQPSK?

7.12 Assuming (7.55a) and (7.55b) and the out-of-band definition for power as

$$P_{ob} = 1 - \frac{\int_{-B}^{B} s(f)df}{\int_{-\infty}^{\infty} s(f)df}$$

draw Figure 7.35. What conclusions can you draw with respect to the applicability of BPSK, QPSK or OQPSK, and MSK as the BT product varies?

7.13 Starting from (7.68a) and (7.68b) arrive at (7.71). Provide reasoning at every step of your derivation.

7.15 Provide arguments for the following:
   a. A two-level eye diagram is the correct representation of I and Q channels data for baseband differential detection.
   b. A four-level eye diagram is the correct representation of I and Q channel data for FM discriminator detector.
   c. A five-level eye diagram is the correct representation of I and Q channel output data for coherent detection.

7.16 Frequency response of the raised cosine pulse is given in (7.70). Find the time domain response, $h(t)$.

REFERENCES

[1] Petrovic, P. M. "Digitized Speech Transmission at VHF Using Existing FM Mobile Radios," *IEEE Trans. on Vehicular Technology,* Vol. VT-31, No. 2, May 1982.

[2] Nettleton, R. W., and H. Alavi. "Power Control for a Spread Spectrum Cellular Mobile Radio System," *IEEE Vehicular Technology Conference*, 1983.

[3] Mehrotra, A. "EE-246 and EE-247," class notes, George Washington University, Department of Electrical Engineering.

[4] Shanmugam, K. S. "Digital and Analog Communication Systems," John Wiley and Sons: New York, NY, 1977.

[5] Taub and Schilling. *Principles of Communication Systems*, McGraw-Hill Book Co.: New York, NY, 1971.

[6] Carlson, A. B. *Communication Systems*, McGraw-Hill Book Co.: New York, NY, 1968.

[7] Shimbo, O., and C. Loo. "A Simple Formula for Threshold Characteristics of FM Signals," *IRE Proc.*, July 1968, pp. 1241–1242.

[8] Lee, W.C.Y. *Mobile Cellular Telecommunications Systems*, McGraw-Hill Book Co.: New York, NY, 1987.

[9] Hirade, K., and K. Murota. "A Study of Modulation for Digital Mobile Telephony," *29th IEEE Vehicular Technology Conference*, March 1977.

[10] Chuang, J. C-I. " The Effects of Delay Spread on 2-PSK, 4PSK, 8-PSK, and 16-QAM in a Portable Radio Environment," *IEEE Vehicular Technology Conference* 1987, pp. 21.4.1–21.4.4.

[11] Murota, K., K. Kinoshita, and K. Hirade. " Spectrum Efficiency of GMSK Land Mobile Radio," *IEEE Vehicular Technology Conference* 1981, pp. 23.8.1–28.8.5.

[12] Asakawa, S., and F. Sugiyama. "A Constant Spectrum Constant Envelope Digital Modulation," *IEEE Trans. on Vehicular Technology*, Vol. VT-30, No. 3, Aug. 1981.

[13] Goode, S. H. " A Comparison of Gaussian Minimum Shift Keying to Frequency Shift Keying for Land Mobile Radio," *IEEE Vehicular Technology Conference*, 1984.

[14] Proakis. *Digital Communications*, McGraw-Hill Book Co.: New York, NY, 1987.

[15] Bhargava, V. J., et al. *Digital Communications by Satellite*, a Wiley-Interscience Publication, 1981.

[16] Rhodes, S. "FSOQ, a New Modulation Technique That Yields a Constant Envelope," *National Telecommunication Conference* 80, pp. 51.1-1–51.1-7.

[17] Ziemer, R. E., and R. L. Peterson. *Digital Communications and Spread Spectrum Systems*, Macmillan Publishing Company: New York, NY, 1985.

[18] Murota, K., and K. Hirade. "GMSK Modulation for Digital Radio Telephony," *IEEE Trans. on Communication*, Vol. COM-29, July 1981, pp. 1044–1050.

[19] Simon, M. K., and C. C. Wang. " Differential Detection of Gaussian MSK in a Mobile Radio Environment," *IEEE Trans. on Vehicular Technology*, Vol. VT-33, No. 4, Nov. 1984, pp. 307–320.

[20] Feher, K. " Modems for Emerging Digital Cellular-Mobile Radio System," *IEEE Trans. on Vehicular Technology*, No. 2, May 1991, pp. 355–365.

[21] Le-Ngoc, T., and K. Feher. " Power and Bandwidth Efficient Modulation Techniques," *IEEE Vehicular Technology Conference*, 1982, pp. 89–93.

[22] Ziemer, and C. R. Ryan. "Minimum-Shift Keyed Modem Implementations for High Data Rates," *IEEE Communication Magazine*, Oct. 1983, pp. 28–37.

[23] de Buda, R. "A Comparison of Two Digital Modulation Schemes," *IEEE Vehicular Technology Conference*, 1975.

[24] EIA/TIA IS-54, Dual-Mode Mobile Station-Base Station Compatibility Standard, Dec. 1989.

[25] Sklar, B. "A Structured Overview of Digital Communications-Tutorial Review-Part I," *IEEE Communication Magazine*, 1983, pp. 4–17.

[26] Liu, C., and K. Feher. "Noncoherent Detection of p/4-QPSK Systems in a CCI-AWGN Combined Interference Environment," *IEEE Vehicular Technology Conference*, 1989, pp. 83–94.

[27] Guo, Y., and K. Feher. "Performance Evaluation of Differential p/4-QPSK Systems in a Rayleigh Fading/Delay Spread/CCi/AWGN Environment," *IEEE Vehicular Technology Conference*, 1990, pp. 420–424.

[28] Ono, S., et al. " Digital Cellular System With Linear Modulation," *IEEE Vehicular Technology Conference*, 1989, pp. 44–48.

[29] Goode, S. H., et. al. "A Comparison of Limiter-Discriminator, Delay and Coherent Detection for p/4 QPSK," *IEEE Vehicular Technology Conference*, 1990, pp. 687–694.

[30] Hoover, D. "An Instrument for Testing North American Digital Cellular Radios," *Hewlett Packard Journal*, April 1991.

[31] Noordanus, J. "Digital Radio Transmission in Mobile Automatic Telephone Systems (MATS)," *IEEE Vehicular Technology Conference*, 1983, pp. 544–548.

[32] Jager, F., and C. B. Dekker. "Tamed Frequency Modulation, a Novel Method to Achieve Spectrum Economy in Digital Transmission," *IEEE Trans. on Communication*, May 1978.

[33] Sundberg, C-E. "Continuous Phase Modulation," *IEEE Communication Magazine*, 1986, pp. 25–38.

[34] Pasupathy, S. "Minimum Shift Keying: A Spectrally Efficient Modulation," *IEEE Communication Magazine*, 1979.

[35] Haykin, S. *Communication Systems*, John Wiley & Sons: New York, 1978.

[36] Rhodes, S. "Effects of Hardlimiting on Bandwidth Transmissions with Conventional and Offset-QPSK Modulation," *IEEE National Telecommunications Conference*, 1972.

[37] Messaros, R. B., and F. J. Seifert. "Performance of GMSK Transmission Under Typical Indoor and Outdoor Channel Conditions Using Coherent and Noncoherent Reception," *IEEE Vehicular Technology Conference 1990*, p. 494.

Chapter 8
Diversity and Combining

8.1 INTRODUCTION

In mobile communication, reception suffers from fast fading in urban environments as well as in the heavily wooded areas. Fast fading is characterized by deep fades occurring within a fraction of a wavelength and is caused by multiple reflections from fixed and mobile structures. In heavily built-up areas, waves arriving from different directions with different amplitudes and phases are random to cause the resultant signal amplitude to follow a Rayleigh distribution over relatively small distances.

The fast fading is removed by averaging over distances of tens of wavelengths, but the slow fading still exists. This is caused by variations in both the terrain profile and the general nature of the environment. The average signal strength can be 40–50 dB below the free-space path loss and is found to have a log-normal distribution with a standard deviation of 4–8 dB. Slow fading can be improved by increasing the transmitted power and is therefore not of interest in this chapter. Fast Rayleigh fading, which cannot be improved by increasing the signal power (and where diversity helps), will be discussed here. The diversity reception is one way to improve the reliability of communication without increasing either the transmitter's power or the channel bandwidth. Reliability is being used here in the communication sense rather than with reference to the reliability of equipment. For emergency voice applications, reliability on the order of 100% is desired. For other normal voice services, holes in the coverage may be acceptable as the messages usually have enough redundancy that short fades and flutter generally do not matter. The redundancy can also be increased by the use of techniques such as repeating the voice message. However, a natural redundancy that is inherent in speech is not present in data communication, and thus for data applications the system is required to have a high reliability. Therefore, the basic idea behind diversity reception is that if two or more independent samples of a random process are taken, then these samples fade independently. The probability that all the samples will simultaneously be below a given level is much smaller than the individual level being below that level. In fact, the probability that all m samples will simultaneously be below a certain level is p^m, where p is the probability that a single sample is below the level. Thus, by properly combining samples one can improve the performance more than if the receiver acts on a single sample.

Diversity, as discussed above, improves communication performance without an increase of transmitter power or bandwidth. However, this ideal may not be achievable in all cases; that is, an increase in bandwidth is sometimes required, and in some cases full diversity gain may not be achieved. A competitor for diversity is coding, but one pays the price in terms of bandwidth. Redundancy may be another choice, but it is regarded as another form of coding. There are different ways by which diversity can be achieved: time, frequency, space, angle (direction), field component, rake (multipath), and polarization.

Among these techniques, space diversity is of main interest to us, and thus will be extensively treated in this chapter. In order to gain complete advantage of diversity, combining has to be done at the receiving end. Combiners are designed such that the input signal levels, after phase and time delay corrections for the multipath effects, do add vectorially while the noise outputs are added randomly. Therefore, on the average the combined output SNR will be greater than that present at the input of a single receiver. Since the probability of having all the uncorrelated inputs fading simultaneously is small, this condition serves to provide an additional reliability.

Combining can be done at radio frequency (RF) or intermediate frequency (IF) (predetection), or after signal detection. Based on where the combining takes place, we categorize combining as either predetection combining or postdetection combining. The advantages and disadvantages of each type are a subject of considerable controversy. It is, however apparent that for predetection combining, less (and perhaps more complicated) hardware is required since the combined receivers in this configuration utilize common circuitry beyond the IF combiner. This is however, not the case with a postdetection combiner. Since combining is done at the baseband level, all stages of a receiver are duplicated. When considering noise, predetection combining is advantageous, because the noise contributed by the extra receiver stages required for postdetection combining degrade the system performance. For linear modulation, it does not matter whether the combining is done before or after the detection. However, for nonlinear modulation such as FM, the predetection combining yields a higher output SNR, than postdetection combining. If combining is done at the IF or RF level, it will also be necessary to bring the signals in phase before combining. It does not matter whether combining is done at the IF, RF, or after detection, but there is generally no limit to the number of independently fluctuating signals that can be combined after proper processing.

Among various techniques of interest for combining are selective or switched combining, which can be used at the mobile, and maximal ratio or equal gain combining at the cell site. In view of the above discussion, we shall discuss different diversity schemes in Section 8.2 where we emphasize why techniques other than space diversity are not very useful for cellular radio, followed by a brief treatment of predetection and postdetection combining in Section 8.3. Section 8.4 contains details of combining. Section 8.5 treats data systems, followed by a summary and conclusion in section 8.6 and 8.7, respectively.

8.2 DIVERSITY TECHNIQUES

The advantage offered by diversity is the improved performance without an increase in transmitter power or bandwidth. In all the diversity techniques, it is implied that the received signals have very little or, theoretically, zero correlation. However, as shown in Chapter 3, a correlation coefficient as high as 0.7 can still provide a substantial diversity advantage. We discuss below different diversity techniques.

8.2.1 Time Diversity

If the objective of the diversity is to increase the communication reliability without the increase of power, time diversity is the best possible technique. This scheme is easily applicable to digital systems, where the same bits of information can be transmitted at time intervals well separated compared to the reciprocal of the average fading rate. The fade levels with various repetitions will then be essentially independent so that the appropriate combinations of repetitions will provide good diversity performance. In order to gain advantage of diversity, a time diversity scheme will require storage both at the transmitter and at the receiver. Such a storage requirement, and specifically the diversity combination of the processed signal, is easy for digital rather than for analog systems. Thus, the scheme cannot be adopted in analog cellular radio. Also, time diversity will reduce the data throughput rate. In some cases the data rate has to be increased also. As an example, if the data is discontinuous and if one decides to repeat the data, it may not be necessary to increase the data rate. On the other hand, for continuous data there may not be a choice but to increase the data rate. However, time diversity has several advantages, amongst which is that only a single antenna is required and there is no requirement for co-phasing or duplication of the receiver. In spite of the advantages, time diversity cannot be used easily by mobile radios. This is seen by an example discussed below. We know that the vehicle velocity v and the distance traveled D are related by a time parameter t such that $D = vt$ or $t = D/v$. From samples of a Rayleigh envelope, the distance between independent fades D is equal to $\lambda/2$, where λ is the wavelength of the signal. Thus, $t = \lambda/2v$.

Therefore, time separation between independent fades must be

$$t \geq \lambda/2v \quad (8.1)$$

Since the time separation between independent fades is inversely proportional to the vehicle velocity v for a static mobile, time separation of infinity is required. In this situation, time diversity will fail theoretically for both analog and digital transmissions. However, at UHF the wavelength is so small that minor movements of people and vehicles ensures that the standing wave pattern is never truly stationary. As discussed in Section 3.6, the time delay spread of the signal received is exponential in distribution and its value lies between 0.25–5 μs in suburban and urban areas. Thus, if the multiple received signals can be brought in phase, a form of time diversity can be achieved due to multipath. However, it is not an easy matter to achieve this in practical systems. The processing used for received signals in the time diversity is majority voting.

8.2.2 Frequency Diversity

As shown in Chapter 3, if the spacing between carriers is larger than the coherence bandwidth, then the independent fading of the two signals is expected. Obviously, the price paid is the increased bandwidth. From discussions in Section 3.6, the coherence bandwidth can be given by

$$B_c \geq 1/2\pi\Delta \quad (8.2)$$

Thus, for a delay spread of the order of 0.5–5 μs, the required coherence bandwidths are 318 kHz and 31.8 kHz. Therefore, signals that are separated by at least these values can be combined. To obtain relatively complete decorrelation, one would want a spacing of the order of magnitude larger than this value. For achieving maximum decorrelation one needs at least 3–5% of the channel center frequency as bandwidth. However, this separation though feasible in cellular surroundings will severely limit the already limited number of channels. Thus, the cost of frequency diversity in mobile surroundings is high. On the other hand, frequency diversity provides a complete redundant path and the full advantage of diversity gain.

8.2.3 Space Diversity

As discussed in Chapter 3, the model for a mobile transmission path consists of an antenna mounted at a high mast essentially removed from the influence of local scatterers, a line-of-sight path to the general vicinity of the mobile in question, and scattering through various obstructions near mobile. This model was analyzed by Clark, who has shown that a high decorrelation between received signals can be obtained if the separations between two electric dipole antennas are greater than $\lambda/2$. Specifically, the correlation coefficient between detected outputs for the signal received by two electric dipoles spaced at a distance L is given by

$$\rho_E(L) = J_0(2\pi L/\lambda) \tag{8.3}$$

where J_0 is the zero-order Bessel function of the first kind. The correlation function is shown in Figure 8.1. Limiting the correlation to 0.7 or less shows the spacing between two antennas to be greater than $\lambda/4$. With this spacing, the advantage of space diversity can be obtained by the mobile. From example 3.1, it is easy to see that if two uncorrelated signal envelopes can be obtained at two antennas, then the probability of simultaneous fades in excess of 20 dB is reduced to 0.01%, while for a single antenna system this number is 1%. The requirement for decorrelation remains the same for signals received at the cell site. However, the conditions are somewhat different. Signals transmitted from the mobile go through scattering in the near vicinity of the mobile itself. Assuming once again that the cell-site antenna is free of obstructions in the near vicinity and a line-of-sight path exists up to a point close to mobile, correlation will be high for antennas spaced at 0.5λ. Spacing

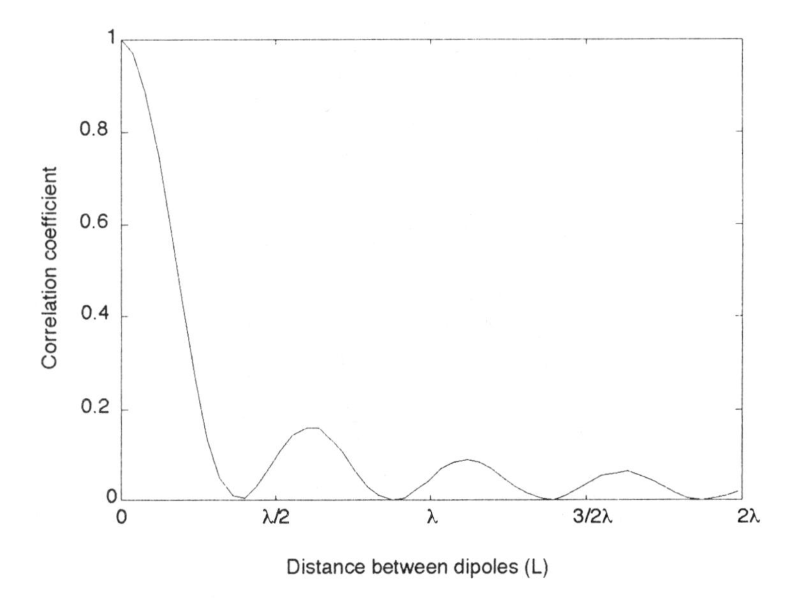

Figure 8.1 Correlation between two electric dipoles.

at the cell-site antenna is a function of the scattering volume A, the horizontal distance R between the scattering volume and the antenna mast, and the electrical phase shifts that the waves produce at the phase center of the antennas [2]. Figure 8.2 shows two antennas with a separation distance D in a direction perpendicular to the direction of the arrival of the rays from the scattering volume. Assuming that the scattering volume forms the same angle at the electrical-phase centers of the two antennas, then the phase difference due to path difference between two electrical waves L_1 and L_2 is given by

$$\Delta_1 = (2\pi/\lambda)(L_1 - L_2) \tag{8.4}$$

Similarly, the phase difference due to the path length difference at the second antenna is

$$\Delta_2 = (2\pi/\lambda)(L_1' - L_2') \tag{8.5}$$

Thus, the differential phase shift between the two antennas is

$$\Delta = \Delta_1 - \Delta_2 = (2\pi/\lambda)[(L_1 - L_2) - (L_1' - L_2')] \tag{8.6}$$

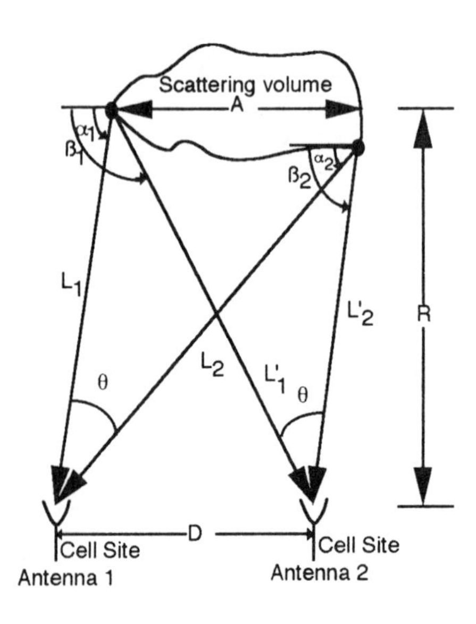

Figure 8.2 Dual space-diversity reception configuration. After [2, p. 419].

Assuming the distance $R \gg D$ and $R \gg A$, the angles Θ seen by the cell-site antennas are

$$\theta = (A/R) \ll 1 \text{ rad} \tag{8.7}$$

to the first order approximation $\alpha_2 \cong \alpha_1$ and $\beta_2 \cong \beta_1$. Hence, the rays L_1' and L_2' are parallel and L_1' and L_2' are parallel. Reference [2] has shown that (8.7) with α_1 and β_1 near $\pi/2$ can be approximated as

$$|\Delta| \cong (2\pi/\lambda) D\theta \tag{8.8}$$

Thus, for uncorrelated fading $(2\pi/\lambda)D\theta \gg 2\pi$, or $D \gg \lambda/\theta$.

Assuming operation at 850 MHz, which corresponds to $\lambda = 13.8$ in and an antenna beamwidth at 30 deg (sector antenna), the ratio λ/θ is 26.4 in. Thus, an antenna separation of about 5–10 times above the computed value would be adequate for independent fading of the signal.

Based on experimental data of Zysman and Lee, the normalized value of antenna separation has been defined as [3]

$$\eta = \frac{\text{antenna height}}{\text{antenna spacing}} \tag{8.9}$$

Figure 8.3 provides an upper bound of correlation versus η when the base antennas are pointed in different directions. The data was taken for vertically polarized antennas in suburban areas when the transmission path was kept at three miles. The same curves can, however, be used for a larger length of transmission. A linear regression curve was fitted for the data at each angle at which the measurements were conducted. A system designed with these curves provides a safe separation distance between the cell-site antennas for a given correlation coefficient. The curve shows very little difference between the values of the correlation coefficient between the angular orientations of 0 deg and 30 deg. Thus, along broadside orientation the value of the correlation coefficient between signals received by two antennas will remain about the same. As shown in the curve, the worst case correlation value (higher value of ρ) is obtained when the signals are received at a 90-deg angle with respect to the antenna broadside. We illustrate the use of these curves with an example below.

Example 8.1

For the configuration in Figure 8.4, find the antenna separation at a frequency of 850 MHz so that the value of the correlation coefficient is always less than 0.7.

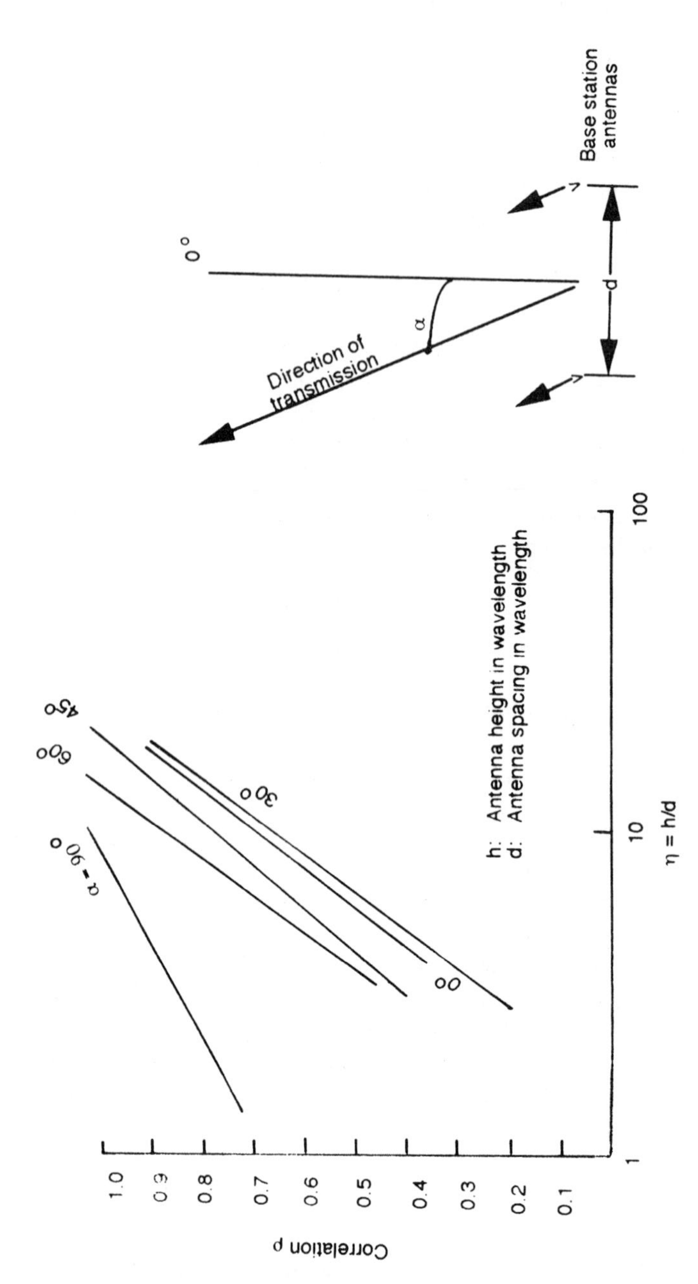

Figure 8.3 Correlation versus antenna height and spacing. Source: [3, p. 291].

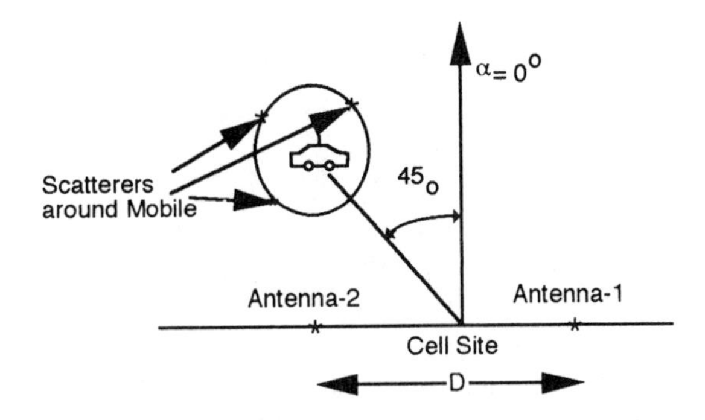

Figure 8.4 Mobile offset at 45 deg with antenna broadside.

Assume that the antenna's phase centers are located at 100-ft above the tower base. At $\alpha = 45$ deg, and for correlation ρ of 0.7, the normalized value of antenna separation (from Figure 8.3) is approximately 9. Thus, $\eta = h/d = 9$ or $d =$ 100 ft/9 ≅ 11 ft. Therefore, the required separation between two cell-site antennas is 11 ft. In the above configuration with 30-deg beamwidth, η is nearly equal to 10. For $h = 100$ ft, the value of d is nearly equal to 10 ft. This value is in the same range as we get from (8.8).

8.2.4 Angle Diversity (Direction Diversity) [4]

Assume the radiation pattern of two antennas to be $G_1(\alpha_n)$ and $G_2(\alpha_n)$ for a radio wave coming from a direction α_n, as shown in Figure 8.5(a). Thus, a wave of

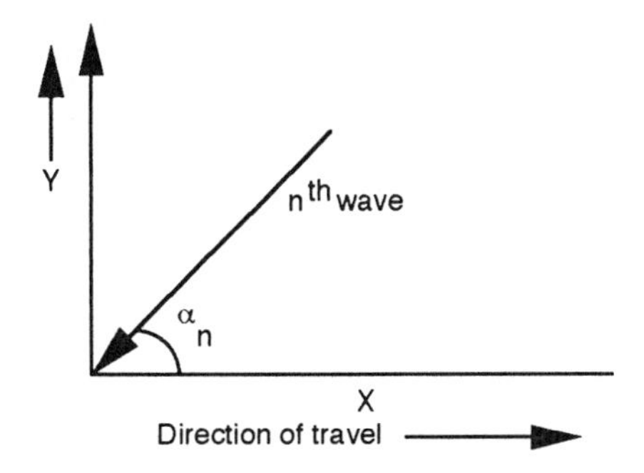

Figure 8.5(a) An incident wave at field point.

amplitude E_0 and phase ϕ_n coming from a direction α_n produces a voltage at the receiver antenna output as

$$E_1 = E_0 \sum_{n=1}^{N} G_1(\alpha_n) e^{j\phi_n}$$
$$E_2 = E_0 \sum_{n=1}^{N} G_2(\alpha_n) e^{j\phi_n} \tag{8.10}$$

It is assumed here that the two antenna gains are at least partially nonoverlapping. Signals received in these antennas will be decorrelated to the extent of the non-overlapping part of the gains. The cross covariance of the two outputs is

$$R_{12} = \langle E_1 E_2^* \rangle = E_0^2 \sum_{n=1}^{N} \sum_{m=1}^{N} \langle G_1(\alpha_n) G_2(\alpha_m) e^{j(\phi_n - \phi_m)} \rangle \tag{8.11a}$$

Due to statistical independence of ϕ and α, we can write

$$R_{12} = E_0^2 \sum_{n=1}^{N} \sum_{m=1}^{N} \langle G_1(\alpha_n) G_2(\alpha_m) \rangle \langle e^{j(\phi_n - \phi_m)} \rangle \tag{8.11b}$$

The second term produces an average of zero except when $n = m$. Thus,

$$R_{12} = N\, E_0^2 \, [\langle G_1(\alpha_n) G_2(\alpha_n) \rangle] \tag{8.12}$$

Equation (8.12) assumes that at every field point, the number of incident waves is the same and equal to N. The angle α is also assumed to be uniformly distributed. Thus,

$$R_{12} = N E_0^2 \int_0^{2\pi} p(\alpha) G_1(\alpha) G_2(\alpha) \, d\alpha = \frac{N E_0^2}{2\pi} \int_0^{2\pi} G_1(\alpha) G_2(\alpha) \, d\alpha \tag{8.13}$$

Similarly, the standard deviation for two antennas is

$$\sigma_1^2 = \left\langle \sum_{n=1}^{N} \sum_{m=1}^{N} E_0^2 G_1(\alpha_n) G_1(\alpha_m) e^{j(\phi_n - \phi_m)} \right\rangle$$
$$\sigma_2^2 = \left\langle \sum_{n=1}^{N} \sum_{m=1}^{N} E_0^2 G_2(\alpha_n) G_2(\alpha_m) e^{j(\phi_n - \phi_m)} \right\rangle \tag{8.14}$$

Again, the exponential terms are zero except for $n = m$, for which case σ_1^2 and σ_2^2 are reduced to

$$\sigma_1^2 = E_0^2 \sum_{n=1}^{N} \langle G_1(\alpha) G_1(\alpha) \rangle = E_0^2 \sum_{n=1}^{N} \langle G_1^2(\alpha) \rangle$$
$$= (NE_0^2/2\pi) \int_0^{2\pi} G_1^2(\alpha)\, d\alpha \tag{8.15a}$$

Similarly,

$$\sigma_2^2 = (NE_0^2/2\pi) \int_0^{2\pi} G_2^2(\alpha)\, d\alpha \tag{8.15b}$$

Using (8.13), (8.15a) and (8.15b), the correlation coefficient becomes

$$\rho = \frac{\displaystyle\int_0^{2\pi} G_1(\alpha) G_2(\alpha)\, d\alpha}{\left[\displaystyle\int_0^{2\pi} G_1^2(\alpha)\, d\alpha \int_0^{2\pi} G_2^2(\alpha)\, d\alpha\right]^{1/2}} \tag{8.16}$$

It is seen from the above expression that the correlation coefficient depends on the amount of angle overlap between two antennas. Thus, for an ideal perpendicular loop with radiation patterns cos α and sin α in the horizontal plane where α lies between 0 to 2π, the correlation coefficient equals zero. A loop and a whip antenna installed at the same point having a voltage gain of 1 and 1.2 cos(α) or 1.2 sin(α) will also produce a correlation coefficient of zero. Thus, two perpendicular loops or a loop and a whip can be used by a mobile.

Test setup and the results of two antennas installed parallel to one another on the top of the roof are shown in Figures 8.5(b) and Figure 8.5(c) [10]. The radiation pattern in the horizontal plane changes due to mutual coupling between elements. The radiation pattern of two antennas is given by

$$G_{1,2}(\phi) = \left| 1 - \frac{Z_{12}}{Z_{ii}} \exp^{(j\beta d \cos\phi)} \right| \tag{8.17}$$

where d is antenna spacing, z_{12} is the mutual impedance, and z_{ii} ($i = 1, 2$) is the self-impedance. The incoming waves are concentrated in horizontal plane (i.e., $\theta = 0$). The solid line in Figure 8.5(c) shows the plot of (8.17). The dotted line

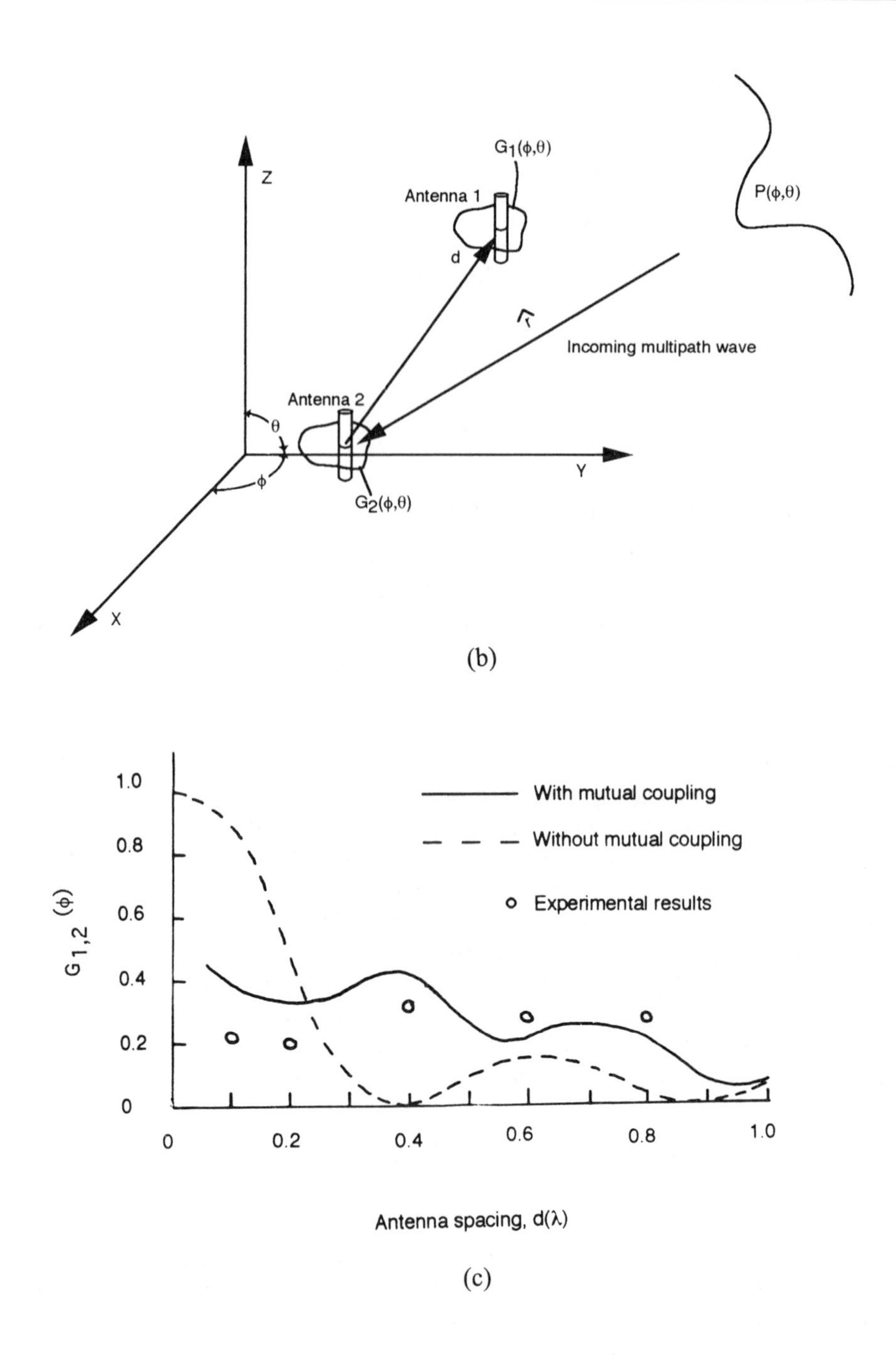

Figure 8.5 (cont.) (b) Antenna configuration. (c) Theoretical and measured results. Source: [10, p. 33

shows the results without mutual coupling. The experimental results agree with the solid line more than the dotted line.

8.2.5 Field Component Diversity

In fading fields, the total electromagnetic energy density remains constant, although the electric and the magnetic fields fade independently. An antenna system that can separate out the electric and magnetic fields and then combine them appropriately can act as a diversity system. Unfortunately, such a system can be used only at low frequencies. If it consists of separate antennas sensitive to electric and magnetic fields, a combined signal proportional to energy density $D = 1/2[\epsilon|E|^2 + \mu|H|^2]$ can be formed. One such composite aerial with hybrids to resolve components of the different fields is shown in Figure 8.6. The antenna consists of two mutually orthogonal semi-loops above a ground plane. The loops are fed to the hybrids, where the sum port provides an output proportional to the electric field, and the difference port provides an output proportional to the magnetic components H_x and H_y. The problem in the above implementation is that the squaring process causes distortion for linear modulated signals and it is undesirable to take the square roots of the output fields. Therefore, a more practical approach will be to use the composite antenna to provide the uncorrelated signals and then combine these signals to gain advantage of diversity combining. Based on this, two types of combining are possible: an incoherently combined signal of the form $V_0 = E_z + H_z + H_y$ and a coherently combined signal of the form:

$$V_1 = |E_Z| + |H_x| + |H_y| \tag{8.18}$$

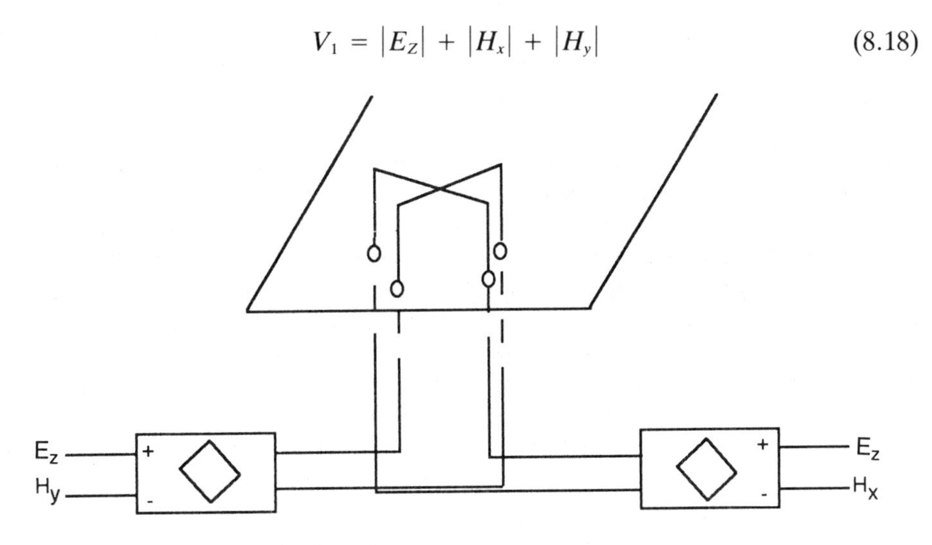

Figure 8.6 Two hybrids to resolve the components of electromagnetic fields.

8.2.6 Polarization Diversity

Two orthogonal signals, when independently affected by the channel characteristics, can be used to gain advantage of diversity at the receiver. However, the fading of these signals is far from being independent in the UHF band and thus cannot be used for combining at the mobile. Also, for a specified transmitted power, a 3-dB reduction in signal power at the mobile is seen in each of the fields.

8.2.7 RAKE Diversity

This is a special form of time diversity that applies to a wideband system with time reference. If the time reference is provided, then the different components of the multipath can be separately identified as distinct echoes of the signal separated in time. These separately identified components of the received signal can be brought in phase for diversity combining. However the design of such a system, which can detect the different components of the multipath accurately and align them in phase for combining, is difficult to design. It should be noted that the design itself is based on the transmission medium, and thus as the medium changes, so does the combining process. The RAKE diversity is being considered for CDMA system by Qualcomm.

8.3 PREDETECTION AND POSTDETECTION COMBINERS

In the previous section we discussed different methods to provide multiple copies of a signal. Depending upon the type of application, one of these methods will be chosen. The problem at the receiver is to combine them in a suitable form to achieve the maximum gain from combining. There are two forms of combining into which all forms of combining can be categorized, namely predetection and postdetection. For linear modulation, it does not matter whether the combining takes place before or after the detection. However, for nonlinear modulations such as FM the difference between these two can be large. For FM modulation, predetection combining is an advantage for those cases where the CNR of the individual channels is below the threshold. Predetection combining, if carried out properly, will bring the output SNR above threshold due to FM improvement, while postdetection combining may not be able to do so due to additional receiver losses. Due to nonlinear characteristics of FM modulation, Rayleigh distribution at the input will not lead to a Rayleigh distribution at the output. Thus, Rayleigh distribution does not apply for postdetection combining. We describe below different schemes for predetection combining.

Figure 8.7(a) has a phase shifter inserted in one of the antenna ports. The phase shifter is adjusted by a suitable dc voltage derived from a differential phase

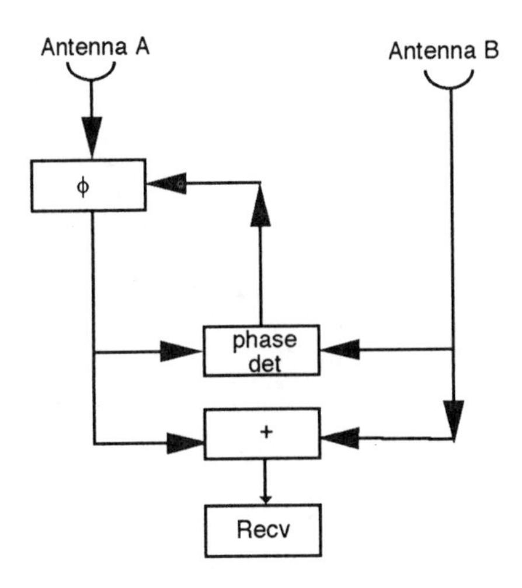

Figure 8.7(a) Phase shifter in one-antenna port.

between two RF signals until the two signals are in phase. The advantage of this scheme is a simple implementation, while a disadvantage is that at a high frequency operation the phase detector may not be able to follow the fast fades. Also there is no simple way of extending this scheme to three or more branches.

The second scheme, based on the hybrid combiner, is shown in Figure 8.7(b). Before the signal reaches the first hybrid, a phase shifter is inserted at one of the inputs of the first hybrid to provide quadrature phase shift at the hybrid input. This will also provide equal amplitude signals at its output ports. The sum port output signal is then passed through a second phase shifter, which brings two signals in phase thereby providing output at the sum port only. The difference port signal is summed to zero. The sum signal now goes to a receiver for detection. The use of two hybrids to make the signals co-phased with each other make the system complex. There is also a loss of power at the difference port. An active hybrid can be substituted in its place, but the intermodulation may become a problem.

The third scheme, shown in Figure 8.7(c), is based on the phase sweeping principle. Here the phase shifter inserted at one of the two antenna output ports is continuously swept between 0 and 2π at a frequency f_s, which is twice the highest modulation frequency present on the carrier. The phase variation can be written as

$$\phi(t) = 2\pi f_s t \tag{8.19}$$

The carrier signals from the two antennas must therefore come in phase at least twice per modulation cycle and will satisfy the sampling theorem. Thus a peak

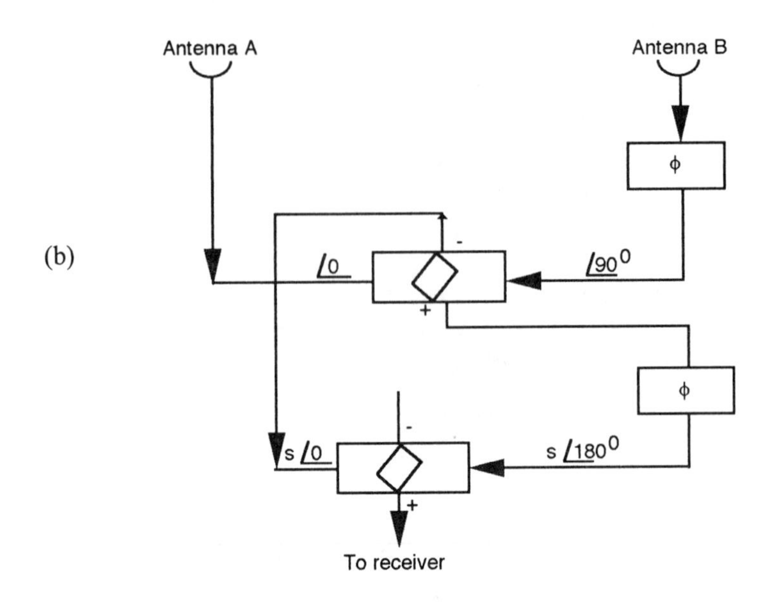

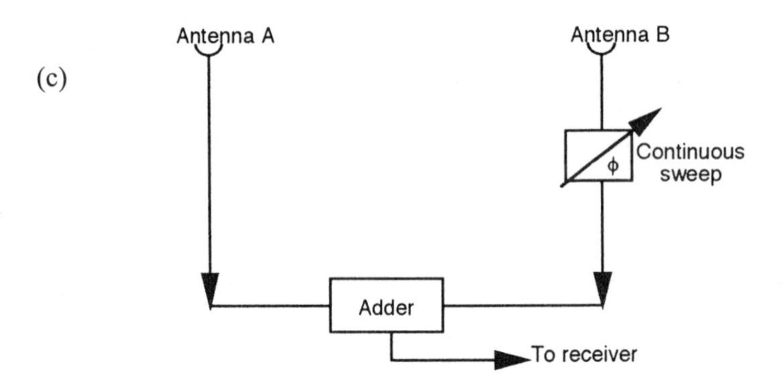

Figure 8.7 (cont.) (b) Hybrid combiner; (c) phase sweeping network.

detector, a suitable filter, and an AM demodulator are expected to provide diversity improvement. The way in which the sweeping takes place is important to avoid serious spectrum spreading. The best choice is to sweep the phase linearly as shown in (8.19). This can be done by using repetitive sawtooth phase modulation and is equivalent to frequency translating the whole of the spectrum passing through the phase shifter by $1/T$ where T is the repetition period of the waveform. Although in theory the scheme works on FM signals, a majority of the work based on this scheme is on AM modulations. Lastly, the most important scheme is based on a class of heterodyne phase strippers, which brings the output in phase for predetection combining [5]. Here, the phase of the output signal is independent of the phase of the input. To illustrate this principle, consider Figure 8.7(d) where the input is

$$S_1(t) = A(t)\cos[\omega_c t + \phi(t) + \phi_o] + A_P \cos(\omega_p t + \phi_o) \tag{8.20}$$

where the first term is due to modulation and the second is due to pilot tone, which is close to the carrier but not close enough to produce interference with the carrier. The principle of operation is that the random phase ϕ_o, due to channel, should affect the carrier as well as the pilot in the same way and therefore can be canceled by a suitable design. The signal $S_1(t)$ is multiplied by a local oscillator signal $S_2(t)$ whose frequency is assumed to be below the carrier and the resultant signal is passed through a narrow bandpass filter F1, which is centered at a frequency equal to the difference of the pilot tone and local oscillator. Then,

$$S_2(t) = B\cos(\omega_o t + \theta) \tag{8.21}$$

and

$$S_3(t) = A_F \cos[(\omega_p - \omega_o)t + \phi_o - \theta] \tag{8.22}$$

Since F1 is a narrow bandpass filter, F1 rejects all modulation associated with carrier. The filter, which is supposed to be wide enough to pass the modulation sidebands to the output, has a center frequency given by the difference between the center frequency of F1 and the pilot tone. Thus, the signal at the output of the filter F2 is given by

$$\begin{aligned} S_4(t) = K\{&A(t)\cos[(\omega_c - \omega_p + \omega_o)t + \phi(t) + \theta] \\ &+ a_P \cos[(\omega_P - \omega_P + \omega_o)t + \theta]\} \end{aligned} \tag{8.23}$$

The phase ϕ_o, introduced due to random variations of the channel, is canceled from the output and the phase of the pilot θ is inserted and the modulation part is preserved. Therefore for multiple channel operation, the output of each diversity channel has the same phase and thus are coherent to each other provided the channel inserted phase varies slowly. In other words, the fading rate should be slow compared with the bandwidth of F1. Figure 8.7(e) illustrates the diversity scheme based on this principle. Here, the predetection coherent combining is possible as the output of each channel multiplier has the same phase.

(d)

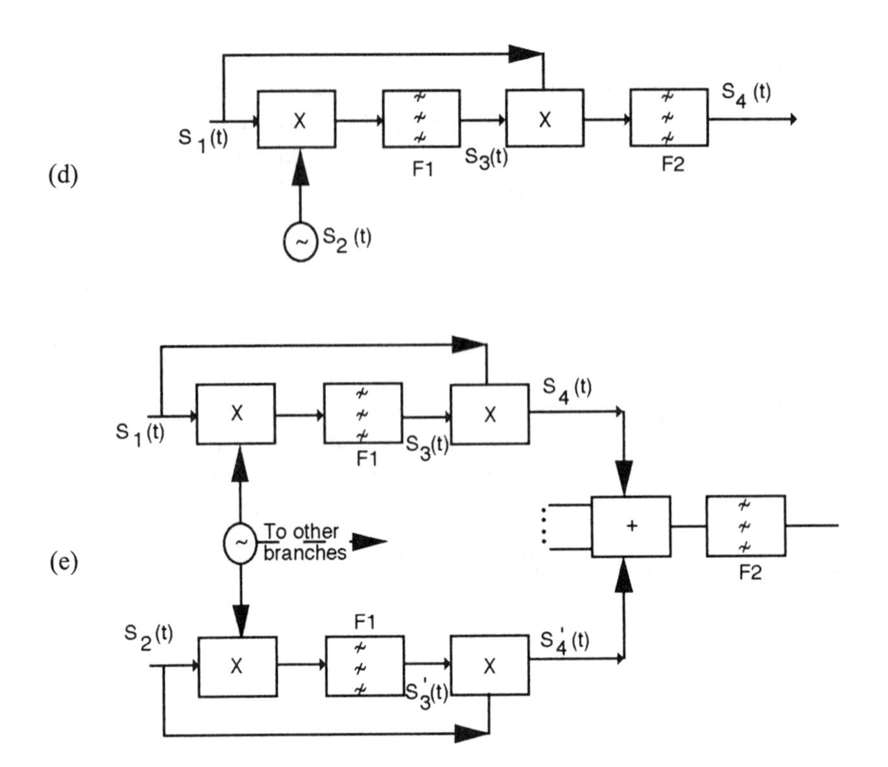

(e)

Figure 8.7 (cont.) (d) Double heterodyne phase stripper; (e) predetection combiner using phase stripper principle.

8.4 COMBINING TECHNIQUES

In general, the term "diversity system" refers to a system that has two or more closely similar copies of the desired signal. More generally, one may have N such copies $f_1(t), f_2(t), \ldots, f_N(t)$, for which

$$f(t) = f_1(t) + f_2(t) + \ldots + f_N(t) = \sum_{i=1}^{N} f_i(t) \tag{8.24}$$

where each of the signals

$$f_i(t) = s_i(i) + n_i(t) \tag{8.25}$$

Thus, $s_1(t), s_2(t), \ldots, s_N(t)$ are the desired message parts, and $n_1(t)$, $n_2(t), \ldots, n_N(t)$ are the noise components. Since the channels are fading, the amplitudes of the received signals are different. Thus, we can rewrite the above equation in the general form as

$$f(t) = a_1(t)f_1(t) + a_2(t)f_2(t) + \ldots + a_N(t)f_N(t) = \sum_{i=1}^{N} a_i(t)f_i(t) \quad (8.26)$$

where $a_i(t)$ is associated with the received $f_i(t)$ and is proportional to the channel gain. Since the summed signal is a linear combination of different copies of the received signal, we classify this combining as linear. Essentially the analysis of the combiners is carried based on the following assumptions:

- The signals are locally coherent, implying that the fading is slower than the lowest modulation frequency present in the signal.
- Noise in each branch is independent and additive.
- The noise components are locally incoherent and have zero means.

Based on linearity there are basically four types of diversity combining: scanning, selection, maximal ratio, and equal gain. Some variations of these basic types also exist, but most of those can easily be identified and their effects can be included in the basic results. We briefly describe the basic properties and operating principles of these combining techniques before we take up their detailed study.

Scanning diversity combining is a techniques of the switched type; that is, at any instant of time, only one of the coefficients $a_1(t)$ is equal to 1. A selector device scans the channels in a certain order until it finds a signal above the threshold and then uses that signal until the signal drops below the threshold. At that point, scanning resumes and a new channel is selected. Figure 8.8(a) shows the configuration of a scanning diversity combiner. Here, only one circuit for a short-term measurement of power is required irrespective of the number of channels actually being used. Short-term average means that the duration of power measurement is short compared to the signal wavelength.

In order to avoid frequent switching from channel to channel, the preset threshold should be low. Sometimes the setting of the threshold provided in one geographical area may not be suitable when the mobile is moved to other locations. Thus, the threshold setting based on long-term average power may be desirable. The modified configuration based on long-term average is shown in Figure 8.8(b). Here, the long-term average is computed over tens of wavelengths and the setting of the attenuator determines how far below the long-term average the signal is allowed to fall before switching occurs. This form of switching is generally applied to the case of two antennas supplying a single receiver through a switch, and that is why this form of diversity combining is also known as antenna selection diversity.

Selection diversity combining is also a switched system in the sense that one of the number of possible inputs is allowed into the receiver. The system is in some sense similar and picks out the best out of the N copies of the received signals

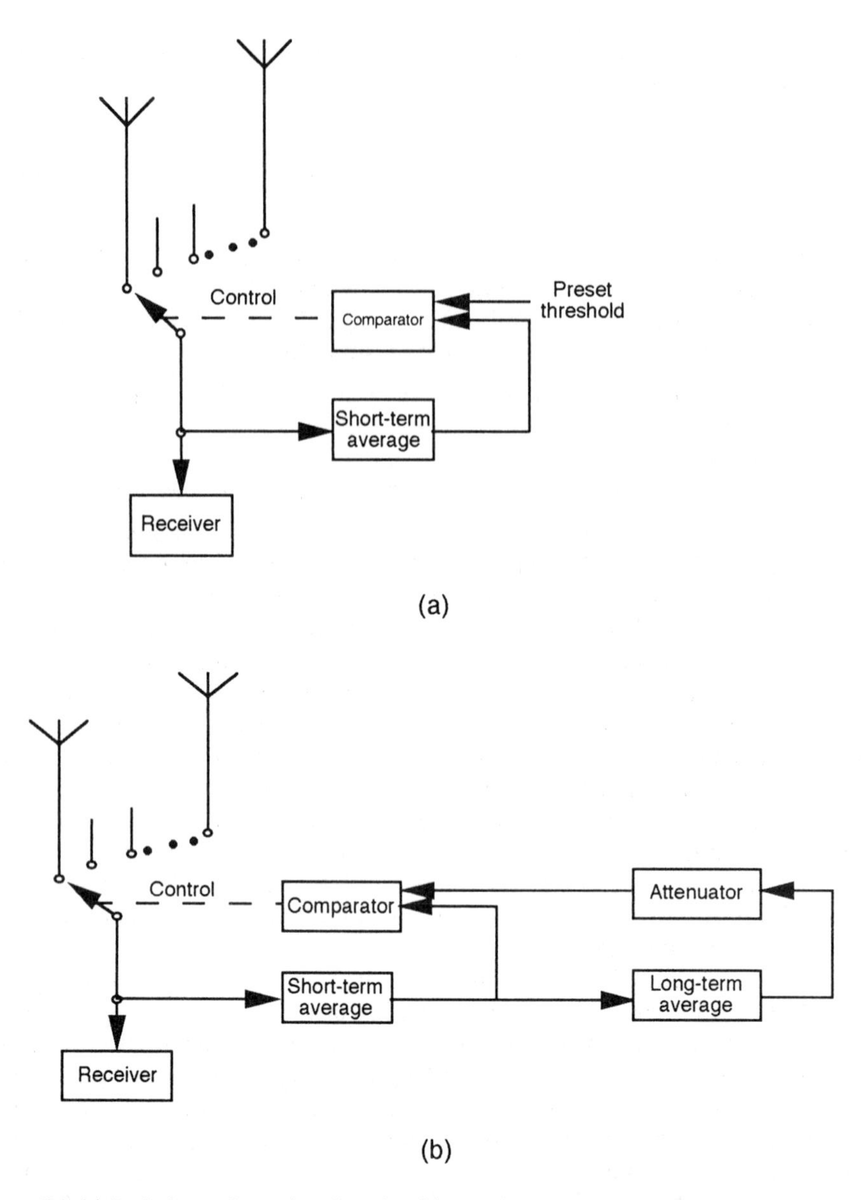

Figure 8.8 (a) Basic form of scanning diversity. (b) scanning diversity—variable thresholds.

$f_1(t), f_2(t), \ldots, f_N(t)$ and uses that alone. Others do not contribute towards the output $f(t)$. Thus, in this case the switching coefficient a_i associated with the signal $f_i(t)$ has a value of unity and all others have zero values. In other words,

$$\begin{aligned} a_i &= 1 \quad j = i \\ a_i &= 0 \quad j \neq i \end{aligned} \tag{8.27}$$

The scheme for three-channel selection diversity is shown in Figure 8.9. Once again the selection can either be done at the RF or at the baseband. If switching is done at the IF or RF, the switching transients may be a problem. The major advantage of this scheme is the best selection of a channel on an instantaneous basis and the disadvantage lies in continuous monitoring of signals in all channels.

A maximal ratio diversity combiner system is defined by the property that among all diversity systems, it yields the maximum SNR to the output signal $f(t)$. If P_i denotes the SNR of $f(t)$, then the maximal ratio combiner realizes

$$P = \sum_{i=1}^{N} P_i \tag{8.28}$$

(i.e., the maximum power ratio realizable from any linear combination is equal to the sum of the individual power ratios). Here, the coefficients in (8.26) are weighed

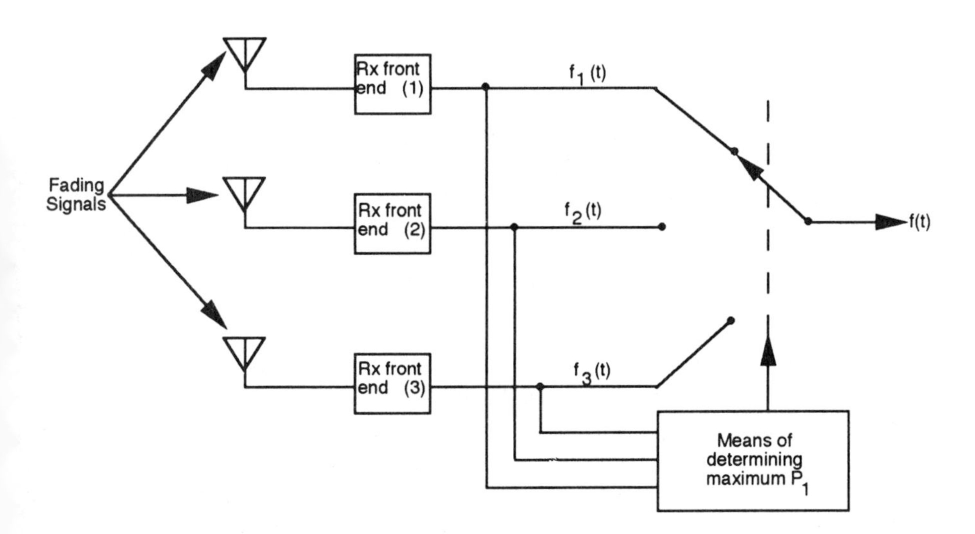

Figure 8.9 Selection diversity (general scheme).

in proportion to their signal and inversely proportional to the mean square noise in that channel, that is,

$$a_i = f_i(t)/\overline{n_i^2} \tag{8.29}$$

In this condition a maximum gain is realized. The system requires co-phasing of the individual signals before combining (coherent), therefore the technique is complex as it requires weighing, co-phasing, and summing of these signals. Its main advantage is that it is possible to produce an output with an acceptable SNR even though none of the signals on the individual channels themselves are acceptable. Figure 8.10 shows a typical implementation of the maximal ratio diversity combiner. Due to its complexity, the system can only be used at the cell site and not at the mobile.

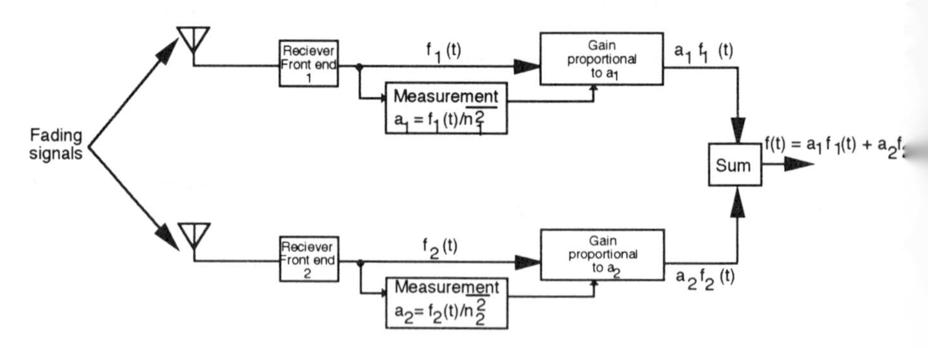

Figure 8.10 Maximal ratio diversity.

Equal gain diversity combining is the simplest type of linear diversity combining. Signals are simply added together to produce the output signal $f(t)$. With all coefficients set to unity, (8.26) is reduced to

$$f(t) = \sum_{i=1}^{N} f_i(t) \tag{8.30}$$

Figure 8.11 is the implementation of two-channel equal gain combining. The blocks with the designation of variable gain have the same gain value. An equal gain combiner operating as a predetection combiner will include a variable phase shift at the front end. Similar to the case of a maximal ratio combiner, the possibility of producing an acceptable signal from a number of unacceptable signals exists and the performance is only marginally inferior to that of a maximal ratio combiner. We elaborate below these combining techniques.

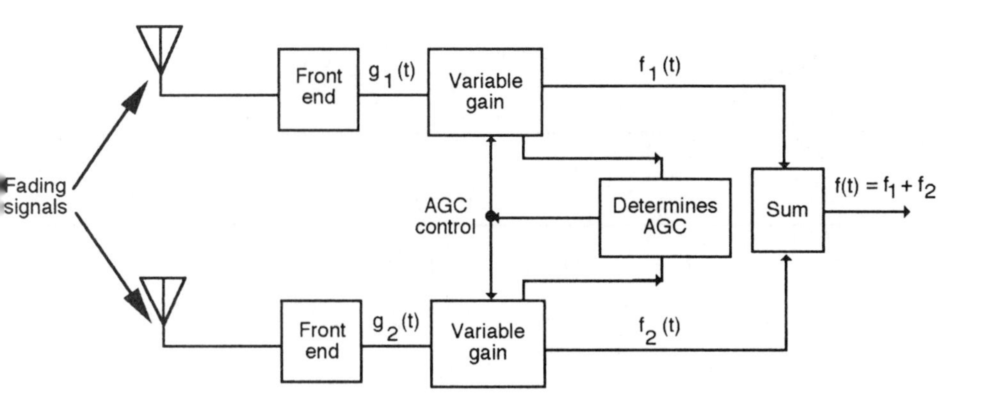

Figure 8.11 Equal gain diversity.

8.4.1 Details of the Scanning Diversity Combiner

As indicated above, this form of combining requires only one measurement circuit. A simplified form with two antennas that can be conveniently used at the mobile receiver is shown in Figure 8.12(a). Here, the instantaneous envelope of the received signal is measured: if it is below a predetermined threshold, the antenna switch is activated, thus selecting a second antenna input. If the newly selected branch is above the threshold, the switch will stay in its new position. Otherwise, either it can stay in this new position or can revert back to the first antenna port. The scheme is shown in Figure 8.12(b). Performance is controlled by the choice of threshold and the time required to detect fading and activating the switch. As

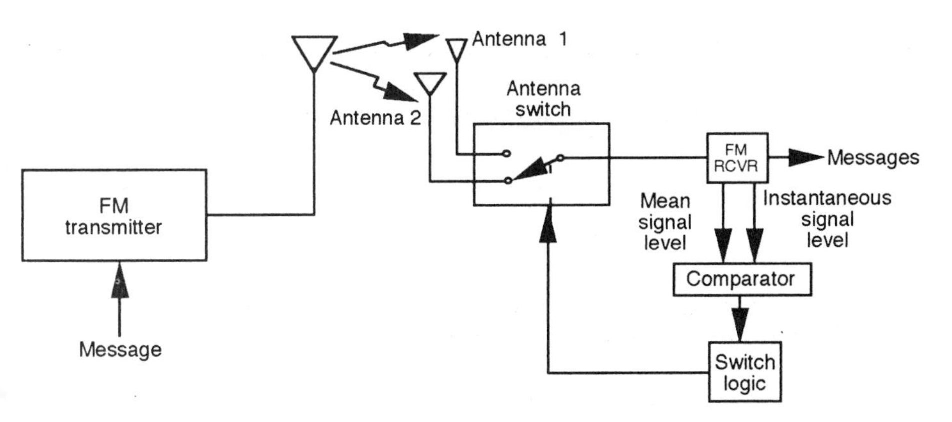

Figure 8.12(a) Switched diversity system.

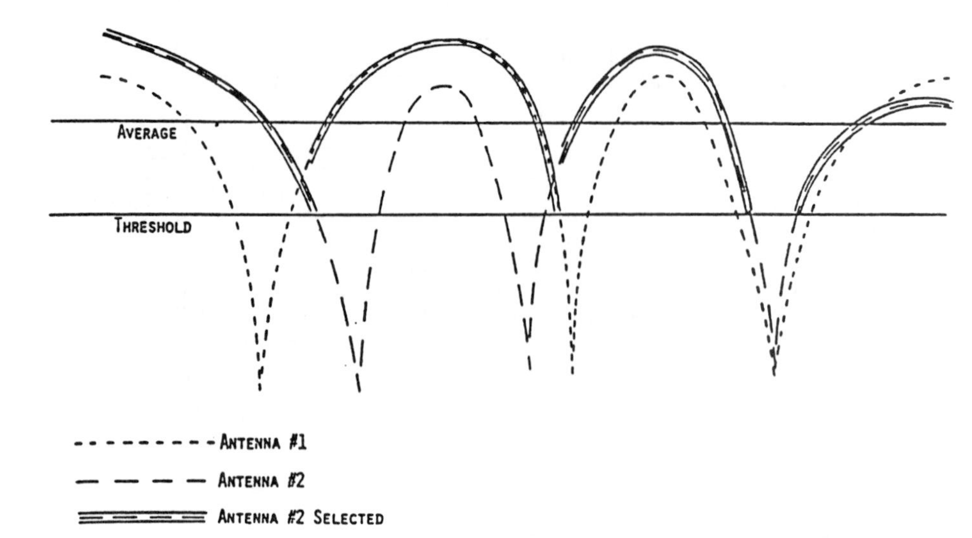

Figure 8.12(b) Switch and stay principle.

stated in the previous section, this switching will cause transients that will appear in the baseband as pulses whose width is proportional to the reciprocal of the IF bandwidth. We derive below the performance of this system under a Rayleigh fading assuming a switch and stay strategy. The density function for the Rayleigh process is

$$\begin{aligned} p(r) &= \frac{r}{b_0} e^{-r^2/2b_o} \qquad r \geq 0 \\ &= 0 \qquad\qquad\quad r < 0 \end{aligned} \tag{8.31}$$

where $\langle r^2 \rangle = 2b_o$. Thus, the cumulative distribution for r is given by

$$P_r(A) = q_A = P_r(r \leq A) = 1 - e^{-A^2/2b_o} \tag{8.32}$$

and

$$P(r > A) = P_A = 1 - P(A) = e^{-A^2/2b_o} \tag{8.33}$$

where A is the threshold value of the signal. It is easy to see that the cumulative distribution function, if the switching is done at any other threshold B, is given by

$$\begin{aligned} P_r(r \leq B) &= P(B) - q_A + q_A P(B) \qquad B > A \\ &= q_A P(B) \qquad\qquad\qquad\quad B \leq A \end{aligned} \tag{8.34}$$

The composite level $r(t)$ is stochastic and is shown in Figure 8.12(b). A plot of $P_r(r < B)$ is shown in Figure 8.13. Both the experimental curve and the theoretical curves for $A = -4$ dB, -10 dB, and -13 dB are plotted. If the threshold is kept at A itself, the $P(r \leq A)$ can be computed by (8.34) as

$$P_r(r < A) = q_A \qquad B = A \tag{8.35}$$

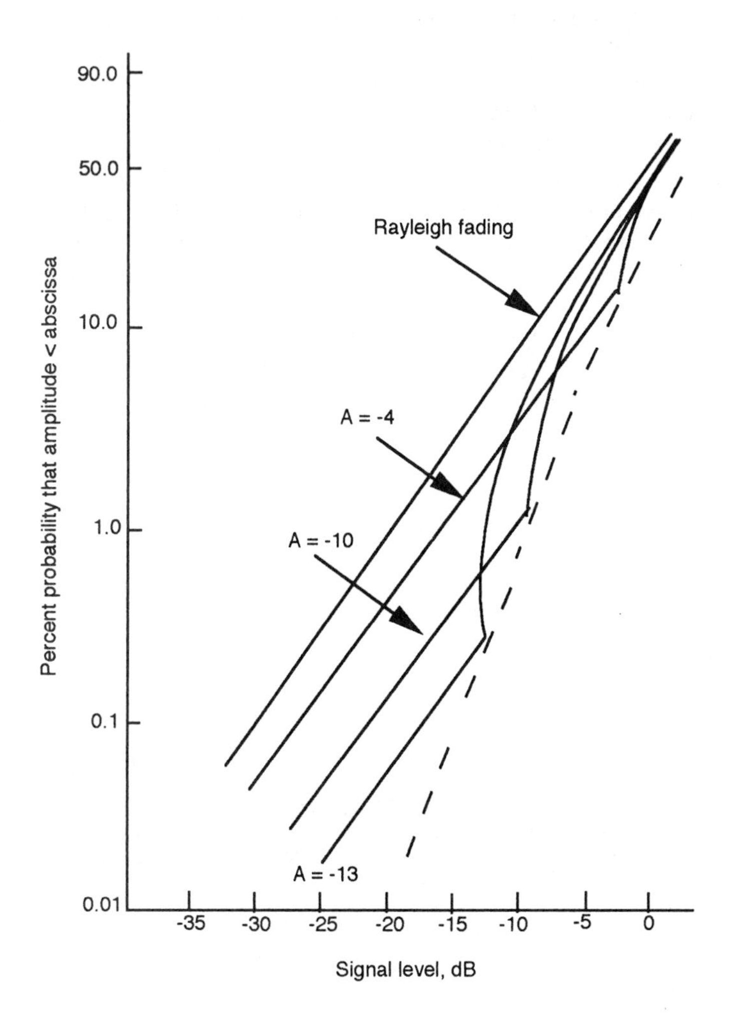

Figure 8.13 Cumulative probability distribution curves of two-branch switching diversity systems (switch and stay). Source: [11].

For comparison purposes, the envelope distribution with a threshold equal to A is drawn along with the case of a single channel under Rayleigh fading. As seen in Figure 8.13, below the threshold the CDF has a Rayleigh curve. Above the threshold the CDF quickly merges with a single Rayleigh cumulative distribution function again.

A modified form of this diversity is known as feedback diversity. The two-branch diversity scheme with two antennas at the base is shown in Figure 8.14. It is assumed that the mobile base path (control channel) is reliable, using some form of base receiver diversity. In this scheme, the base transmitter will be connected to one of the transmitters until the received signal at the mobile falls below the present threshold. At this time the mobile, through the mobile base station path, informs the base of its deteriorating signal condition and the base will switch to a new antenna. The base will keep this new antenna until the level at the mobile falls below the threshold again, when the antenna will be switched. The performance of this scheme is close to the scanning diversity except for a slight degradation, which is caused by delay in switching, and is the sum of the round-trip propagation time and the time delay corresponding to the response time of the control channel. If the delay is high, the signal at the mobile could continue to fade below the threshold before the antenna at the cell site switches. At 900 MHz and a vehicle speed of 60 miles per hour, the expected signal drop is of the order of only 1–2 dB. Thus, the scheme can be effectively used for diversity combining.

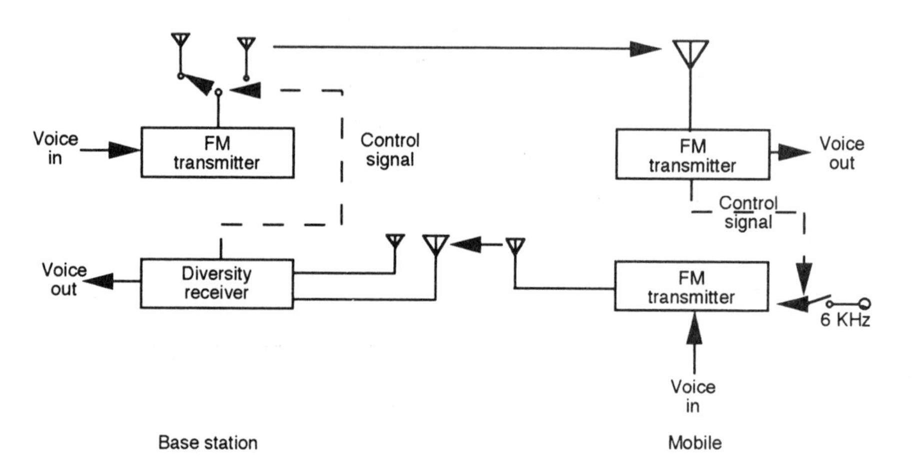

Figure 8.14 Feedback diversity.

8.4.2 Selection Combining

In this scheme, the channel with the highest instantaneous CNR is selected such that the output SNR is equal to the best among the incoming channels. However, in reality the system cannot function on a truly instantaneous basis. So, to be practical the internal time constants of the selection combiner are made substantially shorter than the reciprocal of the signal fading rate. This depends upon the available bandwidth. The basic configuration is shown in Figure 8.15, where one of the *M* receivers having the highest CNR is connected to the output. In addition to switching transients, continuous monitoring of the signals on all branches is required and thus it is difficult for mobile use. This scheme is more suitable for the cell site. The preset threshold should be low enough in this scheme to prevent unnecessary switching. As shown in Figure 8.15, one of the *M* receivers having the highest baseband SNR is connected to the output. This is regarded as a baseband selection. For selection at RF, the signals at the antenna can be sampled and the antenna system with the best instantaneous CNR can be selected.

Assuming that each diversity branch is independent and Rayleigh distributed with equal mean signal power b_o, the density function of the signal envelope is given by

$$p(r_i) = r_i/b_o e^{-r_i^2/2b_o} \tag{8.36}$$

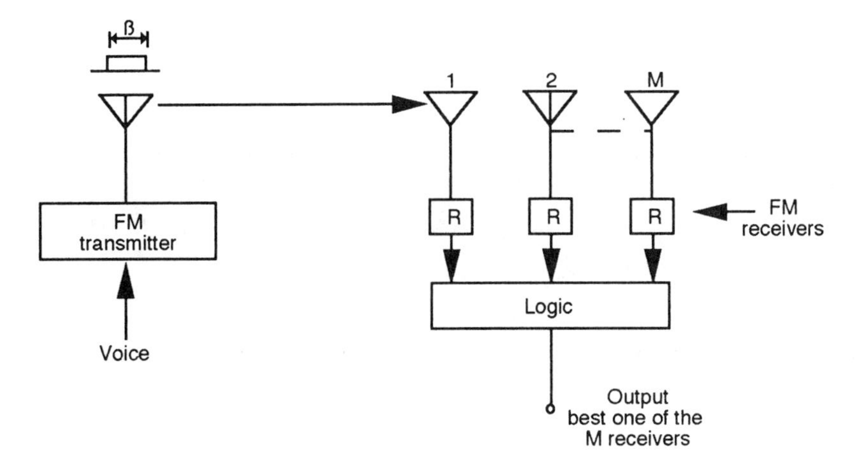

Figure 8.15 Principles of selection diversity.

where $2b_o$ = mean signal power per branch = $\langle r_i \rangle$ and r_i^2 = instantaneous power in the i_{th} branch. Let the mean noise power per branch $E(n_i^2) = 2N_i$ and let γ_i equals the instantaneous signal power per branch divided by the mean power per branch

$$\gamma_i = r_i^2/2N_i \tag{8.37}$$

Let Γ_i equal the mean signal power per branch divided by the mean noise power per branch

$$\Gamma_i = \frac{2b_o}{2N_i} \tag{8.38}$$

Thus,

$$\gamma_i/\Gamma_i = r_i^2/2b_o \tag{8.39}$$

Therefore the density function for γ_i is given by

$$p(\gamma_i) = \frac{r_i}{|J|b_o} e^{-\gamma_i/\Gamma_i} \tag{8.40}$$

where J is the Jacobian and is equal to $(r_i/b_o)\Gamma_i$. Thus

$$p(\gamma_i) = \frac{1}{\Gamma_i} e^{-\gamma_i/\Gamma_i} \tag{8.41}$$

Here we have assumed that the mean power per branch is constant. Thus, the probability that the *CNR* γ_i is less than or equal to γ is given by

$$P(\gamma_i \leq \gamma) = \int_0^\gamma \frac{1}{\Gamma_i} e^{-\gamma_i/\Gamma_i} \, d\gamma_i = (1 - e^{-\gamma/\Gamma_i}) \tag{8.42}$$

Therefore, the probability that the CNR in all diversity branches will be less than or equal to γ is given by

$$P(\gamma_1, \gamma_2, \ldots, \gamma_M \leq \gamma) = \prod_{i=1}^{M} (1 - e^{-\gamma/\Gamma_i}) \tag{8.43}$$

Assuming Γ_i, the mean CNR, to be same in all the diversity branches, then

$$P(\gamma_1, \gamma_2, \ldots, \gamma_M \leq \gamma) = (1 - e^{-\gamma/\Gamma})^M = P_M(\gamma) \tag{8.44}$$

Thus, the probability that at least one branch will exceed the threshold value of γ (CNR) is given by

$$\gamma = 1 - P_M(\gamma) \tag{8.45}$$

The plot of $P(\gamma_i > \gamma)$ for different values of M and the number of diversity branches are shown in Figure 8.16. The biggest value of gain occurs for the two-branch diversity combiner. The values of diversity advantage for different values of avail-

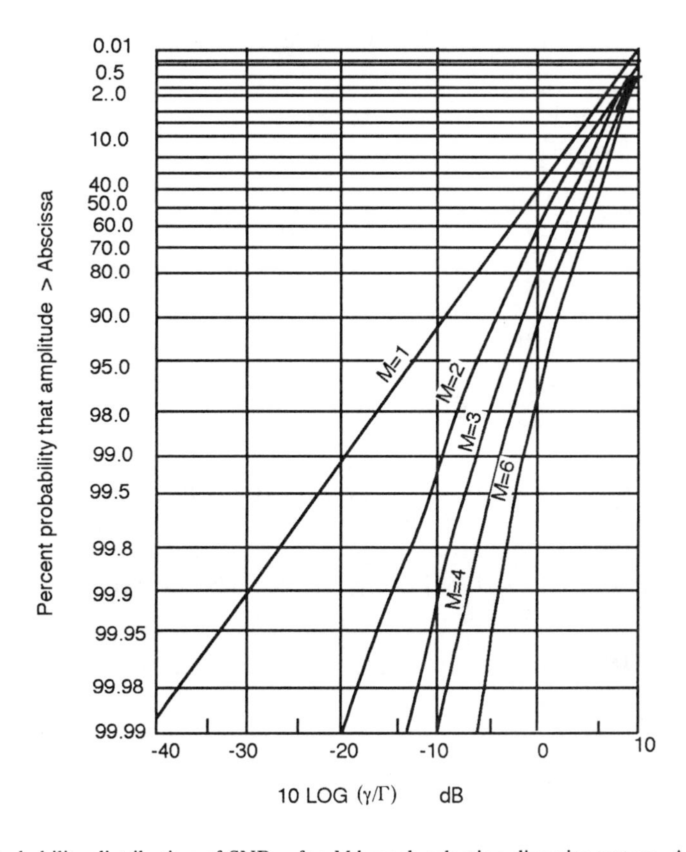

Figure 8.16 Probability distribution of SNR γ for M branch selection diversity system. After [8].

ability and for different values of M can be directly read from Figure 8.16. It should be noted that the gains are reduced as the value of M is increased. To find the mean CNR of the selected branch, we find the density function from (8.44) as

$$p_M(\gamma) = M/\Gamma\,(1 - e^{-\gamma/\Gamma})^{M-1}\,e^{-\gamma/\Gamma} \tag{8.46}$$

The mean value of CNR is given by

$$E[\gamma] = \overline{\gamma} = \int_0^\infty \gamma M/\Gamma\,(1 - e^{-\gamma/\Gamma})^{M-1}\,e^{-\gamma/\Gamma}\,d\gamma \tag{8.47}$$

If $x = \gamma/\Gamma$, then $dx = (1/\Gamma)d\gamma$. Thus,

$$\overline{\gamma}/\Gamma = M\int_0^\infty x(1 - e^{-x})^{M-1}\,e^{-x}dx \tag{8.48}$$

Let $y = 1 - e^{-x}$ or $x = -\ln(1 - y)$. Then $e^{-x}\,dx = dy$. The above equation is reduced to

$$\overline{\gamma}/\Gamma = M\int_0^1 [-\ln(1 - y)]y^{M-1}dy \qquad |y| < 1 \tag{8.49}$$

Thus,

$$\begin{aligned}
\overline{\gamma}/\Gamma &= M\int_0^1 y^{M-1}\left(\sum_{K=1}^{\infty} y^K/K\right)dy = M\sum_{K=1}^{\infty}\frac{1}{K}\int_0^1 y^{K+M-1}dy \\
&= M\sum_{K=1}^{\infty} 1/K(K+M) = \sum_{K=1}^{\infty}[1/K - 1/(M+K)] \\
&= \sum_{K=1}^{M}[1/K - 1/(M+K)] + \sum_{K=M+1}^{\infty}[1/K - 1/(M+K)] \\
&= \sum_{K=1}^{M} 1/K - [1/(M+1) + 1/(M+2) + \ldots + 1/2M] \qquad (8.50) \\
&\quad + [1/(M+1) + 1/(M+2) + \ldots + 1/2M + 1/(2M+1) + \ldots] \\
&\quad - [1/(2M+1) + 1/(2M+2) + \ldots] \\
&= \sum_{K=1}^{M} 1/K
\end{aligned}$$

Thus, the mean SNR increases slowly with M, as shown in Table 8.1. For framed digital transmission, antenna switching noise can be eliminated by carrying out the switching during the preamble or a postamble portion, which is the part of framed signal. This scheme has been adopted by business for cordless telephones, as discussed in Chapter 10.

8.4.3 Maximal Ratio Combining [6–9]

As expressed in Section 8.4, the M signals are weighted proportional to their signal voltage-to-noise power ratios and then summed. The resulting signal is

$$r_M = \sum_{i=1}^{M} a_i r_i(t) \tag{8.51}$$

Since noise in each branch is weighted according to noise power,

$$\begin{aligned}\overline{n^2(t)} &= \sum_{j=1}^{M}\sum_{i=1}^{M} a_i a_j \overline{n_i(t)\, n_j(t)} \\ &= \sum_{i=1}^{M} a_i^2 \overline{n_i^2(t)}\end{aligned} \tag{8.52}$$

From (8.38), $\overline{n_i^2(t)}$ is equal to $2N_i$. Thus, the average noise power is

$$N_T = 2\sum_{i=1}^{M} |a_i|^2 N_i \tag{8.53}$$

Table 8.1
SNR Versus the Order of Diversity

M	γ/Γ	$10 \log \gamma/\Gamma$
1	1.0	0
2	1.5	1.76
3	1.833	2.63
4	2.083	3.2
5	2.283	3.6

The resulting SNR at the output is

$$\gamma_M = \frac{1}{2}\frac{\left|\sum_{i=1}^{M} a_i r_i(t)\right|^2}{\sum_{i=1}^{M} |a_i|^2 N_i} \tag{8.54}$$

We seek to maximize the value of γ_M. This is readily done by using the Schwartz inequality

$$\left|\sum_{i=1}^{M} a_i r_i\right|^2 \leq \left(\sum_{i=1}^{M} |r_i|^2\right)\left(\sum_{i=1}^{M} |a_i|^2\right) \tag{8.55}$$

From [7], if $a_i = r_i/\sqrt{N_i}$ the maximum output SNR is obtained. Thus,

$$\begin{aligned}\gamma_M &= \frac{1}{2}\frac{\left(\sum_{i=1}^{M} r_i^2\right)\left(\sum_{i=1}^{M} r_i^2/N_i\right)}{\sum_{i=1}^{M} |a_i|^2 N_i} \\ &= 0.5\sum_{i=1}^{M}\frac{r_i^2}{N_i}\end{aligned} \tag{8.56}$$

Assuming the mean noise power is the same for each branch of the diversity combiner and equals N, then

$$\gamma_M = 1/2 \sum_{i=1}^{M} \frac{r_i^2}{N} = \sum_{i=1}^{M} \gamma_i \tag{8.57}$$

Thus, the SNR at the combiner output equals the sum of the branch SNR. Hence,

$$\gamma_i = \frac{1}{2}\frac{r_i^2}{N} = \frac{1}{2N}(x_i^2 + y_i^2) \tag{8.58}$$

where x_i and y_i are independent Gaussian random variables of equal variance b_0 and zero mean. Therefore, γ_i is a chi-square distribution of $2M$ Gaussian random variable with variance

$$b_o/2N = \Gamma/2 \tag{8.59}$$

Thus, the density function of the combiner output SNR is given by

$$p(\gamma_M) = \frac{\gamma_M^{M-1}\, e^{-\gamma_M/\Gamma}}{\Gamma^M (M-1)!} \qquad \gamma_M \geq 0 \tag{8.60}$$

The probability that $\gamma_M \leq \gamma$ is given by

$$P[\gamma_M \leq \gamma] = \int_0^\gamma \frac{\gamma_M^{M-1}\, e^{-\gamma_M/\Gamma}}{\Gamma^M (M-1)!} d\gamma_M = \frac{1}{\Gamma^M (M-1)!}\int_0^\gamma \gamma_M^{M-1} e^{-\gamma_M/\Gamma} d\gamma_M \tag{8.61}$$

or

$$\begin{aligned} P(\gamma_M \leq \gamma) &= \frac{1}{\Gamma^M (M-1)!}\left[-\gamma_M^{M-1}\Gamma e^{-\gamma_M/\Gamma}\Big|_0^\gamma + \int_0^\gamma \Gamma\, e^{-\gamma_M/\Gamma}(M-1)\gamma_M^{M-2}\, d\gamma_M\right] \\ &= -\frac{e^{-\gamma/\Gamma}(\gamma/\Gamma)^{M-1}}{(M-1)!} + \frac{1}{\Gamma^M (M-1)!}\int_0^\gamma \Gamma\, e^{-\gamma_M/\Gamma}(M-1)\gamma_M^{M-2} d\gamma_M \end{aligned}$$

Integrating the second term $(M - 1)$ times, we get

$$\frac{1}{\Gamma^M (M-1)!}\int_0^\gamma \Gamma^{(M-1)}(M-1)!\, e^{-\gamma_M/\Gamma}\, d\gamma_M = 1 - e^{-\gamma/\Gamma}$$

Combining all the terms, we obtain

$$1 - e^{-\gamma/\Gamma}\left[\frac{(\gamma/\Gamma)^{M-1}}{(M-1)!} + \frac{(\gamma/\Gamma)^{M-2}}{(M-2)!} + \ldots + \frac{(\gamma/\Gamma)^0}{0!}\right] = 1 - e^{-\gamma/\Gamma}\sum_{K=1}^{M}\frac{(\gamma/\Gamma)^{K-1}}{(K-1)!}$$

Thus,

$$P(\gamma_M > \gamma) = e^{-\gamma/\Gamma}\sum_{K=1}^{M}\frac{(\gamma/\Gamma)^{K-1}}{(K-1)!} \tag{8.62}$$

Figure 8.17 shows the plots of $P(\gamma_M > \gamma)$ for different orders of diversity M. Values of $P(\gamma_M > \gamma)$ for different availabilities can be read from the figure. From (8.57), the mean SNR is given by

$$E(\gamma_M) = E\sum_{z=1}^{M}\gamma_i \tag{8.63}$$

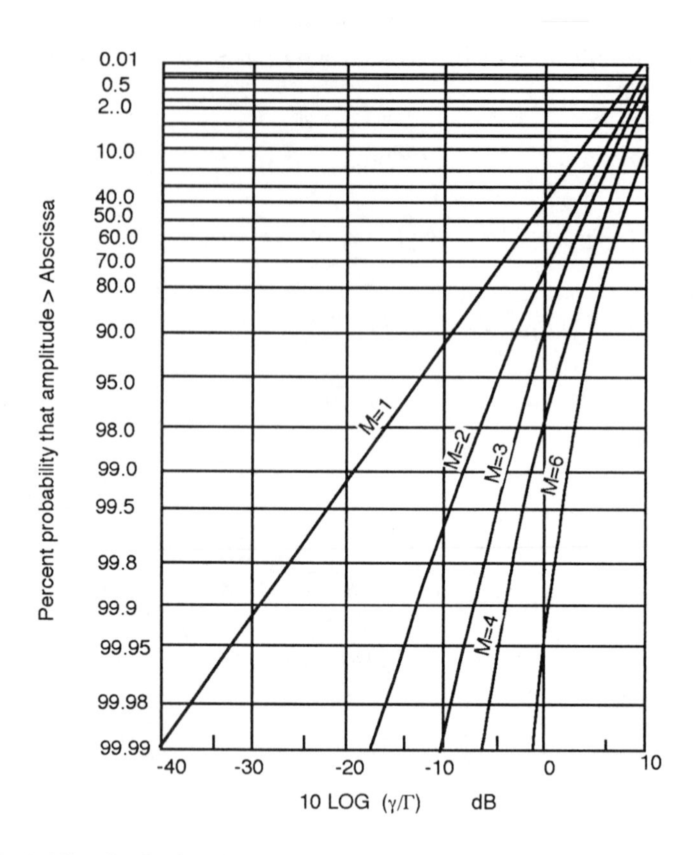

Figure 8.17 Probability distribution of SNR γ for M branch maximal ratio diversity system. After [8].

Since $E(\gamma_i) = \Gamma$, we get

$$E(\gamma_M) = M\Gamma \tag{8.64}$$

The average value of γ_m increases linearly with the order of diversity. This increase in average SNR at the output is a lot slower for selection diversity.

8.4.4 Equal Gain Combining [8–9]

Maximal ratio predetection combining gives the best performance, but in practice it is difficult to achieve because of the requirement of correct weighing factors.

Performance of equal gain combining is only slightly inferior to the maximal ratio combiner, but the implementation complexity is drastically reduced. In this approach, the weighing factor in all branches is the same. Thus (8.54), with all a_i set to 1, reduces to equal gain combining with the resulting output SNR as

$$\gamma_M = \frac{1}{2} \frac{\left(\sum_{i=1}^{M} r_i \right)^2}{\sum_{i=1}^{M} N_i} \tag{8.65}$$

where $|r_i|$ are the individual signal envelope factors. With

$$Y_T = \frac{1}{\sqrt{2N_T}} \sum_{i=1}^{M} |r_i| \tag{8.66}$$

where

$$N_T = \sum_{i=1}^{M} N_i$$

(8.65) is reduced to

$$\gamma_M = y_T^2 \tag{8.67}$$

Unfortunately, the closed form solution of the above equation is difficult to obtain. Reference [2] provides the distribution function for output SNR versus the percent outage rate assuming equal noise powers in all branches. This is shown in Figure 8.18. As seen from the figure, the biggest improvement over Rayleigh fading is obtained for the two-branch combining case ($N = 2$). The performance of this scheme lies in between maximal ratio and selection combining, as shown in Figure 8.19, and is in general only marginally below the maximal ratio combining. However, as in the case of the maximal ratio combiner, the average SNR increases linearly with M. The mean value of SNR is given by

$$\overline{\gamma}_M = \frac{1}{2} \frac{\overline{\left(\sum_{i=1}^{M} r_i \right)^2}}{\sum_{i=1}^{M} \overline{N_i}} \tag{8.68}$$

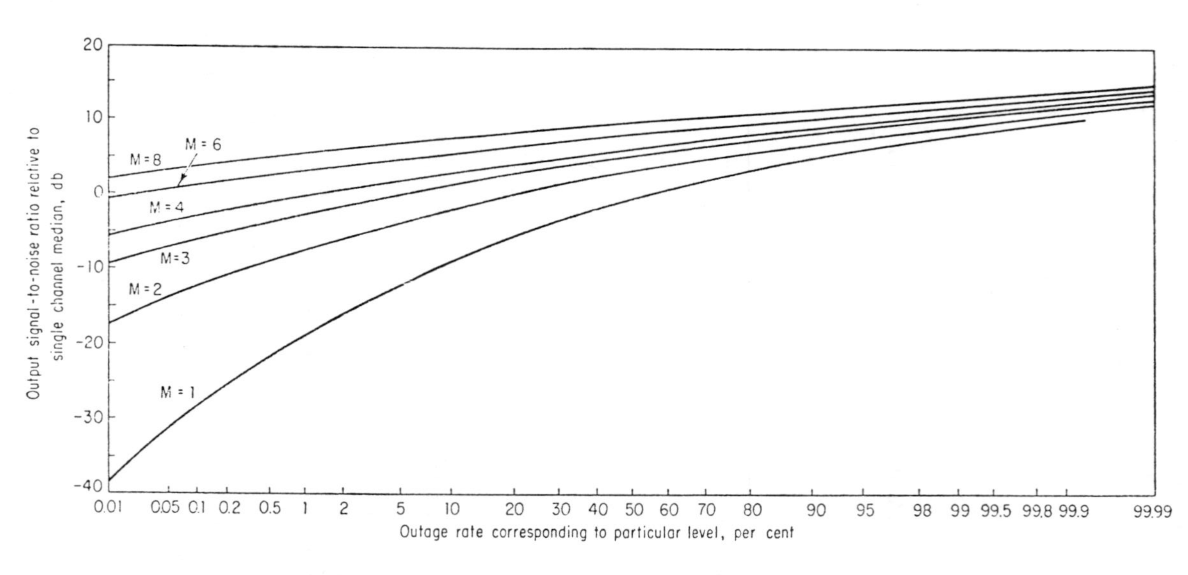

Figure 8.18 Probability distribution for predetection equal gain combining. After [2, p. 455].

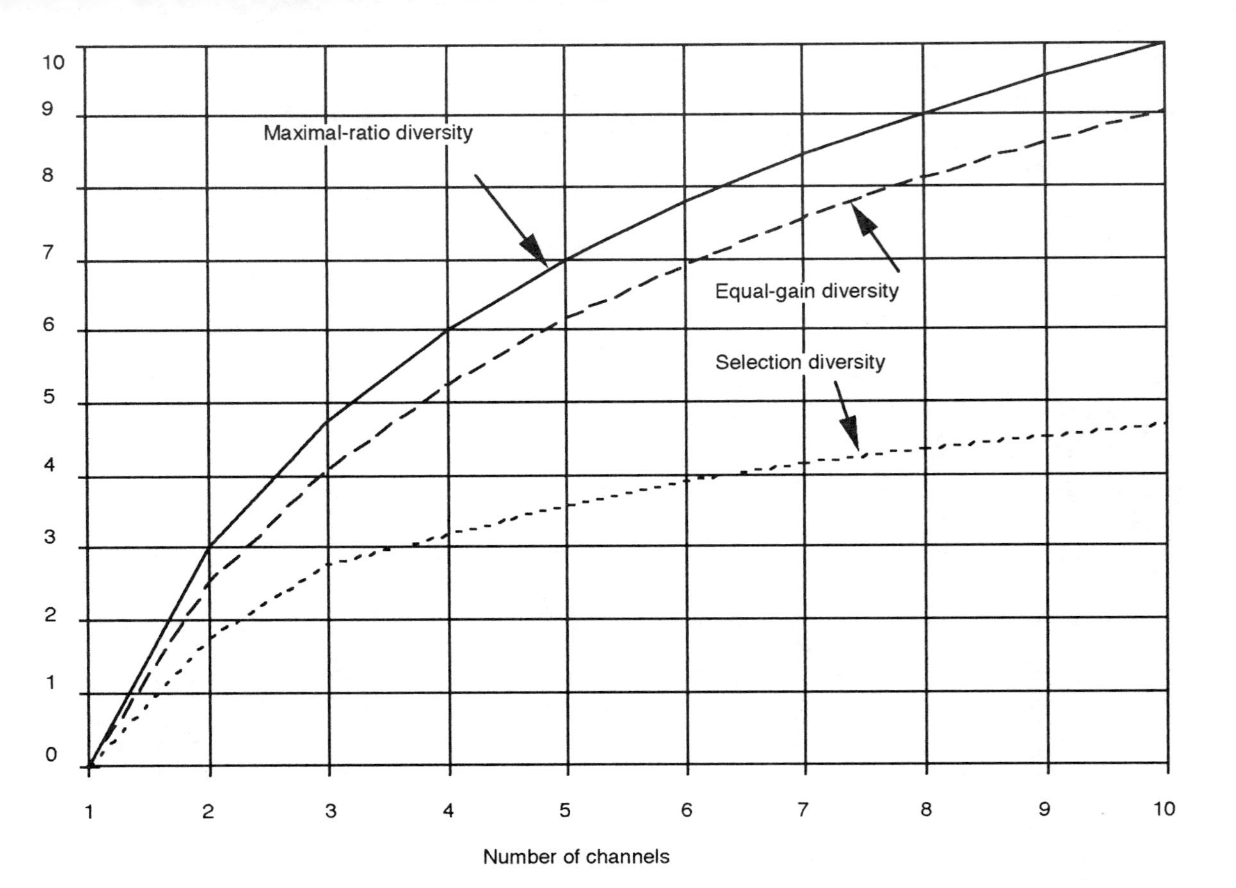

Figure 8.19 Diversity improvement in average SNR. Source: [6, p. 1086].

Assuming noise power to be same in each branch, (8.68) can be rewritten as

$$\overline{\gamma}_M = \frac{1}{2NM} \sum_{i,j=1}^{M} \overline{r_i r_j} \tag{8.69}$$

since $\overline{r_i^2} = 2b_o$, and $\overline{r_i} = \sqrt{\pi b_o/2}$. Also, assuming $\overline{r_i r_j} = \overline{r_2 r_j}$ for $i \neq j$, we can rewrite the above equation as

$$\overline{\gamma}_M = \gamma_o \left[1 + (M - 1) \frac{\pi}{4} \right] \tag{8.70}$$

8.5 EFFECTS OF DIVERSITY ON FSK MODULATION

In this section we shall apply the technique of selection and maximal ratio combining to a noncoherent FSK system. In the previous section we have evaluated different combining techniques for analog signals, where we have applied the average output SNR as an evaluation criterion under diversity. These techniques apply equally for digital systems except the performance measure should be replaced by BER. Thus, the objective for these cases is to find the reduction in BER due to combining. The error rate equation for noncoherent FSK has been derived in Chapter 7 and is given by

$$P_e(\gamma) = 1/2(e^{-\gamma/2}) \tag{8.71}$$

Here, γ is equal to the signal energy/bit divided by the noise spectral density, which is equal to E_b/N_o. The Rayleigh density function is

$$p_\gamma(\gamma) = 1/\gamma_0 \, e^{-\gamma/\gamma_0} \tag{8.72}$$

where $\gamma_o = \Gamma =$ the mean carrier power divided by the mean noise power.

In order to find the error rate under Rayleigh fading, we integrate the probability of error equation over all possible values of γ after weighing the integral by the PDF of γ, or

$$P_{e,\text{fading}} = \int_0^\infty P_e(\gamma) p_\gamma(\gamma) \, d\gamma \tag{8.73}$$

Thus, for noncoherent FSK,

$$P_e = \int_0^\infty \frac{1}{2} e^{-\gamma/2} \frac{1}{\gamma_0} e^{\gamma/\gamma_0}\, d\gamma = \frac{1}{2}\gamma_0 \int_0^\infty e^{-\gamma(1/2+1/\gamma_0)}\, d\gamma = \frac{1}{2+\gamma_0} \tag{8.74}$$

Asymptotic behavior is obtained as the value of γ_0, mean SNR, is increased (high CNR). Thus, the asymptotic error probability for noncoherent FSK is given by

$$P_e \approx 1/\gamma_0 \tag{8.75}$$

Thus, P_e is inversely proportional to the CNR, as opposed to the exponential change in the nonfading case. This will impose an additional design constraint when the system operates in fading channel operation. Techniques to improve the system performance are diversity reception and combining, which we shall discuss now. The probability density function for two-branch selection combining from (8.46) is given by

$$p(\gamma) = M/\Gamma(1 - e^{-\gamma/\Gamma})^{M-1} e^{-\gamma/\Gamma} \tag{8.76}$$

Since, Γ = mean CNR = γ_0, and for a two-branch selection system $M = 2$, we get

$$p(\gamma) = 2/\gamma_0(1 - e^{-\gamma/\gamma_0})e^{-\gamma/\gamma_0} \tag{8.77}$$

The BER of the system output is the integral of P_e over all values of γ, weighted by the density function of (8.78).

$$\begin{aligned} P_{e,2} &= \frac{1}{2}\int_0^\infty e^{-\gamma/2} \frac{2}{\gamma_0}(1 - e^{-\gamma/\gamma_0})e^{-\gamma/\gamma_0}\, d\gamma \\ &= \frac{1}{\gamma_0}\int_0^\infty [e^{-\gamma(1/2+1/\gamma_0)} - e^{-\gamma(1/2+2/\gamma_0)}]\, d\gamma \\ &= \frac{4}{(\gamma_0+2)(\gamma_0+4)} \end{aligned} \tag{8.78}$$

or,

$$P_{e,2} \approx 4/\gamma_0^2 \tag{8.79}$$

The density function for two-branch maximal ratio combining from (8.60) for noncoherent FSK is

$$p(\gamma) = \frac{\gamma_M^{M-1} e^{-\gamma_M/\Gamma}}{\Gamma^M (M - 1)!} \qquad \gamma \geq 0 \tag{8.80}$$

Proceeding similar to the selection combining case, the error probability for noncoherent FSK under two-branch maximal ratio combining is

$$P_{e,2} = \frac{1}{2} \int_0^\infty e^{-\gamma/2} \frac{\gamma e^{-\gamma/\gamma_0}}{\gamma_0^2} d\gamma = \frac{2}{(2 + \gamma_0)^2} \tag{8.81}$$

Asymptotically,

$$P_{e,2} = 2/\gamma_0^2 \tag{8.82}$$

As an example for a specific value of γ_0, if the BER $= 1 \times 10^{-3}$ under Rayleigh fading using two-branch selection diversity, the BER will become

$$P_{e,2} = 4 \times 10^{-6} \tag{8.83}$$

For two-branch maximal ratio combining

$$P_{e,2} = 2 \times 10^{-6} \tag{8.84}$$

Thus the BER is drastically improved by using either diversity techniques. Also, the BER is two times better for maximal ratio combining than selection diversity.

Table 8.2
Comparative Average SNR

Number of Channels	*Number of dB by Which Maximal Ratio Exceeds*		
	Equal Gain	*Selection*	*One Channel*
2	0.49	1.25	3.01
3	0.67	2.14	4.77
4	0.76	2.83	6.02
8	0.9	4.69	9.03
.			
.			
.			
∞	1.05	∞	∞

8.6 COMPARISON OF SELECTION, MAXIMAL RATIO, AND EQUAL GAIN COMBINING TECHNIQUES

The average improvement in output SNR for different schemes and for different values of the diversity branches M in Rayleigh fading are shown in Figure 8.19. In Section 8.4 the value of combining gain for maximal ratio combining is $10 \log_{10} M$, for equal gain combining is $10 \log_{10}[1 + (M - 1)\pi/4]$, and for selection combining

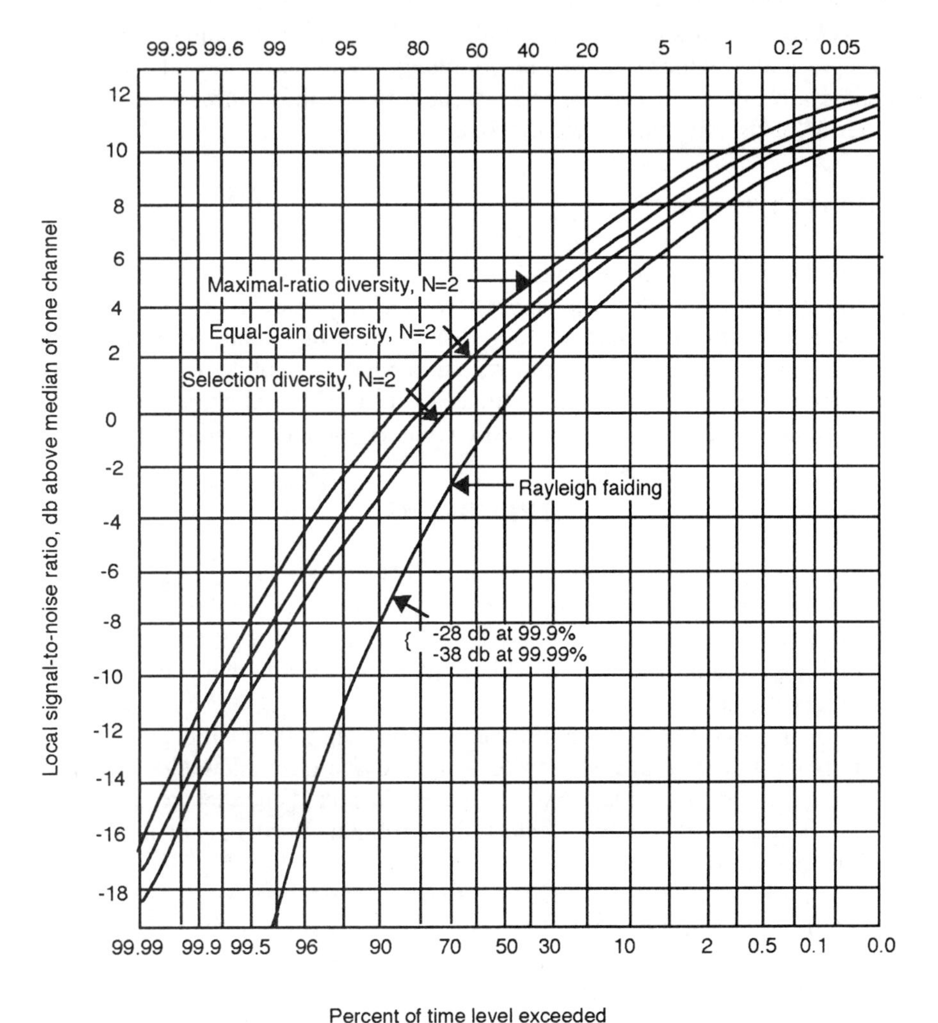

Figure 8.20 Comparison—dual diversity distribution.

is 10 $\log_{10} \Sigma(1/K)$. The values of excess gain of maximal ratio combining with respect to equal gain, selection, and single channel under Rayleigh fading are shown in Table 8.2. As shown in the table, the difference between maximal ratio and equal gain combining can never exceed 1.05 dB. Diversity improvement for the two-channel case ($N = 2$) is shown in Figure 8.20. Here, a reference is taken at the 50% point of the Rayleigh distribution curve. All three schemes of diversity, namely maximal ratio, equal gain, and selection diversity appear to be close to each other. However, the offset gain between these and the single channel non-diversity case appears to be substantial. Thus, for the two-channel diversity case it does not matter which scheme one adopts; give and take, they all provide about the same advantage. However, the offsets between these three increases as the number M of the diversity branch increases.

8.7 CONCLUSIONS

In the first part of this chapter we discussed the different techniques for diversity, where we have logically discarded most of the techniques in favor of space diversity. In the second half of the chapter, four basic combining techniques were discussed where scanning combining has been recommended for mobile use and equal gain combining is a preferable choice for the cell site. In the next chapter we shall take up the studies of signal processing and coding as applied to cellular radio.

PROBLEMS

8.1 Find the time separation required at a 850 MHz cellular system between samples of the received signal when the mobile is moving at the constant velocity of 10, 30, 60, and 80 mi/hr. Draw also the plot of the velocity versus time separation between the vehicle. What conclusions can you draw by these observations?

8.2 Explain physically the significance of coherence bandwidth of the system. Explain why it would be more easy to provide frequency diversity in urban areas than in rural areas. Assuming a cell has fixed channels, can it be paired for frequency diversity? Neglect the problem of cochannel and adjacent-channel interference.

8.3 The normalized distances between two electric dipoles are 0.1, 1.5, 2.5, and 3.5 if the maximum value of the correlation coefficient does not exceed 0.1. What is the recommended spacing between dipoles.

8.4 Can two directional antennas mounted at the automobile be used for diversity advantage? Show that an omnidirectional antenna and whip antenna having a voltage gain of 1.2 cos θ, if installed at the automobile, will act as an angle

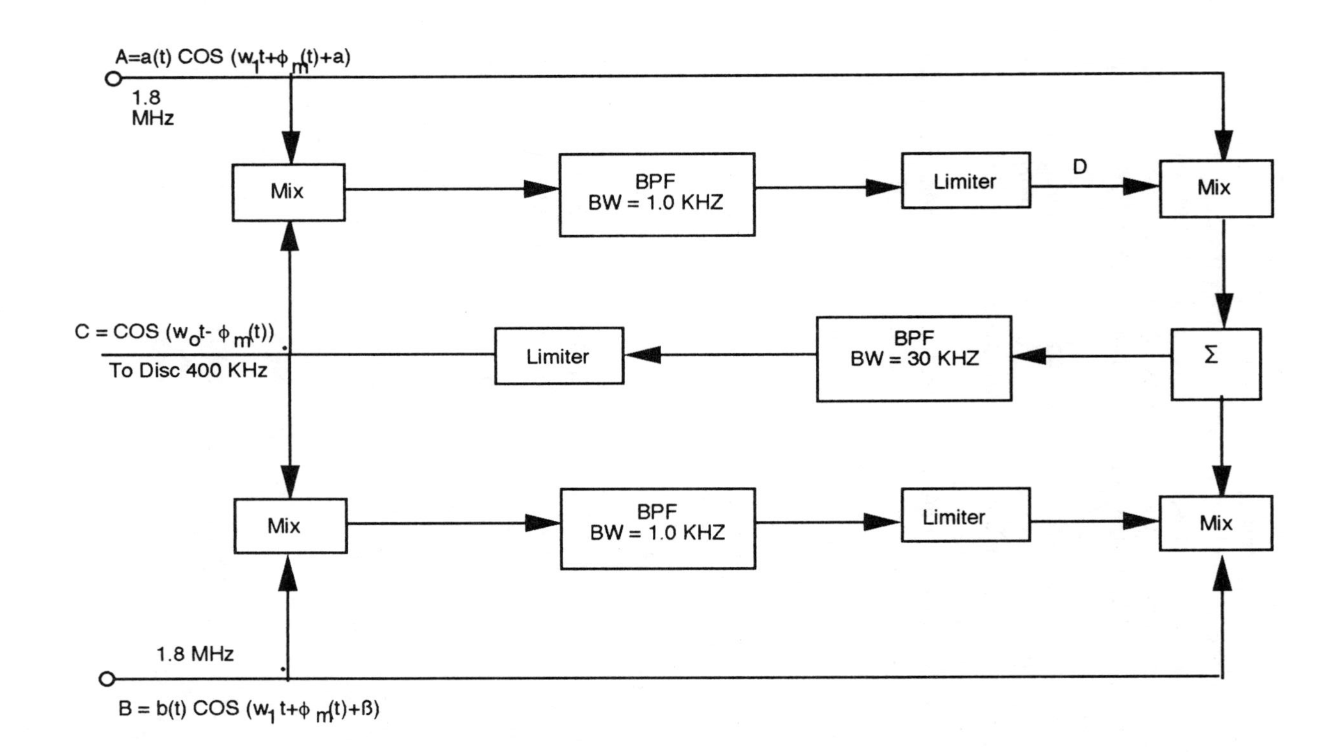

Figure 8.21 Two-branch equal gain diversity predication diviner.

diversity combiner. Show in the block diagram how to connect the two antennas to the receiver in this configuration.

8.5 Enumerate the difficulties encountered in providing a RAKE diversity scheme. Draw a block diagram implementation of the scheme assuming that you have overcome all the difficulties and have decided to implement the scheme at 900 MHz. List all the critical areas of the design and the precautions you have to take.

8.6 The two-branch equal gain diversity predication combiner is shown in Figure 8.21. Show that the two inputs to the summing network are in phase with each other (i.e., the random phases are stripped out and only the modulation component of the phase remains).

8.7 Explain with respect to Figure 8.13 that as the switching point moves above the threshold, the curve approaches the Rayleigh distribution curve. Explain also with the help of (8.34) why the curve follows parallel to the Rayleigh distribution curve below the threshold with the signal power offset equal to the threshold value at A. Draw the probability distribution curve below threshold for $A = -12$ dB and $A = -15$ dB.

8.8 For feedback combining, prove that the expected signal drop due to time delay in switching the cell-site transmitter antenna is less than 2 dB for a vehicle traveling at a speed greater than 60 mi per hr.

8.9 Explain why is selection diversity difficult to use at the mobile. Derive the expression for $P(\gamma < \Gamma_k)$, $k = 1, 2, \ldots, M$ when the instantaneous CNR is far below the average CNR in all the branches. Plot the result for $M = 2$ and 4. What conclusions can one draw from this special case of selection diversity?

8.10 Why is maximal ratio best among all the combining techniques? Compare the result of scanning and selection combining with maximal ratio combining for $M = 2$, 3, and 4 at the CNR of -30 dB, -20 dB, -10 dB, 0 dB, and $+8.0$ dB. Comment on why it is difficult to adopt this scheme at the mobile. Would you recommend this scheme for the cell site?

REFERENCES

[1] Clarke, R. H. "A Statistical Theory of Mobile-Radio Reception," *Bell System Technical Journal*, July-Aug. 1968.

[2] Schwartz, M., et al. *Communication Techniques and Techniques*, McGraw-Hill Book Co.: New York, NY, 1966.

[3] Lee, W. C. Y. "Mobile Radio Signal Correlator Versus Antenna Height and Spacing," *IEEE Trans. on Vehicular Technology*, Aug. 1977, pp. 290–292.

[4] Awadalla, K. H. "Direction Diversity in Mobile Communication," *IEEE Trans. on Vehicular Technology*, Aug. 1981, pp. 121–122.

[5] Parsons, J. D., M. Henze, P. A. Ratliff, and M. J. Withers. "Diversity Techniques for Mobile Radio Reception," *IEEE Trans. on Vehicular Technology*, Aug. 1976, pp. 75–84.

[6] Brennan, D. J. "Linear Diversity Combining Techniques," *IRE Pro.*, June 1959, pp. 1075–1101.
[7] Rustako, A. J., Y. S. Yeh, and R. R. Murray. "Performance of Feedback and Switch Space Diversity 900 MHz FM Mobile Radio Systems With Rayleigh Fading," *IEEE Trans. on Communication*, Nov. 1973, pp. 1257–1275.
[8] Jakes, W. C. "A Comparison of Specific Space Diversity Techniques for Reduction of Fast Fading in UHF Mobile Radio Systems," *IEEE Trans. on Vehicular Technology*, Nov. 1971, pp. 81–92.
[9] Panter, P. F. *Communication Systems Design*, R. E. Krieger Publishing Company: Florida, 1972.
[10] Ebine, Y., and Yoshihide Yamada. "Vehicular-Mounted Diversity Antennas for Land Mobile Radios," *IEEE Conference on Vehicular Technology*, 1988.
[11] Lee, W. C. Y. *Mobile Communications Engineering*, McGraw-Hill Book Co.: New York, 1982, p. 304.

Chapter 9
Signal Processing and Coding

9.1 INTRODUCTION

Two different generations of cellular radio are in existence today, old analog systems based on FDMA technology and the newer version based on digital TDMA principles. The same holds for cordless telephones where both analog (CT-1) and digital systems are in existence (CT-2, DECT). There is also a strong possibility that future personal communication services in the United States will be based on digital CDMA technology. In this chapter we shall cover these systems from the point of view of signal processing and data coding.

First, the analog cellular telephone system employs a digital signaling technique for various stages of call setup and completion. Different digital messages are required for routine operation and control of mobile equipment. These messages are categorized into two groups:

1. A continuous digital message stream on the forward (cell site to mobile) setup channels to establish the land originated call.
2. A discontinuous burst message stream, which originates from the mobile or from the cell, either over the reverse setup channel or over speech channels.

These messages are used to originate mobile-to-land calls or provide acknowledgment for a land-to-mobile call. The reverse setup channel used for this is shared among all mobile users in a cell, and thus a definite contention probability exists.

To obtain synchronization to these discontinuous messages, transmissions contain a prefix known as "dotting" or an alternating 101010.... sequence. A dotting sequence is recognized by the receiver as a fixed-frequency tone, which initializes the clock phase that is subsequently updated in a phase-locked loop (PLL). The data following synchronization is generally Manchester-coded with FSK or FFSK modulations and lasts tens of milliseconds. Since channels following Rayleigh fading cause bursts of errors, suitable combating techniques in the form of block or convolutional codes and repeats of messages are included. These messages are stored in the receiver where the decision, including majority voting, takes place. Both categories of signals will require word synchronization.

On the signal processing side with analog FM modulation, the average power at the baseband is proportional to the mean-squared frequency deviation f^2_{rms} produced at the transmitter and since the talker volume varies widely, some form of level control is necessary to optimize the system performance. If the level control is not provided at the transmitter, a weaker talker will suffer degradation proportional to reduction in frequency deviation. This reduction in signal-to-impairment ratio (SIR) affects the perceived signal quality, and thus the subjective rating of the channel. On the other hand, a loud talker will cause excessive frequency deviation at the transmitter, resulting in adjacent-channel interference at the receiver. Use of a compandor-compressor in the transmitter and compandor-expandor in the receiver controls the speech level variability on clipping distortion and thus the frequency deviation generated in the modulator. Assuming that the signal power spectrum remains uniform over the whole baseband, then, due to the parabolic behavior of noise power density, higher frequency components of signal at the receiver output will have a lower signal-to-noise ratio (SNR). The use of a preemphasis filter circuit in the transmitter shapes the signal spectrum such that the higher frequency is amplified relative to the lower frequency. In order to bring the spectrum back to the original level, the receiver uses an inverse deemphasis network at the baseband.

Due to multipath propagation of the signal from the base to the mobile, the data signals in digital cellular are likely to be "smeared out" in time. In other words, the radio channel will introduce intersymbol interference (ISI). To reliably decode the signal, an estimate of the channel transfer function is needed so that an equalizer can be used to compensate for ISI. For a mobile user, the transfer function of the channel may change rapidly. Thus, the channel characteristics have to be estimated for each time slot. To measure the dynamic channel transfer function, a constant training sequence is included in every data burst to enable the repeated evaluation of the channel. The equalizer knows the transmitted training sequence and also what it has received, and thus can estimate the transfer function of the channel. To minimize channel errors due to multipath propagation, different block and convolutional codes have been selected. Among block codes, the Bose-Chaudhuri-Hocquenghem (BCH) and Reed-Solomon (RS) codes are important. Since channel errors occur in the burst form, interleaving is used at the expense of increased delay.

Narrow band TDMA has been adapted both in Europe and the United States. In order to contain a toll-quality digital speech spectrum within the allocated band, a maximum data rate of coded speech can be 16 Kbps. A residual excited linear prediction (RELP) coder is used in GSM and a vector sum excited linear predictive (VSELP) coder has been chosen in the U.S. adaptive differential PCM based on a speech rate of 32 Kbps has also been adapted in some mobile systems.

In view of the above discussions, we divide this chapter into three major areas: signal processing, block coding and speech coding. The section on signal processing deals with time and frequency domain representation of baseband waveform, application of PLL as an element for frequency synthesizer and timing circuits, nonlinear equalizers, and speech processing associated with FM modulation.

In the section on coding, we discuss block coding. Lastly, Section 9.4 deals with two different algorithms; namely, residual excited linear predictive coding (RELP) and code excited linear prediction (CELP).

9.2 SIGNAL PROCESSING

9.2.1 Characteristics of Baseband Signals

One of the requirements for analog cellular radio is the transmission of discontinuous signals over a voice channel. Discontinuous message streams are employed to accomplish the following. The mobile is commanded from the cell site to switch to a new voice channel when a call is in progress. On the reverse side, the mobile may request for special services during a call. This is accomplished by the mobile briefly interrupting the voice message and transmitting a discontinuous digital message. This discontinuous data stream requires that the received signal should contain timing information that allows efficient clock recovery for bit synchronization. Biϕ coded data helps in clock recovery.

If a common receiver circuitry is selected for both voice and data, it is also necessary that the spectrum density for data be above the voice band. In the AMPS system in the U.S., both FM speech and FSK-modulated data are passed through a discriminator before the speech and data are recovered. Accounting for the above two factors, in most cellular systems raw NRZ data are first Manchester coded before FSK modulation. Some other desirable characteristics of the baseband data signals are that:

- Signals should be immune to noise and interference;
- The complexity of the system should be minimal to make it reliable and easy to maintain.

Thus, it is important to investigate the properties of NRZ and Biϕ signals. Common types of NRZ and Biϕ waveform are shown in Figure 9.1.

The simplest among these waveforms is the nonreturn to zero (NRZ) level. Here, 1 is represented by a certain voltage level $+A$ and 0 is represented by a voltage level $-A$. Thus for a long sequence of ones or zeros, signals will not have any transition and will cause the clock recovery circuit to drift with time. Synchronization of the recovered clock with incoming data becomes unreliable and errors in sampling the incoming pulses can even occur at a high SNR. On the other

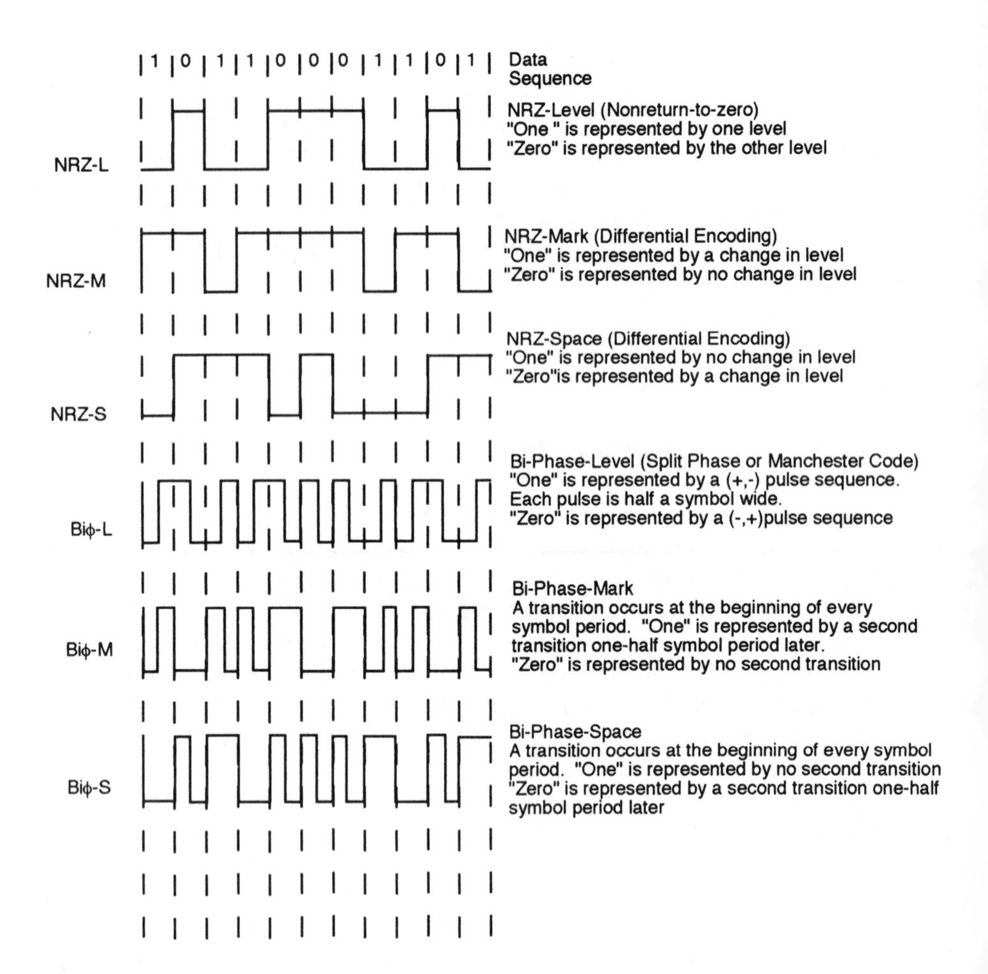

Figure 9.1 Binary PCM waveform as related to a binary input data sequence. After [3, p.11].

hand, Biϕ signals (Biϕ -L, -M, and -S) assure at least one transition per data bit irrespective of 0 or 1 transmission, which makes the clock acquisition and tracking easy.

9.2.1.1 Power Spectral Density of Baseband Data

It has been shown that the power spectral density (PSD) for a purely random binary signal source; that is, one that emits an elementary signal in a given signaling interval independent of those transmitted in previous intervals, is given by [1–4]:

$$s(f) = \frac{1}{T_s^2}\sum_{n=-\infty}^{+\infty}\left|\sum_{i=1}^{N} p_i s_i\left(\frac{n}{T_s}\right)\right|^2 \delta\left(f - \frac{n}{T_s}\right) + \frac{1}{T_s}\sum_{i=1}^{N} p_i(1 - p_i)|s_i(f)|^2$$
$$- \frac{2}{T_s}\sum_{i=1}^{N}\sum_{k=1}^{N} p_i p_k \mathrm{Re}[s_i(f)\bar{s}_k(f)] \qquad i \neq k,\ k > i \tag{9.1}$$

where T_s is the period of each elementary signal; $s_i(f)$ is the Fourier transform of the ith elementary signal; $\bar{s}(f)$ is the complex conjugate of $s_i(f)$; p_i is the probability of the ith elementary signal; and $\delta(f)$ is the Dirac delta function. For binary signaling, $N = 2$, $p_1 = p$, and $p_2 = 1 - p$. The above equation is reduced to

$$s(f) = \frac{1}{T_s^2}\sum_{n=-\infty}^{\infty}\left|ps_1\left(\frac{n}{T_s}\right) + (1 - p)s_2\left(\frac{n}{T_s}\right)\right|^2 \delta\left(f - \frac{n}{T_s}\right)$$
$$+ \frac{1}{T_s}p(1 - p)|s_1(f) - s_2(f)|^2 \tag{9.2}$$

Let us use the above equation to find the PSD of NRZ and Biϕ signals. For the power spectrum of NRZ signaling,

$$\begin{aligned} s_1(t) &= A \qquad 0 \le t \le T \\ s_2(t) &= -A \qquad 0 \le t \le T \end{aligned} \tag{9.3}$$

Thus, the Fourier transform is

$$s_1(f) = A\int_0^T e^{-j2\pi ft}dt = -\frac{A}{\pi f}e^{-j\pi fT_s}\sin(\pi fT_s) \tag{9.4}$$

Since $s_1(t) = -s_2(t)$, we have

$$s_2(f) = -s_1(f)$$

Thus, the composite PSD is

$$s(f) = \frac{1}{T_s^2}\sum_{n=-\infty}^{\infty}\left|-\frac{1}{2}\frac{A}{\pi(n/T_s)}e^{-j\pi\frac{n}{T_s}T_s}\sin\left(\frac{n\pi}{T_s}T_s\right)\right.$$
$$\left. + \frac{1}{2}\frac{A}{\pi n/T_s}e^{-j\pi\frac{n}{T_s}T_s}\sin\left(\frac{n\pi T_s}{T_s}\right)\right|^2 \delta(0)$$
$$+ \frac{1}{4T_s}\left|-2\frac{A}{\pi f}e^{-j\pi fT_s}\sin(\pi fT_s)\right|^2 \tag{9.5}$$

Since the first two terms of the series are zero,

$$
\begin{aligned}
s(f) &= \frac{A^2}{\pi^2 f^2 T_s} \left| e^{-j2\pi f T_s} \right| \sin^2(\pi f T_s) \\
&= \frac{A^2}{\pi^2 f^2 T_s} \sin^2(\pi f T_s)
\end{aligned}
\tag{9.6}
$$

Thus, the composite PSD is

$$
\frac{s(f)}{E_s} = \frac{\sin^2(\pi f T_s)}{(\pi f T_s)^2} \tag{9.7}
$$

where $E_s = A^2 T_s$ is the power of the rectangular pulse. For the power spectral density of Biϕ signaling,

$$
\begin{aligned}
s_1(t) &= A \qquad & 0 \le t \le T_s/2 \\
s_2(t) &= -A \qquad & T_s/2 \le t \le T_s \\
s_1(t) &= -s_2(t)
\end{aligned}
\tag{9.8}
$$

Proceeding similar to the NRZ case and making use of (9.2), we get

$$
\frac{s(f)}{E_s} = \frac{\sin^4 \pi f T_s/2}{(\pi f T_s/2)^2} \tag{9.9}
$$

where $E_s = A^2 T_s$ is the power of the rectangular pulse. A normalized plot of $s(f)/E_s$ versus fT_s for NRZ and Biϕ signals is shown in Figure 9.2, where it is seen that choosing 10-Kbps signaling over a voice band of 3 kHz (i.e., $fT_s = 0.3$) will interfere with the speech when signaling is done by NRZ waveform. However, for Biϕ signaling, most of the energy is beyond the normalized $fT_s = 0.3$, and the peak of $s(f)/E_s$ occurs at $fT_s = 0.75$. Hence, it is preferable to use Biϕ signaling to minimize interaction of speech with data. Also note that the Manchester coding provides a solution to the baseband dc wander problem typically found in digital communication systems. Biϕ coding is used in AMPS, TACS, and Japanese systems.

9.2.1.2 Manchester Biϕ Encoding and Decoding

Due to the advantages of Manchester coding, in most cellular systems of the world data at the input to the modulator is first converted from NRZ to Biϕ before feeding to the modulator and the corresponding conversion from Biϕ to NRZ takes

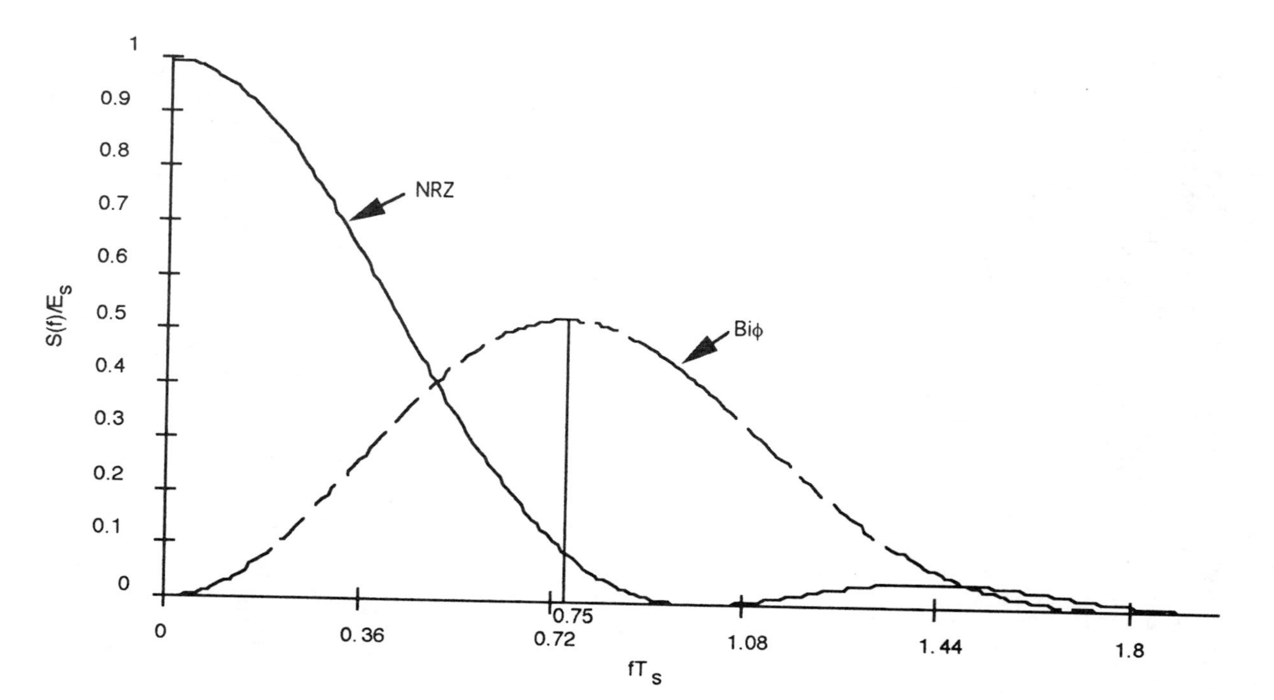

Figure 9.2 Spectral densities of NRZ and Manchester-coded waveforms. Source: [19].

place in the receiver. A block diagram along with waveform at different stages of the generation is shown in Figure 9.3(a) and (b) [1–4]. Essentially, when "exclusive ORed" with the phase-coherent 10-kHz bit clock provides, NRZ data Manchester-coded data at C. The clock phase is chosen so that an NRZ binary one becomes a zero-to-one transition in the middle of each NRZ bit. Correspondingly, an NRZ binary zero becomes a one-to-zero transition in the middle of each NRZ bit.

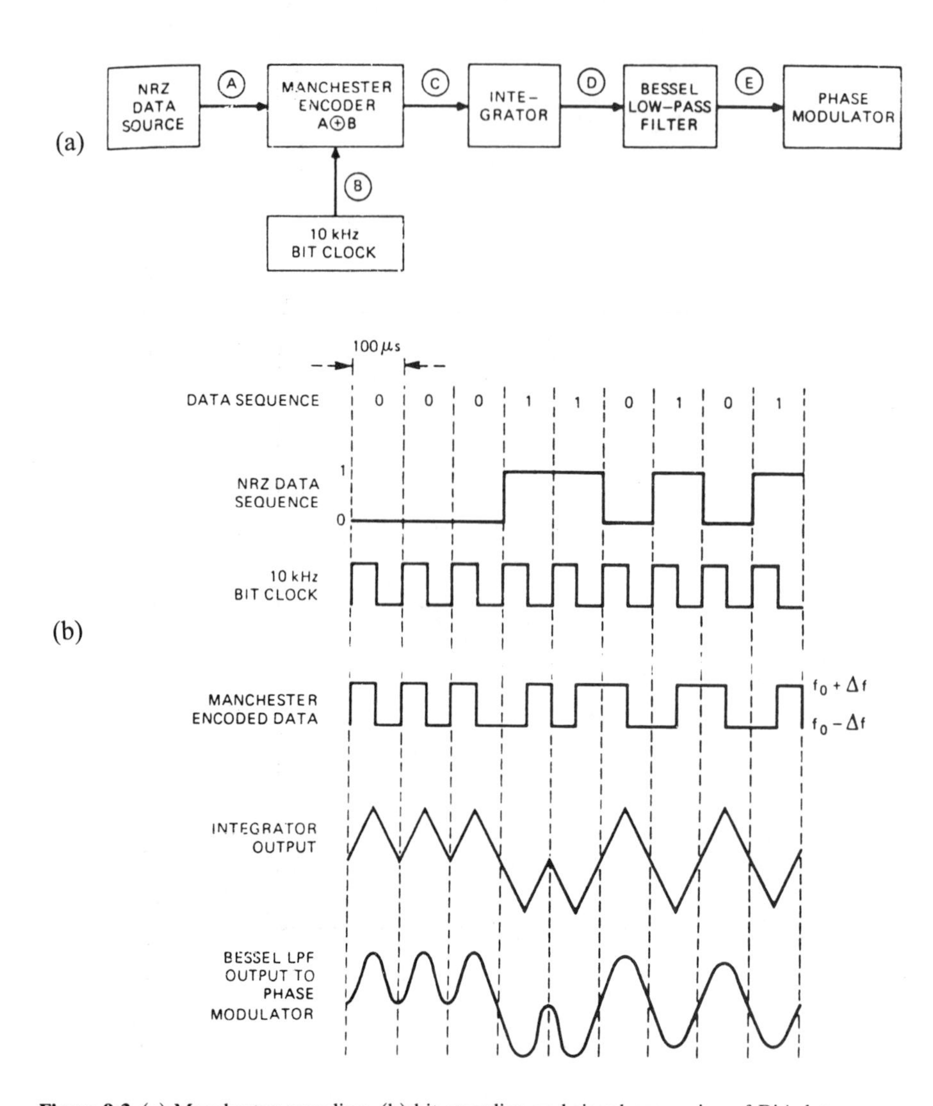

Figure 9.3 (a) Manchester encoding, (b) bit encoding and signal processing of Biϕ data.

A block diagram for a Manchester decoder is shown in Figure 9.4 [1–4]. The clock generator circuit extracts an almost noiseless 10-kHz phase-coherent clock waveform at B. The generator employs a narrowband phase-locked loop to extract the 5-kHz Fourier component of the Manchester bit stream. This waveform is then frequency doubled to obtain the desired 10-kHz bit clock. The recovered and synchronized clock and Manchester encoded stream at A are "exclusive ORed"

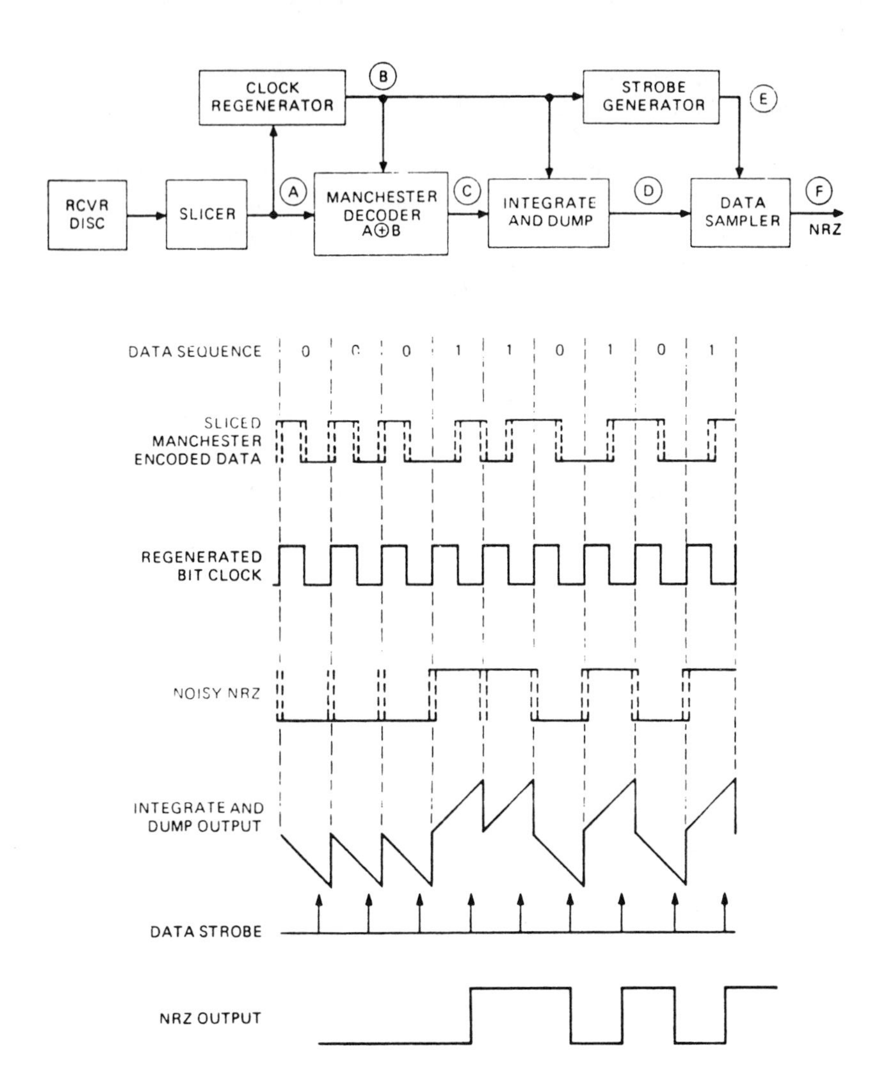

Figure 9.4 Manchester decoding of the received bit stream. Source: [19].

with the bit clock at B to provide the NRZ output waveform at C. Note that the phase noise of the data at A or the phase error of the clock at B produce unwanted spikes at the borders of each NRZ bit. The receiver is a matched filter type where an integrator and a dump are used. At the end of the integrating interval, the accumulated energy is dumped before the next pair of Manchester symbols are processed. The waveform at D is strobed to the sampler before each dump by the data strobe E. The data sampler retains the output from each strobed sample to produce the clean NRZ output at F.

9.2.2 Baseband Filtering Requirements

A sharp pulse passing through a band-limited channel produces differential attenuation and causes stretches in the received pulses. Consequently, the pulses appearing at the receiver predetection filter output are dispersed over a period longer than the symbol interval T_s, as shown in Figure 9.5. Here, the first pulse is stretched out to second and third pulses, while the effect of the second and third is felt on the fourth pulse.

To control the spectral splatter and thus the intersymbol interference (ISI), filtering is applied to the baseband pulses. In order to combat the effect of ISI, one can also apply equalization at the receiver or, instead of sending rectangular pulses with infinite rise and fall times, pulses can be shaped at the transmitting end. To understand this phenomenon, let us consider the baseband model shown in Figure 9.6.

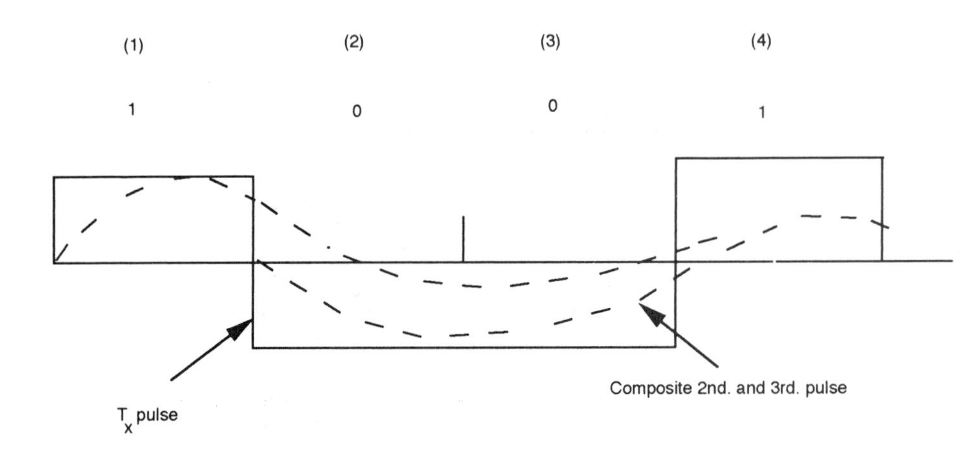

Figure 9.5 Stretched pulses.

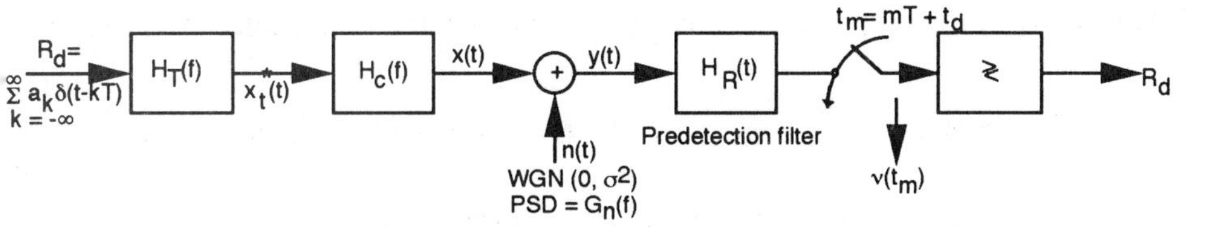

Figure 9.6 A typical transmitter and receiver.

The input to the transmitter filter with low-pass transfer function $H_T(f)$ or impulse response $h_T(t) = F^{-1}[H_T(f)]$ is a series of symbol pulses $a_k = \pm 1$. During the kth signaling interval $(k - 1)T \leq t \leq kT$, pulse can be represented by

$$R_d(t) = \sum_{k=-\infty}^{\infty} a_k \delta(t - kT) \tag{9.10}$$

The transmitted signal $x_t(t)$ is

$$x_t(t) = \sum_{k=-\infty}^{\infty} a_k \delta(t - kT) * h_T(t) = \sum_{k=-\infty}^{\infty} a_k h_T(t - kT) \tag{9.11}$$

The receiver predetection filter output can be represented as:

$$\begin{aligned} v(t) &= y(t) * h_R(t) \\ &= \sum_{k=-\infty}^{\infty} A a_k p_r(t - t_d - kT) + n_0(t) \end{aligned} \tag{9.12}$$

where A is a scale factor chosen such that $p_r(0) = 1$, $n_0(t)$ is the noise component, and $p_r(t - t_d)$ is the pulse shape at the predetection filter, $H_R(t)$, output. Delay t_d is relative to $x_t(t)$ and is due to prior filters. Thus, the received pulse $Ap_r(t - t_d)$ is the result of various convolutions, which are represented as

$$h_T(t) * h_c(t) * h_R(t) \tag{9.13}$$

where $p_r(t - t_d)$ is the response of the system when the input pulse is $x_t(t)$ for $k = 0$.

If the maximum value of $p_r(t - t_d)$ occurs at $t = t_d$, then sampling points can be chosen such that

$$t_m = mT + t_d \tag{9.14}$$

The samples at the output of the receive filter are written as (sampled A/D output).

$$V_m = v(t_m) = \sum_{k=-\infty}^{\infty} Aa_k p_r(mT + t_d - t_d - kT) + n_0(mT + t_d)$$
$$= Aa_m p_r(0) + \sum_{\substack{k=-\infty \\ k\neq m}}^{\infty} Aa_k p_r[(m - k)T] + N_m \tag{9.15}$$

where $m = \ldots, -1, 0, 1, 2, \ldots$ and $N_m = n_0(t_m) = n_0(mT + t_d) = n(t) * h_R(t)\big|_{t=t_m}$. The noise sample $n_0(t_m)$ is Gaussian, and $G_n(f)|H_R(f)|^2$ is wideband such that N_k and N_m; $m \neq k$ are uncorrelated and thus independent.

Let us assume that the noise term is zero and investigate the effects of ISI only. The peak value of ISI from (9.15) is given by

$$\text{ISI}\big|_{P-k} = \sum_{k=-\infty}^{\infty} |p_r(m - k)T| \tag{9.16}$$

In the absence of noise and ISI, the mth transmitted bit can be decoded by $v(t_m)$ or $V_m = Aa_m p_r(0)$. One can note that $Aa_m p_r(0)$ is directly related to the transmitted pulse a_m and thus the decoding will provide the information element during the pulse interval T.

Thus, ISI can be completely eliminated provided that we can design the overall system such that the received pulse $p_r(t = nT)$ is

$$p_r(nT) = V_n(t) = 1 \qquad n = 0$$
$$= 0 \qquad n \neq 0 \tag{9.17}$$

The above condition is satisfied if one can satisfy the following equation by Nyquist in the transform domain

$$\sum_{k=-\infty}^{\infty} P_r(f + k/T) = T \qquad |f| \leq 1/2T \tag{9.18}$$

where $P_r(f) = F[p_r(t)]$.

The above equation requires abrupt transition from the response of T at $f = \pm 1/T$ to 0 for $f > 1/T$. This is shown in Figure 9.7(a). Pulse, $p_r(nT) = (\sin n\pi)/(n\pi)$ is a slowly decaying function, thus it stays high for large values of n (high sidelobes for high n) and is likely to cause errors with drifts of sampling times. This is shown in Figure 9.7(b). From Figure 9.7(a), one can conclude that the theoretical minimum system bandwidth required to detect $1/T$ symbols/sec without

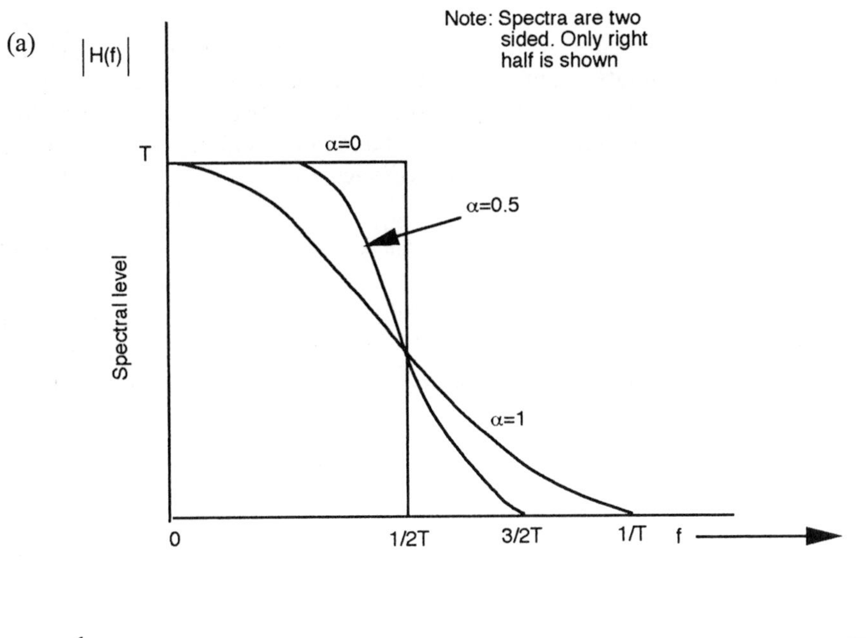

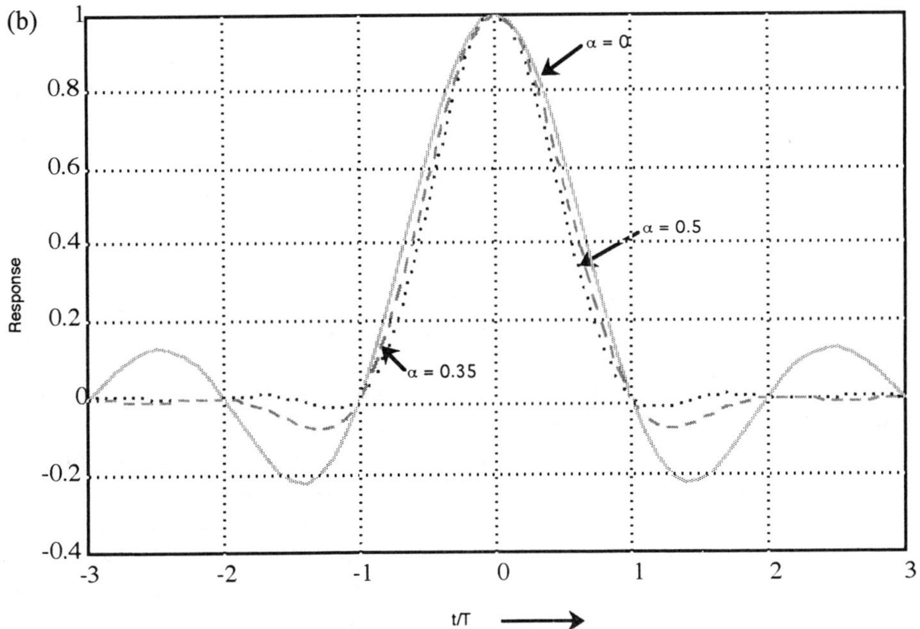

Figure 9.7 (a) Raised-cosine spectra; (b) pulses with raised-cosine spectra.

intersymbol interference is $1/2T$ Hz. For this, a rectangular filter in the frequency domain is required. This filter is difficult to approximate, but if the filter bandwidth is increased, the approximation can be easier. The modification to the rectangular filter in the frequency domain is characterized by a term called filter roll-off factor, α. If $W_0 = 1/(2T)$ is the minimum required bandwidth and W is the designed bandwidth of the system, then the filter roll-off factor is

$$\alpha = (W - W_0)/W \tag{9.19}$$

Thus, for $W = 1/(2T)$, $\alpha = 0$, and for $W = 1/T$, $\alpha = 1.0$. This roll-off factor specifies the required excess bandwidth divided by the filter's 6-dB bandwidth. A family of spectra that satisfies the Nyquist (9.18) pulse-shaping criteria is a raised-cosine pulse, defined as

$$H(f) = \begin{cases} T & |f| \leq 1/2T - \alpha \\ 1/2T\left(1 + \cos\dfrac{\pi|f| - 1/2T + \alpha}{2\alpha}\right) & 1/2T - \alpha \leq |f| \leq 1/2T + \alpha \\ 0 & |f| > 1/2T + \alpha \end{cases} \tag{9.20}$$

The inverse transform $h(t)$ is given by

$$h(t) = \frac{\cos 2\pi\alpha t}{1 - (4\alpha t)^2}\left[\sin\frac{(\pi r_b t)}{\pi r_b t}\right]; \qquad r_b = 1/T \tag{9.21}$$

The normalized frequency and time domain response of raised-cosine pulse with different values of α are shown in Figure 9.7(b). Note that the sidelobe response is higher for lower values of α. With higher α, sidelobe levels decrease. Thus, if α is properly chosen in the system design, a slight unstability in the receiver sampling point may not contribute to large ISI.

The North American digital mobile cellular system (NADMCS) specifications, IS-54, has recommended a square-root raised-cosine filter with $\alpha = 0.35$. By using this filter both at the transmitter and at the receiver, matched filtering problems are solved. Impulse response of a square-root raised-cosine filter with α of 0.35 is shown in Figure 9.8. Note that the impulse response does not have nulls at symbol decision points. This indicates the presence of ISI at the receiver. However, the ISI is removed because the receiver has another square-root raised-cosine filter. This is due to the cascade response of two square-root raised-cosine filters, which results in a raised-cosine response with no ISI.

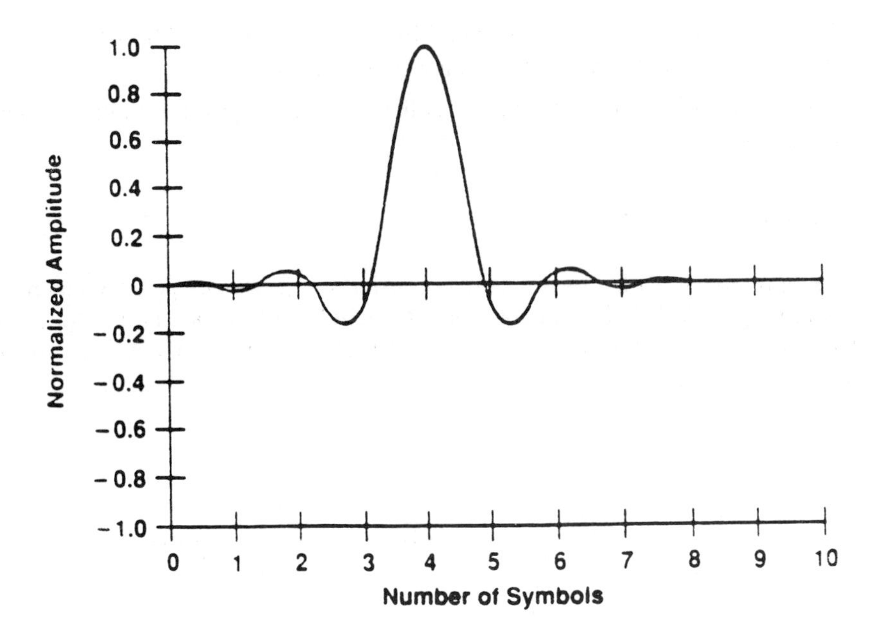

Figure 9.8 Response of a square-root cosine filter with an α of 0.35 [24].

9.2.3 Fundamentals of Phase-Locked Loop [5–6]

Applications of PLL are many in cellular radio. Thus this section describes the fundamentals of PLL before we discuss applications to cellular radio. For further details, refer to [4–6]. Some important applications of PLL are:

- For timing recovery, to generate a single-frequency tone at the output of the VCO. The required frequency f_o of this tone should equal the average symbol rate at the modem input.
- For carrier recovery, to track the phase of the input signal while at the same time minimizing the effects of noise.
- As a tunable (automatic tracking) narrowband filter. To illustrate this principle, let us take an example of S band filter where the doppler shift can be of the order of ± 75 kHz. Use of an ordinary receiver will require a minimum bandwidth of 150 kHz. A PLL-based receiver can track the signal with a bandwidth of nearly 3 Hz. Thus, the noise improvement factor due to the use of PLL is $(150 \times 10^3)/3 = 50$ kHz $\rightarrow \approx 47$ dB.
- As a frequency discriminator. PLL as a frequency discriminate is superior in performance than conventional discriminator.

- PLL as a synthesizer, frequency multipliers, and dividers. Almost all modern oscillators are designed as frequency synthesizers.

From a cellular radio point of view, we shall apply the PLL for the timing recovery and as a synthesizer.

Basic Components of PLL

A block diagram of a PLL of an arbitrary order along with its phase conventions is shown in Figure 9.9. Different components include a phase detector, a loop filter (LPF), and a voltage controlled oscillator (VCO) whose frequency is controlled by an external signal.

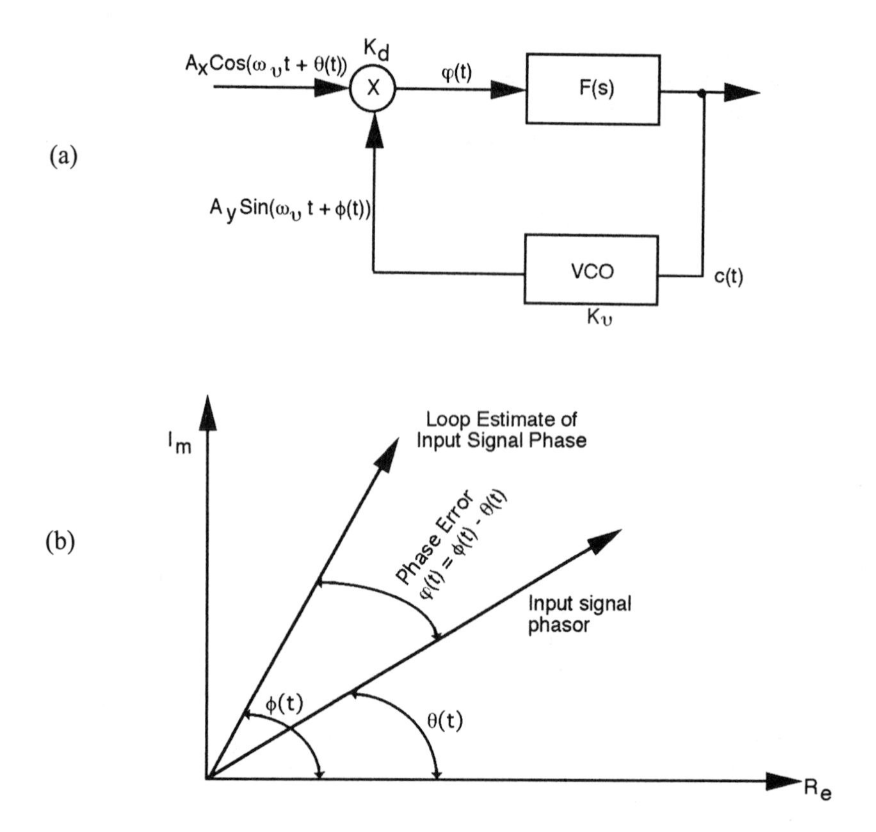

Figure 9.9 (a) Configuration of a PLL; (b) notations in PLL.

Sinusoidal Phase Detector

We assume that the phase detector is an ideal multiplier followed by a low pass filter which removes the sum frequency component at the low pass (LP) filter output. Then, from Figure 9.9(a),

$$\varphi(t) = \frac{k_m A_x A_y}{2} \sin[\theta(t) - \phi(t)] + \frac{k_m A_x A_y}{2} \sin[2\omega_v t + \theta(t) + \phi(t)] \quad (9.22)$$

After LP filtering,

$$\varphi_{LP}(t) = K_d \sin[\theta(t) - \phi(t)] \quad (9.23)$$

where $k_d = k_m A_x A_y/2$, where k_d is the phase detector constant and k_m is the multiplier constant. For minimum error voltage at the output of the LP filter,

$$\sin[\theta(t) - \phi(t)] \bmod 2\pi = 2\pi n \quad (9.24)$$

for small phase error $\theta(t) = \phi(t)$. In other words, the local input to the phase detector is in quadrature with the external signal. From (9.24) one can note that the input or the VCO can abruptly change by 2π and will not be detected by the phase detector. However, such changes are detrimental and are known as clicks. In the derivation of the above equation we have assumed a sinusoidal phase detector, as shown in Figure 9.10. Some other phase detectors used in PLL are triangular and sawtooth, as shown in Figure 9.11 [5, 6].

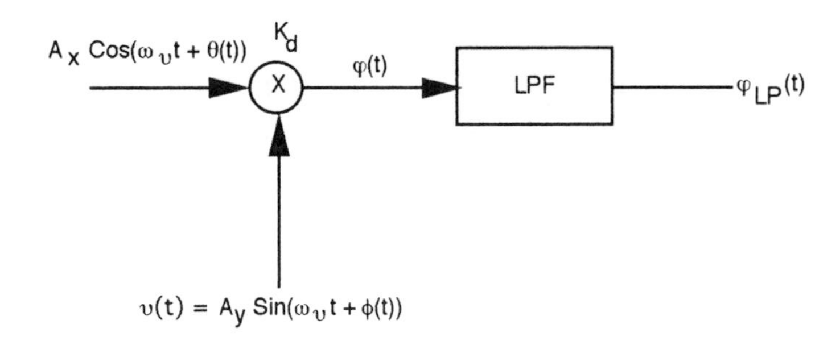

Figure 9.10 Sinusoidal phase detector.

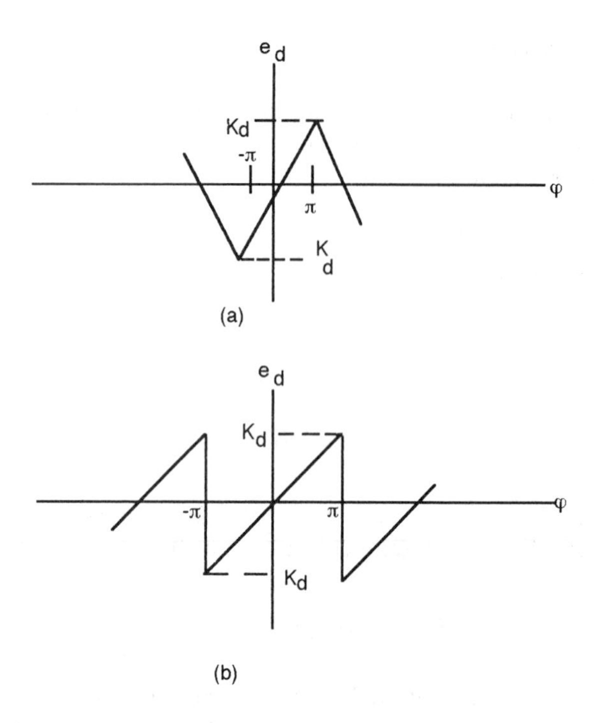

Figure 9.11 Phase detector characteristics: (a) triangular, (b) saw tooth.

Ideal Voltage Controlled Oscillator (VCO)

A VCO model is shown in Figure 9.12. Assuming $c(t)$ is the error signal from the LP filter and the VCO gain is K_v (rad/sec/V), then in time domain the VCO behavior can be represented by

$$\frac{d}{dt}[\omega_v t + \phi(t)] = \omega_v + \frac{d\phi(t)}{dt} \tag{9.25a}$$

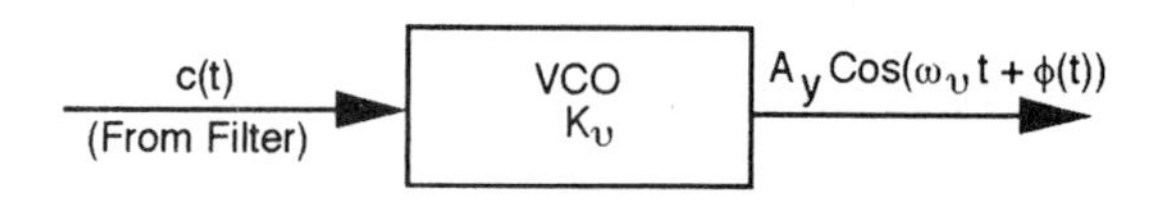

Figure 9.12 VCO model of PLL.

and

$$k_v c(t) = \frac{d\phi(t)}{dt} \tag{9.25b}$$

where $k_v c(t)$ represents the change in VCO frequency due to control voltage $c(t)$. This represents in Laplace domain,

$$k_v c(s) = s\phi(s) \qquad \phi(s) = k_v \frac{c(s)}{s} \tag{9.26}$$

From (9.26), one can conclude that with respect to input, VCO acts as an integrator. In other words, the VCO transfer function is $1/s$. Also, the PLL can't have an order below one. Here, the gain constant k_v can be modeled as a part of the loop gain of the filter. A loop filter is simply a lowpass filter. For analog PLL, we can consider this to be a conventional Butterworth type , a Chebychev type, or any of the other types.

Modes of Operation

There are two operating modes of PLL: tracking and signal acquisition. In the tracking mode of operation, the VCO operates at the same frequency as an external signal and the phase error is within the range of the phase detector. Signal acquisition mode is the transient behavior of PLL when the loop is trying to synchronize with the external signal. We shall only discuss the behavior of PLL under tracking conditions. For the signal acquisition mode of operation, readers are referred to references [5–6].

PLL as a closed-loop system is shown in Figure 9.13. Assuming that the phase detector is sinusoidal and the error signal is small, we can write the transfer functions at three points of the loop as shown below.

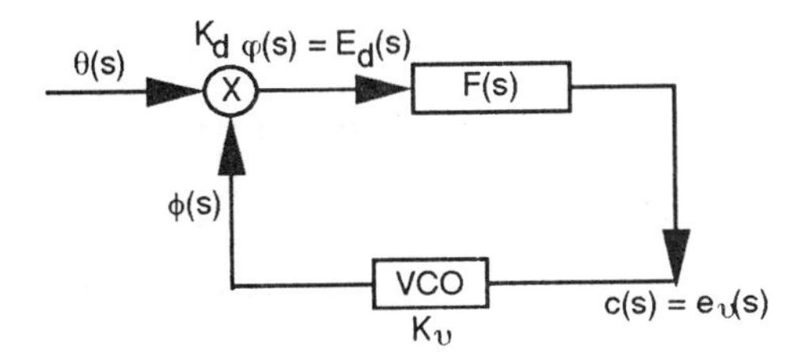

Figure 9.13 Block diagram of PLL.

$$H(s) \triangleq \frac{\phi(s)}{\theta(s)} = \frac{k_d k_\nu F(s)}{s + k_d k_\nu F(s)} = \frac{\frac{k_d k_\nu}{s} F(s)}{1 + \frac{k_d k_\nu}{s} F(s)} = \frac{G(s)}{1 + G(s)} \tag{9.27a}$$

$$H_e(s) \triangleq \frac{E_d(s)}{\theta(s)} = \frac{s}{s + k_d k_\nu F(s)} = 1 - H(s) \tag{9.27b}$$

$$H_\nu(s) \triangleq \frac{E_\nu(s)}{\theta(s)} = \frac{s k_d F(s)}{s + k_d k_\nu F(s)} \tag{9.27c}$$

where $G(s) = (k_d k_\nu F(s))/s$ is an open-loop transfer function. The above are general equations for PLL and can be applied to any order PLL. We shall limit ourselves to first and second order only.

PLL Analysis

Assuming $F(s) = 1$, we achieve the first-order PLL. As explained above, due to the presence of VCO, which has a transfer function of the form $1/s$, the PLL is at minimum of order one. The transfer function $F(s)$ of the filter for the second-order PLL can be of three different forms: lag-lead network, lag filter, and perfect integrator. Table 9.1 presents a summary of relevant design parameters. Filter configurations, transfer functions, and the open loop phase and gain plots of these are shown below.

The lag-lead network configuration is shown in Figure 9.14. The transfer function is

$$F(s) = (1 + s\tau_2)/(1 + s\tau_1) \tag{9.28}$$

Substituting the value of $F(s)$ in $G(s)$ (9.27a), the open-loop transfer function becomes

$$G(j\omega) = \frac{k(1 + j\omega\tau_2)}{j\omega(1 + j\omega\tau_1)} \tag{9.29}$$

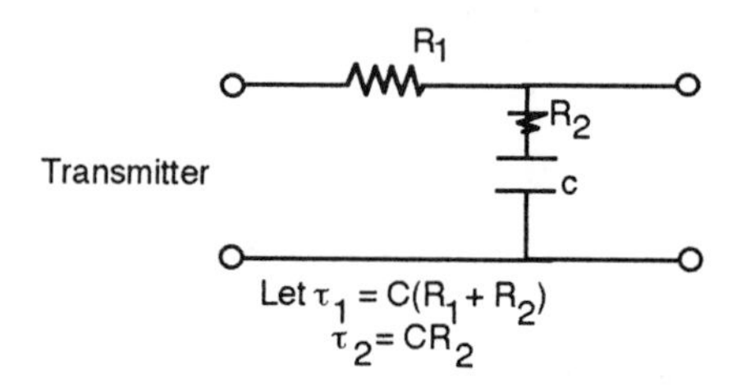

Figure 9.14 Lag-lead network.

Table 9.1
Transfer Functions and Parameters for First- and Second-Order Phase-Locked Loops

Loop Filter, F(s)	*Natural Frequency, ω_n (rad/s)**	*Damping Factor*	*Closed-Loop Transfer Function, H(s)*	*Error Transfer Function, 1 − H(s)*	*Single-Sided Bandwidth (Hz)*
Case 1 1 (first order):	K	—	$\frac{K}{s + K}$	$\frac{s}{s + K}$	$\frac{K}{4}$
Case 2 $\frac{s\tau_2 + 1}{s\tau_1 + 1}$ (passive, second order)	$\sqrt{\frac{K}{\tau_1}}$	$\frac{\omega_n}{2}(\tau_2 + K^{-1})$	$\frac{(2\zeta\omega_n - \omega_n^2/K)s + \omega_n^2}{D(s)}$	$\frac{s^2 + \omega_n^2 s/K}{D(s)}$	$\frac{K\tau_2(1/\tau_2^2 + K/\tau_1)}{4(K + 1/\tau_2)}$
Case 3 $\frac{s\tau_2 + 1}{s\tau_1}$ (active, second order)	$\sqrt{\frac{K}{\tau_1}}$	$\frac{\tau_2\omega_n}{2}$	$\frac{2\zeta\omega_n s + \omega_n^2}{D(s)}$	$\frac{s^2}{S(s)}$	$\frac{1}{2}\,\omega_n\left(\zeta + \frac{1}{4\zeta}\right)^{\dagger}$
Case 5 $\frac{1}{s\tau + 1}$ (lag, second order)	$\sqrt{\frac{K}{\tau}}$	$\frac{1}{2\sqrt{K\tau}}$	$\frac{\omega_n^2}{D(s)}$	$\frac{s^2 + 2\zeta\omega_n}{D(s)}$	$\frac{K}{4}$

After [4, p. 265].
*$K = K_v K_d$.
†For a second-order loop with $\zeta = 0.5$, $B_L = 0.5\omega_n$; with $\zeta = 1/\sqrt{2}$, $B_L = 0.53\omega_n$. B_L is the single-sided noise bandwidth in hertz, and the dimensions of ω_n are rad/s.

where $k = k_d k_v$. The open-loop gain and phase plots of $G(j\omega)$ are shown in Figure 9.15. Let us illustrate the behavior of the open-loop transfer function with a numerical example.

Example 9.1

From Table 9.1, $\omega_n = \sqrt{k/\tau_1}$, the natural frequency of oscillation and $2\xi\omega_n = (1 + k\tau_2)/\tau_1$. Let $k = 10^5$, $\omega_n = 100$ rad/sec, and $\xi = 0.7$ (damping constant). Then, $\omega_n = 100 = \sqrt{10^5/\tau_1}$, or $\tau_1 = 10$ sec. Since $2 \times 0.7 \times 100 = (1 + 10^5\tau_2/10)$, or $\tau_2 = 14 \times 10^{-3}$ sec, where, from Figure 9.15 ω corresponding to a loop gain of 0 dB $= k\tau_2/\tau_1 = (10^5 \times 14 \times 10^{-3})/10 = 140$ rad/sec. The phase shift for an open-loop gain of 0 dB is $-90 - \tan^{-1}(\omega\tau_1) + \tan^{-1}(\omega\tau_2) = -90 - \tan^{-1}(140 \times 10) + \tan^{-1}(140 \times 14 \times 10^3) = -90 - 90 + 63 = 117$ deg. Thus, the phase margin (the difference between -180 deg and the phase-shift for an open-loop gain of 0 db) is high and the loop will be stable. We should note from the above example that for a greater safety margin, τ_2 should be large.

A network configuration for a lag filter is shown in Figure 9.16. The transfer function is

$$F(s) = 1/(s\tau + 1)$$

The open-loop transfer function is given by

$$G(s) = \frac{k}{s(s\tau + 1)}$$

Thus,

$$G(j\omega) = \frac{k}{j\omega(1 + j\omega\tau_1)} \tag{9.30}$$

The open-loop gain and phase plots are shown in Figure 9.17.

Example 9.2

Let $k = 10^6$, $\omega_n = \sqrt{k/\tau} = 10^3$ rad/sec. (See Table 9.1.) Then, $\tau_1 = k/\omega^2 = 10^6/10^6 = 1$ sec.

$$\xi = \frac{1}{2\sqrt{k\tau_1}} = \frac{1}{2\sqrt{10^6 \times 1}} = \frac{1}{2000} = 5 \times 10^{-4}$$

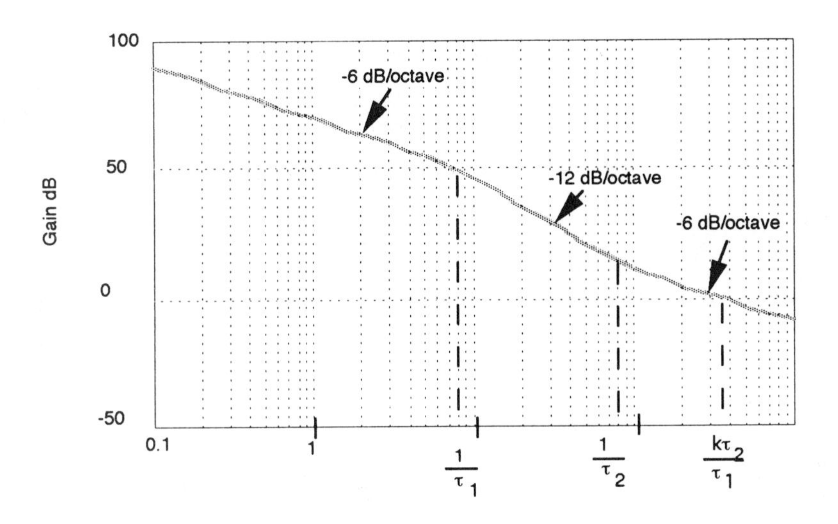

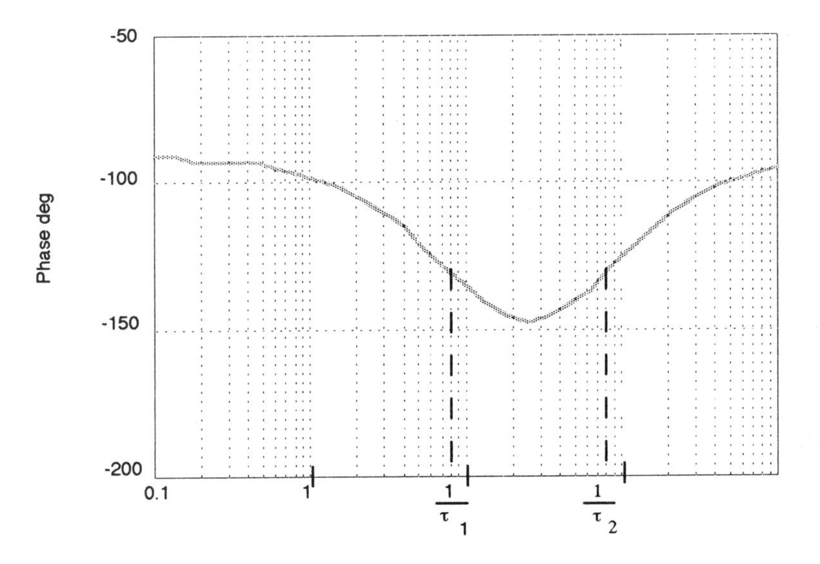

Figure 9.15 Bode plot for lag-lead network.

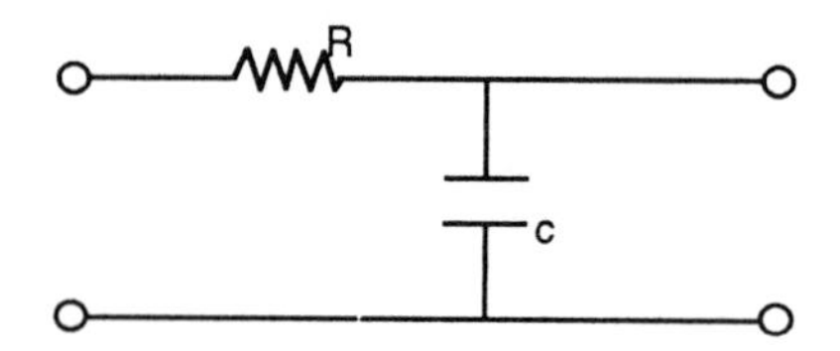

Figure 9.16 Lag filter.

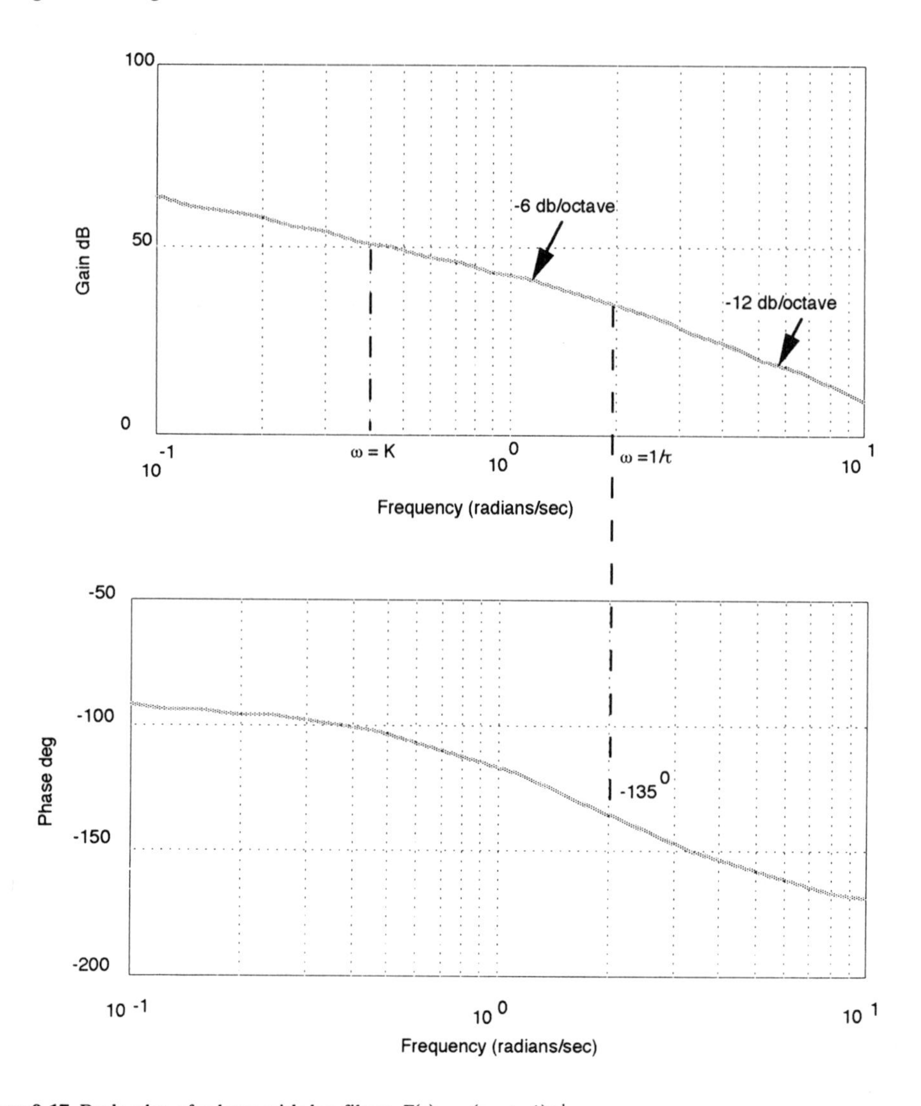

Figure 9.17 Bode plot of a loop with lag filter; $F(s) = (s\tau + 1)^{-1}$.

(See Table 9.1.) Phase shift at $\omega = \omega_n$ from (9.30) is

$$-90 - \tan^{-1}(\omega_n t_1) = -90 - \tan^{-1}(10^3 \times 1) = -180 \text{ deg}$$

where the phase margin is zero, thus a slight spurious phase shift will be sufficient to make the loop unstable.

The network configuration for a perfect integrator is shown in Figure 9.18.

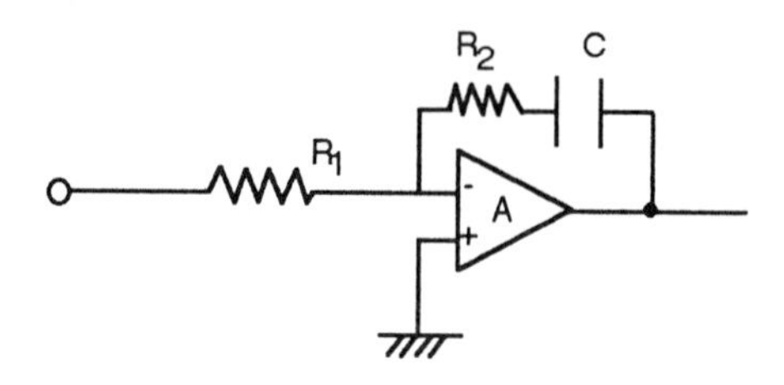

Figure 9.18 Perfect integrator.

The transfer function is

$$F(s) = (s\tau_2 + 1)/s\tau_1 \qquad \tau_2 = R_2C \qquad \tau_1 = R_1C \tag{9.31}$$

The transfer function $F(s)$ assumes an ideal operational amplifier with the gain value, $A = \infty$. The open-loop transfer function is given by

$$G(s) = \frac{k(1 + s\tau_2)}{\tau_1(s^2)}$$
$$G(j\omega) = \frac{k(1 + j\omega\tau_2)}{\tau_1(j\omega)^2} \tag{9.32}$$

Open-loop gain and phase are shown in Figure 9.19.

A summary of important characteristics of first and second-order PLL is shown in Table 9.1. With these as background information of PLL, we shall discuss the use of PLL as a synthesizer and clock recovery elements in cellular radio.

Frequency Synthesizer [5–6]

In recent years, the shortage of available radio spectrum has imposed constraints of spectral purity and effective use of channels. A PLL-based frequency synthesizer

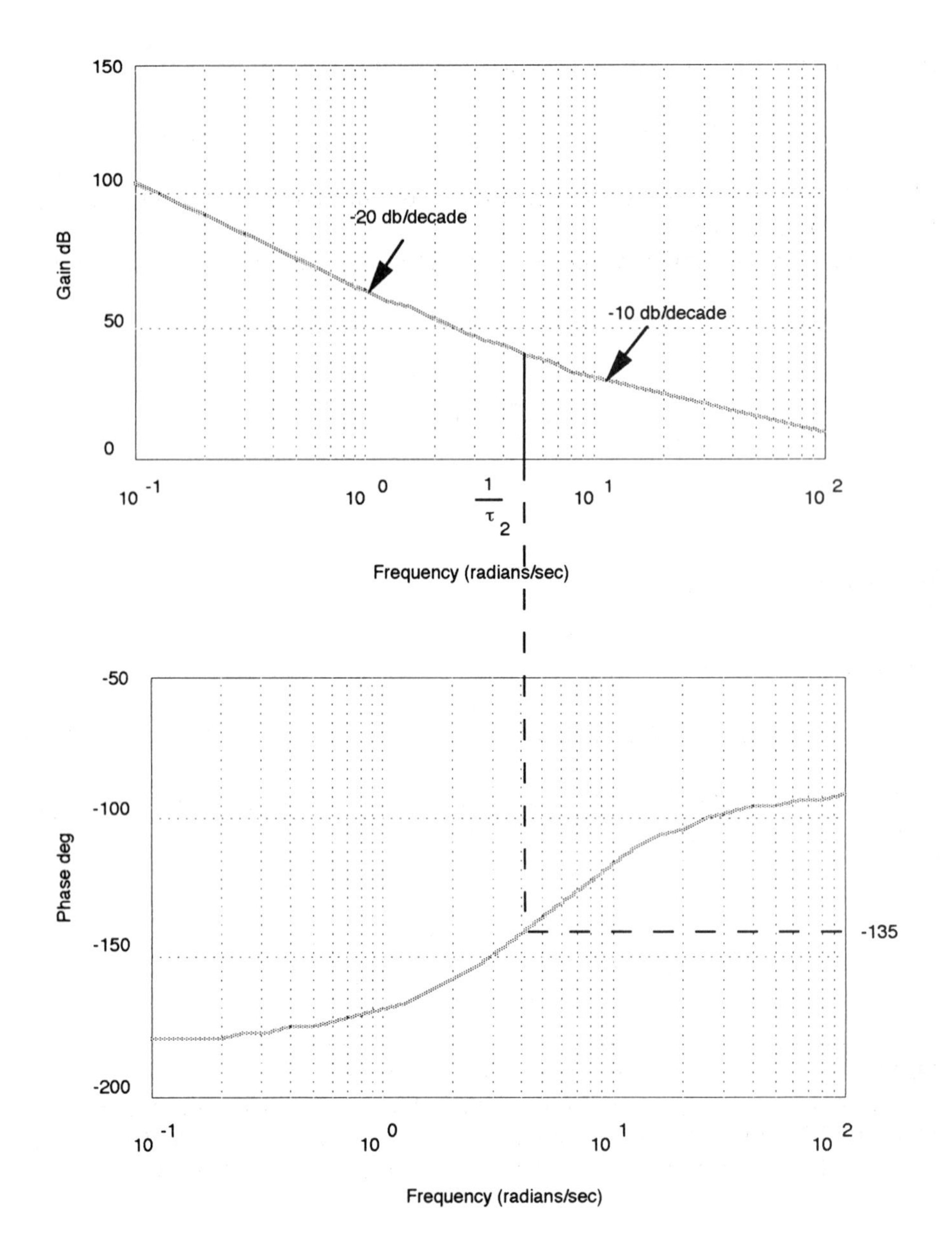

Figure 9.19 Open-loop gain (top) and phase (bottom) plots of a perfect integrator.

provides a single stable reference source of all channel center frequencies that are spectrally pure. The economic advantage of most frequency synthesizers results from the large number of high-performance oscillators that are eliminated. A synthesizer must have the following properties: programmability, switching time, resolution, and spurious signals and noise. Almost all modern frequency synthesizers are also controlled by a set of dc levels on the control wires. The advantages of this technique are the remote control, optimum layout of the circuit to minimize spurious radiations, and more rapid and reliable control. Switching time (i.e., the time between a command to a new frequency and the time when the output is useful) is an important parameter. Some applications also require that the phase be continuous. The switching speed in many of the synthesizers is proportional to the difference between the initial and final frequencies; for others, switching speed is determined by the number of frequency control bits that must be changed. Resolution is the minimum frequency difference between any two adjacent output frequencies. In U.S. cellular radio, this frequency difference is 30 kHz. In the European digital system (GSM), this difference is 200 kHz. For most analog cellular systems of the world this is either 20 kHz, 25 kHz, or 30 kHz. Since the frequency synthesizer's output is not a perfect sinusoid, corruption of the desired output signal, usually referred to as the "carrier," may be divided into coherent spurious signals generated in various nonlinear operations such as mixing, and noncoherent noise outputs due to internal circuit noises. Because synthesizer output is normally amplitude-leveled to a very high degree of accuracy, the primary source of noise at the output is phase noise.

In this section we shall discuss PLL-based synthesizer realizations. A block diagram of the PLL basic synthesizer is shown in Figure 9.20. From the basic knowledge of the PLL, $f_1/n_1 = f_2/n_2$. Thus, the output selection is provided by changing the divider integers n_1 and n_2. For the minimum increment in output frequency of (Δf), the loop bandwidth (B_L) must be less than Δf in order that

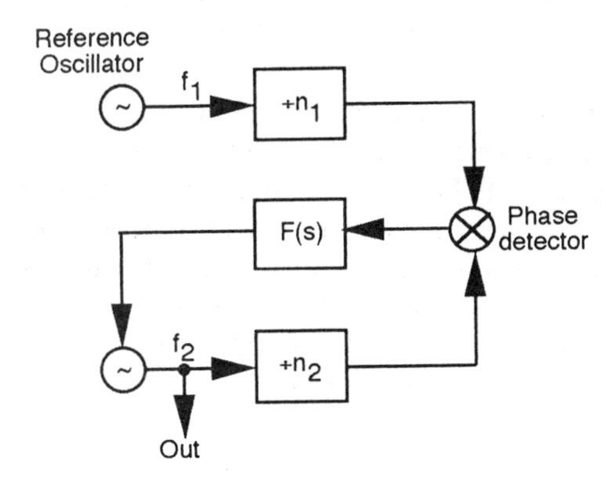

Figure 9.20 Basic PLL synthesizer.

ripples of Δf do not pass through. However, smaller bandwidth B_L makes the acquisition time T_{acq} large. Also, small-loop bandwidth means more phase jitter and undesirable characteristics of a synthesizer. In order to avoid these conflicting requirements, PLL is modified as shown in Figure 9.21. For this configuration the output frequency is

$$f_2 = \frac{n_2 f_1}{n_1 m} \tag{9.33}$$

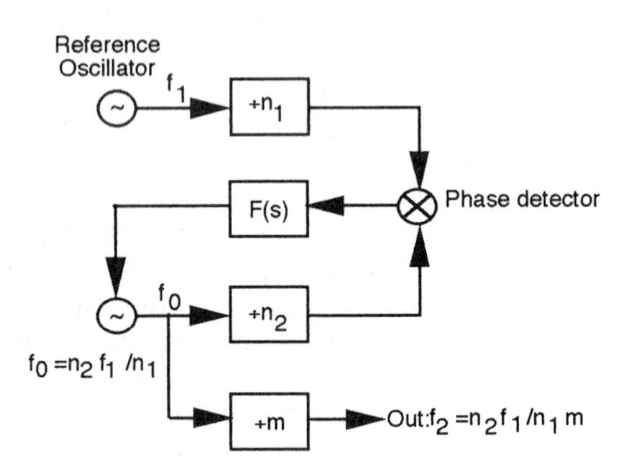

Figure 9.21 Modified phase-locked loop.

The frequency step size is

$$\Delta f = \left(\frac{(n_2 + 1)f_1}{n_1 m} - \frac{n_2 f_1}{n_1 m}\right) = \frac{f_1}{n_1 m} \tag{9.34}$$

Since VCO is now working at a higher frequency, $n_2 f_1/n_1$, the loop bandwidth is m times larger, which in turn reduces the jitter by a factor of m. Thus, the configuration is a more practical choice for a frequency synthesizer.

One practical configuration of a frequency synthesizer is shown in Figure 9.22. Here, the reference frequency oscillator multiplies and divides by N_1 and N_2. $N_1 f_{ref}$ is the input to the phase detector (PD2), where it is mixed with the output frequency f_o and produces the sum and difference frequencies. Assuming that the mixer allows only the sum frequency component, we get

$$\frac{f_o + N_1 r_{ref}}{N} = \frac{f_{ref}}{N_2} \tag{9.35a}$$

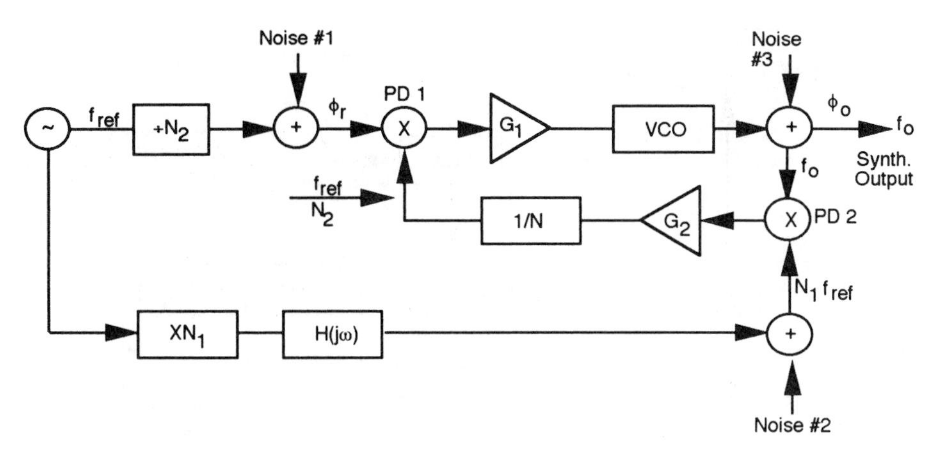

Figure 9.22 Single-loop frequency synthesizer.

or,

$$f_o = \left(\frac{N}{N_2} - N_1\right) f_{\text{ref}} \qquad \frac{N}{N_2} > N_1 \tag{9.35b}$$

Thus, the resolution or the step size is given by

$$\Delta f_o = \left(\frac{N+1}{N_2} - N_1\right) f_{\text{ref}} - \left(\frac{N}{N_2} - N_1\right) f_{\text{ref}} = \frac{f_{\text{ref}}}{N_2} \tag{9.36}$$

The AMPS system makes use of the above PLL configuration for both their cell-site and mobile frequency synthesizers. See Figures 9.23 and 9.24, respectively. At the cell site, VCO generates the synthesizer output frequency at f_o. A portion of the VCO output power is fed to the mixer where it is heterodyned against f_o = 228.02250 MHz, which is derived out of a quartz crystal. The output difference frequency is divided by a programmable divider, which is set by a 10-bit binary number. The integer N lies between 737 and 1,402, which is selected by the dc voltage on the control lines. A stable 7.50-kHz reference oscillator, f_2, is compared with the divider output frequency of $(f_1 - f_0)/N$. Any phase error is fed back to the VCO in the form of a dc control voltage, which changes the output frequency. When in lock, the output frequency $f_o = f_1 - Nf_2$. Thus, the output oscillator stability solely depends upon the stability of oscillators f_1 and f_2. It is obvious that the output frequency of the synthesizer can be changed in steps of f_2 or 7.5 kHz by assigning different values of N, which is controlled remotely. Since f_o is in the range of 217.5–222.5 MHz band, which is a quarter of the output frequency (actual

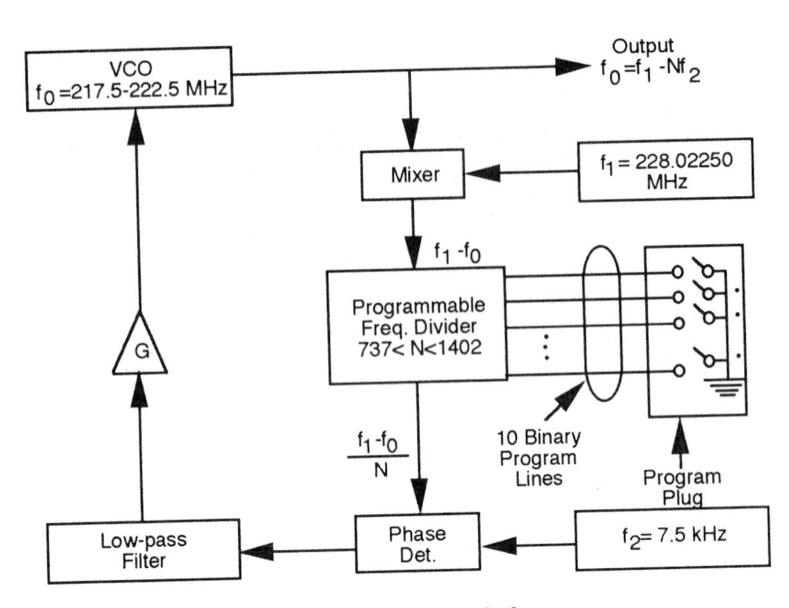

Figure 9.23 AMPS cell-site frequency synthesizer. Source: [20].

band 800–900 MHz), the 7.5-kHz frequency steps are multiplied by four to produce the output channel frequencies, which are 30-kHz apart.

As an example, if one wants to tune to the 134th channel whose center frequency is 870.030 MHz, the output frequency of the synthesizer f_o is f_o = $228.0225 - N \times 7.5 \times 10^{-3} = 870.030/4$, where $N = 1{,}402$. Thus, the frequency synthesizer must be programmed by setting the number $N = 1{,}402$.

The second example of the synthesizer is the one with a mobile transceiver. The design is based on 666-channel frequencies on digital command from ten parallel binary lines, which are set remotely by the cell site. A relatively unstable VCO generates the output carrier frequency f_o. A portion of the output carrier power enters a mixer where it is heterodyned against the third harmonic of a master crystal oscillator operating at f_{3rd} = 43.82 MHz. The difference frequency, $f_o - f_1$, which is in the range of 5.7–9.1 MHz, is divided down by an integer N in a programmable frequency counter. The divider output frequency, $(f_o - f_1)/N$ nominally at 5 kHz, is compared with a stable 5.0-kHz source f_2. This reference is obtained by dividing down the reference oscillator output by 549 × 4, as shown in Figure 9.24. The phase error feeds back to the VCO in the form of a dc control voltage, thus keeping the total loop in phase lock. When in lock, the output frequency is given by

$$\frac{f_o - f_1}{N} = f_2 = 5 \text{ KHz} \qquad f_o = f_1 + Nf_2 \tag{9.37}$$

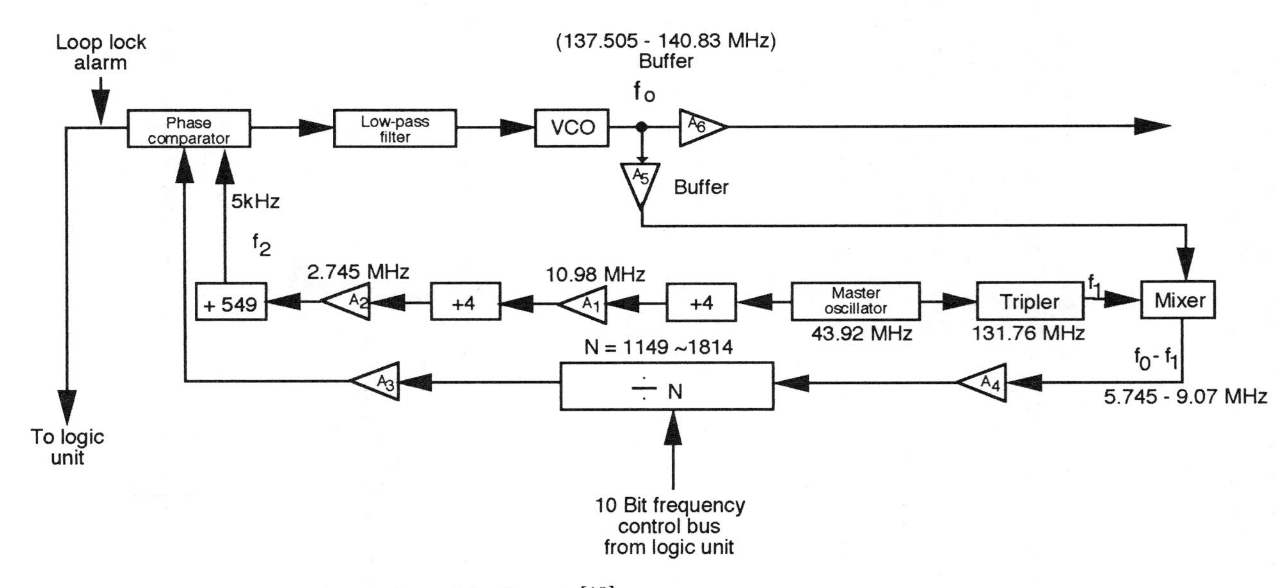

Figure 9.24 Frequency synthesizer at the mobile. Source: [19].

Thus, the synthesizer has the same frequency stability as that of a master oscillator. Obviously, the different channel frequencies can be set remotely by the cell site by appropriately setting up the dc control voltage on a 10-bit frequency control bus.

Timing Recovery

All digital receivers need to have their demodulator synchronized to the incoming digital symbol transitions in order to achieve optimum demodulation. Thus, the mobile data receiver must also reconstitute a stable bit clock derived from the incoming data stream, which is contaminated by Rayleigh fading and noise. There are basically two types of timing recovery techniques: deductive and inductive. The other name of deductive timing recovery is an open-loop synchronizers. These circuits directly recover the clock from the data, as shown in Figure 9.25. On the other hand, inductive timing recovery does not process the received signal to get a timing tone as in the open-loop case, but rather uses a closed-loop technique such as that shown in Figure 9.26. PLL finds wide application in this area of realizations.

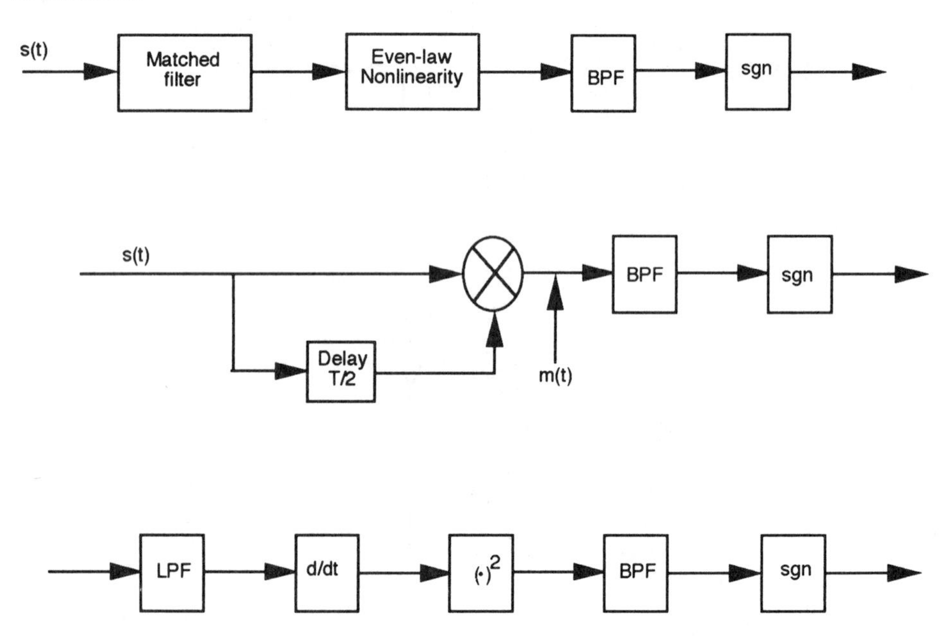

Figure 9.25 Open-loop bit synchronizers.

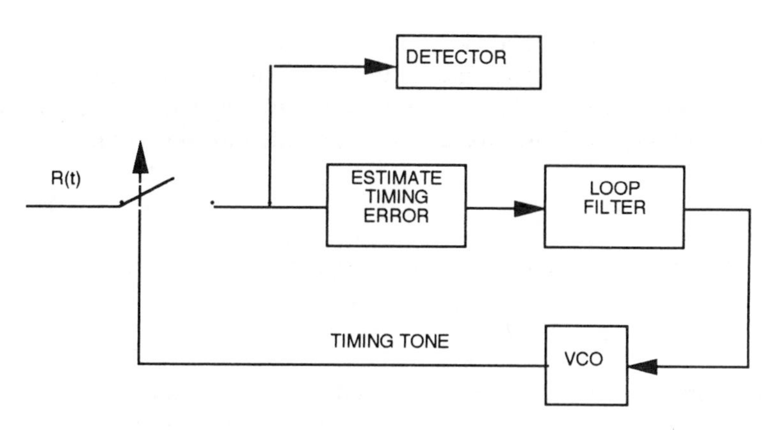

Figure 9.26 Closed-loop timing recovery system (inductive).

Open Loop Recovery Process: Spectral Line Method

Consider a baseband pulse amplitude modulation (PAM) system represented as

$$R(t) = \sum_{k=-\infty}^{\infty} A_k p(t - kT) \tag{9.38}$$

The process is cyclostationary; that is, the moments vary in time and are periodic with period T, the symbol time. We operate on $R(t)$ to form

$$Z(t) = f[R(t)] \tag{9.39}$$

where f is a memoryless nonlinearity. When $R(t)$ itself contains spectral line at T, then the timing recovery is linear. However, in most cases the mean value of $R(t)$ is zero and the higher moments of $R(t)$ are periodic. In this case, f is nonlinear and the recovery process is known as nonlinear spectral line method.

Linear Spectral Line Method

When the mean value of data symbol is nonzero, the baseband PAM system may contain a line at the baud rate. In order to find this, the signal can be partitioned into independent (deterministic) components and data dependent, which is a zero-mean stochastic process. For the PAM waveform under consideration, we can decompose $R(t)$ into the following form:

$$R(t) = E(A_k) \sum_{m=-\infty}^{+\infty} p(t - mT) + \sum_{m=-\infty}^{+\infty} [A_m - E(A_k)]p(t - mT) \tag{9.40}$$

where the first term is independent of data A_k and is periodic with period T, and can be recovered with a bandpass filter (BPF). Its periodicity implies a fundamental at the symbol rate of $2\pi/T$. The second term is zero mean and results in jitter on the recovered timing tone. An illustration is provided for linear timing recovery.

Example 9.3

Let the received pulse be a 50% duty cycle square pulse represented as

$$p(t) = 1 \qquad 0 < t < T/2 \tag{9.41}$$

and zero otherwise. Thus, $E(A_k) = 1/2$, and the signal has clock at the symbol rate. The deterministic part of the waveform is shown in Figure 9.27.

Nonlinear Spectral Line Method

As stated above, often the mean of $R(t)$ is zero, and the second and higher moments are non-zero and periodic. In this case, the clock can be derived by passing the received signal through a memoryless nonlinearity. The output in this case has a deterministic mean value and the timing signal can be derived by filtering the output through a bandpass filter. Consider the above example of a PAM signal. Passing the signal through a second-order nonlinearity and taking the expected value of the output, we get

$$E[|R(t)|^2] = \sigma_A^2 \sum_{m=-\infty}^{m+\infty} |p(t - mT)|^2 \tag{9.42}$$

We assume that the correlation function of data symbols are white; that is,

$$E[A_m A_n^*] = \sigma_A^2 \delta_{m-n} \tag{9.43}$$

Some random components in $|R(t)|^2$ will also pass through BPF and result in timing jitter. In this case, any nonzero excess BW (above Nyquist) will be sufficient to guarantee the baud-rate clock. There are three types of nonlinearity of interest:

$$\begin{aligned} z(t) &= f[x(t)] = x^2(t) \\ &= \ln \cosh [x(t)] \\ &= |x(t)| \end{aligned} \tag{9.44}$$

where cosh () is the natural log of hyperbolic cosine. The nonlinearity ln cosh $[x(t)]$

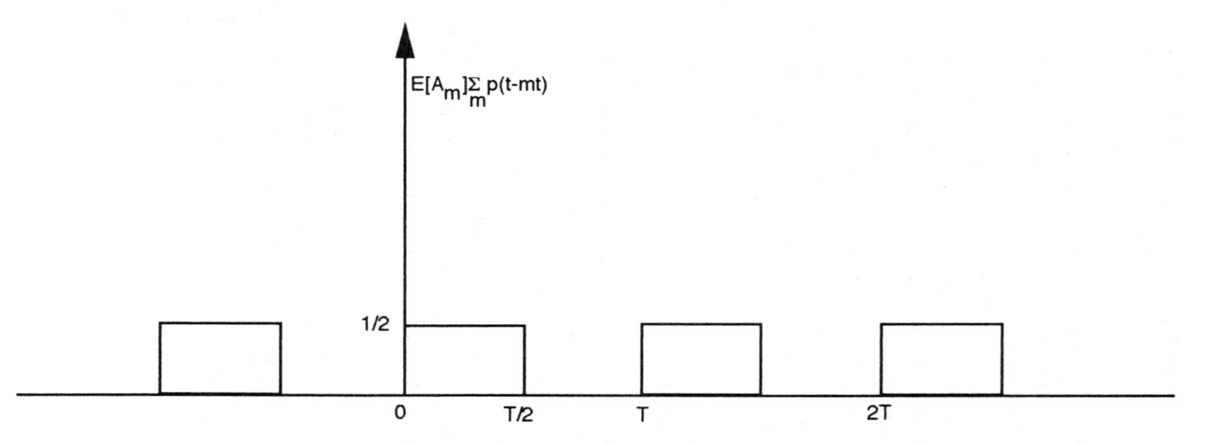

Figure 9.27 The deterministic part of a binary on-off signal where the pulse shape is a 50% duty cycle square pulse.

is the optimum of nonlinearity as it behaves like $x^2(t)$ for small values of $x(t)$, and also like $|x(t)|$ for large values of $x(t)$. In other words,

$$\begin{aligned} \ln \cosh\,[x(t)] &\approx |x(t)|/2 \qquad |x(t)| \geq 1 \\ &\approx x^2(t)/2 \qquad |x(t)| < 1 \end{aligned} \tag{9.45}$$

While the best type of nonlinearity is ln cosh[$(x(t)$], it has been shown that the square law provides near-optimum performance for rounded pulse while the hard limiter is best for square pulses.

Figure 9.28 shows the block diagram of the square-law circuitry for clock recovery for the GMSK signal. The low-pass filter, squarer, and zonal filter form a suboptimum configuration. Here the low-pass filter preceding the nonlinearity approximates the matched-filter operation of the optimum synchronizer. BPF (zonal filter) output is approximately a sine wave, which is fed to the PLL. The zonal filter is centered at the frequency corresponding to data rate. The PSD at the squarer output for a B_tT of 0.3 and the modulation index h of 0.6 is shown in Figure 9.29. The zonal filter eliminates the dc component of the PSD and filters the continuous component, causing the PSD to assume a bandpass shape rather than the low-pass it assumes at the squarer output. The discrete component of the PSD remains unchanged when passed through zonal filter. The zonal filter output is then applied to a PLL. In the steady-state condition, VCO will follow the discrete component, which is the recovered clock. Some jitter will also be present due to continuous component and the additive noise.

As a second application of the above theory we shall discuss the experimental clock recovery system for data channels in AMPS. This is an open-loop nonlinear clock recovery process. The test signal formats for the forward set-up channel in the forward blank and the burst test signal for voice channel are shown in Figure 9.30(a). In the forward set-up channel data is transmitted continuously, while in

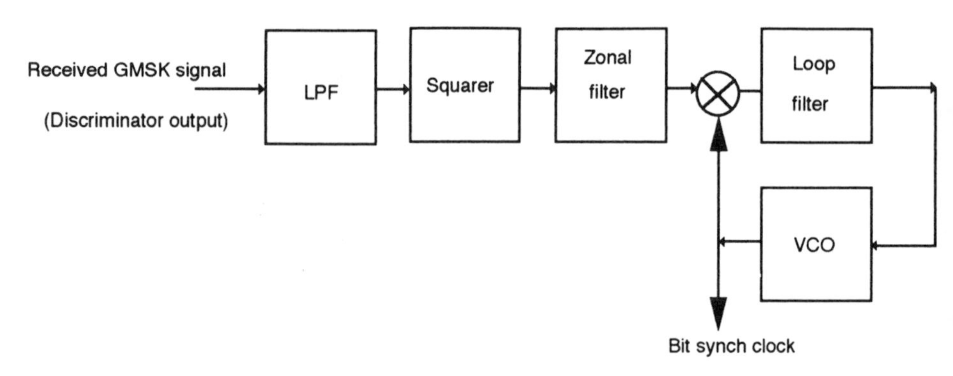

Figure 9.28 Block diagram of a square-law bit synchronization.

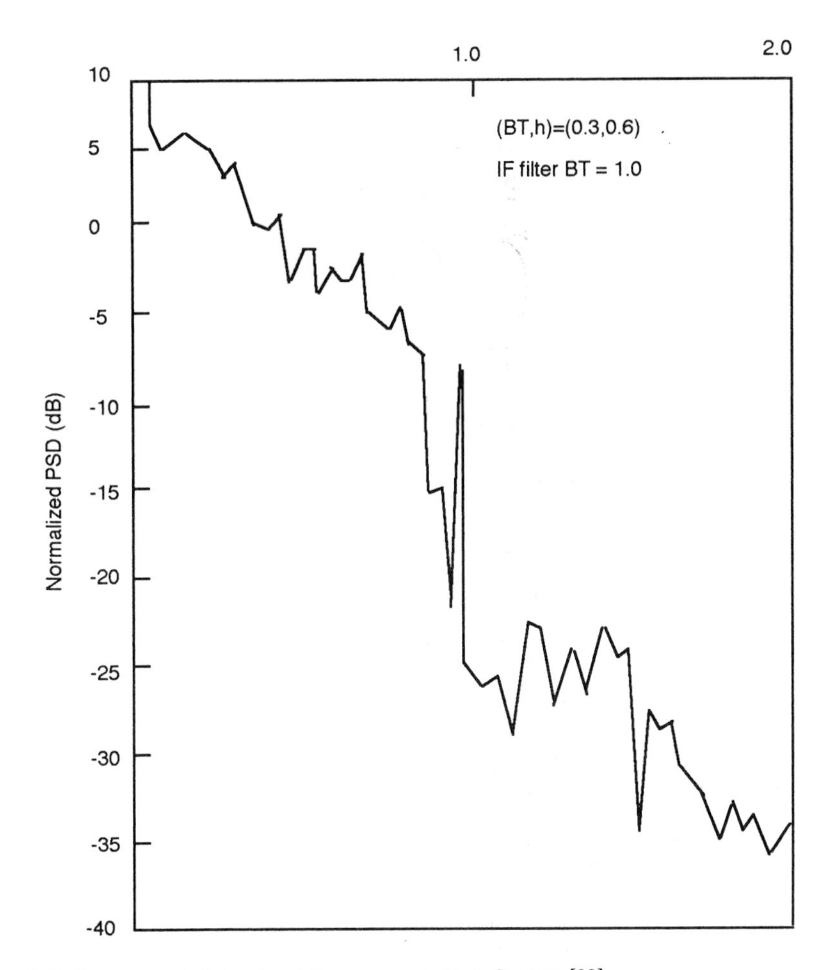

'igure 9.29 Power spectral density at the squarer output. Source: [22].

he forward blank-and-burst mode, data are sent interleaved with voice. The test ignal format for the forward setup channel in AMPS consists of the following equence.

1. A 10-bit dotting sequence 1010101010 for bit synchronization acquisition.
2. An 11-bit Barker sequence, 11100010010 for word acquisition, and an 88-bit message data repeated five times. Test signal format for the forward blank-and-burst channel consists of a 37-bit dotting sequence for bit synchronization acquisition, followed by an 11-bit Barker sequence for word synchronization, followed by a 40-bit random code message.

The waveforms at different points of the system are shown in Figure 9.30(b).

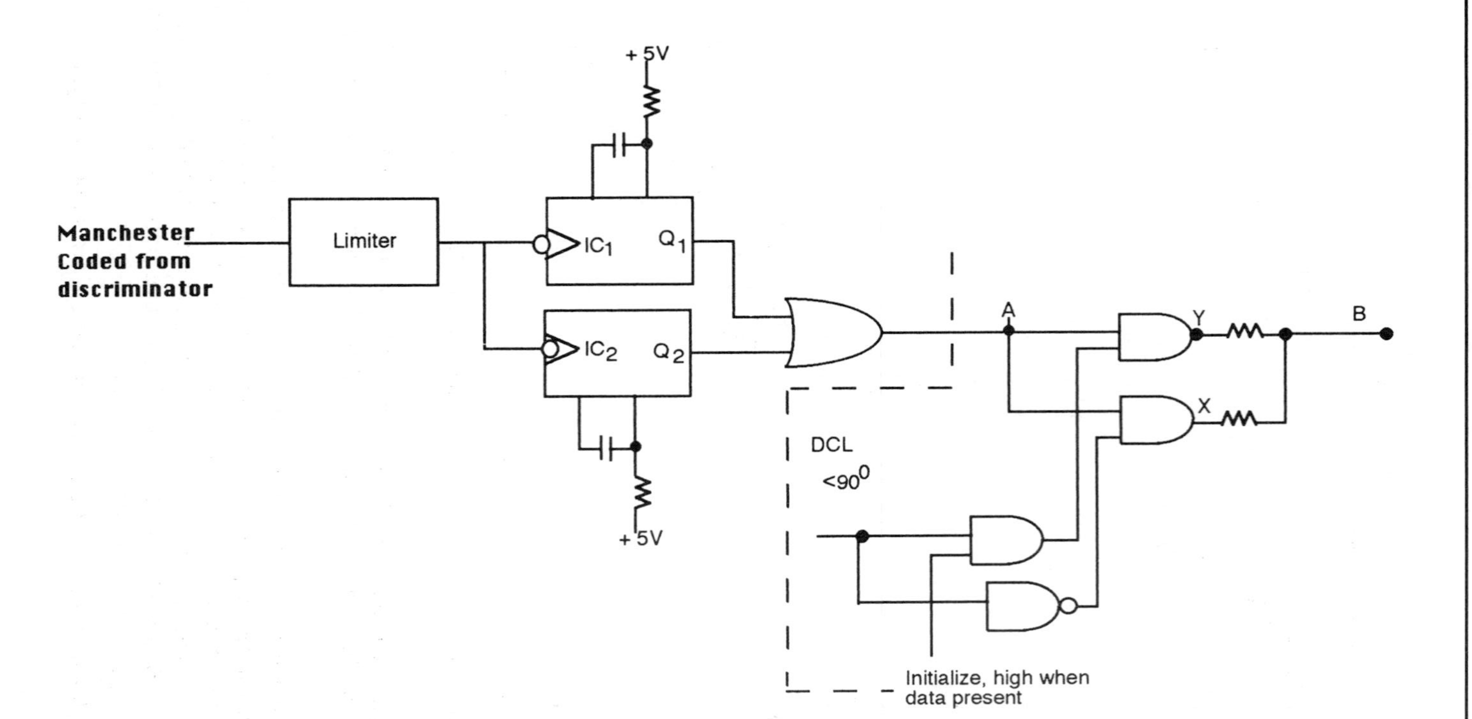

Figure 9.30(a) Digital bit clock recovery system.

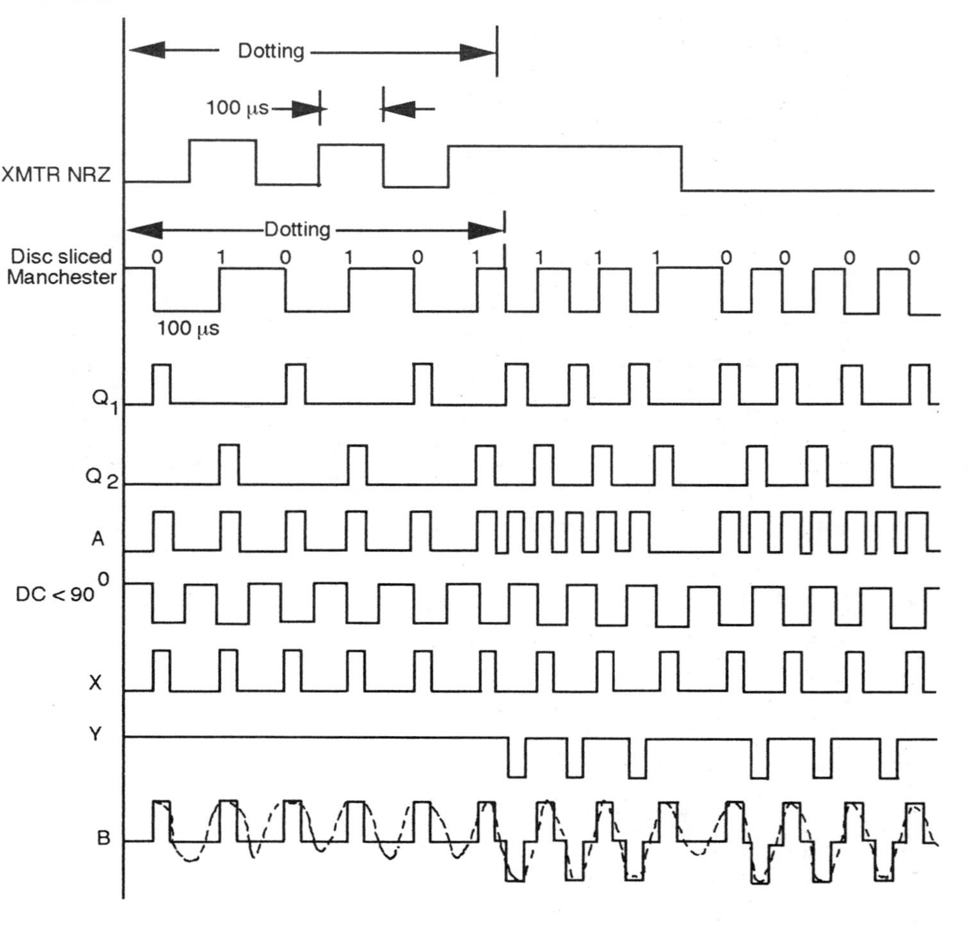

Figure 9.30(b) Waveforms for the digital bit clock recovery system.

The discriminator output which is Manchester-coded data, is fed to the two ICs, IC_1 and IC_2. These are two monostable multivibrators that together produce a 10-kHz pulse at point A when the input is the dotting sequence (10101010. . .) and a 20-kHz square wave when the input code sequence is all ones or all zeros. With the help of steering logic, the unipolar pulses of 20-kHz rate are converted into bipolar signals. This is done by inverting each alternate pulses. X and Y signals when mixed together at the output B provide a stable 10-kHz clock source.

9.2.4 Barker Code Recovery

For receiver output to make sense, it needs to be synchronized with the frame structure of the incoming signal. Usually the start of an N-bit frame is indicated by one or more bits periodically inserted at the beginning of each frame. Sometimes even a single bit may be enough. However, if synchronization is required at a low value of SNR, L-bit code words may be necessary for frame synchronization. A good choice for the frame synchronization is the Barker code due to its unique correlation properties. The scheme locates the synchronization code word by correlating successive L-bit segments of the received bit sequence with the locally stored synchronization word. A good synchronization code word is one that has the property that the absolute value of its correlation sidelobes are small. The set of Barker codes is shown in Table 9.2. It has been proven that there is no Barker code of odd length greater than 13, and if there is any other of even length, the length must be the square of an integer. It has been shown that the Barker codes of lengths three and seven are maximal sequences.

Table 9.2
Known Barker Codes

Length	*Code*
2	00 or 01
3	001
4	0001 or 0010
5	00010
7	0001101
11	00011101101
13	0000011001010

A code of length n is a sequence of symbols c_j for $j = 0, 1, \ldots, n - 1$ taking values of $+1$ and -1 with the autocorrelation property $\phi(k) = n$ if $k = 0$, and $|\phi(k)| \leq 1$ if $k \neq 0$, where $\phi(k)$ is a discrete (aperiodic) autocorrelation function and is given by

$$\phi(k) = \sum_{j=0}^{n-1} c_j c_{j+k} \tag{9.46}$$

Thus, when k is not zero, the only allowed value of $|\phi(k)|$ is 1. An example of autocorrelation for seven-bit Barker code now follows.

The seven-bit code is first written as the sequence $C_1, C_2, \ldots, C_7$:

$$1 \quad 1 \quad 1 \quad 0 \quad 0 \quad 1 \quad 0 \quad \rightarrow \quad + \quad + \quad + \quad - \quad - \quad + \quad -$$

Applying (9.46),

- $k = 0, \phi(0) = 7; k = 1, \phi(1) = 0; k = 2, \phi(2) = -1;$
- $k = 3, \phi(3) = 0; k = 4, \phi(4) = -1; k = 5, \phi(5) = 0;$
- $k = 6, \phi(6) = -1$

The plot of the autocorrelation function $\phi(k)$ is shown in Figure 9.31.

A block diagram of a Barker code detection scheme as used in AMPS is shown in Figure 9.32. The 11-bit Barker sequence detector delivers an output pulse whenever the Barker pattern is recognized. Under dynamic conditions of operation, the first detectable Barker sequence is used to initialize the timing of the window generator. Thereafter, only those "wordsync" flags that occur coincidentally with the output of the window generator are passed on to the data sink as a valid word-sync flag. If the synchronization is correct, successive pulses from the 11-bit Barker sequence (output designated as c) will coincide with the word from the window generator.

9.2.5 Equalizer

As discussed in Chapter 3, due to multipath the mobile signal is dispersed in times. Due to the smearing of signals in time, a radio signal introduces ISI. To reliably decode the signal, an estimate of the channel transfer function is required, which can then be used by an equalizer to compensate ISI. Conceptually, the job of the equalizer is to take the different time-dispersed components, weight them according to the channel characteristics, and sum them up after inserting appropriate delay between components so that a replica of the transmitted signal is restored. The problem in cellular radio becomes more complex due to the dynamic nature of the channel. As the mobile moves through multipath surroundings, the equalizer must continually adapt to the changing channel characteristics, which vary at a rate proportional to the wavelength of the signal. If the operational mode is TDMA, this will necessitate measuring the channel impulse response from one slot to another. A constant training sequence may be included in every data burst to make this fast and repeated evaluation. The equalizer knows the transmitted training sequence. It also knows what it has really received, and thus can make an estimate

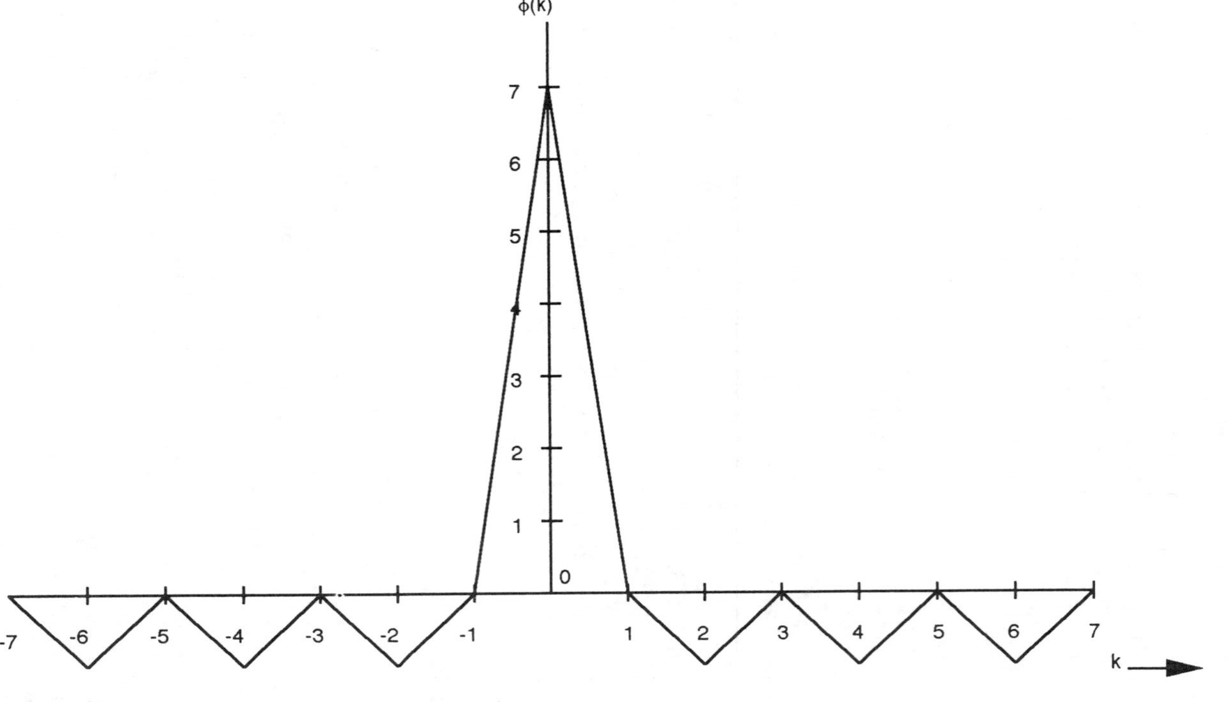

Figure 9.31 Autocorrelation of a seven-bit Barker code.

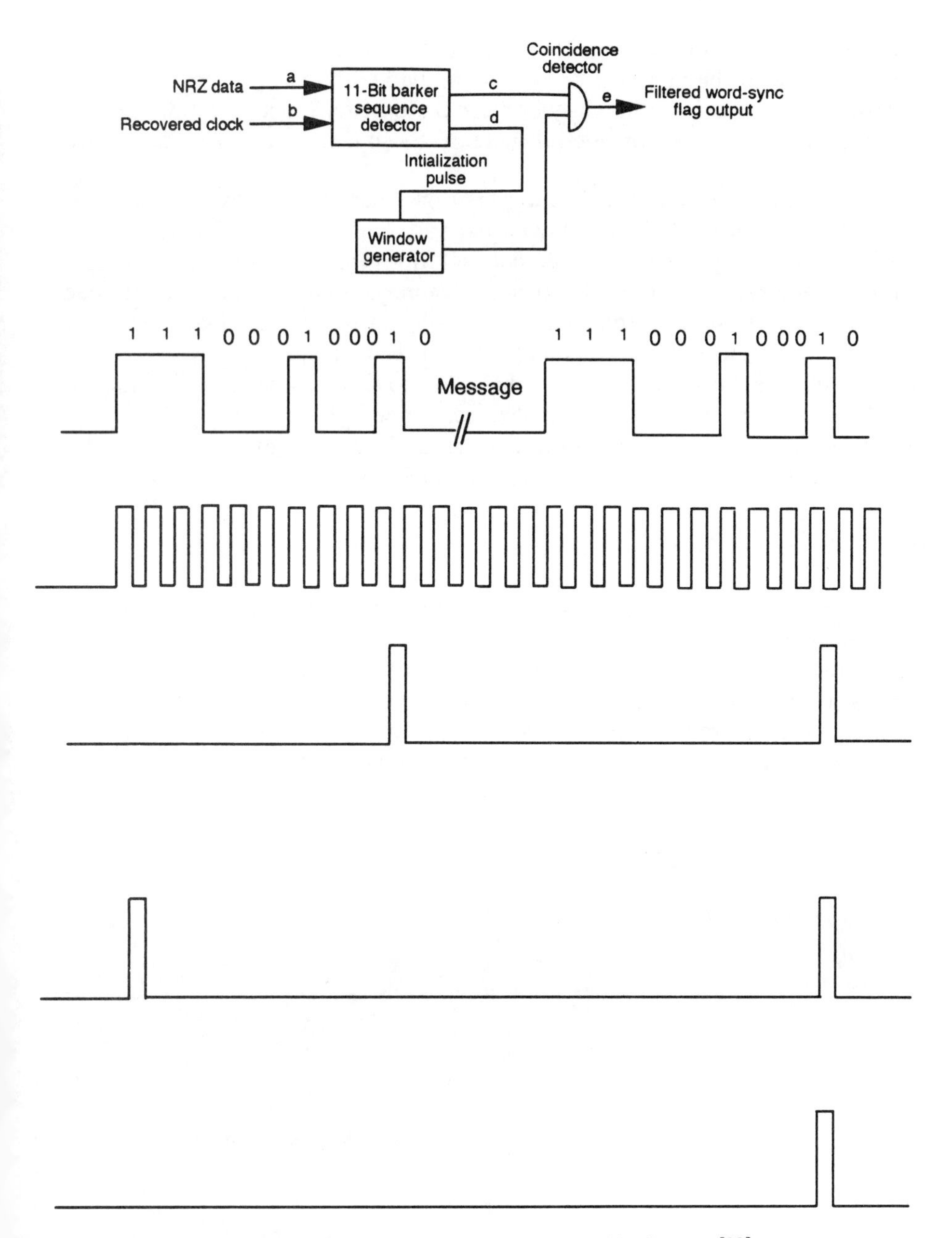

Figure 9.32 System for detection of the Barker word-sync preamble. Source: [22].

of the channel transfer function. If the multipath is severe, it may even be necessary that the equalizer be adaptive during the individual burst of transmission. Thus, an adaptive equalizer continuously updates the transfer function estimate, making sure that the decision error doesn't increase too much during the channel transmission.

Two types of nonlinear equalizers, one based on maximum likelihood sequence estimation (MLSE) and the other based on decision feedback estimation (DFE), have been proposed as candidate solutions for this application. The use of nonlinear equalization offers significant advantages under severe channel distortions since it avoids the noise enhancement, which would otherwise arise if a linear equalizer were used. Both these equalizers are used to adapt to the dynamic changes of the channel and operate under two different modes of operations. Training mode enables a proper setting of the equalizer parameters by processing the preamble at the beginning of each time slot. The tracking mode continuously adjusts the equalizer parameters to match with the channel variations during the time slot.

9.2.5.1 Maximum Likelihood Sequence Estimation Equalization (MLSE)

A block diagram of the (MLSE) adaptive Viterbi receiver is shown in Figure 9.33. The MLSE principally scans all possible data sequences that could have been transmitted by a digital data source, computes the corresponding receiver input sequences, compares them with the actual input sequence received by computing the weighting parameters (known as metric), and selects the data sequence that shows the highest a posteriori probability (highest metric) to be the one transmitted.

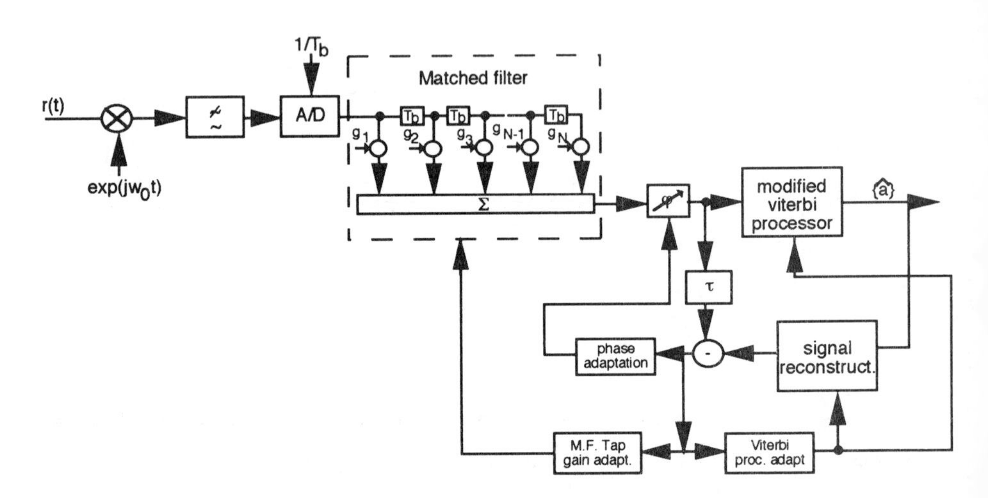

Figure 9.33 Block diagram of the Viterbi (MLSE) adaptive receiver. Source: [7, p. 391].

The process of metric computation is the same as that used in Viterbi decoding of convolutional codes [15]. The MLSE algorithm is almost unrealizable due to the large number of computations for even relatively short binary sequences. The MLSE technique has become realizable after recognizing that the Viterbi algorithm can effectively be used not only to decode convolutionally coded sequences, but also to deal with the unintentional coding produced due to ISI. The MLSE receiver considered here consists of matched filter (MF) with N taps and a Viterbi processor. Here, the received signal is sampled and each sample is filtered through a transversal filter to approximate the matched filter response. The use of MF makes the receiver insensitive to the carrier and clock phases, provided the MF coefficients are correctly adjusted and the time span of the MF is long enough to include all the channel impulse response. The choice of the number of taps, N, is important as it relates directly to the maximum echo delays that are expected to be equalized and also to the complexity of the Viterbi processor.

It has been found that a choice of 5–7 for N is satisfactory under most practical conditions [7–10]. The channel impulse response can be derived by correlating the time slot with the modulating sequence. This makes it possible to set up the MF tap gains, which are the complex conjugate of the channel impulse response. The complexity and the processing effort depend exponentially on the bit periods during which the complex impulse response has to be computed. If the processing is excessive, the channel impulse response has to be truncated, which in some cases will degrade the system performance. The output of the MF equalizer is fed to the Viterbi processor. The metric computation in the Veterbi processor does not require any squaring operation. Also, an all-digital implementation of the block is possible. As shown in Figure 9.33, the receiver is continuously adjusted when the CIR changes during the time slot. The gradient algorithm can be used to minimize the mean square error by varying the carrier phase, φ, the MF tap gains, and the Viterbi parameters.

9.2.5.2 Decision Feedback Equalizer

The equalizer shown in Figure 9.34 consists of a feedforward part that receives the input and a feedback part that is fed by the past detected symbols. Both these filters consist of a finite impulse response (FIR) filter with their tap spacings equal to bit duration, and their respective outputs are added to form the equalizer output before threshold detection. The tap coefficients of both FIR filters can be computed by solving a set of linear equations or by iterative adjustment by processing the training sequence and subsequently the information symbols. For iterative tap coefficient adjustment, either slowly converging but less complex gradient algorithms or more advanced and more complex but faster converging algorithms can be used. It has been pointed out in the literature that the equalization performance

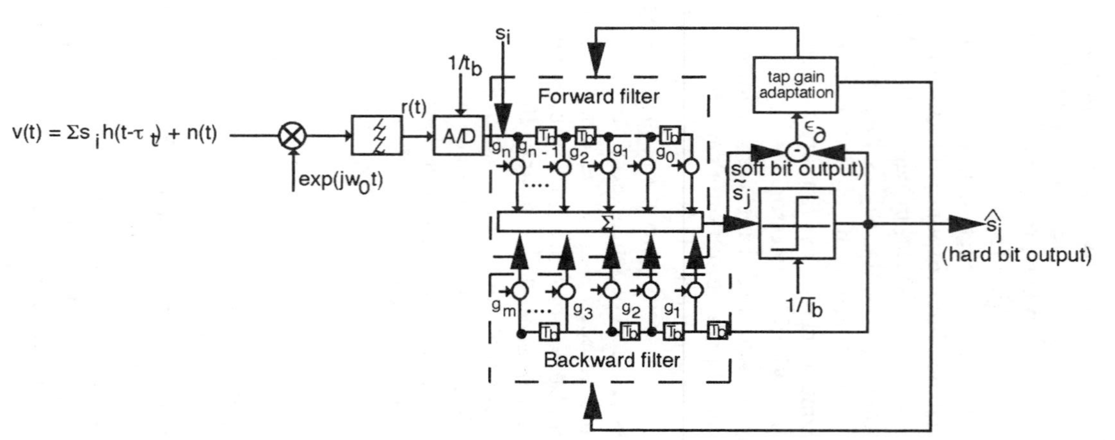

Figure 9.34 Block diagram of the DFE adaptive receiver. Source: [7, p. 391].

is highly dependent on the capabilities of the coefficient adaptation algorithm used, which is the main processing load of the equalizer [7–11].

The number of taps is generally between 2–5, depending on the maximum delay to be equalized. The adaptation of the taps has to take place during the training as well as during the actual interval when the messages are received. A long preamble may be required during the training sequence, especially for a low SNR. After the initial preamble, the tap gains are adjusted to minimize the mean square error (LMS algorithm) during the message part of the time slot. The error, as shown in Figure 9.34, is defined as the difference between the analog sample at the input to the decision device and the corresponding transmitted binary digit and can be expressed as

$$\epsilon_n = \tilde{S}_n - S_n \tag{9.47}$$

In terms of the forward and backward filter coefficients, the error can be expressed as:

$$\epsilon_n = \int \gamma(t) r(t + nT_b) dt - \sum_{k=1}^{\infty} g_k \hat{S}_{n-k} - s_n \tag{9.48}$$

where

$$g(t) = \sum_{t=1}^{\infty} g_i S(t - iT)$$

is the impulse response of the backward filter, whose tap gains are g_k. The first two terms represent τ_n in (9.47). The forward filter impulse response is $\gamma(-t)$ and is defined over the whole range $|t| < \infty$. Thus, the minimization problem is

$$\min_{\gamma(t),\ b_k} E[\epsilon_n^2] \tag{9.49}$$

As the number of taps increases, not only do these filters become more complex but also the tap gain adjustment becomes more involved.

9.2.6 Voice Processing in FM Systems [13]

A block diagram of the voice processing circuitry is shown in Figure 9.35. The circuit consists of four voice processing sub-blocks in the transmitter: compandor (compressor), preemphasis, deviation limiter, and bandpass filter. The corresponding processing blocks in the receiver are: deemphasis, filtering, and compandor.

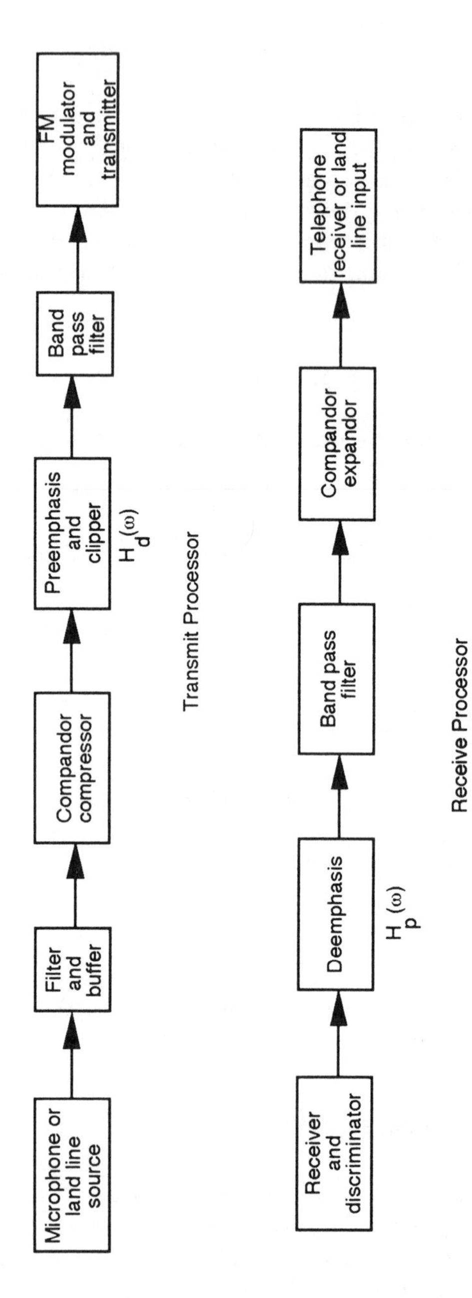

Figure 9.35 Transmit and receive processor in analog FM in AMPS.

In this section we shall describe the basic characteristics of and reasons for using these blocks in different analog cellular radios of the world. The use of filtering and amplitude limiting is done in the transmit processor to control the splatter of spectrum into the adjacent channels and the filtering is used in the receiver to control the effects of noise.

The differentiator preemphasis and integrator deemphasis improve the performance in the presence of channel impairments by preemphasizing the higher frequency components in the transmitter signal and correspondingly deemphasizing those components in the receiver. The composite effect of preemphasis and deemphasis is that the original voice signal remains unchanged and receiver noise is reduced. The compandor is used to provide important improvements in analog FM system by controlling the effects of speech-level variability on clipping distortion and frequency deviation generated by the modulator.

9.2.6.1 Compandor

A well known characteristic of FM modulation is that the baseband signal power is directly proportional to mean-squared frequency deviation f^2_{rms} produced in the transmitter modulator. In order to maximize the performance of an FM system, it is essential to control the rms deviation, which in turn is controlled by the speaker volume. The compandor is a device that monitors the speaker volume so that the rms deviation applied to the modulator is controlled. It has been shown that in the absence of some type of control in the modulator, talker and microphone variability will result in a log-normal distribution for f^2_{rms}, which has a standard deviation around 5 dB. There are two types of problems that can result due to variability of microphone and the talker. Weak talkers will suffer a degradation in the receiver voice band signal-to-impairment ratio, which is directly proportional to the reduction in f^2_{rms}. On the other hand, a loud talker will produce distortion in the transmitter due to clipping. Clipping is used in the transmitter to limit the peak instantaneous frequency deviation so that the adjacent-channel interference effects can be controlled. It has been verified that the ratio of about 13 dB between peak and rms frequency deviation is a reasonable balance between clipping distortion and impairments introduced by noise, random FM, and interference.

This compression and expansion for FM in cellular systems is implemented at a syllabic rate. The syllabic compandor and expansion are matched-pair devices with controlled time constants. Both are variable gain devices, as shown in Figure 9.36. The signal dependent gain of the compressor is matched by a complementary signal-dependent loss of the expandor so that the speech may be transmitted without any level changes. This matching is done by stabilizing the operating point of each of the devices and the proper time constants for its gain control. The compressor has a nominal attack time of 3 ms and a nominal recovery time of 13.5 ms as defined by CCITT. The nominal reference level to the compressor is a 1,000-Hz acoustic tone at the expected nominal speech volume level, which provides ±2.9-kHz peak frequency deviation at the transmitter output in the U.S. AMPS system. As shown in the block diagram of the 2:1 compressor (Figure 9.36), for a low value of e_i output, e_o is low. Reverse action takes place for high-volume speech input when finally the gain is set to a lower value by the feedback signal at the amplifier input. This output signal is passed through a half-wave rectifier and a filter with 20-ms time constant, which is used to increase the gain of the amplifier.

Burst noise lasting less than 20 ms does not change the gain setting of the forward adjustable gain of the amplifier. Also, due to sufficiently slow response time, the compressor does not respond to level changes that occur significantly faster than the time constant of 20 ms, but will only react to the slower syllabic variations. The bandpass filter that precedes the compressor serves to band limit the speech signals so that the out-of-band speech energy does not influence the gain setting of the compressor. The characteristics of a 2:1 compressor are shown in Figure 9.37. Since the slope of the compressor is 1/2, for every 2-dB level change at the input signal, power at the output changes by 1 dB. The receiver expandor

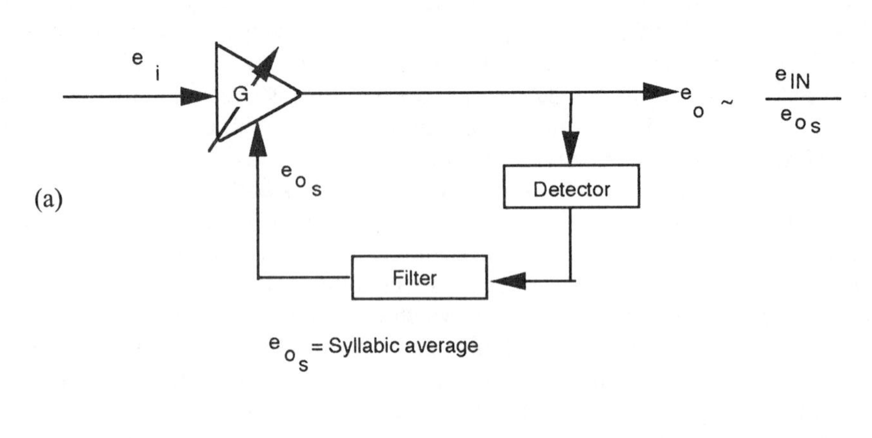

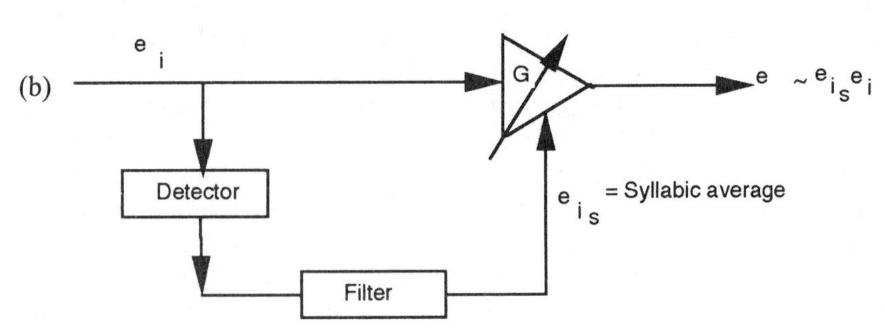

Figure 9.36 Compandor operation (a) 2:1 compressor, (b) 2:1 expandor.

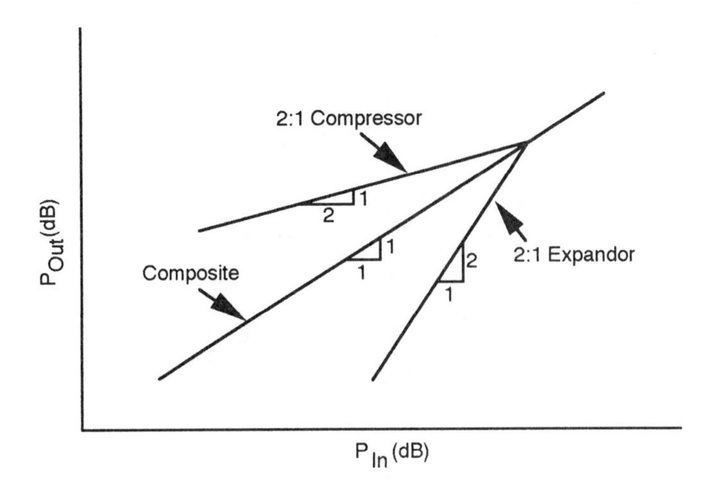

Figure 9.37 Input/output characteristics of compandor.

not only removes transmitter predistortion created by the compandor compressor, but also suppresses receiver noise and interference relative to speech signals, thereby quieting the receiver. The expandor is nominally at a high loss state in the absence of speech signal. High-level burst noise is larger in magnitude than levels corresponding to nominal talkers and are eliminated in the FM discriminator itself due to the amplitude limiter and thus may not effect the operation of the expandor. The composite characteristics of the compressor and expandor leave the original power level of the speech unchanged.

9.2.6.2 Pre-emphasis and De-emphasis

A detailed analysis of the noise response of the FM discriminator suggests that the noise spectral density at the output varies as a square of frequency:

$$G_{no} = \frac{\alpha^2 \omega^2 \eta}{A^2} \qquad |f| \le B/2 \tag{9.50}$$

where $\eta/2$ is the two-sided noise spectral density, A is the amplitude of the modulated signal, and B is the bandwidth of the speech signal. Thus, the total noise power at the baseband filter output, as shown in Figure 9.38, will be

$$N_o = \frac{\alpha^2 \eta}{A^2} \int_{-f_M}^{f_M} 4\pi^2 f^2 df = \frac{8\pi^2}{3} \frac{\alpha^2 \eta}{A^2} f_M^3 \tag{9.51}$$

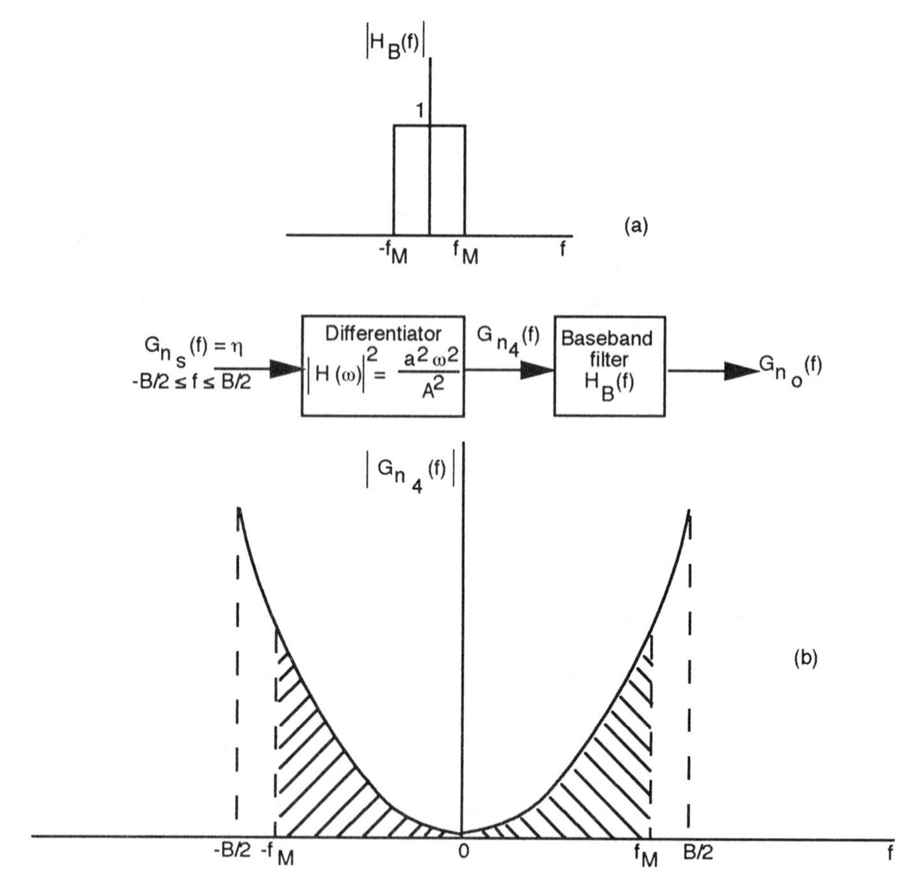

Figure 9.38 (a) Noise density at the baseband filter output; (b) power spectral density at the output of a FM demodulator.

Due to the natural characteristics of speech, the PSD is relatively high at low frequency and falls off at the higher end of the spectrum. In other words, most of the speech power is concentrated below 1,600 Hz and there is relatively less power around 3 kHz. Thus, the recovered speech will have a lower SNR.

One way to reduce the effect of noise at the receiver output is to include a filter after the discriminator, which will de-emphasize the noise at $1/\omega^2$ rate. This is desirable, but if nothing else is done this de-emphasis filter will degrade the high-frequency performance of speech also. The way to circumvent this problem is to include a pre-emphasis network in the transmitter, which will emphasize the speech and correspondingly de-emphasize the speech in the receiver in a way that the composite effects of both networks are nullified. Thus, the transmitter filter must have a transfer function that is the inverse of the receiver transfer function, that is,

$$H_d(\omega) = 1/H_p(\omega) \tag{9.52}$$

The pre-emphasis filter has no effect on channel noise. However, noise is reduced by the de-emphasis network in the receiver. The FM system configuration showing the exact placement of the pre-emphasis and de-emphasis networks is shown in Figure 9.39.

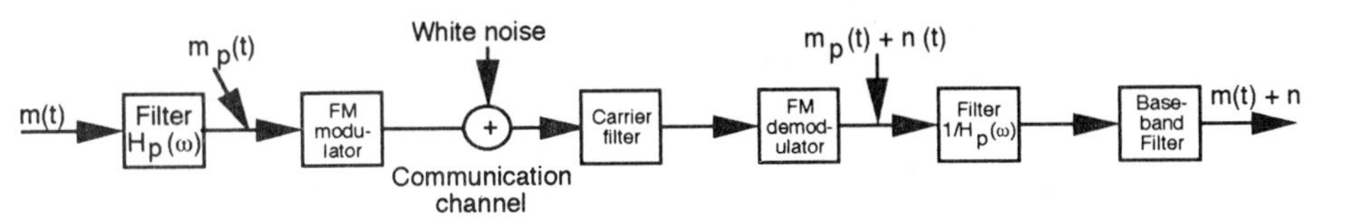

Figure 9.39 Pre-emphasis and de-emphasis networks in an FM system.

Since the signal is unaffected by these two networks in cascade, the signal-to-noise improvement due to these is simply the ratio of total noise power (9.51) in the baseband without de-emphasis and with de-emphasis network. This ratio, R, can easily be computed:

$$R = \frac{\left(\frac{\alpha}{A}\right)^2 (4\pi^2\eta) \int_{-f_M}^{f_M} f^2 df}{\left(\frac{\alpha}{A}\right)^2 (4\pi^2\eta) \int_{-f_M}^{f_M} \frac{f^2}{|H_p(f)|^2} df} \tag{9.53}$$

The characteristic of $H_p(f)$ has a rising response of +6 dB per octave between 300 Hz and 3 kHz, as shown in Figure 9.40. The corresponding de-emphasis network response is dropping with the same slope of − 6 dB per octave. Thus, these networks simply boost the response in the transmitter and correspondingly reduce the response in the receiver. This arrangement of filters can easily provide signal-to-noise improvement nearly equal to 5 dB.

9.3 BLOCK CODING

Channel coding is required to combat the bursty nature of the channel and occasional random errors. Techniques used in cellular radio to combat channel errors include using block and convolutional codes along with interleaving. Block codes

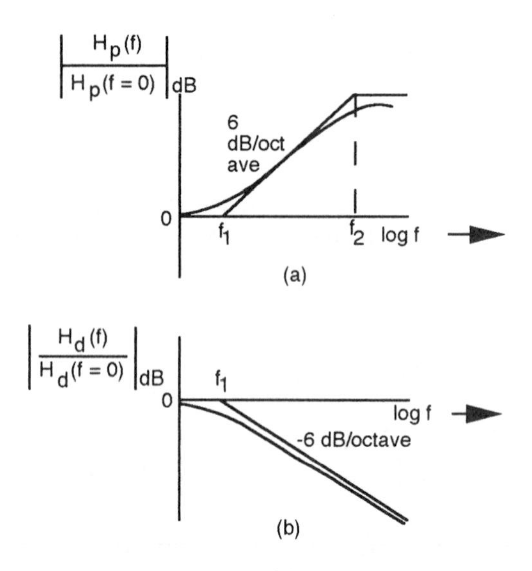

Figure 9.40 Normalized frequency response of the (a) pre-emphasis and (b) de-emphasis networks.

used are BCH and Reed-Solomon (block coding) codes. We shall discuss these block codes in this section along with some background information.

9.3.1 Shannon Bound

The assessment of various modulation techniques for use in mobile communications, bandwidth, power, and their tradeoffs are important. For a Gaussian memoryless channel of nominal bandwidth, B Hz, received signal power, P, and noise spectral density, N_o, the channel capacity is given by Shannon's formula.

$$C = B \log_2(1 + P/N_o B) \text{ bits/sec} \tag{9.54}$$

If R bits/sec is the actual data rate transmitted over a channel with capacity C, then for $R < C$, some coding, modulation, demodulation, and decoding exists to yield an arbitrarily small error probability. For $R > C$, the error probability will be bounded away from zero.

$$\begin{aligned} C_\infty &= \lim_{B\to\infty} C = \lim_{B\to\infty} \log_2\left(1 + \frac{P}{N_o B}\right)^B \\ &= \lim_{B\to\infty} \frac{1}{\ln 2} \ln 2\left(1 + \frac{P}{N_o B}\right)^B = \frac{P/N_o}{\ln 2} \end{aligned} \tag{9.55}$$

As

$$\lim_{n \to \infty} \ln(1 + x/n)^n = x/n$$

Thus, $E_b/N_o = P/N_oR = P/N_oC_\infty = \ln 2 = 0.693 = -1.59$ dB, which is the Shannon limit. Below -1.6 dB, there is no error-free performance, even if B is arbitrarily large or R is vanishingly small. Close to this value, the performance is likely to be expensive. Real systems operate above -1.6 dB. Probabilistic decoding at the receiver is the most attractive way to attain Shannon bound. Probabilistic decodings are sequential decoding and Viterbi decoding. Coding gain is obtained at the expense of an increase in transmission BW. Performance of coding for different schemes is shown in Figure 9.41 and Table 9.3.

Table 9.3
Coding Gain for Several Coded Systems

System	*Required Value of E_b/N_o at $P_b = 10^{-5}$ (dB)*	*Coding Gain (dB)*
Ideal PSK (no coding)	9.6	—
BCH (15, 7)	8.7	0.9
Golay (24, 12)	7.6	2.0
BCH (128, 112)	7.5	2.1
Threshold ($r = 3/4$)	7.4	2.2
Viterbi ($K = 7$, $r = 1/2$)		
Hard quantization	6.5	3.1
Soft quantization	4.5	5.1
Sequential $K = 41$, $r = 1/2$		
Hard quantization	4.4	5.2

Some basic fundamentals of coding include gain over uncoded system due to hard and soft decision, code rate, redundancy, weight and distance, throughput of the coded system, and coded gain. We shall discuss these below.

9.3.2 Hard and Soft Decisions

In an uncoded system, the correlator output is compared with the decision threshold to declare if the binary 1 or 0 is present. We call this the binary quantization and say that a hard decision has been made on the correlator output as to which level was actually transmitted. With $Q = 2$ (hard decisions), a channel is regarded as a binary symmetric channel (BSC) with $P(0/0) = P(1/1)$ denoting the probabilities

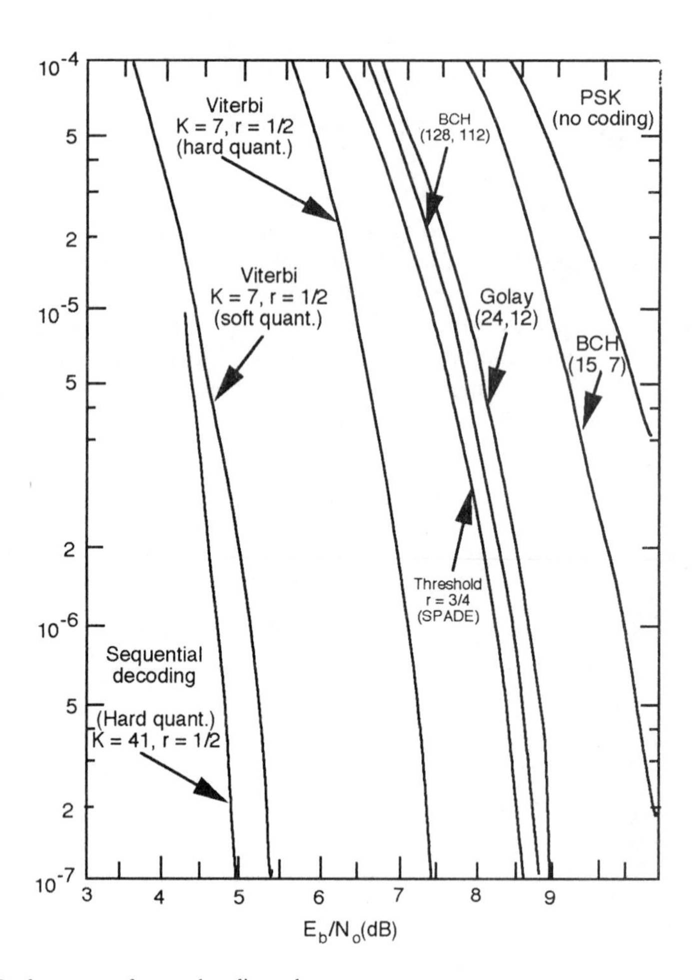

Figure 9.41 Performance of several coding schemes.

of correct reception, conditioned on a 0 and 1 having been transmitted, respectively. On the other hand, the probability of making an incorrect decision is given by $P(0/1)$ and $P(1/0)$, which correspond to events that the receiver declares the received bit to be 0 when 1 is transmitted and declares the bit to be 1 when 0 is being transmitted. The binary symmetric channel is represented in Figure 9.42. For coded signal, it is desirable to keep an indication of how reliable the decision was. A soft-decision demodulator first decides whether the output voltage is above or below the decision threshold and then computes a "confidence" number, which specifies how far from the decision threshold the demodulator output is. This process also

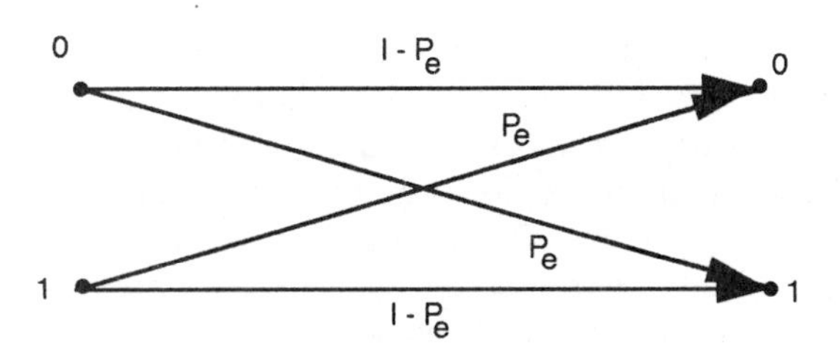

Figure 9.42 Binary symmetric channel.

makes it possible to predict the BER by observing the signals over a shorter period of time.

An example of three-bit quantization is shown in Figure 9.43. The input to the demodulator is binary, while the output is 8-ary delineated by one decision threshold and four pairs of confidence thresholds. The information available to the decoder in this case is considerably increased as demonstrated below by an example of an antipodal signal. The channel resulting from three bit quantization on a Gaussian channel is called the binary input, 8-ary output, and is shown in Figure 9.44.

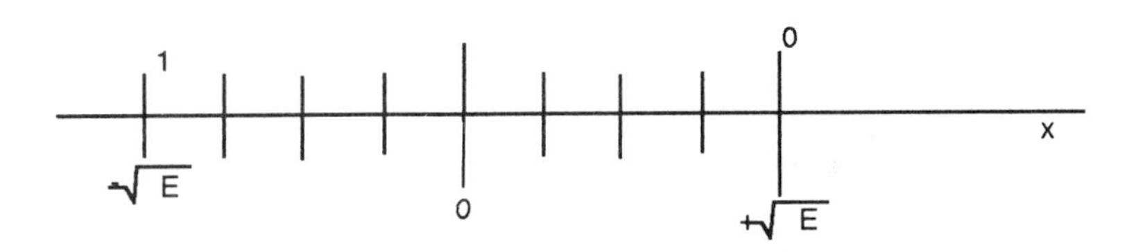

Figure 9.43 Confidence thresholds for antipodal signal.

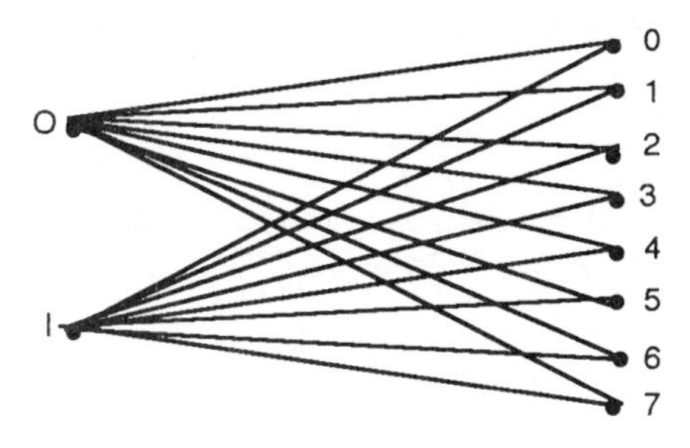

Figure 9.44 Eight-level soft quantized DMC produced by a three-bit quantizer on a Gaussian channel.

Each of these regions is assigned a number between 0 and 7, inclusive, as shown in Figure 9.45. When the received signal lies between $\pm\sqrt{E/2}$, we consider the channnel to be performing poorly. The region of poor performance has been designated as a flunk region. Since the threshold for a flunk is one-half the nominal magnitude of a received bit, we shall call this $\alpha = 0.5$ flunk. Whenever the magnitude of the received bit is less than or equal to $\sqrt{E/2}$, one can count the flunk regardless of whether it was a correctly received bit or not. We describe below the relationship between $\alpha = 0.5$ flunks and the BER. We first derive the expected value of $\alpha = 0.5$ flunk rate as a function of SNR. The PDF of the received signal is given by

$$\begin{aligned} f(x) &= f(x/-\sqrt{E} \text{ was sent}) \times \Pr(-\sqrt{E} \text{ was sent}) \\ &\quad + f(x/\sqrt{E} \text{ was sent}) \times \Pr(\sqrt{E} \text{ was sent}) \\ &= N(-\sqrt{E}, \sigma^2) \times 0.5 + N(\sqrt{E}, \sigma^2) \times 0.5 \end{aligned} \tag{9.56}$$

where $N(\mu, \sigma^2)$ is defined as

$$\frac{1}{\sqrt{2\pi\sigma^2}} e^{-[(x-\mu)^2/2\sigma^2]} \tag{9.57}$$

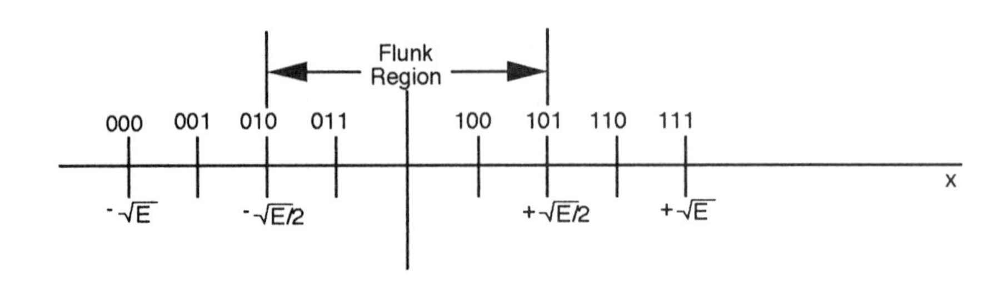

Figure 9.45 Three-bit quantization of an antipodal signal.

Figure 9.46 shows the hatched area of the curve that corresponds to $\alpha = 0.5$ flunk. The probability of $\alpha = 0.5$ flunk error is given by

$$\frac{1}{\sqrt{2\pi\sigma^2}} \left[\frac{1}{2} \int_{-\sqrt{E}/2}^{\sqrt{E}/2} e^{-(x+\sqrt{E})^2/2\sigma^2} dx + \frac{1}{2} \int_{-\sqrt{E}/2}^{\sqrt{E}/2} e^{-(x-\sqrt{E})^2/2\sigma^2} dx \right] \tag{9.58}$$

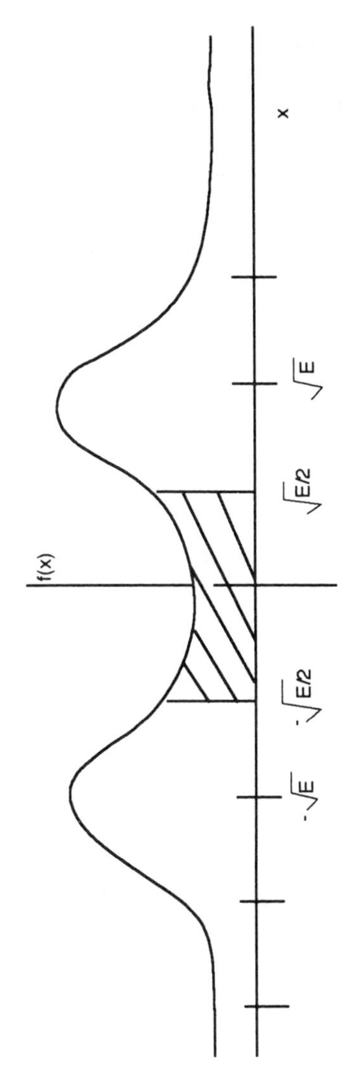

Figure 9.46 Probability distribution $f(x)$.

Due to symmetry, the first term can be combined with the second term, and the probability of a flunk is

$$Q(\sqrt{E}/2\sigma) - Q(3\sqrt{E}/2\sigma)$$

Making the substitution $N_o = 2\sigma^2$ reduces the above equation to

$$Q\sqrt{E/2N_o} = Q\sqrt{4.5\ E/N_o} \tag{9.59}$$

The BER probability can similarly be arrived at by integrating between 0 and $\pm\infty$ as shown below:

$$\frac{0.5}{\sqrt{2\pi\sigma^2}}\int_0^{\infty}\frac{e^{-(x+\sqrt{E})^2}}{2\sigma^2} + \frac{0.5}{\sqrt{2\pi\sigma^2}}\int_{-\infty}^{0}\frac{e^{-(x-\sqrt{E})^2}}{2\sigma^2} = Q(\sqrt{2E/N_o}) \tag{9.60}$$

Plots of $\alpha = 0.5$ flunk and BER are shown in Figure 9.47. These curves clearly show that one can predict the BER by monitoring the $\alpha = 0.5$ flunk curve for a short duration. Thus, the decoder can monitor the BER accurately.

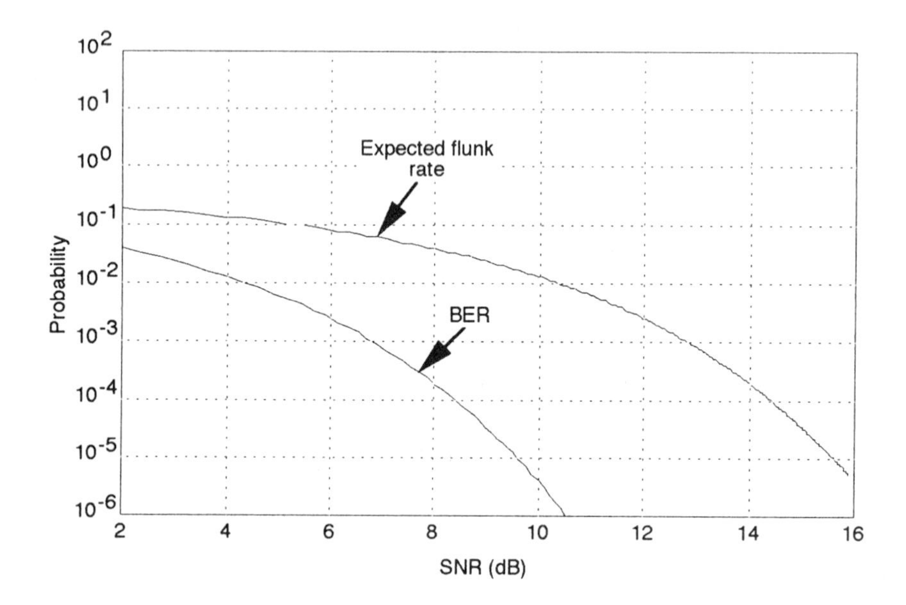

Figure 9.47 True BER and expected flunk rate versus SNR.

9.3.3 Code Rate, Redundancy, Weight, and Distance [14–15]

In the case of block codes, the data is segmented into blocks of k messages or information bits and each block can represent any one of 2^k distinct messages. The encoder adds $(n - k)$ bits and forms a block n bits long, which are called code bits or channel symbols. These $(n - k)$ added bits are known as redundant bits, parity bits, or check bits and they carry no information. The code is referred as (n, k) code. The ratio $(n - k)/k$ within a block is called the redundancy of the code and the ratio of data bits to total number of bits, k/n is called the code rate. The code rate is the portion of a code bit that constitutes information. Thus for 3/4 rate code, there are three bits of information for every four bits of code word. Thus, the redundancy in this case is 33% and the bandwidth expansion is 4/3. On the other hand, for 1/2 rate code, each code bit carries 1/2 bit of information, both the redundancy and the bandwidth expansion is 100%. The Japanese cellular system uses (43, 31) and (11, 7) BCH codes for forward and reverse channels. Thus, the code rates are 31/43 and 7/11 and the redundancies are 38% and 57%, respectively. A block diagram of coder and decoder, including channel modulator and demodulator, is shown in Figure 9.48.

Besides the code rate k/n, an important parameter of a code word is its weight, which is simply the number of nonzero elements that it contains. When all the M code words have equal weight, the code is called a fixed weight or a constant-weight code. Let us consider two code words C_j and C_i in an (n, k) block code. The distance between two code words is defined as the measure of the difference between them or the number of positions where they differ. This measure is called

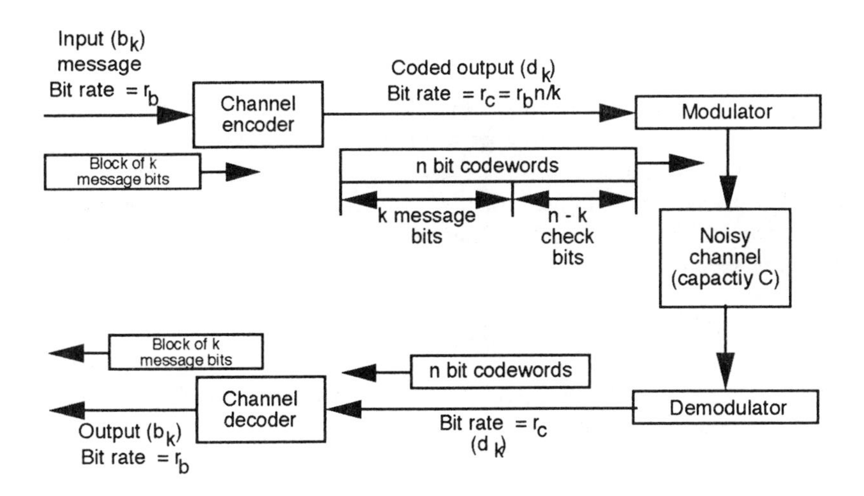

Figure 9.48 Block diagram of a general coder/decoder.

the Hamming distance $d(j, i)$. Clearly, $d(j, i)$ for $j \neq i$ satisfies the condition $0 < d(j, i) < n$. The smallest distance value of $d(j, i)$ among all code words is called the minimum distance of the code and is denoted as $d_{min}(j, i)$. Minimum distance is important as it represents the weakest link in a chain and provides the minimum capability, which provides the strength of the code. The Hamming distance for (40:28) and (48:36), used in the forward and reverse channels of AMPS, has a minimum distance of 5.

Example 9.4

If two codes j and i, are represented as

$$j = 1\,0\,0\,1\,1\,0\,1\,0\,1\,1\,1$$
$$i = 1\,1\,0\,0\,1\,1\,0\,1\,0\,1\,0$$

Thus, $d(j, i) = 7$. By the property of modulo-2 addition, we conclude that the sum of two binary vectors is another vector whose binary ones are located in those positions in which the two vectors differ. Thus, in the above example,

$$j + i = 0\,1\,0\,1\,0\,1\,1\,1\,1\,0\,1$$

From this, one can note that the Hamming distance between two code vectors equals the Hamming weight of their sum, that is, $d(j, i) = W(j + i) = 7$.

9.3.4 Throughput and Coding Gain

Two major coding schemes used are the block code and convolutional code. Block code designated as $B(n, k)$ contains k bits of information out of the total n bits transmitted. On the other hand, for convolutional code designated as $C(n, k)$, for k bits fed to the encoder input block, n coded bits appear at the encoder output. Thus, the throughput R_{th}, which is defined as a fraction of the information bit rate, can be expressed in both cases as

$$R_{th} = Rk/n \tag{9.61}$$

where R is the data rate without coding. Another important parameter of the coding is its gain, defined as

$$CG = (E_b/N_o)_{unc} - (E_b/N_o)_{cod} \tag{9.62}$$

Here, E_b is the average energy per bit and N_o is the noise power density. E_b/N_o is related to CNR as

$$\text{CNR} = P_T/N = E_b R/N_o B \tag{9.63}$$

Comparing E_b/N_o with and without coding provides coding gain. C-450 uses (15:7) BCH code and, as shown in Table 9.3, provides a coding gain of 0.9 dB to BPSK data (9.6–8.7 = 0.9dB). Usually a coding gain of 3–5 dB is required and increases as the error rate performance is improved. Coded versus uncoded performance for coherent PSK with (127, 92) and (24, 12) is shown in Figure 9.49. As seen in the curve, coded performance is the worst as compared to the uncoded case at low values of BER. Improved performance is only seen after a certain value of E_b/N_o, known as the threshold value. Below threshold, the additional bits due to coding

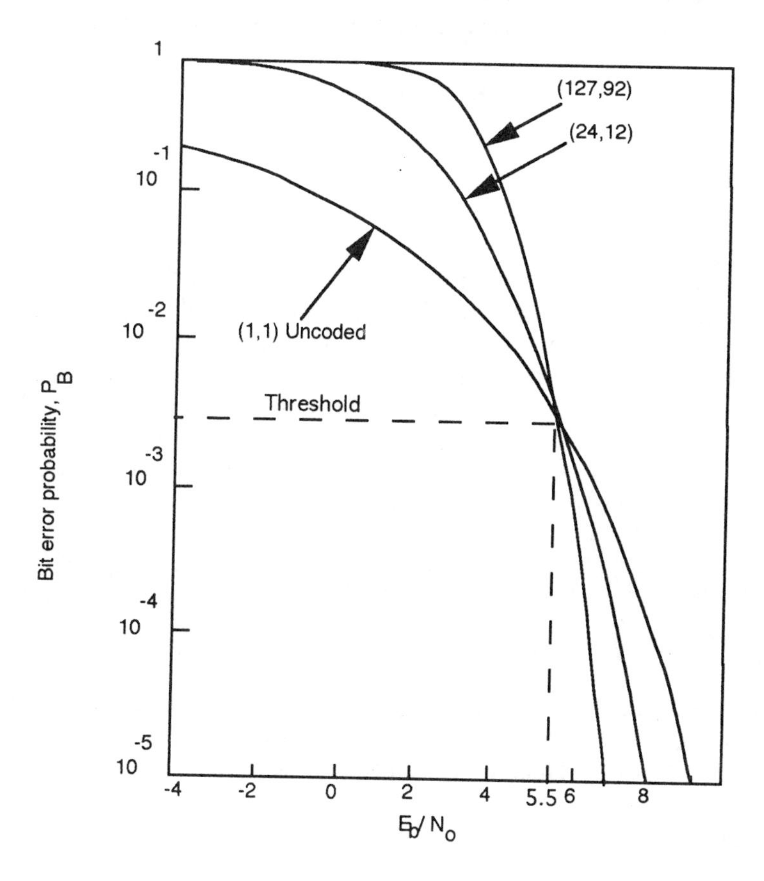

Figure 9.49 Coded and uncoded performance of coherent PSK. After [6, p. 267].

simply act as excess baggage without the ability to improve the performance. This can be conceptualized, as in this region excess overhead bits simply result in reduced energy per bit compared to the uncoded case. Once the threshold is exceeded, performance improvement of the code more than compensates for the reduction of energy per coded bit. Thus in the above example, once E_b/N_o exceeds the threshold value of about 5.5 dB, the positions of these curves are reversed. Coding gain, as seen from these curves, is nearly equal to 2.5 dB and 1.5 dB at a BER of 10^{-5}. Let us take up an example to clarify this point.

Example 9.5

We take up an example of (63, 52) BCH code with two error correcting capabilities to illustrate how to estimate the BER after decoding. Let E_b/N_o be the SNR per bit without coding and E_s/N_o be the SNR of the coded bit or symbol. Then,

$$E_s/N_o = E_b R/N_o \tag{9.64}$$

or

$$E_s/N_o = E_b/N_o + 10 \log_{10} R \tag{9.65}$$

where E_s/N_o is the coded bit energy-to-noise density ratio at the receiver. Since $E_s/N_o < E_b/N_o$, more symbols received will be in error than in the uncoded system. However, if the coding is done properly to combat channel degradation, a large number of channel errors will be compensated for by error correcting capability, resulting in a superior performance than with the coded system.

Since $R = k/n = 52/63$, and assuming $E_b/N_o = 8$ dB, then $E_s/N_o = 8 + 10 \log_{10}(52/63) = 7.2$ dB. From Figure 9.50, the BER prior to decoding is nearly equal to 6×10^{-4}. The coded word error probability is

$$\begin{aligned} P_{CW} &= \sum_{i=3}^{63} \binom{63}{i} p_s^i (1 - p_s)^{63-i} \approx \binom{63}{3} (6 \times 10^{-4})^3 \, (1 - 6 \times 10^{-4})^{60} \\ &= 8.3 \times 10^{-6} \end{aligned} \tag{9.66}$$

The most likely case of the coded error occurs with three errors. Since the Hamming distance is 5 ($t = 2$, $D = 2 \times t + 1$), the received word is erroneously corrected with two additional errors. Thus, there can be 5 decoded bits in errors out of 63 or at most 5 errors out of 52 information digits. Thus, an estimate of probability of bit error after decoding is

$$\frac{5}{63} P_{CW} = 6.6 \times 10^{-7} \tag{9.67}$$

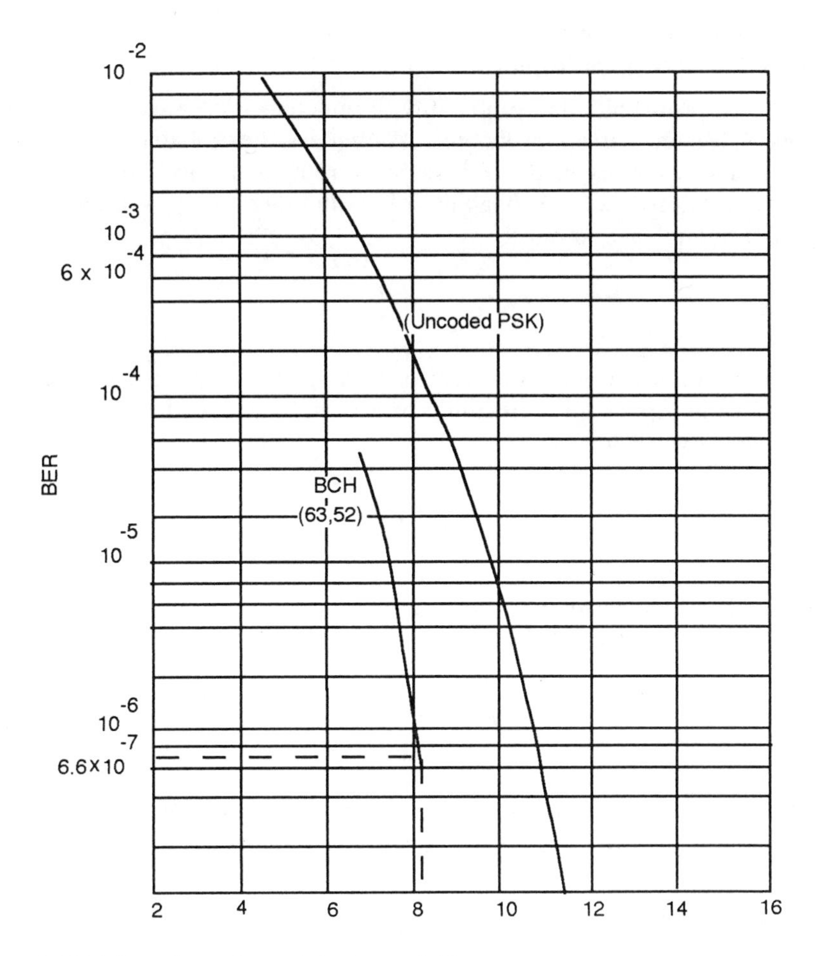

Figure 9.50 BER for coded and uncoded coherent PSK.

Without coding, the SNR required to achieve this BER is nearly equal to 10.5 dB. With coding, the SNR required is nearly equal to 8 dB. Thus, a savings of 2.5 dB due to coding. Therefore, the transmitter power can be reduced by 2.5 dB to achieve the equivalent BER performance of the uncoded system.

9.3.5 Message Requirements in Coded Communication

In the coded communication, there are various message parameters of interest:

- Message error probability;
- Probability of undetected message error;

- Missed message probability;
- Spurious message probability.

Message error probability and probability of undetected message are discussed in detail below. Missed message probability is defined as the probability that a message is rejected due to errors in the checksum or redundant bits. Spurious message probability is defined as the probability of a valid message generation in a given time period due to random noise.

Message Error

For the coded information, word error or message error rate plays an important part. For a code capable of correcting t or less bits in error, the message or word error rate P_M is given by

$$P_M = \sum_{j=t+1}^{n} \binom{n}{j} p^j (1 - p)^{n-j} \tag{9.68}$$

where p is the channel symbol rate. The parameter t is related to the minimum distance between code words by

$$t = \left[\frac{D_{\min} - 1}{2}\right]_I \tag{9.69}$$

where $[]_I$ is the integer of the bracketed quantity. For example, for a block code (36, 25) capable of correcting a single error, the message error rate becomes

$$P_M = \sum_{j=2}^{36} \binom{36}{j} p^j (1 - p)^{36-j} \tag{9.70}$$

For small bit error probability p, the first term dominates. Thus, (9.70) can be approximated as

$$P_M \approx \binom{36}{2} p^2 (1 - p)^{34} \tag{9.71}$$

As a second example of AMPS, since $D_{\min} = 5$, the parameter $t = 2$ from (6.69). Thus,

$$P_M \approx \binom{40}{3} p^3 (1 - p)^{37}$$

for forward link with $n = 40$ and

$$p_m \approx \binom{48}{3} p^3 (1 - p)^{45}$$

for the reverse link. From the message error probability, it is always difficult to calculate the bit error probability, as it depends on the particular code and decoder. However, for systematic code the bit error probability can be estimated by:

$$P_b = \frac{1}{n} \sum_{j=t+1}^{n} \alpha_j \binom{n}{j} p^j (1 - p)^{n-j} \tag{9.72}$$

where α_j is the average number of symbol errors remaining in the corrected received sequence given that the channel caused j symbol errors. The value of $\alpha_j = 0$ for $j \leq t$. When $j > t$, α_j can be bounded by noting that when more than t errors occur, the decoder can at most correct t errors. In other words, it can at best correct t errors and at worst add t errors. Thus,

$$j - t \leq \alpha_j \leq j + t; \qquad j > t$$

We shall look at an example to prove this point.

Probability of Undetected Error

Another important performance measure for the error correcting code is the probability of undetected error in the message code word. It is defined as

$$P_{nd} = \sum_{j=1}^{\substack{n \neq 2 \text{ (for even } n) \\ [n-1]_I \text{ (for } n \text{ odd)}}} \binom{n}{2j} p^{2j} (1 - p)^{n-2j} \tag{9.73}$$

We shall compute the probability of undetected error by taking up an example of (4, 3) even parity code.

Example 9.6

For the even parity code word (4, 3) shown in Table 9.4, compute the probability of undetected error. Assume the probability of channel symbol error $p = 10^{-2}$. Since this code will detect all single and triple error, the probability of undetected

Table 9.4
Even Parity (4, 3) Code

Message	*Parity Bit*	*Code*	*Word*
000	0	0	000
001	1	1	001
010	1	1	010
011	0	0	011
100	1	1	100
101	0	0	101
110	0	0	110
111	1	1	111

error will be when two or four bits are in error in a code word. Thus, the probability of undetected error in a code word is given by

$$\begin{aligned} P_{nd} &= \binom{4}{2} p^2(1 - p)^2 + \binom{4}{4} p^4 = 6p^2(1 - p)^2 + p^4 \\ &= 6(10^{-2})^2(1 - 10^{-2})^2 + (10^{-2})^4 = 5.9 \times 10^{-4} \end{aligned} \tag{9.74}$$

For block codes used for error detection only, the decoder fails to detect an error in the block only when the error sequence in the received code word is such that it looks like another valid code word. If A_j is the number of code vectors of weight j within an (n, k) linear code, the numbers $A_1, \ldots, A_n$ are called the weight distribution of the code. For BSC, the probability of undetected error can be arrived at by the weight distribution of the code as follows:

$$P_{nd} = \sum_{j=1}^{n} A_j p^j (1 - p)^{n-j} \tag{9.75}$$

where p is the error transition probability in BSC. If the minimum distance is $d_{\min}$, the values of A_1 to $A_{d\min} - 1$ are all zero.

Example 9.7

For (6, 3) error detection code, calculate the probability of undetected error in BSC with an error transition probability of 10^{-2}. In this example, there will be four coded words with weight 3 and three coded words with weight 4. Thus, the probability of undetected error is given by

$$\begin{aligned} P_{nd} &= 4p^3(1-p)^3 + 3p^4(1-p)^2 \\ &= 4(10^{-2})^3(1-10^{-2})^3 + 3(10^{-2})^4(1-10^{-2})^2 \\ &= 3.9 \times 10^{-6} \end{aligned} \tag{9.76}$$

We shall now discuss two important techniques, namely majority voting and interleaving for burst error randomization used in cellular radio.

9.3.6 Majority Voting Process (Time Diversity)

As discussed in Section 8.2.1, time diversity can be used to improve the system performance under fading channel. Majority voting is a simple unweighted form of time diversity. It relies on the fact that if a given data bit is received at a time when the signal is under fade, it is likely to be received in error. However, this situation is highly unlikely to recur if the bit is repeated at a later time. If the time intervals between the repeated bits of information are comparable with the reciprocal of the average fading rate, the fade levels associated with the various repetitions will essentially be independent and appropriate combinations of repetitions will give a diversity improvement. The process best known and most widely used for processing signals is the time diversity technique known as majority voting. It assumes that each word is transmitted J times (J an odd number). Each received bits of J words must be aligned and a majority voting process can be used to determine each valid message bit. The process is as follows. If $(J + 1)/2$ or more bits among the repeated bits are 1 or 0, then the received bit will be assumed as 1 or 0 accordingly. The resulting majority-voted message words then constitute the improved message stream. The improved bit-error rate is given by

$$\langle P_e' \rangle = \sum_{k=(J+1)/2)}^{J} C_k^j \langle P_e \rangle^k (1 - \langle P_e \rangle)^{J-k} \tag{9.77a}$$

where $\langle P_e \rangle$ is the BER in fading conditions, and

$$C_k^J = \frac{J!}{(J-k)!k!} \tag{9.77b}$$

The value of J in AMPS is 5 for the forward and the reverse control channels.

9.3.7 Interleaving for Burst Error Randomization

When overall average error rates are low but errors are known to occur in bursts, interleaving can be applied to distribute the errors randomly among data words. Figure 9.51 shows N-bit data words transmitted as successive groups of N consecutive data bits. A burst of errors will cause a large number of errors in a single word. However, if the words are interleaved before transmission, a burst of errors will tend to place single bit errors in each of a number of different interleaved words. Given an (n,k) cyclic code it is possible to construct (In, Ik) cyclic interlaced code by arranging I code vectors of the original code into I rows and transmitting them column by column. Let us illustrate this case by an example of a (15, 7) BCH code generated by polynomial $g(x) = x^8 + x^4 + x^2 + x + 1$. This code is cyclic and has the following parameters:

- Block length: $2^m - 1$, where $m = 4$;
- Number of information bits: $k \geq n - mt$, where $k = 7$, $t = 2$.

Thus, the code vector consists of fifteen bits, message length is seven bits long, and the number of bits that can be corrected in a code words is two. Let us construct a (75, 35) interleaved code with interleaved distance $I = 5$, with a burst error correcting capability of 10. Table 9.5 shows the arrangement of code words where each row has a code word fifteen bits long, where there are seven message bits and eight parity bits. These five code words of a length of fifteen bits each are generated by using the generating polynomial $g(x)$ given above. Transmission of these code words will be by columns. From the inspection of the table it is seen that two successive symbols from the same code word at the output of the encoder are separated by five symbols in the transmitted sequence. In this example, the code words are fifteen bits long for two error correcting and if a burst of errors spans no more than five symbols of the transmitted sequence, then it can only affect a single symbol of any given code word, which can be easily corrected. This "spreading out" of the errors in a burst from the point of view of the reconstructed code word at the output of the deinterleaver is the underlying principle of symbol interleaving. In general, if each code word in an array has an error correcting capability of t errors, then a burst of errors spanning $I \cdot t$ encoded symbols can be corrected by using an interleaving I code word where I is the interleaving distance.

Table 9.5
Interleaved Code Words

1	6		21	26		51	56	71
2	7		22	27		52	57	72
3	8		23	28		53	58	73
4	9		24	29		54	59	74
5	10		25	30		55	60	75

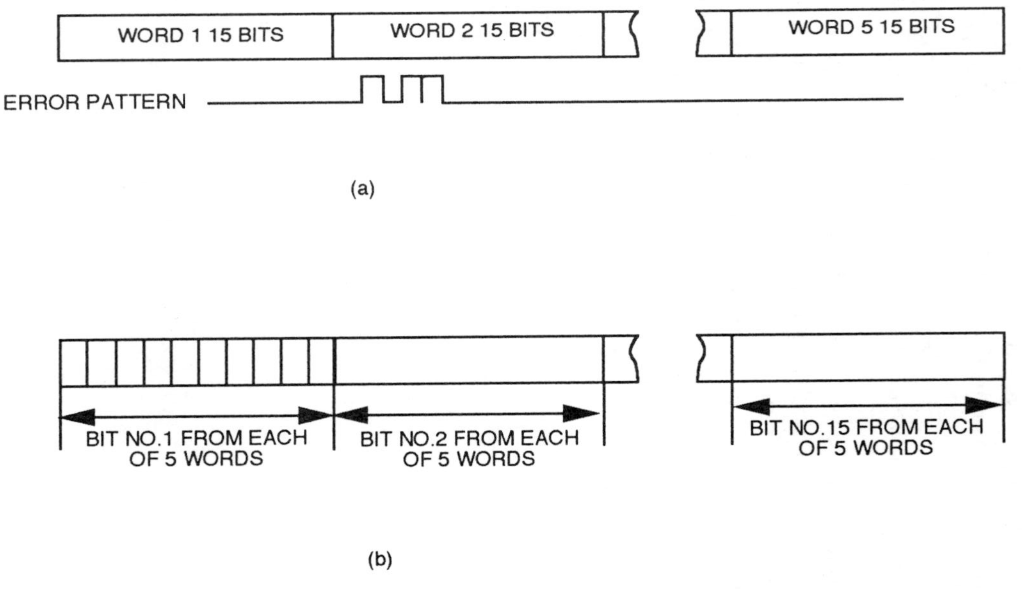

Figure 9.51 Transmission of five interleaved messages: (a) noninterleaved data words; (b) interleaved data words.

Assuming a burst of errors effecting bits 22–29. Since each row has a maximum of two errors, a (15, 7) BCH will be able to correct errors and the received message word will be error free.

As discussed above, the interleaving/deinterleaving function is to convert a nonrandom channel to one that is random. At the transmitter, the interleaving occurs after the encoding process. At the receiver, the deinterleaving takes place prior to the decoding, as shown in Figure 9.52.

Bose-Chaudhuri-Hocquenghem (BCH)Codes [14–15]

BCH codes are a powerful class of codes that have well-defined decoding algorithms and allow quite a selection of block lengths, code rates, error-correcting capability, and alphabet sizes. The most common codes use a binary alphabet, and an encoder output block length of $n = 2^m - 1$. Here, m is a positive integer and assumes values of 1, 2, ... Table 9.6 provides the minimum distance (d_B) of a BCH code and the generating polynomial $g(x)$ for different values of n (total number of bits in the code word, n, up to 31) and k (number of information bits). The coefficients of $g(x)$ are presented as octal numbers arranged so that when converted to binary digits the rightmost digit corresponds to the zero-degree coefficient of $g(x)$. BCH code is important for large block size, where it outperforms all other block codes with the same block length and code rate. Block, bit, and the probability of undetected error P_{nd} can be found from (9.68), (9.72), and (9.75) by finding t_B from (9.69).

$$t_B = \left[\frac{d_B - 1}{2}\right]_I \tag{9.78}$$

Table 9.6
BCH Codes Up to $n = 31$

n	k	d_{BCH}	$g(x)$
7	4	1	13
15	11	1	23
	7	2	721
	5	3	2467
31	26	1	45
	21	2	3551
	16	3	107657
	11	5	5423325
	6	7	313365047

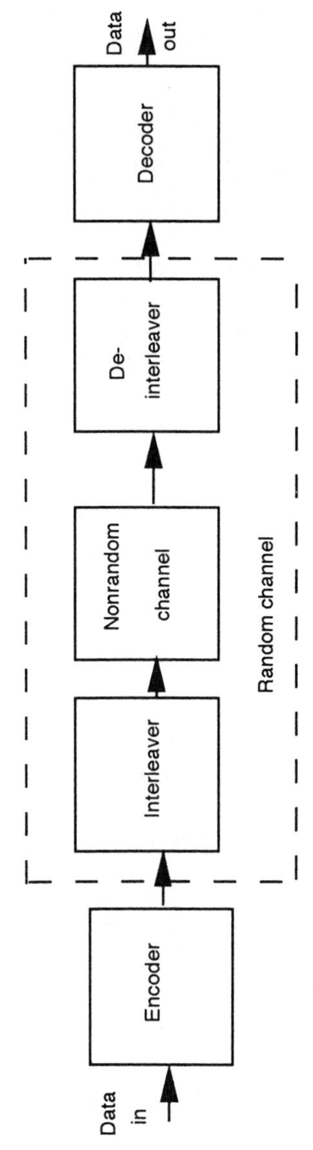

Figure 9.52 Coding using interleaving and deinterleaving.

Block error rate is the direct application of (9.68) with $j = t_{BCH}$ found from the above equation. BER probability, which depends upon the coefficient α_j, is always difficult to estimate as it depends upon the code and the design of the decoder. However, for small channel error rate, approximation is usually obtained using only the first few terms of (9.72). Substituting, $\alpha_i = i$, (9.72) provides the estimate of BER P_b as

$$P_b = \frac{1}{n} \sum_{i=t_B+1}^{n} i \binom{n}{i} p^i (1 - p)^{n+i} \tag{9.79}$$

The probability of undetected error P_{nd} requires estimation of A_j, the number of code vectors of weight j. However, this can be estimated for the first few terms, which provides a good estimate for low channel error rate. The decoding procedure generally uses hard decision decoding due to complexity of soft decision. However, if soft decision is implemented, it provides approximately a 2-dB gain over hard decision decoding. Plots of channel error rate (9.66) versus BER (9.79), with $\alpha_j = j$) for block length $n = 127$ bits capable of correcting five, ten, and fifteen channel errors are shown in Figure 9.53. It should be noted that, for the same channel error rate performance, the improved BER performance is obtained as the number of encoded data bits, k, are decreased. We now turn to an example illustrating a block code generation.

Example 9.8

Using the generator polynomial $g(x) = 1 + x + x^3$, $g(x) = 13$ in Table 9.6, generate a systematic code vector from a (7, 4) code word set for the message vector $m = 1011$. $m(x) = 1 + x^2 + x^3$, $n = 7$, $k = 4$, $n - k = 3$. Thus, $x^{n-k}m(x) = x^3(1 + x^2 + x^3) = x^3 + x^5 + x^6$. Dividing $x^3m(x)$ by $g(x)$ we can write

$$x^3 + x^5 + x^6 = \underbrace{(1 + x + x^2 + x^3)}_{q(x)} \underbrace{(1 + x + x^3)}_{g(x)} + \underbrace{1}_{r(x)} \tag{9.80}$$

Thus, the code vector $U(x)$ is

$$\begin{aligned} U(x) &= r(x) + x^3m(x) = 1 + x^3 + x^5 + x^6 \\ U &= (\underbrace{100}_{r(x)} \quad \underbrace{1011}_{m(x)}) \end{aligned} \tag{9.81}$$

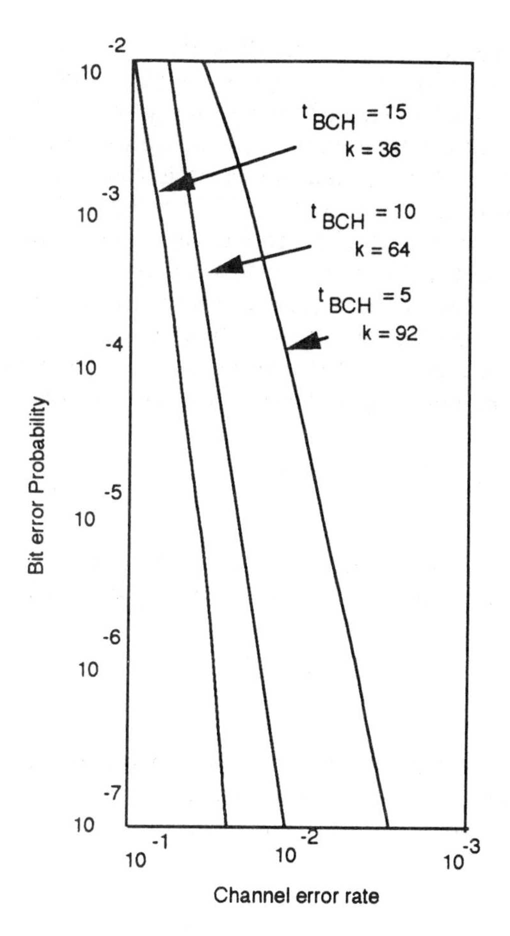

Figure 9.53 Bit error probability versus error rate performance of several length $n = 127$, BCH codes. After [14, p. 87].

Reed-Solomon Coding [14–15]

Reed-Solomon (RS) codes have been successfully used for mobile communications in both bursty and random error environments. RS codes are a class of nonbinary linear block codes that are also a subset of BCH codes. An (n, k) RS code has k information symbols, n coded symbols, $n - k$ redundant symbols (or parity check symbols), and a minimum distance $d_{\min} = n - k + 1$. The rate k/n code can correct t symbols in the n symbol block where $t = (d_{\min} - 1)/2 = (n - k)/2$. An RS code with an alphabet of 2^m symbols has $n = 2^m - 1$ and $k = 2^m - 1 - 2t$ where

$m = 2, 3, \ldots$. The advantage of an RS code can be seen by considering the following example. Consider a binary $(n, k) = (15, 9)$ code. The entire n-tuple space amounts to $2^n = 2^{15} = 32{,}768$ binary words, of which $2^k = 2^9 = 512$ message binary words. Thus, 1/64 of the n-tuple are code words. Now consider a nonbinary RS code with $(n, k) = (15, 9)$ code where each symbol is comprised of $m = 3$ bits. The n-tuple space amounts to $2^{nm} = 2^{15\times 3} = 3.518437 \times 10^{13}$ binary words, of which $2^{km} = 2^{9\times 3} = 134{,}217{,}728$ are message words. Thus, the ratio of code words to possible n-tuple is $1/(2.62144 \times 10^5)$, a very large number when compared to binary code transmission. The ratio will also decrease as the symbol size m increases. When a small fraction of the n-tuple space is used for code words, a large $d_{\min}$ can be achieved and thus a larger error correcting capability. Weights α_k, which satisfies the minimum distance $d_{\min}$ of an RS code, is given by [14–15]:

$$\alpha_k = \binom{2^m - 1}{k}(2^m - 1) \sum_{i=0}^{k-2t-1} (-1)^i \binom{k-1}{i} 2^{k-i-2t-1} \tag{9.82}$$

for $k \geq 2t + 1$. $\alpha_0 = 1$; $\alpha_k = 0$ for $1 \leq k \leq 2t$. From (9.72), decoded symbol error probability P_{sym} in terms of channel symbol error probability P_{sym} is given by

$$P_{\text{sym}} \cong \frac{1}{2^m - 1} \sum_{i=t+1}^{2^m - 1} i \binom{2^m - 1}{i} p_s^i (1 - p_s)^{2^m - 1 - i} \tag{9.83}$$

The weight α_i (9.72) is replaced by i. This is a good approximation for $i > t$. The bit error probability can be upper bounded by the symbol error probability for specific modulation types. For MFSK modulation with $M = 2^m$, the relationship between P_{sym} and P_b is given by

$$P_b/P_{\text{sym}} = 2^{(m-1)}/(2^m - 1) \tag{9.84}$$

Figure 9.54 shows the bit error probability versus channel symbol error probability obtained by using the above equation for $n = 31$ code capable of correcting channel errors $t = 1, 2, 4$, and 8.

9.4 SPEECH CODING [16–18]

In this section we shall describe two new vocoders, namely the residual excited linear prediction (RELP) vocoder and the code excited linear prediction (CELP) coder. These coders are based on analysis-by-synthesis techniques. These techniques model the speech signal as some excitation signal and a vocal tract filter model, which processes the excitation. The vocal tract filter is developed by application of some form of linear predictive coding (LPC) and a set of parameters

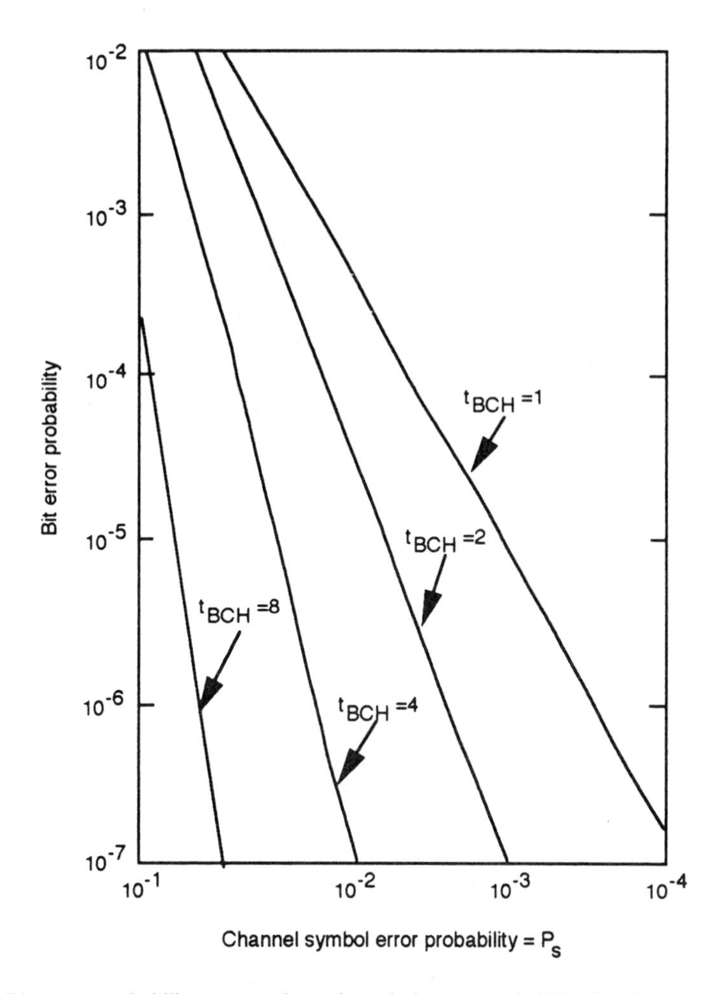

Figure 9.54 Bit error probability versus channel symbol error probability for 32 orthogonal signal modulation and $n = 31$, error correcting Reed-Solomon coding. After [14, p. 91].

chosen for the efficient transmission of the filter description. The excitation is handled by one of a variety of methods, leading to different names like RELP and CELP. Before we describe these two coders, let us establish the requirements in speech coding from the cellular point of view.

9.4.1 Transmission Rates in Speech Coding

Figure 9.55 shows a spectrum of speech coding transmission rates currently of interest. It highlights the dichotomy between nonspeech-specific waveform coders

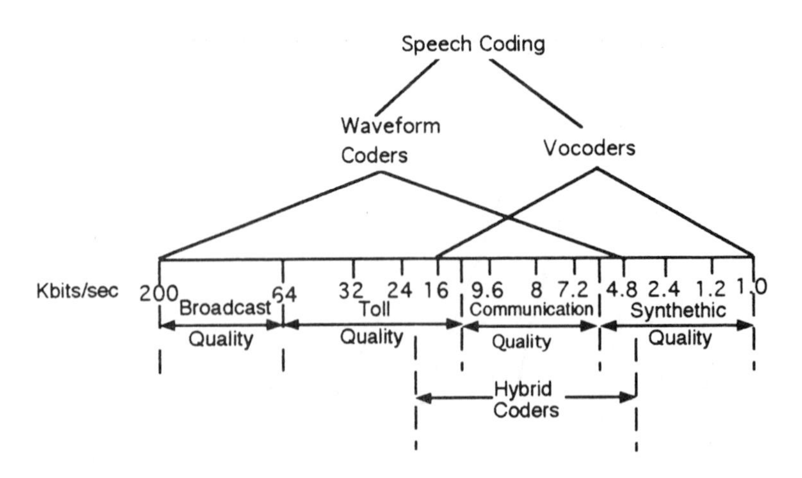

Figure 9.55 Classification of speech coders. After [23].

that need relatively higher transmission rates, and speech-specific vocoders for digitization at relatively lower bit rates. Figure 9.55 also indicates the overlapping region where hybrid coding is done. Hybrid coding utilizes suitable combinations of waveform coding techniques and vocoder techniques. The figure displays the quality of speech reproduction that can presently be attained at a prescribed bit rate. The quality characterizations are: broadcast, toll, communication, and synthetic.

Commentary or broadcast quality refers to a wide bandwidth (between 0 and 7,000 Hz) high-quality speech with no perceptible noise and requires a data rate in excess of 64 Kbps. Toll quality describes speech as heard over the switched telephone network (200–3,200 Hz with an SNR greater than or equal to 30 dB and harmonic distortion less than or equal to 2–3%). At the present time we know of no way, with moderate complexity, to achieve toll quality at rates much below 16 Kbps. Communication quality speech is highly intelligible but has noticeable distortion compared to toll quality, some detectable distortion, and perhaps less talker recognition. Communication quality speech is presently achieved below 16 Kbps, and specifically the data-speed range is 9.6–7.2 Kbps. Here, the complexity is a function of data rate and goes up as the rate is decreased. Synthetic quality speech, while 80–90% intelligible, has substantial degradation. These coders are realized below the data rate of 4.8 Kbps and sound automated. Talker recognition is substantially degraded and the coder performance is talker dependent.

In the design of speech coders for mobile radio channels, aspects such as the efficient use of the radio spectrum, the speech quality, and the hardware complexity, play important roles. The efficient use of radio channel is determined by the bit rate and the modulation method applied. With modulations such as DQPSK and MSK, presently, speech at 16 Kbps can be transmitted in a channel with the bandwidth of 30 kHz with an adjacent-channel interference of about −60 dB, which is the requirement in cellular radio. Here, in the United States, it is desirable to choose 30-kHz channel bandwidth from a compatibility point of view. Thus, the maximum bit rate available for the speech coder is 16 Kbps.

A simple waveform coding scheme is pulse code modulation (PCM), where the speech signal is sampled, quantized, and encoded. To establish telephone quality, 12 bits per sample are required at a sampling rate of 8 kHz. By using companding (logarithmic PCM), 8 bits per sample are sufficient. These high bit rates are not attractive for cellular radio. Better results are obtained with differential coders, in which a dynamic range of compression can be applied, such as adaptive delta modulation (ADM) and adaptive differential PCM (ADPCM). The reason for these coders is to get a better signal-to-quantization noise performance over straight PCM. Though the advanced variants of these systems can still provide intelligible speech at 16 Kbps, the speech quality at this bit rate should be considered as marginal. At lower bit rates, the signal quality does not meet the requirement. In vocoders, the speech signal is analyzed and represented by excitation (pitch, voicing) and vocal tract parameters. About 10–20 parameters are required, which have to be updated every 10–25 ms, resulting in a bit rate lower than 5 Kbps. Despite the fact that we presently have at our disposal some refined techniques for pitch measurement and voiced-unvoiced detection, we have to consider the underlying model as a gross simplification of the human speech production mechanism with respect to the excitation part. For this reason, the quality of vocoders is not high enough to be adapted in cellular radio application. Also, increasing the rate above 5 Kbps does not pay off. In the midrange of bit rates, from about 5 Kbps to 16 Kbps, the best speech quality is obtained by hybrid coders that utilize suitable combinations of waveform coding techniques and vocoder techniques. Between several types of hybrid coders, there are significant differences in speech quality and complexity. A simple hybrid coding scheme that can yield nearly telephone quality, and which can be realized with a few integrated digital signal processors, is residual excited linear prediction (RELP) coding. This belongs to a class of coders known as analysis-synthesis coders and is based on linear predictive coding (LPC). A new class of coders, also based on LPC are termed as analysis-by-synthesis coders. This reduces the bit rate over analysis-synthesis coders but requires more complex processing. Multiple pulse excited linear predictive coder (MPLPC), regular pulse excited linear predictive coder (RPLPC), code excited linear predictive coder (CELPC), stochastically excited linear predictive coder (SELPC), and self-excited vocoder (SEV) are some of the coders that have been developed in this class. Next, we describe RELP coders and code excited linear prediction (CELP) coders as they have been adapted for digital cellular in the United States and Europe.

9.4.2 Residual Excited Linear Prediction Vocoder [16–18]

In this section, we discuss the RELP vocoder based on [17]. In the RELP vocoder, the vocal tract is characterized in the same way as in the pitch excited technique. However, instead of excitation parameters such as pitch, amplitude, and voiced/unvoiced being extracted and transmitted, the residual signal is encoded and sent to the receiver. The concept of residual encoding is similar to that of the voice excited vocoder (VEV), but the speech signal processing is fundamentally different. Since no pitch extraction is done in the RELP vocoder, the system is robust in any operating environment. The RELP vocoder is shown in Figure 9.56. The coding of the residual signal is done by adaptive delta modulation (ADM). Since the coding of the unfiltered residual would require large bandwidth, a low-pass filtering is done to limit the bandwidth to 800 Hz. This makes it possible to reduce the transmission rate of the residual to about 5 Kbps. At the receiver, this process is reversed where the spectral flattening is done to recover the high-frequency harmonics of the signal. This spectrally flattened signal mixed with an appropriate amount of white noise is used as the excitation signal to drive the LPC synthesizer, which is formed with the received prediction coefficients. There are five different elements of this encoder-decoder. They are an LPC analyzer and a residual encoder at the transmitter, a residual decoder, a spectral flattener, and an LPC synthesizer at the receiver. We discuss these subblocks below.

LPC Analysis and Residual Encoding [16–18]

The encoder in Figure 9.56 is segmented into two sections: LPC analysis, and encoding by adaptive delta modulation. The function of the LPC analyzer is two-fold. Firstly, it computes the prediction coefficients that characterize the vocal tract transfer function and, secondly, it generates one-step prediction error or the residual error signal. Before the signal is fed to the analyzer box it is low-pass filtered, A/D converted and pre-emphasized. The low-pass analog filter is elliptic, having the cutoff frequency of 3.2 kHz and the stopband decay of 85-dB per octave. The low-pass filtered signal is A/D converted at the sampling rate of 6.8 Kbps. The digitized samples are pre-emphasized such that the 3-dB breaking point occurs at 100 Hz. These samples are stored in the memory for analysis. By pre-emphasizing the speech signal, the higher frequency content of the signal is increased, thereby reducing the effect of low-pass filtering. The other advantage of pre-emphasizing is the reduction of spectral dynamic range. Thus, coding of the pre-emphasized signal is more accurate.

Essential elements of the coder are predicted error energy, LPC coefficients, and the error signal r_n. For LPC analysis, autocorrelation techniques can be used because it is less prone to instability of the synthesizer filter. The other method

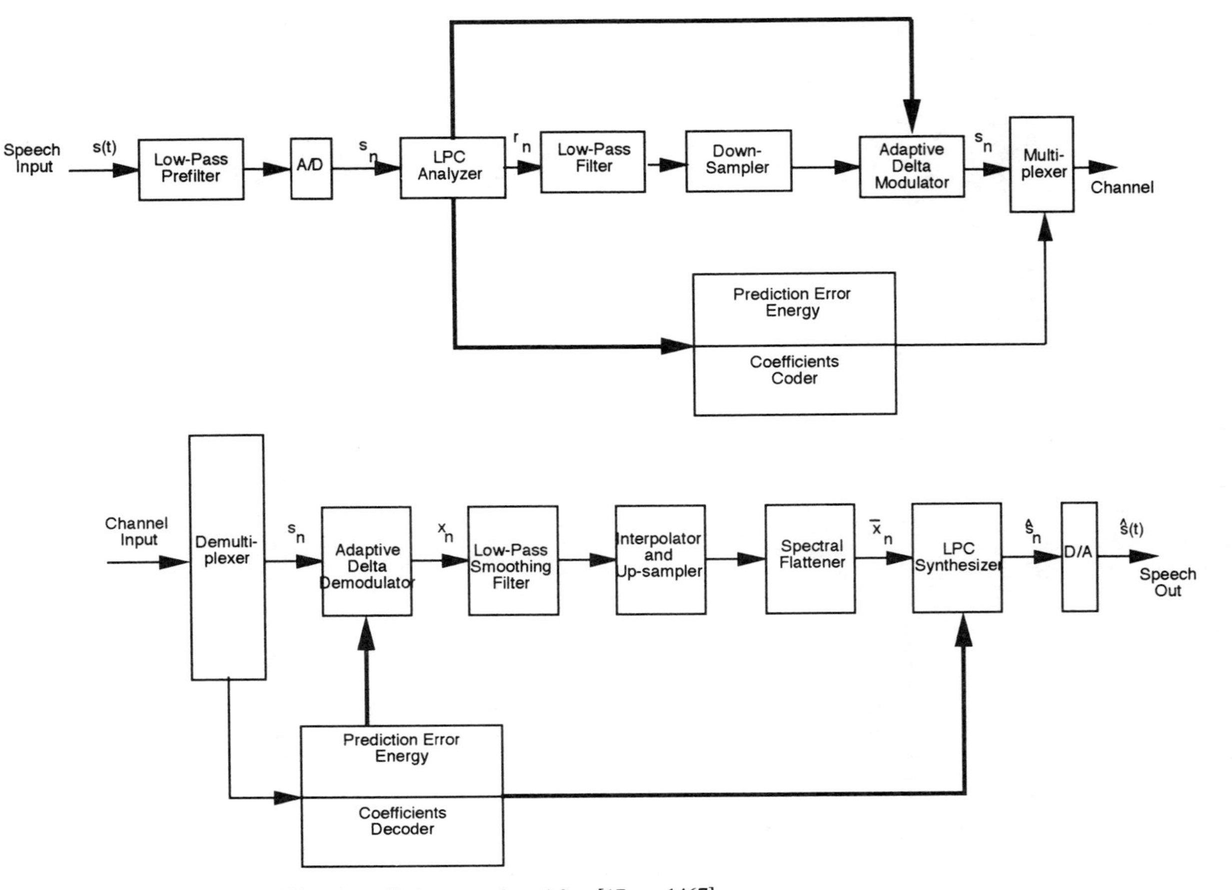

Figure 9.56 Residual excited linear prediction vocoder. After [17, p. 1467].

generally used is the co-variance method. In the autocorrelation method, the preprocessed speech samples are windowed by the Hamming window, represented as

$$W_n = 0.54 - 0.45 \cos \frac{2\pi n}{N} \tag{9.85}$$

The windowed signal, represented as s_n, contains N samples. The length of the window is made up of 196 sample points and the length of the analysis block is made up of 136 points. Thus, the LPC analysis block, say B_k, consists of 196 windowed samples that are made up of 30 samples of the previous block, B_{k-1}, 136 samples of B_k, and 30 samples of the next block. For analysis of next block, B_{k+1}, the window is moved by 136 samples. This overlapping provides a smoothing effect in LPC analysis and avoids missing data samples that fall in the window null. This results in enhanced quality of the overlapped windowed speech. These windowed speech samples are used to predict a sampled speech signal $s(nT)$ at discrete time $t = nT$ by linearly adding past p samples as

$$\hat{s}_n = \sum_{k=1}^{p} a_k s_{n-k} \tag{9.86}$$

where $\hat{s}_n$ is the predicted value of $s(nT)$, and (a_k) is a set of real constants that represent the predictor coefficients. The predictor coefficients are determined by minimum mean-square error analysis. The error between the predicted and real speech samples is

$$r_n = s_n - \sum_{k=1}^{p} a_k s_{n-k} \tag{9.87}$$

The transformed equation is

$$R(z) = A(z)S(z)$$

where (9.88)

$$A(z) = 1 - \sum_{k=1}^{p} a_k z^{-k}$$

where $R(z)$ and $S(z)$ are the z transform of r_n and s_n, respectively. The minimization of rms energy over all n results in a set of simultaneous linear equations in the autocorrelation coefficients and are given by

$$\sum_{k=1}^{p} a_k R_{|i-k|} = R_i \qquad i = 1, 2, 3, \ldots, p \tag{9.89}$$

where

$$R_i = \sum_{n=o}^{N-1-|i|} s_n s_{n+|i|}$$

where $R_i = R_{-i}$ for all i. The error energy is given by

$$E = \sum_{n=0}^{N-1} r_n^2 \tag{9.90}$$

The solution of a_k requires inversion of the matrix $\overline{R}_{|i-k|}$. This matrix is inverted by using the efficient Levinson-Durbin recursive procedure to solve for reflection coefficient (k_i), which is a transformed version of the prediction coefficients (a_k). These reflection coefficients, plus prediction error energy (E) and the residual error signal (normalized error signal E/R_o) are encoded by PCM and transmitted at the rate of 50 Hz or every 20 ms, as discussed below.

Before encoding the signal, it is necessary to band limit the signal by low-pass filtering to reduce the transmission rate. In order to cover the range of the fundamental frequency of speech and its higher harmonics, the bandwidth of the low-pass filter is kept at about 800 Hz. The low-passed residual must be downsampled before ADM encoding. The ADM encoder is shown in Figure 9.57. The low-passed residual signal to be transmitted is compared with its estimated value, which is obtained by increasing or decreasing the previous estimate of the sample by one step size. The increase or decrease of the size of the residual sample is based on the sign of the difference between the signal and its estimate. The sign information, one bit per sample, is transmitted over a binary channel to the receiver, where this sign bit is used to construct the estimate of the original signal. The process of generating the sign information is controlled by

$$e_n = sgn(r_n - x_n) \tag{9.91a}$$

where $x_n = 0.99x_{n-1} + e_{n-1}\Delta_{n-1}$ and $\Delta_n = [\alpha E^j]\delta_n$, where $\delta_n = \beta_n\delta_{n-1}$, and

$$\beta_n = f(e_n, e_{n-1}, e_{n-2}, e_{n-3}, e_{n-4}) \tag{9.91b}$$

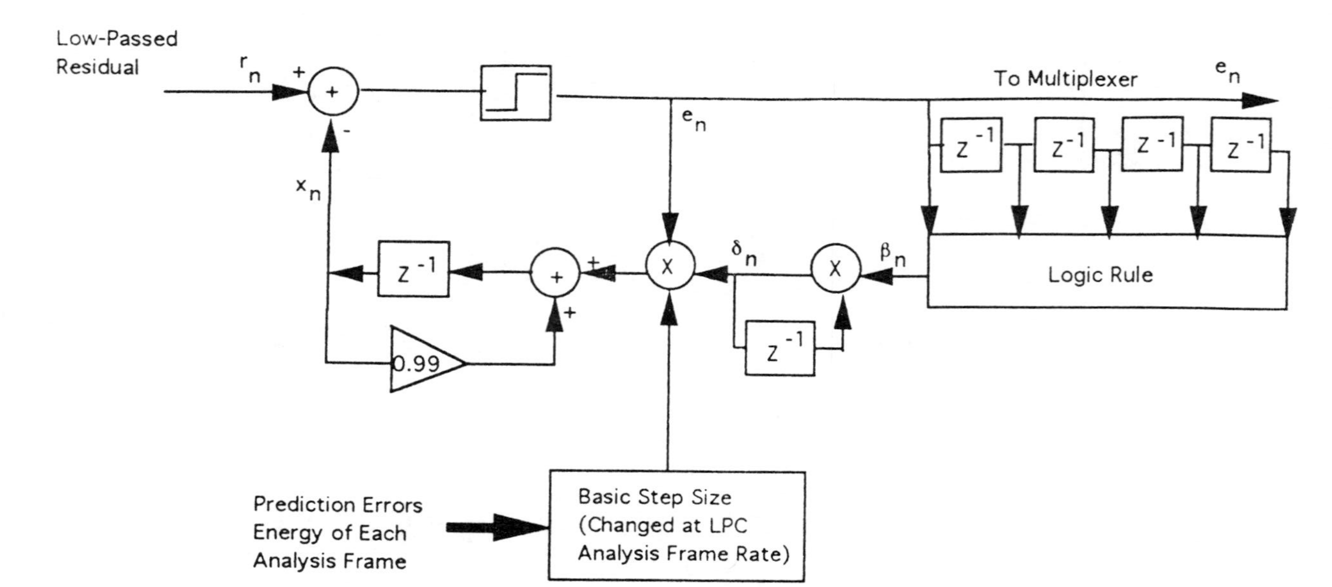

Figure 9.57(a) ADM with hybrid companding encoder.

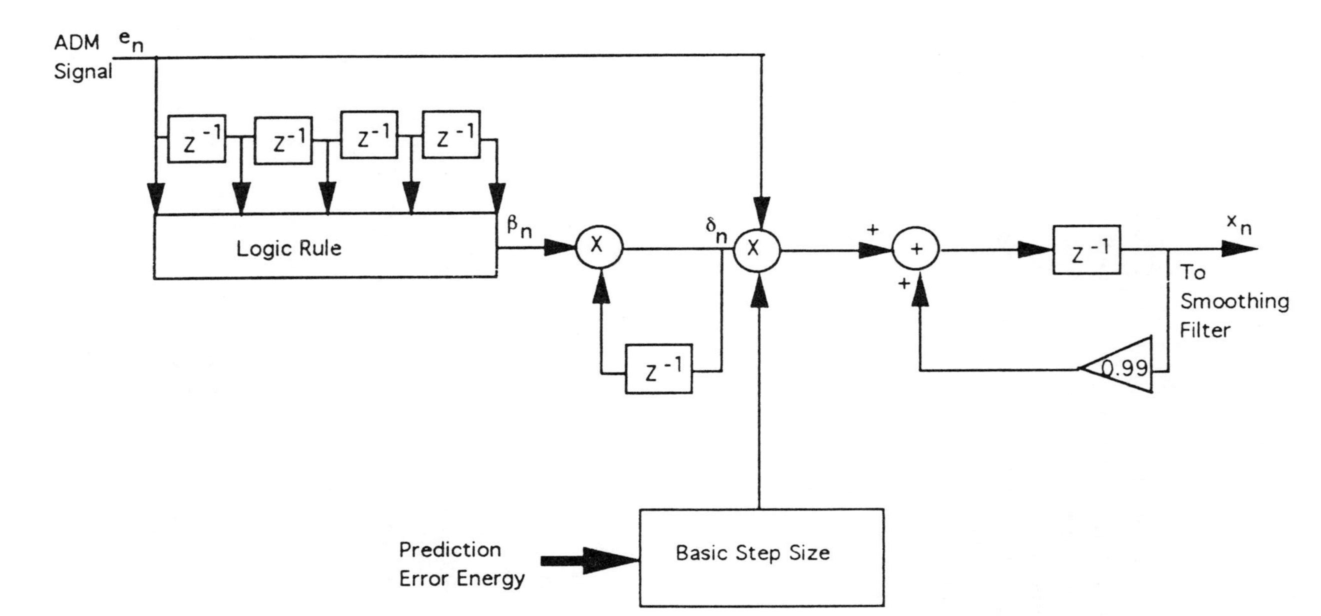

Figure 9.57(b) ADM with hybrid companding decoder. After [17, p. 1469].

where $0 < \delta_{min} \leq \delta_i \leq \delta_{max}$ for all i, Δ_n is the nth step size, and β_n is a multiplication factor. The minimum or the original step size Δ_o of the quantizer is given by

$$\Delta_o = \alpha E^{(j)} \tag{9.92}$$

Here α is a scale factor and $E^{(j)}$ is the average residual energy in the jth analysis block. Therefore, the actual step size Δ_n depends upon the residual energy $E^{(j)}$, the multiplication factor β_n, and the previous step size Δ_{n-1}. The factor β_n depends upon the present and the previous four sign bits, and it ranges from 0.66 to 1.5 [17]. The ADM encoder employs hybrid companding; that is, it uses both syllabic and instantaneous companding. Using the companding in ADM is advantageous, particularly for transmission of a speech signal or its residual, because of its large dynamic range. The process of decoding is the inverse of the encoding where the estimate of the signal is obtained by the received signal bits, the residual energy, and the fixed logic rule. The ADM decoder output is smoothed out by the low-pass filter, as shown in Figure 9.57(b). Here, the interpolation is used to regenerate the higher frequency components of the signal, which were removed due to down-sampling at the transmitter. The signal is then passed through a spectral flattener, which we describe below.

Spectral Flattening

Before the residual signal is applied to the LPC synthesizer, the high-frequency components must be recovered. The high frequency components of the residual signal are generated by the nonlinear device (full-wave rectification), as shown in Figure 9.58. By double differentiating the output, the energy in the harmonics of the residual signal is enhanced. The high-pass filtering ensures only high-frequency components of the signal at the output. The upper path provides the low-frequency baseband residual signal undistorted. Both signals of the upper and the lower arms when summed together provide the excitation signal, whose spectrum is approximately flat.

LPC Synthesis

The input signal to the synthesizer is spectrally flattened residual mixed with random white noise generated from a local noise generator. The amount of random noise in the excitation signal is controlled by the following formula:

$$\chi_n = \overline{X_n}(1 - R) + (E/N)^{1/2} R \Lambda_n \tag{9.93}$$

where $R = ERRN/k$ and $0.05 \leq R \leq 1$, where χ_n denotes the synthesizer excitation

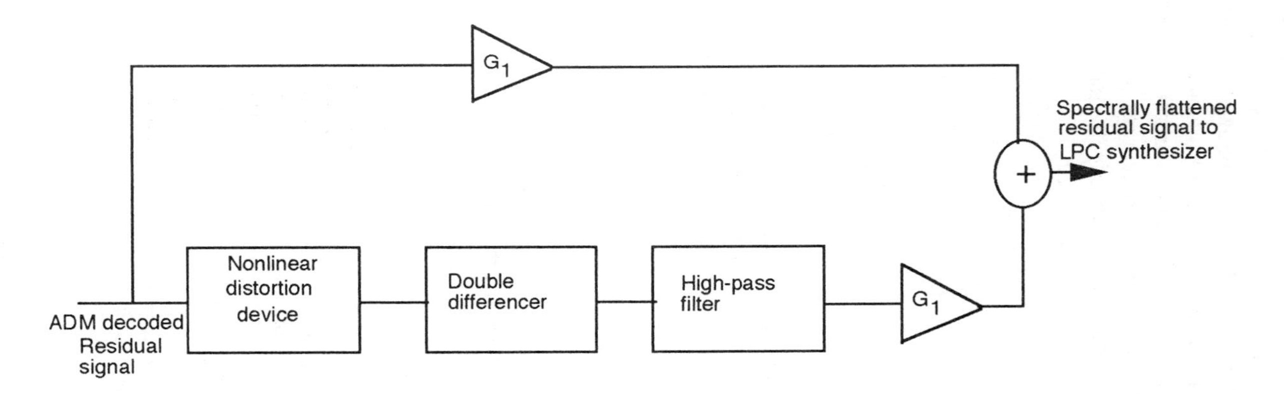

Figure 9.58 Spectral flattener with baseband signal preserved.

signal with random noise, X_n the spectrally flattened residual signal, E the residual energy, N the number of samples in one analysis frame, Λ_n the output of the random number generator, and $ERRN$ the normalized residual energy. The amount of random noise energy is controlled by both the scale factor k and the variation of the lower bound of R. In order to feed the correct amount of excitation energy to the LPC synthesizer so that the short-term average power of the synthetic speech is the same as that of the input speech at the transmitter, gain adjustments as shown below are done.

$$\hat{\chi}_n = (E/W)^{1/2} \chi_n \tag{9.94}$$

where W is given by

$$W = \sum_{n=0}^{N-1} \chi_n^2$$

Note that E and W are computed in each LPC analysis and synthesis block and that the residual energy E is also used in a syllabic companding of ADM coding of the residual signal. A summary of the transmission rate and other parameters is shown in Table 9.7.

Table 9.7
Summary of RELP Decoder

Parameter	*Description*
Transmission rate:	
Residual	6,800 bps
Coefficient	2,250 bps
Gain	200 bps
Normalized energy	200 bps
Frame synchronization	150 bps
Total data rate	9,600 bps
Input bandwidth	3.2 kHz
Input sampling rate	6.8 kHz
Window	Hamming
LPC analysis method	Autocorrelation
Cutoff frequencies of low-passed residual	800 Hz
ADM with hybrid companding spectral flattener	Baseband preservation

9.4.3 CELP Speech Coder

A relatively new speech coding technique, known as CELP has been developed recently. The scheme produces an excellent speech quality at a low data rate at the expense of high computational complexity. Thus, a modified form of CELP coding that reduces the computational process known as VSELP has been adapted in North American digital cellular systems. The VSELP algorithm reduces the code book search process. Here we shall discuss briefly the basic CELP scheme. The CELP synthesis structure is shown in Figure 9.59.

The basic synthesizer consists of a gain term followed by two time varying filters $1/A(z)$ and $1/B(z)$ where

$$A(z) = 1 - P_s(z) = 1 - \sum_{k=1}^{p} a_k z^{-k} \tag{9.95}$$

$$B(z) = 1 - P_l(z) = 1 - \sum_{k=-q}^{q} b_k z^{-(M+k)} \tag{9.96}$$

where $P_s(z)$ is the short-term predictor and $P_l(z)$ is the long-term predictor or the predictor based on the fine spectral structure. The number of short-term predictors lies somewhere between 10–16 ($p \approx 10$–16). The long-term predictor has the order $(M + q)$ while the number of nonzero pitch predictor coefficients is $(2q + 1)$. M is the pitch period ($M \approx 20$–160 samples). The value of $M \approx 2.5 - 20$ msec at a sampling rate of 8 kHz. Though the long-term predictor order is $(M + q)$, most of the coefficients are zero. $1/A(z)$ and $1/B(z)$ in the above diagram represent the short and long-term filters. $A(z)$ is updated every 10–30 ms while $B(z)$ is updated every 5–30 ms. The input $v(n)$ from the code book is either arranged in the form

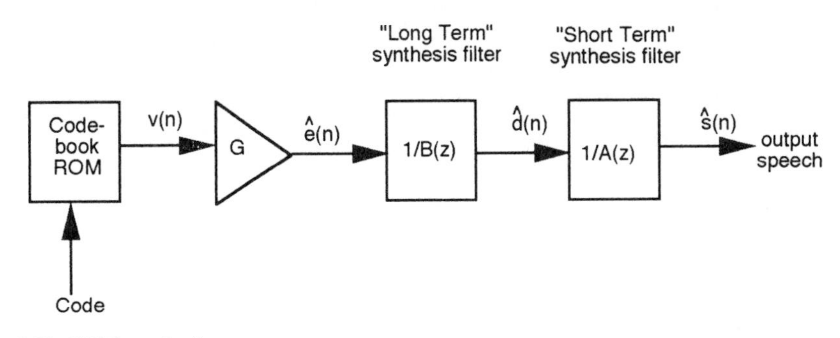

Figure 9.59 CELP synthesizer.

of a tree or trellis coding. The output of the code book $v(n)$, $n = 1, 2, \ldots, 2^N$, is arranged in the form of tree and trellis is shown in Figure 9.60.

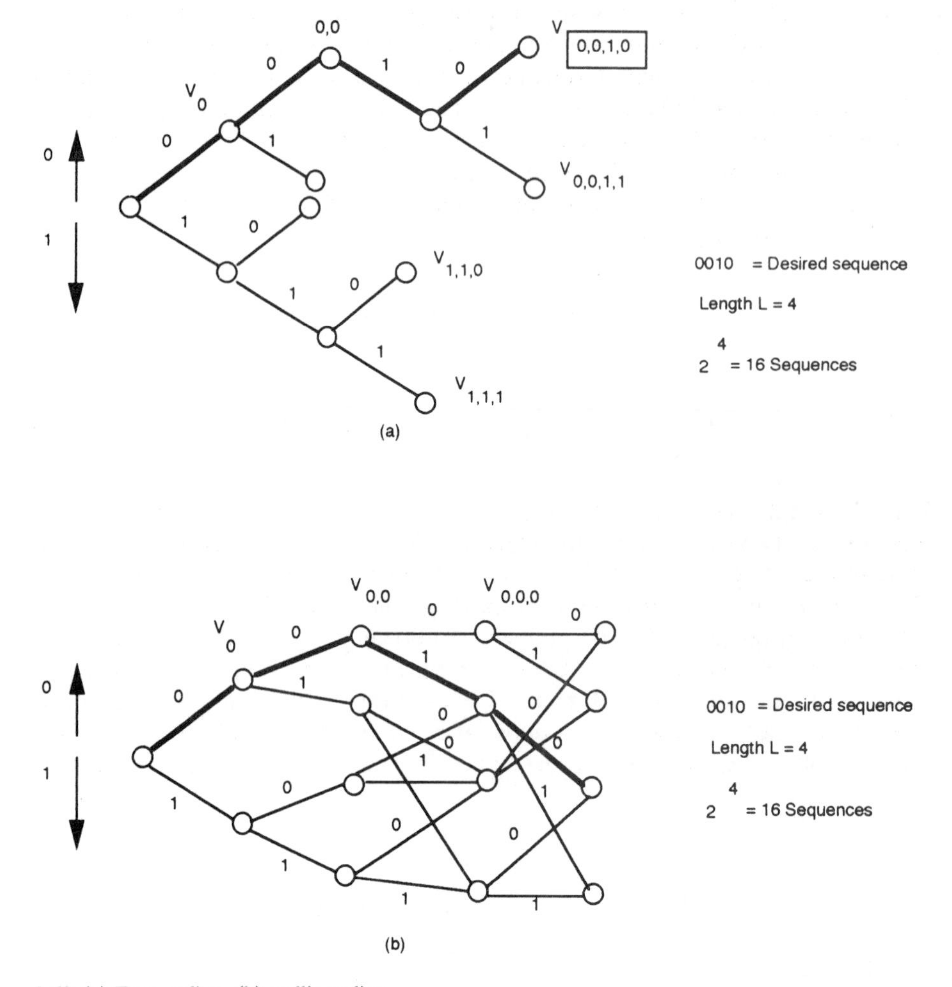

Figure 9.60 (a) Tree coding, (b) trellis coding.

9.5 CONCLUSIONS

We have established that Biϕ coding minimizes the interaction of speech with data, which is a must in cellular applications. To control the ISI, both GSM in Europe and American digital TDMA systems use raised-cosine pulses. PLL as a communication element has been extensively used in the cellular field for frequency-synthesizer and bit clock–recovery applications. Barker code with good correlation properties has been used in AMPS and other analog world systems for frame synchronization. In order to reliably decode the signal in multipath surroundings, an estimate of the channel transfer function is required. Nonlinear equalizers based on maximum likelihood sequence estimation and decision feedback estimation have been recommended for digital cellular radios. Use of compandor and emphasis networks is prevalent in analog cellular radios. Time diversity and interleaving have both been recommended for cellular applications. Majority voting is a simple unweighted form of time diversity, interleaving is used to minimize the effects of burst errors. Due to increased distance between code words, Reed-Soloman coding is recommended. Among voice coding, RELP and CELP, which are based on analysis-by-synthesis techniques, are the preferred choice for digital cellular applications.

PROBLEMS

1. Narrate the advantages and the disadvantages of Biϕ over NRZ in the context of cellular communication.
2. The following $P_r(f)$ are used in binary data transmission with controlled *ISI*. For both these cases, find $p_r(t)$ and the number of received levels.
   a. $4T_b \cos^2(\pi f T_b)$ for $|f| \leq 1/(2T_b)$; 0 elsewhere.
   b. $2T_b \sin(2\pi f T_b)$ for $|f| \leq 1/(2T_b)$; 0 elsewhere.
3. Derive the inverse Fourier transform (9.33), starting from (9.32).
4. A digital synthesizer that produces a maximum output frequency of 1,024 kHz with a frequency resolution of 4 kHz is desired. Compute the capacity and the required increment in the accumulator to produce the following frequencies: 10, 25, and 30 kHz.

REFERENCES

[1] MacDonald, V. H. " The Cellular Concept," *Bell System Technical Journal*, Vol. 58, No. 1, Jan. 1979.

[2] Arredondo, G. A., J. C. Feggeler, and J. I. Smith. "Voice and Data Transmission," *Bell System Technical Journal*, Vol. 58, No. 1, Jan. 1979.

[3] Lindsey, W. C., and M. K. Simon. *Telecommunication Systems Engineering*, Prentice-Hall: New York, NY, 1973.

[4] Zieme, R. E., and R. L. Peterson. *Digital Communications and Spread Spectrum Systems*, Macmillan: New York, NY, 1985.

[5] Gardner, F. M. *Phaselock Techniques*, John Wiley and Sons: New York, NY, 1979.

[6] Sklar, B. *Digital Communications*, Prentice Hall: New York, NY, 1988.

[7] D'Avella, R., L. Moreno, and M. Sant'Agostino. "Adaptive Equalization in TDMA Mobile Radio System," *IEEE Vehicular Technology Conference*, 1987.

[8] Bune, P.A.M. "Effective Low-Effort Adaptive Equalizers for Digital Mobilephone Systems," *IEEE Vehicular Technology Conference*, 1989.

[9] Svensson, L. "Channel Equalization for Digital Mobile Telephone Using Narrow-Band TDMA Transmissions," *IEEE Vehicular Technology Conference*, 1989.

[10] Monsen, P. "Feedback Equalization for Fading Dispersive Channels," *IEEE Trans. on Communication*, Jan. 1971.
[11] Forney, G. D. "Maximum-Likehood Sequence Estimation of Digital Sequences in the Presence of Intersymbol Interference," *IEEE Trans. on Information Theory*, May 1972.
[12] Lee, W. C. Y. *Mobile Communication Engineering*, McGraw-Hill Book Co.: New York, NY, 1982.
[13] Taub and Schilling. *Principles of Communication Systems*, McGraw-Hill Book Co.: New York, NY, 1971.
[14] Odenwalder, J. P. *Error Control Coding Handbook*, Linkabit Corporation, July 1976.
[15] Viterbi and Omura. *Principles of Digital Communications and Coding*, McGraw-Hill Book Co.: New York, NY, 1979.
[16] Viswanathan, V. R., A. Higgins, and W. Russell. "Design of a Robust Baseband LPC Coder for Speech Transmission Over 9.6 kb/sec Noisy Channels," *IEEE Trans. on Communication*, April 1982.
[17] Un, K. C., and D. T. Magill. "The Residual-Excited Linear Prediction Vocoder With Transmission Rate Below 9.6 kb/sec," *IEEE Trans. on Communication*, Dec. 1975.
[18] O'Shaughnessy, D. *Speech Communication*, Addison-Wesley: Reading, MA, 1990.
[19] Fisher, R. E. "A Subscriber Set for the Equipment Test," *Bell System Technical Journal*, Vol. 58, No. 1, January 1979, p. 136.
[20] Ehrlich, N. "Cell-Site Hardware," *Bell System Technical Journal*, Vol. 58, No. 1 January 1979, p. 178.
[21] El-Tanany, M. S., et al. "Data Detection and Timing Recovery for a Noncoherent Discriminator-based GMSK Receiver," *IEEE Transactions on Vehicular Technology*, 1985, p. 246.
[22] Addeo, E. J. "Word Synchronization in Digital Signal Processing for a High-Capacity Mobile Telecommunication System," *IEEE Transactions on Vehicular Technology*, 1977, p. 107.
[23] Flanagan J. L., et al. "Speech Coding," *IEEE Transactions on Communications*, Vol. Com-27, No. 4, April 1979, p. 172.
[24] Hoover, D. M. "Instrument for testing North American Digital Cellular," *Hewlett-Packard Journal*, April 1991, p. 69.

Chapter 10
Microcellular Radio Communication: Cordless Telephones and Personal Communication

10.1 INTRODUCTION

Today, most of us communicate with each other from station to station: I use the telephone on my desk to call other people who have telephones on their desks. This is the way conventional telephone systems have been working over the last several decades. Another system that has emerged over the last ten years is the wireless telephone system. In the wireless world there are three systems that provide diverse service: cellular, cordless telephones, and paging systems.

On the cellular side, there are many mutually incompatible analog systems in existence throughout the world. Some offer handheld units but their weight, battery life, and coverage leave much to be desired. Capacity is also a problem. In order to remove incompatibility and to meet the future demands, both Europe and the United States are coming out with new digital cellular systems. New standards are being specified for a Pan-European digital cellular system that is being put into operation in many big cities in Europe. Similar standardization activity is also going on in North America but these digital systems, which will have to serve both vehicular and hand-held units, will only partially satisfy the future needs of personal communications.

Analog cordless telephone systems, while having the potential of a higher number of units, have limited operating areas and spectrum assignments. Their popularity is due to the fact that people can keep in touch in the garden, up in the attic, and over at the next-door neighbor's. With new improved digital cordless phones for business and residential use, a person will be able to originate calls through "telepoints" in public areas such as airports, shopping centers, railway stations, and other similar places of public interest. This improved version of cordless telephone is a popular item in the U.K. and has been given the name cordless telephone second generation (CT-2). Telepoints will require billing procedures and some level of air-interface standardization. Calls to the cordless telephones as people roam between home, office, and public areas would require networking to location databases as it is being done for cellular radio.

In spite of tremendous progress in the area of telecommunication, none of the above systems are truly personal. Personal communication networks (PCN) and personal communication services (PCS) are the prominent buzz words in the communication industry today. These phrases are supposed to address the present and future requirements of the users, which include mobility, portability, reachability, and ubiquity. Systems having all four characteristics will truly be regarded as PCN and PCS systems.

Mobility refers to the capability to communicate while moving, including high-speed vehicular and low-speed pedestrian traffic. While the present cellular system is mainly designed for vehicular users, they can also be used to provide service to low-speed pedestrian users. Portability refers to the capability to communicate from any location. A requirement for this is a lightweight handset that the subscriber can carry in a pocket. Subscribers should be able to originate and terminate calls independent of their physical location. Reachability refers to the capability of a caller to reach the subscriber without knowing the subscriber's location. Ubiquity refers to coverage everywhere such that users need not be concerned with service discontinuities. Today there are areas of marginal coverage in cellular systems. Special efforts (such as repeaters) have to be made to provide coverage in multistory garage buildings, convention halls, tunnels, and even in some buildings.

In summary, with PCS and PCN we can communicate from person to person, regardless of location. PCS will provide capabilities from simple telephony and paging to more advanced functions such as voice mail, electronic messaging, and other forms of digital transmissions. The global recognition of the value of PCS/PCN has been reflected in the allocation of the 800-MHz band for CT-2 and the 1.7 to 2.2-GHz frequencies for PCS/PCN. CT-2 allocation generally occupies 4 MHz, while PCS allocation will probably occupy 100 MHz or more. The 1992 World Administrative Radio Conference (WARC92) of the International Telecommunication Union (ITU), which ended on March 3 1992 in Torremolinos, Spain, resulted in a worldwide allocation for mobile services in the 1.7 to 2.69-GHz band. This decision has brought all three regions of the world together. As defined by ITU, the future public mobile telecommunications systems (FPLMTS) are systems capable of providing a wide range of services (voice as well as data), including personal communications with regional and international roaming. The FCC in the U.S. has also allocated a total of 220 MHz of bandwidth in 1.85 to 2.2-GHz band for PCS.

In view of the above discussion, we shall divide this chapter into three further sections. Section 10.2 deals with the requirements of microcellular systems which applies to both CTs and personal communication systems. Section 10.3 deals with present cordless telephone systems of the world, and Section 10.4 deals with the future PCN systems in Europe and the United States.

10.2 MICROCELLULAR SYSTEMS

In order to fulfill the rapidly increasing demands for mobile communications, cellular systems with much higher spectrum efficiency than current systems must be offered. Spectrum efficiency exponentially goes up as the size of the cell is decreased. Thus, in the near future microcellular systems using cells with a few hundred meter radius will be introduced into heavy traffic areas. These cells will satisfy the requirements of cordless telephone systems and personal communication systems. The minimum and the maximum cell size within fixed geographic areas or cell boundaries is ultimately limited by technical factors such as terrain roughness, modulation scheme, frequency of operation, and population density. These factors ultimately determine the minimum and the maximum size of the cells.

The minimum cell size is determined by the need to have low traffic congestion and high channel availability, that is, to provide enough capacity to give users a low blocking probability. As the cell radius is reduced by a factor of two, the number of cells is increased fourfold. Unless the number of subscribers also grows by a factor of four, the cost per subscriber will increase. Also, as the cell size decreases, the number of handoffs needed for mobile users will increase. Finally, as cells become very small, the traffic in each cell is reduced, making it more difficult to achieve gains from trunking.

The maximum cell size is set by the economics of providing the service. In any cellular system, the usage of a particular cell must, on average, provide enough revenue to pay for equipment serving that cell, plus a share of any common equipment. On the other hand, the larger the cell, the lower the need for transferring calls to another base station as the mobile unit travels around. However, larger cells require higher power transmission from mobile and base stations. Also, larger cells lead to greater variations in transmission loss due to multipath fading, resulting in greater interference problems and curtailing the frequency reuse plan.

Microcellular systems will be used in the future in office surroundings, residential areas of the major cosmopolitan cities, telepoint services, major highways, inside convention halls, shopping malls, hospitals, and for a host of other applications. The business environment is dominated by PABX, centrex and key system users. A major problem encountered by this segment is that some 60% of the employees spend a large portion of their time away from their workstations without leaving the building. Wireless PABX will allow greater mobility and productivity within the workplace. The residential environment represents the largest potential market. Telepoint service refers to a cellular-like system with smaller, lower power base stations that largely supports pedestrian traffic that carry hand portables capable of outgoing with handoff between base stations. A typical base station layout in the downtown area is shown in Figure 10.1. In the downtown areas, where the traffic is heaviest, base stations may be located at every intersection [1]. However, as the communication traffic gets lighter, fewer base stations are required.

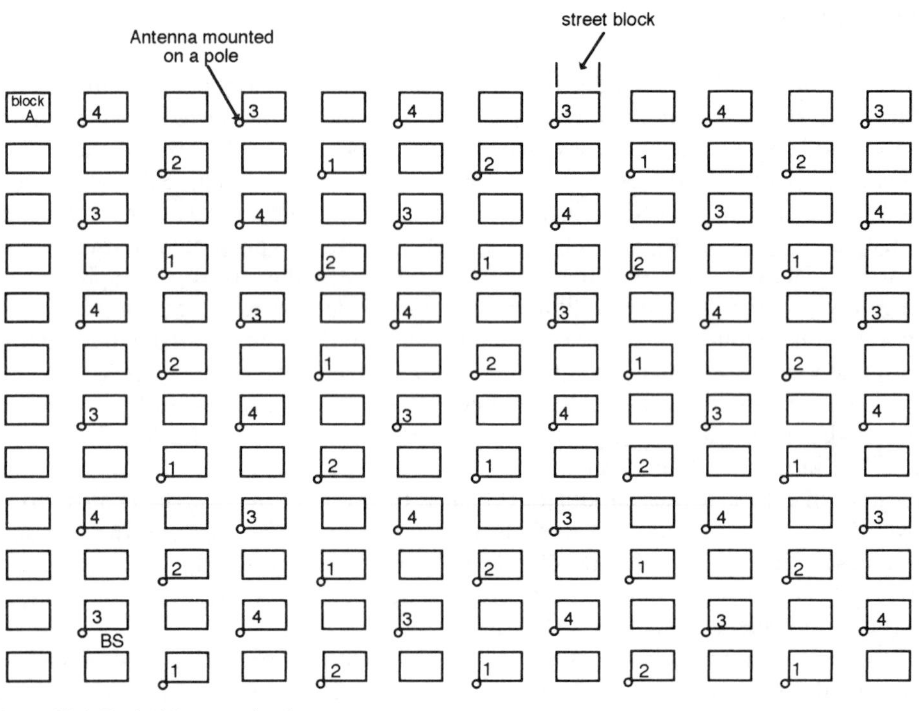

Figure 10.1 Typical base station layout.

As to network ports in residential areas, antenna heights can be about 7–10 meters above ground level, mounted on street lights and flagpoles [2]. Each small antenna symbolizes a radio access port to the local network, as shown in Figure 10.2. The local network provides connectivity, conveys end-to-end signaling, and provides an operations support interface. Estimates indicate that residential area ports separated by about 600 meters could provide nearly continuous coverage for the 5-mw power of the hand portables. Each port is assigned some number of channels, and operation will be based on TDMA or CDMA technology. The low height of the antenna will also limit the time-delay spread. Each port is assigned a frequency-channel pair for full duplex operation according to a frequency reuse factor N, which is the number of channels obtained by dividing the total spectrum by the desired channel bandwidth. The larger N is made, the further apart are cochannel cells, and hence less interference.

In order to provide universal ubiquitous service, radio coverage from network ports should also be available within large buildings where data communication may be the main item of interest. In this case, each floor has two-dimensional radio coverage while the coverage between floors will have a three-dimensional structure [3]. A typical layout of three-dimensional coverage is shown in Figure 10.3(a). In

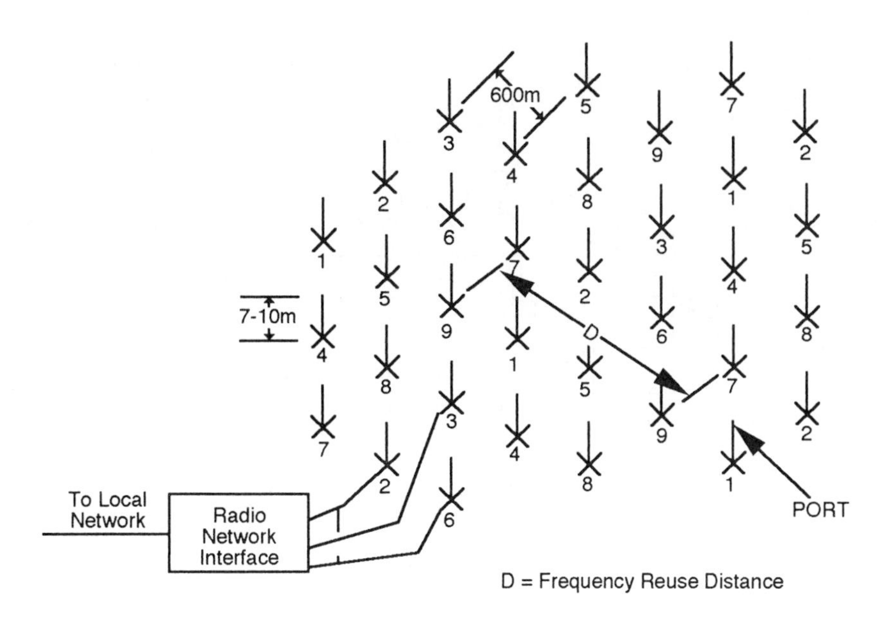

Figure 10.2 Portable radio system. Source: [21, p. 282]

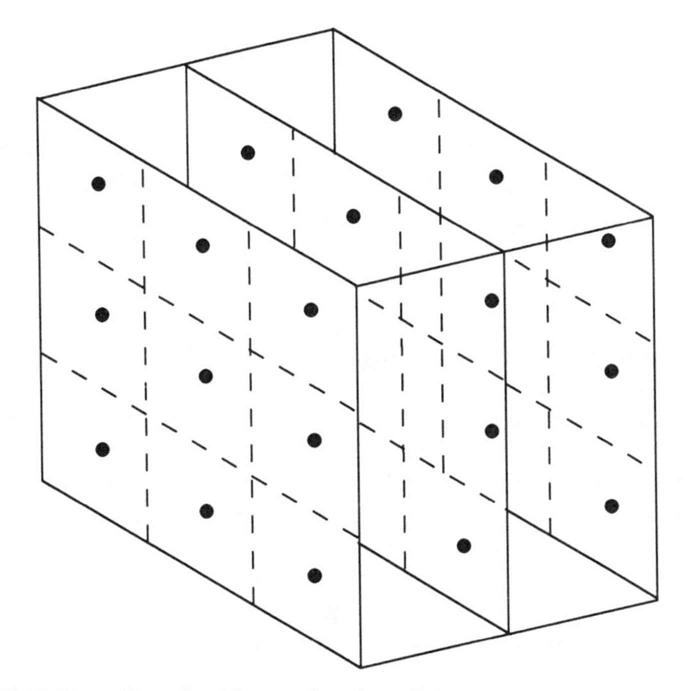

Figure 10.3(a) Three-dimensional layout of a microcellular system.

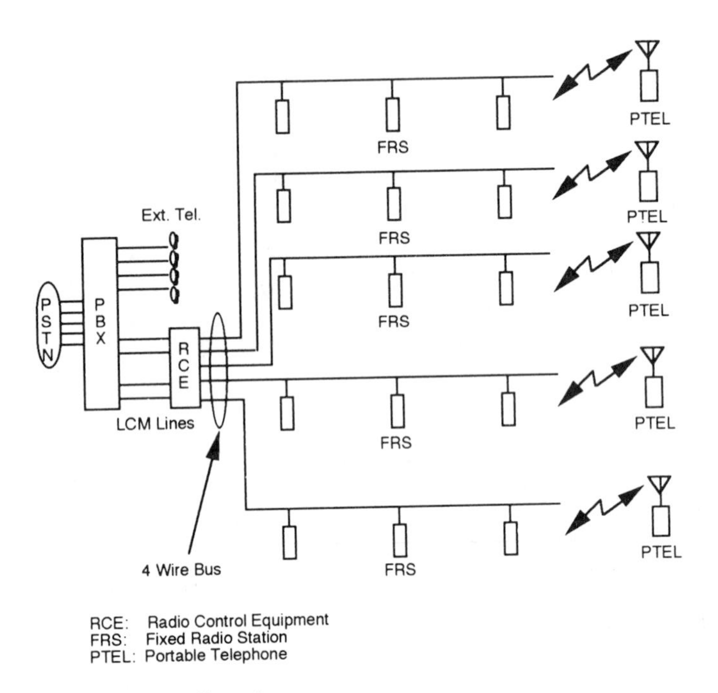

Figure 10.3(b) Office PABX configuration.

order to serve the customers distributed throughout the building, miniaturized (possibly printed circuit board) antennas will be installed. A typical configuration of the equipment setup for portables within the building will have air path to radio ports. Connection to the PSTN will be through PBX. The PBX may also be replaced by centrex service located at a central office. The office PABX configuration is shown in Figure 10.3(b).

10.2.1 Technical and Operational Requirements of a Microcelluar System [5]

For microcellular implementation, control decentralization, spectrum efficiency despite heavily deformed and overlapped zones, and an efficient countermeasure against fading all become important. Decentralization of control includes functions such as location registration and channel assignment, and handoff should be done by subscribers and base stations as much as possible [5], since zone sizes (can be as small as the size of a small room) are so small that microscopic irregularity in wave propagation becomes significant. The irregular deformation of zone shape will render the cochannel interference estimate impossible. Propagation conditions can also change with the construction of a new building in the area or simply due to a new partition within the building. This will also make it impossible to design a cell group structure for frequency reuse. A long duration of fading due to a slow-moving mobile may render the error correction scheme useless. Instead, an inexpensive diversity scheme may have to be designed. In order to resolve the above problems, the following possible techniques must be investigated:

- Channel segregation;
- Multiple base station registration;
- Diversity transmission;
- Proper choice of modulation;
- Performance criterion;
- Propagation model; and
- Multipath model.

10.2.1.1 Channel Segregation

For highly deformed cells, fixed channel assignment to cells is not possible. A decentralized channel assignment, where each cell finds and uses interference-free channels itself is preferable. The basic channel segregation scheme is an improvement over the dynamic channel assignment scheme. In this scheme, each cell acquires its favorite channels by learning through past experience on channel usage. Each base station has a table where a priority function is defined for each channel. Priority function to the channel i is defined as the probability $P(i)$ of a successful transmission over the channel. The algorithm works as follows:

1. When a new call arrives at the base station, the base station selects the channel that has the highest priority.
2. The base station senses the channel, and if found idle, the base station starts communication on the channel and increases the priority function of the channel for the next use.
3. If the channel is sensed busy, the base station looks for the next priority channel and decreases the priority function of the busy channel.
4. If all channels are sensed busy, the call is blocked.

10.2.1.2 Multiple Base Station Registration

This scheme allows the subscriber and the base station to perform location registration procedures without the aid of a central station. The conventional technique is as follows. A subscriber station monitors the received signal level from surrounding base stations. The subscriber selects the base station with the highest signal level. The location registration request signal is sent via the base station with the ID of the base station to the central station. This scheme allows the location registration procedure to be decentralized. The location registration process can be improved further by having multiple cell registration. Since cells are overlapped, multiple registration is possible. The subscriber gives priority to the base stations based on signal level measurements from different cells. In the case of the registration to only one base station, a subscriber cannot communicate with the central station if all the channels of the base station are busy even if the subscriber could communicate to another base station through an idle channels. The scheme allows the channels to be highly utilized.

10.2.1.3 Diversity Transmission

Antenna diversity, as shown in Figure 10.4(a), is a very simple scheme since it requires only one receiver with a simple RF switch. It provides increased capacity, extends the range, and copes with the time dispersion effects. Unfortunately, the scheme works well when the SNR is high and performs poorly when the level is low. Supposedly, a diversity receiver should improve the signal quality while the signal is below a certain level, but unfortunately the scheme generates more noise than a receiver with no diversity. The switching noise generated is due to an abrupt change in amplitude and phase of the signals, as discussed in Chapter 8. There are several methods for overcoming this difficulty:

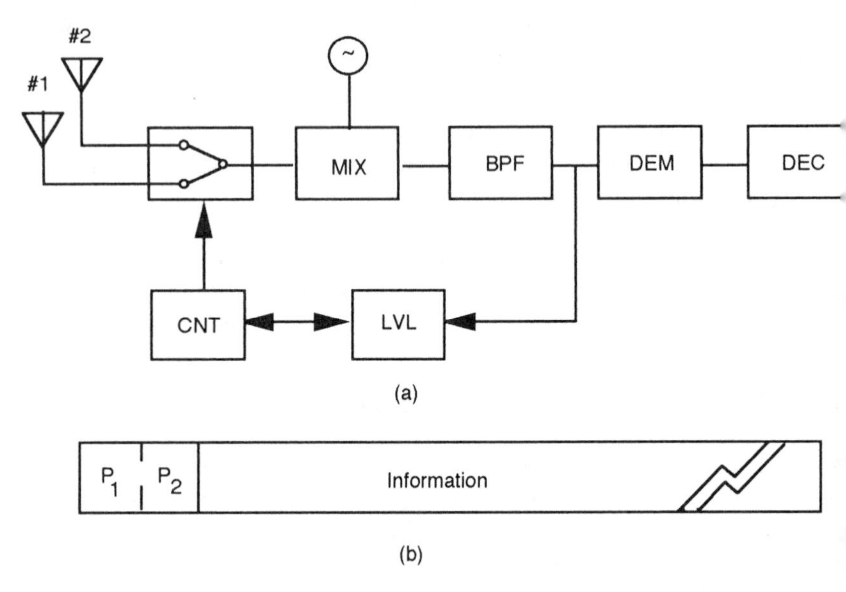

Figure 10.4 Antenna selection diversity: (a) block diagram, (b) signal frame.

- The scheme shown in Figure 10.4(b) switches the antenna during preamble such that the switching transients die down during an active period of data transmission.
- The two schemes shown in Figure 10.5 use two antennas at the base station. The use of diversity at the mobile is essentially done at the base station, which also removes the complexity of the mobile transceiver (a desirable thing to do). In single frequency time-division duplex operation where a single frequency is used for both up and down transmission, the antenna used for downlink transmission is the one that was selected in the previous reception at the base. Here, the base selects the best antenna based on the received signal strength and uses it for the next frame transmission.

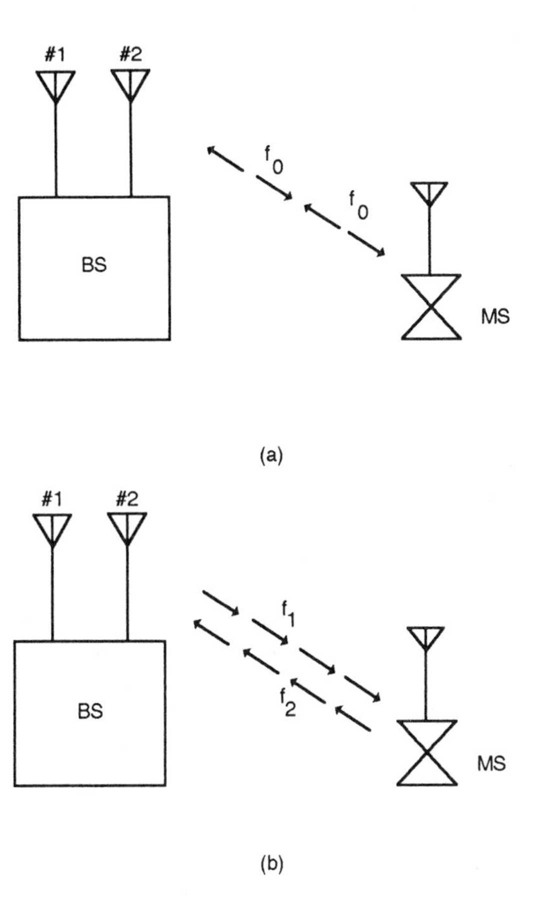

Figure 10.5 Base station antenna selection diversity: (a) single frequency time-division duplex system, (b) dual frequency full-duplex system.

The correlation between up and downlinks is assumed to be high as both transmissions are at the same frequency. For systems with dual frequency operation, coherence between two frequencies is very low as the up-link and down-link channels are purposely widely separated in frequency. In this scheme, test signals are sent from the base station via the two antennas successively at the end of a framed signal. The subscriber measures and compares the two received signal levels, and sends back the result to the base station to select the antenna for the next transmission of the signal.

10.2.1.4 Modulation

Two different choices of modulations are obvious: constant envelope modulation belonging to the MSK family (like GMSK) or a linear modulation belonging to the QPSK family (like $\pi/4$-DQPSK). For higher throughput, $\pi/4$-DQPSK may be a better choice when compared to GMSK. The problem in selecting $\pi/4$-DQPSK is the relative inefficiency of a linear amplifier, as discussed in Chapter 7. Since the power requirements of the mobile are very low for handheld portables, the choice of a linear modulation is quite practical. Since the channel for microcellular system is Rician, fading is relatively infrequent and thus an inexpensive low-speed AGC can be adequate for the mobile.

10.2.1.5 Performance Criterion

The performance of microcellular system can be judged in the same way as that of the cellular radio. A desired signal model is shown in Figure 10.6. The ratio of the desired signal power (S) to the power of the active interfering cochannel signals (I_i) for the duration of the call is given by:

$$\text{SIR} = S/\Sigma_i I_i \tag{10.1}$$

For a TDMA system, the up-link interfering power is due to users assigned to the same frequency at concurrent times at other ports. The summation in (10.1) is over all the cochannel/co-slot active users while the desired user is in communication. The definition of S/I for the downlink is analogous to that for the uplink, except it is calculated for the portable transceiver.

10.2.1.6 Propagation Model

The model that matches well with measurements on microcellular layout in metropolitan cities is a one-dimensional model differentiating between line-of-sight (LOS) and non-line-of-sight (NLOS) paths. The model is represented by [6].

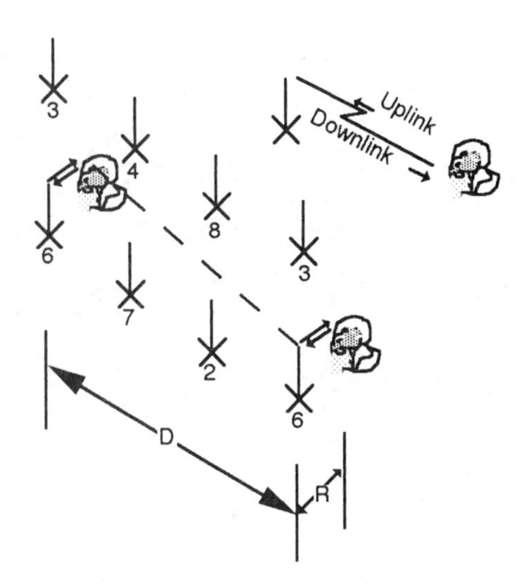

Figure 10.6 Desired signal model. D = frequency reuse distance. R = user distance-to-serving port.

$$L = -20 \log(d^{\alpha}[1 + d/G]^{\beta}) + K \tag{10.2}$$

where d is the distance from the base station; G is the specific distance from the base station, where the exponent increases to β; K is some fixed loss; and α is the attenuation exponent for line-of-sight propagation.

. Measurements conducted on microcells using low omnidirectional antennas in Stockholm show that the exponent changes after some distance from the base station. The layout uses twenty base stations, whose transmitted powers were measured at different streets. Transmitting and receiving antennas were situated at 5 meters and 2 meters, respectively. The frequency of measurements was 870 MHz. Figure 10.7 shows the measurement layout.

Figure 10.8 shows how the signal varies when turning from a LOS street onto an NLOS street. The LOS street is Kommendörsgatan and the NLOS street is Grevgatan for the base station designated as 2. The mobile turns from Kommendörsgatan to Grevgatan for the base station designated as 2. The signal around the corner decreases by 20–25 dB in a short transition distance of 20 meters.

10.2.1.7 Multipath Model

A multipath model of the microcellular is not as severe as in cellular radio. The model that fits well is the Rician fading model represented by the density function $p(x)$ as

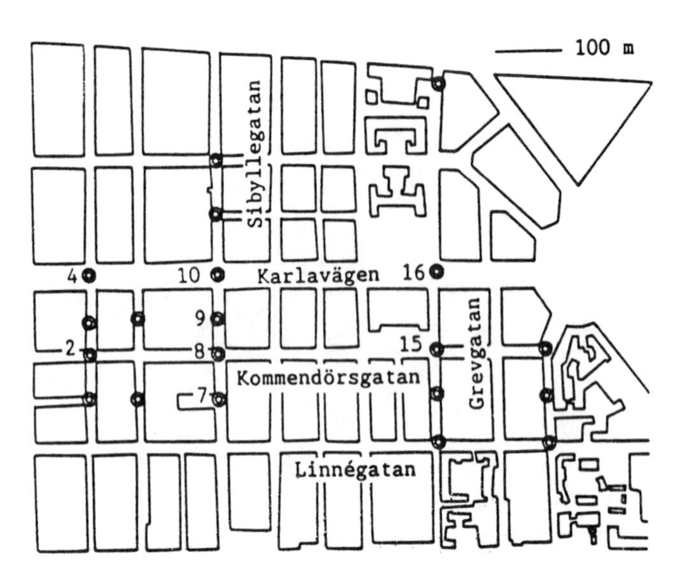

Figure 10.7 Antenna placement on Östermalm. Signal strength levels from 20 antennas were recorded on all of the streets shown. Source: [6, p. 539].

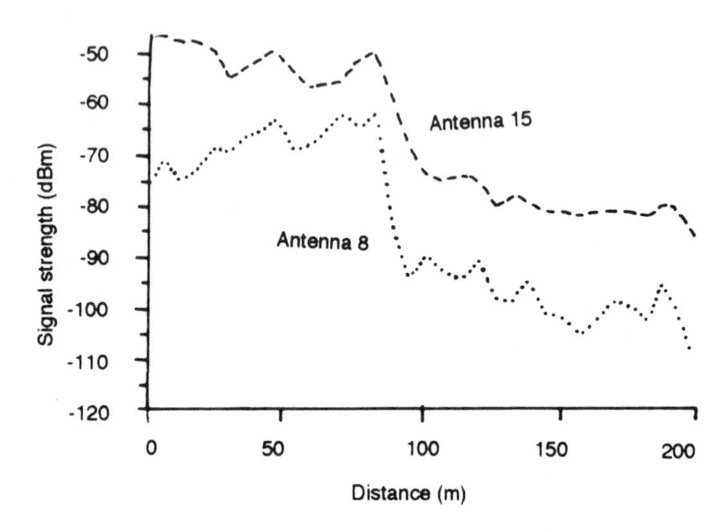

Figure 10.8 Mobile going east on Kommendörsgatan and turning right at Grevgatan. Source: [6, p. 540].

$$p(x) = \frac{x}{\sigma_i^2} e^{-(x^2+A^2)/2\sigma_i^2} I_o\left[\frac{Ax}{\sigma_i^2}\right] \qquad 0 \leq x < \infty, A \geq 0$$
$$K = \frac{A^2}{2\sigma_i^2} \tag{10.3}$$

where $I_o(\bullet)$ is the modified Bessel function of the first kind and zero order; A represents the peak-to-zero value of the specular radio signal, which consists of a LOS signal and scattered time invariant signals received from various paths; K specifies the Rician distribution completely. As K approaches infinity, the Rician density becomes Rayleigh; and σ_i^2 denotes the average power that is received over different paths that vary with time due to moving objects. Rician fading has been discussed in Chapter 2.

10.3 CORDLESS TELEPHONE [7–15]

The basic cordless telephone set is a simple telephone that has its cords replaced by a radio link. The radio path is short, less than 300 meters. There are four major areas of applications: residential, telepoint, wireless-PBX and large cordless business communication, and temporary installations.

Residential

Presently, the CT is used in household surroundings where a cordless telephone and its base station comprise an autonomous communication system. Cordless residential telephones with ranges of 500 to 1,000 ft now number about 30 million in the U.S. and continues to grow at a 20% rate annually. There is no domestic standard for cordless telephones, and given the current mode of operations within a residence, none is required.

Telepoint

In this service, cordless base stations are located at suitable venues, such as railway stations, airports, shopping centers, and so forth, from where a personally owned low-power cordless handset can make a public switched network call. Figure 10.9 shows a typical setup of a public telepoint. It is assumed that the user does not move considerably, and therefore no handover to another telepoint is provided. Since the basic cordless call-box form of telepoint does not offer full national roaming capabilities, it is seen as a natural complement to cellular radio systems by offering personal mobile communications to the less migratory, more localized sector of the traveling market.

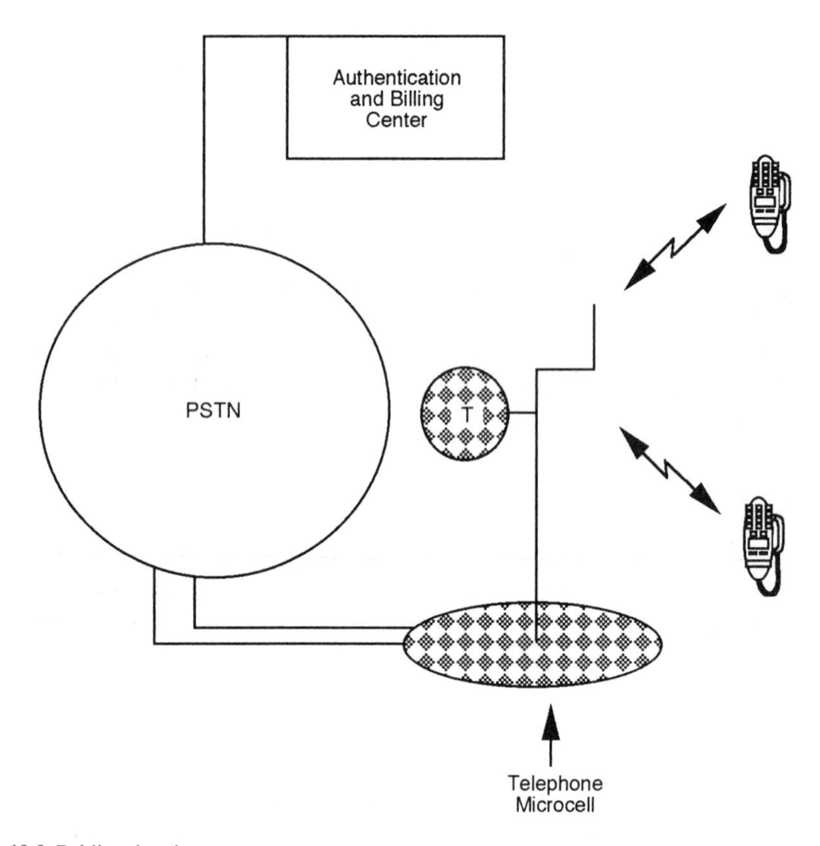

Figure 10.9 Public telepoints.

As a result, telepoint does not have to bear the costs of a comprehensive, centrally controlled and switched cellular network; it is simply a cordless call-box extension of the public switched network with some special requirements for recognition and billing. The telepoint needs access to a central database, which provides the necessary information for authentication and billing data. In many cases this can be provided by PSTN. A cordless telephone architecture consisting of a base switch interfacing the PSTN is shown in Figure 10.10. A benefit for the user is convenience in that no cash money is needed, as billing will be done later to their home or office. Since telepoint service is wireless, it should be almost maintenance free and thus, less costly.

In spite of these advantages, there are also several disadvantages, including:

- Only outgoing calls are allowed (CT1, CT2). A newer generation may permit incoming calls such as DECT 900 used as a cordless PABX.

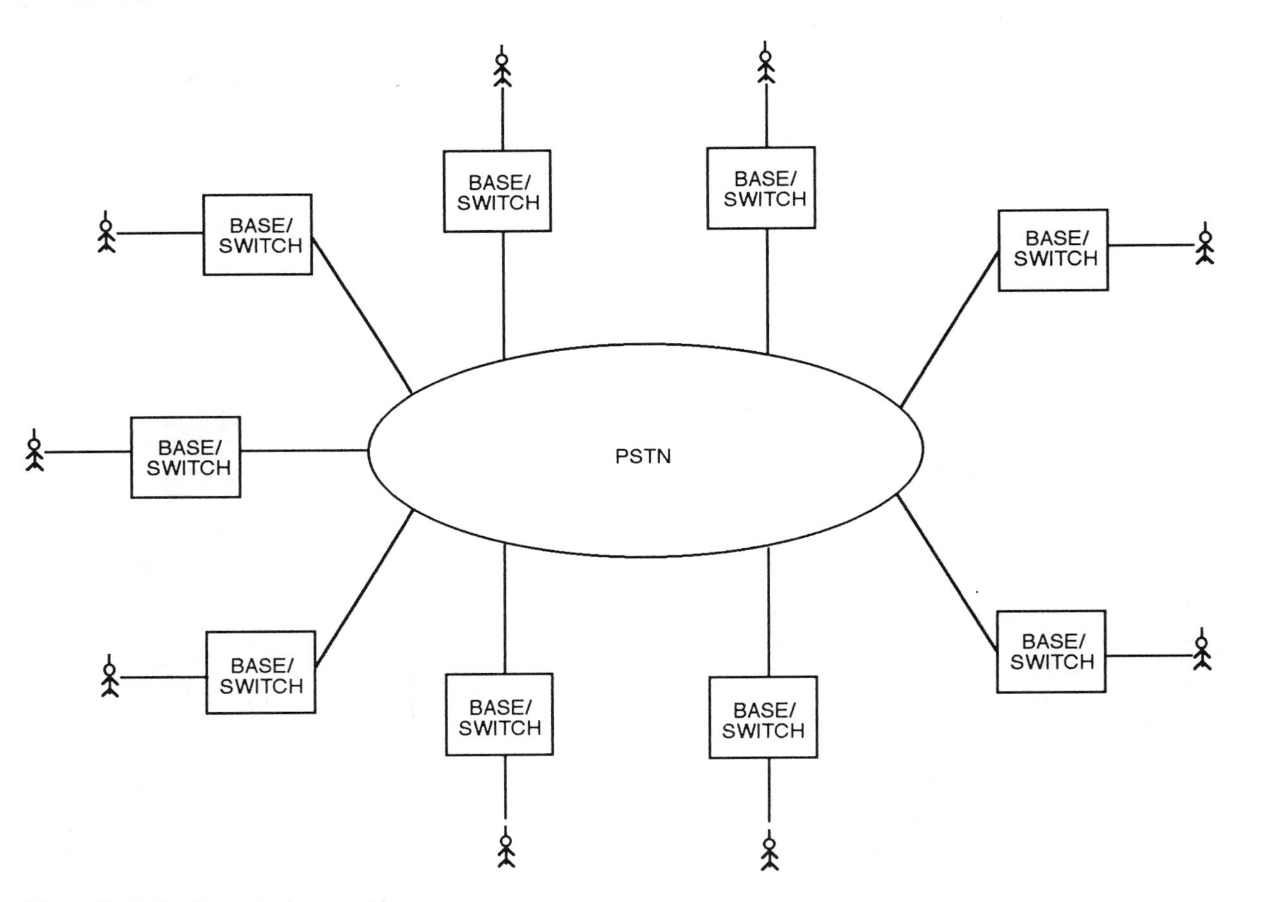

Figure 10.10 Cordless telephone architecture.

- Wide area coverage: it is assumed that telepoint systems installed in public places, such as airports, railway stations, and so forth may not attract enough paying customers. Therefore, in order to have proper return of investment, a telepoint operator would probably need to provide at least nationwide service.

Wireless PBX, and Large Cordless Business Communication

One of the most desired applications of CT is in office business surroundings, as shown in Figure 10.11. Here, portable handsets communicate to small cells, which in turn are tied up to a common control fixed point (CCFP). The CCFP is the controlling facility and will perform networking functions, particularly handover. The combination of PABX/CCFP is sometimes called a wireless PABX. In such a business cordless telecommunication (BCT) system, a user can initiate and also receive both voice and data calls in the coverage area.

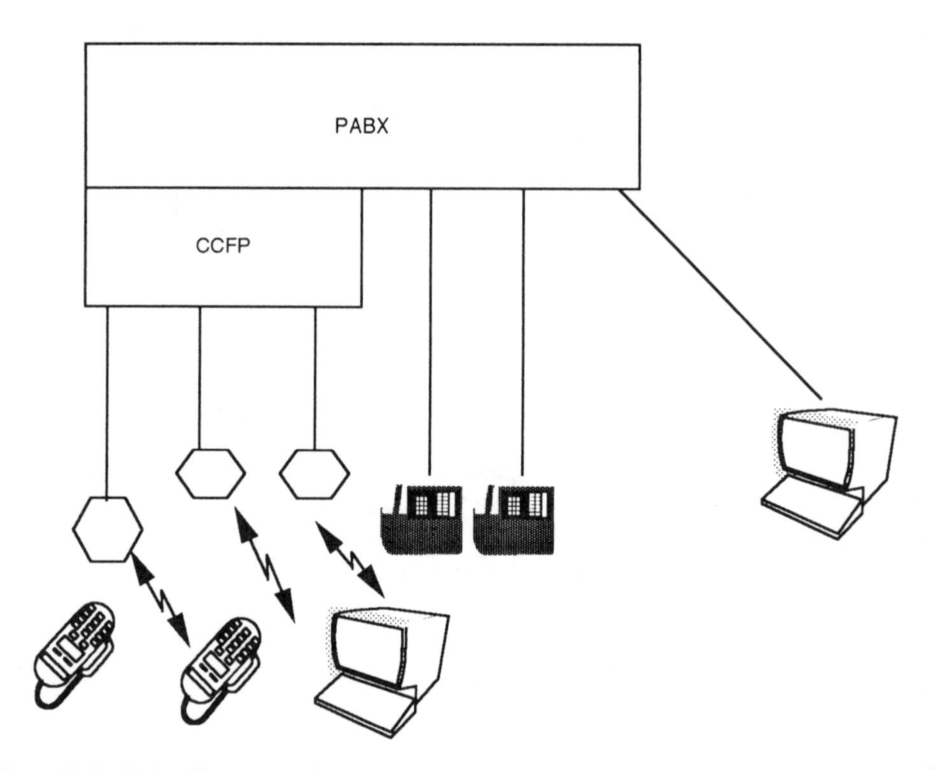

Figure 10.11 CT in office surroundings.

Temporary Installations

Cordless telephone systems can be put into operation temporarily if the wired network is either not available or expensive to get installed for a short duration. Examples of these will be sport competitions, construction sites, exhibitions, and conventions.

10.3.1 Technical Requirements

Since CT is connected to PSTN, the constraints of PSTN also apply. Critical requirements of a PSTN are absolute delay, echo delay, and speech coding rate.

10.3.1.1 Absolute Delay

Absolute delay exceeding 300 ms causes difficulty in user listening in a two-way conversation, and since the user will even expect a cordless-to-cordless international call via a synchronous satellite, which has a one-way delay of nearly 270 ms, the delay attributable to the radio path of the CT must be controlled.

10.3.1.2 Echo Delay

Delayed echo in a two-way PSTN call is a normal phenomenon that may cause difficulties for both users if the loop delay exceeds 20–30 ms. To ensure proper operation in an international call, CCITT recommends that the national end delay should not exceed 12 ms. In order to meet this value, each nation installs echo suppressors at the international gateway. It is reported in the U.K. that in excess of 98% international calls suffer a national end delay of less than 7 ms. Thus, network attachment (PBX plus radio path to CT) delay must be limited to 5 ms. Therefore, an allocation of, say, 2.5 ms each to PBX and radio path to CT is a good choice.

10.3.1.3 Speech Coding Rate

Since CT interfaces with PSTN, the digital speech spectrum must be contained within the bandwidth of the network. Proper choice of speech coding will be required to limit network introduced distortion. CCITT recommends using a 32-Kbps ADPCM speech coder. The ADPCM encoder introduces a negligible encoding delay (nearly equal to zero ms), low implementation complexity, and a high subjective speech quality (mean opinion score (MOS) nearly equal to 4).

10.3.1.4 Channel-Based Requirements

One of the main requirements of CT is the transparency of radio channels to the users. This implies that the CT set should operate in an interference-free channel, rapidly finding a channel out of a large set of 40 or more, if necessary. Thus, sufficient channels need to be available for the user to have a success rate in receiving the calls and also in making outgoing calls from future cordless systems. The number of channels required for a cordless telephone system is determined by four factors:

1. The amount of traffic to be carried in a given area;
2. Efficiency with which channels can be reused;
3. The maximum acceptable call blocking rate;
4. The maximum level of interference that is acceptable on the radio path.

Traffic Estimates [8]

Like cellular radio, traffic (in erlangs) must be estimated before we can select the required number of channels to meet a specified blocking probability. We shall take up an example to illustrate how the total traffic, in erlangs/km^2, was arrived at in a residential neighborhood of London. The raw data gathered in London is:

- Population—139,000;
- Area—12 km^2;
- Residential area—4.8 km^2;
- Number of households—56,068; and
- Density of households—4,700/km^2.

Assuming that in the area only 80% of the households have telephones, then the total number of telephones equals 56,068 × 0.8 = 44,854.4. Further assuming that each subscriber generates 0.07 erlangs of traffic during the peak hour, the total erlangs are: 44,854.4 × 0.07 = 3,140 erlangs. Again, let us assume that out of total Erlangs of traffic, only n% will be carried by CT. Then CT traffic = 3,140 n/100 erlangs. An estimate based on [4], puts the value of n = 10% by the year 2000, which leads to a total traffic of 314 erlangs. Assuming the traffic to be uniformly distributed, traffic per unit area is 314/12 = 26 erlangs/km^2. The area of 12 km^2 does not include areas of parks, rivers, industrial premises, and so forth. Obviously, the above estimate also does not include commercial users. With a reliable cordless telephone system that offers privacy of speech, we would anticipate significantly higher business demand. A figure, including business demand, of 100 erlangs per km^2 is probably not unreasonable in the near future.

Channel Reuse Model

The channel reuse model is based on Figure 10.12. Here, the desired user SIR at the base station is computed by accounting for the signal power in the user band and the interfering power from a single user C. Based on a single interferer the SIR of the channel at the base station is

$$S/I = 10 \log_{10}(r/R)^n \text{ dB} \tag{10.4}$$

where r is the wanted signal range, R is the unwanted signal range, and n is the propagation decay law appropriate to the environment being studied, where $2 \leq n \leq 4$ (typically).

For multiple interfering sources a cellular-type approach can be taken. Postulating six equally spaced interferers gives:

$$S/I = 10 \log_{10}[(1/6)(r/R)^n] \text{ dB} \tag{10.5}$$

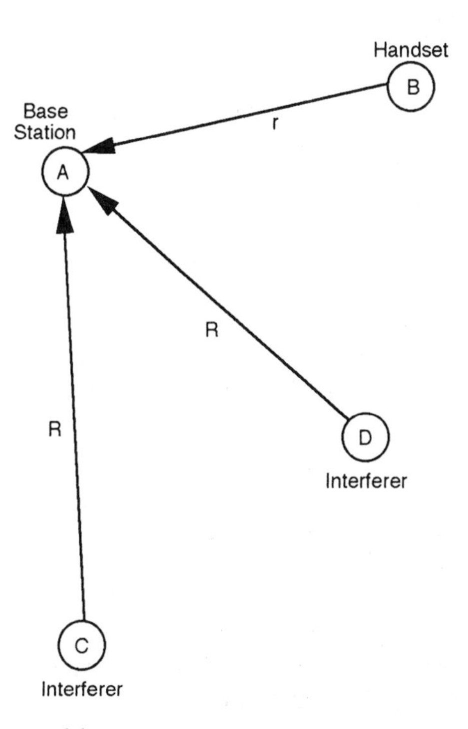

Figure 10.12 Channel reuse model.

The above model is only an approximation as a user operates for both sides of transmission on the same channel (time division duplex), which is unlike cellular where the channel frequencies on both sides of transmission are different. Models based on six interferers are certainly not optimum, as the channel selection in CT channel allocation is made on the instantaneous received signal level by sets and has no other prior knowledge.

Blocking Probability

Knowing the reuse model and the total number of available channels (assuming certain BW and the total spectrum availability), traffic in erlangs/km^2/channel can be arrived at provided the blocking probability can be assumed.

Maximum Level of Interference

For FM modulation, the maximum permissible level of interference is arrived at by knowing the receiver characteristics. As discussed in Chapter 7, the required CNR in the IF bandwidth is nearly equal to 10 dB for an output SNR of nearly 30 dB. A 30-dB output SNR is adequate in thermal noise surroundings. Since the system is interference limited rather than noise limited, adjustments have to be made based on the interference model. Allowing for a 20-dB fading margin, the minimum output SIR desired should be around 50 dB.

10.3.1.5 Other Technical and Operational Requirements

Some other technical requirements for CT system designer and operators are: operation type (simplex, duplex), choice of modulation, choice of multiple access techniques, and channel allocation techniques. We shall discuss each of these.

Operation Type

The choice of duplex operation (ping-pong transmission) over one frequency will make the system controller less complex. Additionally, a duplexing filter in the mobile and base transceiver will not be required. Other advantages of choosing time division duplex mode of operations are:

- Received radio path transmission quality is identical with that transmitted.
- Handset and base unit have identical circuit and component requirements.
- Base-unit antenna diversity optimizes both directions of transmission.

Modulation Type

Since the proper CT operation is strongly based on cochannel interference, the chosen modulation must be resilient to strong interference during the busy-hour traffic. Assuming that the base station will provide the selection of channel based on the lowest interfering power on the available channels, an adjacent-channel interference requirement can be relaxed compared to cellular radio. As shown in Chapter 7, constant-envelope modulation for digital systems can be based on the MSK family of modulations. With a normalized baseband filter BT product of less than 0.5, a 40-dBc adjacent-channel interference power requirement can be met. FM modulation in analog FDMA will also be adequate.

Multiple Access Technique

Among CDMA, TDMA, and FDMA systems, CDMA may be difficult to accept in the liberalized cordless environment. The scheme chosen must be simple, especially in the residential applications. A CDMA system has to be tightly controlled with respect to transmitted power, which in this case will require coordination among base stations. Also, the proper allocation of orthogonal codes to the users will be difficult. This leaves us with the choice of FDMA or TDMA. Here, the choice depends upon implementation complexity, spectrum requirements, regulatory requirements, and most importantly on cost. Table 10.1 provides factors for comparison between FDMA and TDMA.

As shown in the above table, the equalizer for FDMA delay dispersion, is not provided for a wideband channel. For example, a channel bandwidth of 100 kHz

Table 10.1
Comparison of Multiple Access Techniques for CT

Factors	*FDMA/TDD*	*TDMA/TDD*
Dispersion equalizer	No	Yes
Frequency tolerance	Moderate	Moderate
Synchronization conflict between independent systems	No	yes
Dynamic channel allocation	Channel scan	Timeslot scan
Transmission bit rate	32–64 Kbps	2 MBPS
System delay	Low	Medium to high
Echo control	No	Yes
Low-power RF channel combiner	Yes	No
Contiguous frequency band	Desirable	Essential

may not need a dispersion equalizer. This will also reduce the requirement for frequency tolerance of the synthesizers and the bandwidth tightness of filters. Choosing time division duplexing eliminates the need for an RF. duplexer and its moderate delay raises no echo control and absolute delay issues. On the other hand, TDMA/TDD will require a time dispersion equalizer, but its data transmission capacity is also considerably higher than FDMA/TDD. For FDMA systems, frequency bands of operation may or may not be contiguous, while in TDMA/TDD it is absolutely necessary to have contiguous frequency bands.

Channel Allocation Technique

With no central control of cordless base stations, dynamic channel allocation based on interference level measurements on channels are essential before allocation. Deciding which channel to use can be done by the base station by monitoring the interference on each channel to find the first channel that satisfies the *SIR* requirement. Since the interference level is based on traffic, the adjustment to threshold may be needed to provide optimum traffic flow.

10.3.2 Cordless Telephone Versus Cellular Radio [15–22]

The present-day cordless telephone is primarily a microcellular-based system and differs from a cellular system in many ways, as shown in Table 10.2. Terminals for cordless systems are almost exclusively handheld portables, whereas the technology for cellular radio allows for both car-mounted and portable terminals. On the other hand, the technology for cordless systems must allow handheld portables, which includes providing small-size instruments and a long battery life.

The total telephone traffic in an office building may result in densities up to 10,000 erlang/km^2, although this will happen only in a very limited area. Obviously, traffic distribution has to be assessed in three dimensions. Modern cellular systems are designed for a maximum (two-dimensional) traffic of 1,000 erlang/km^2. The range of the radio link (i.e., the distance between a portable handset and the fixed station) is very short, usually less than 300m. Cellular base stations may cover distances up to several miles.

10.3.3 Cordless Telephone First Generation CT-1

Cordless telephone, as the name implies, replaces the cord to the handset with a radio link, serving the user with a high degree of mobility. CT-1 from the Conference of European and Telecommunications Administration (CEPT) is based on the first-generation technology similar to the present analog cellular telephones.

Table 10.2
Comparison of Cordless and Cellular Telephones

Use	*Cellular*	*Cordless*
Typical user model	Cellular phone service primarily used when on the move	Universal: used as cordless phone with residential service and with wireless PBX; used as wireless public service when on the street
Radiated power	High power (1–10W)	Low power (5–10 mW)
Coverage	Several miles	Less than 300m
Radio channel	Fast fading Rayleigh, both Doppler and time spread present based on vehicular speed and areas of applications	Slow fading with negligible Doppler and time dispersion (< 200 ns)
Cell type	Regular cells	Irregular cells
Handover	Centrally controlled, mobile assisted in digital systems	Not a feature with present systems, but may be desired for future developed systems
Coverage	Continuous; not complete in rural areas	Public service likely to be spotty and concentrated in urban areas; also office and home usage
Traffic capacity requirement	Not very high	Very high traffic in office applications
Diversity	Antenna with error detection and correction	Antenna diversity, error detection only

The method of speech transmission is analog FM, and there is limited co-ordination between fixed network elements and wireless telephones. Mobile is low power and limited to 10 mW. The typical characteristics of CT-1 are shown in Table 10.3. The most serious drawback of the first-generation cordless telephone is the operating range, which is limited to tens of meters from a single base station that it can only access. The other limitation is the vulnerability to interference from other cordless phones. Most first-generation cordless telephone are channelized and have access to a single or at most a few channels. The user cannot do anything to avoid interference if the same channel is used nearby. For those that have access to a few channels, almost all rely on manual channel selection by the user to avoid interference. The system mainly serves residential users.

10.3.4 Cordless Telephone Second Generation CT-2

To overcome the limitations of first-generation technology and to satisfy the exploding demand for wireless access to communication networks, the industry has been

Table 10.3
Characteristics of CT-1

Specification	*CT-1 (CEPT)*
Frequency	900 MHz
Access method	Channelized
Modulation	Analog FM
Power	10 mW
Carrier	40 (channels)
Channels per carrier	1
Carrier separation	50 kHz
Handoff	No
Market	Residential

working on second-generation technology. Since CT-1 was designed to communicate with a single base station, no compatibility specification was necessary. In CT-2, a standard air interface will allow mobiles to interface with different base stations. CT-2 allows digital speech transmission and it will dynamically select the best available radio channel. The speech coder is the CCITT standard ADPCM that operates at 32 Kbps. Digital speech allows encryption techniques for privacy. CT-2 is based on a digital technology using combined frequency division/time division transmission techniques. This time division duplexing operation is shown in Figure 10.13. This second-generation telephone system has emerged in Europe, notably in the U.K. to provide telepoint service. In the current mode of operation, the system does not support handoff from cell to cell. Alerting for incoming calls is available only through a paging service extension of CT-2. Compared with the current CT-1 telephones, CT-2 sets can be used in areas of greater density without causing interference and are thus suitable for the business markets also.

Since CT-2 interfaces with many base stations, telepoint participants must conform to a common air interface (CAI) standard, which will enable users to experience complete transparency. The characteristics of CT-2 are shown in Table 10.4. Because power is limited to 10 mW, the operating range of phone is relatively small (nearly 100m) and interference to other users is also somewhat limited. The dynamic channel allocation ensures that the best use is made of the available spectrum at any time. The system can operate in high traffic density surroundings such as offices. When the subscriber wants to make a call, the radio searches around 40 channels and chooses the one that is most free of interference. The user then sends the paging signal on this channel. The base station continuously scans these channels for the paging signal. When the paging message is received by base station, dialing information can be transmitted and the call can progress in the normal way.

Carriers are spaced 100-kHz apart and each conveys one conversation using time division duplex for two-way conversation. The frame duration shown in Figure 10.14 is of 2ms duration. For 1-ms, the mobile transmits to the base station and

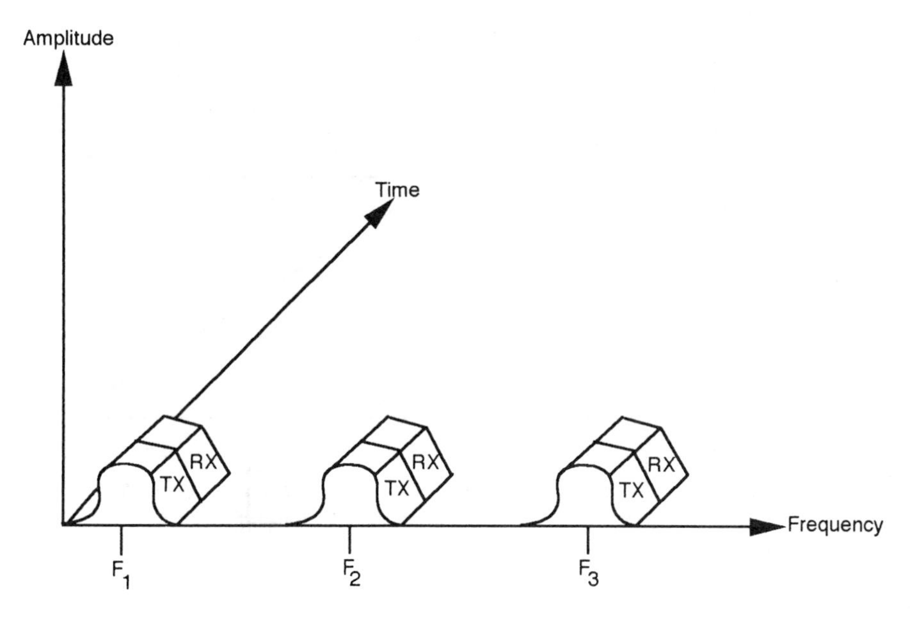

Figure 10.13 Second-generation cordless phones.

Table 10.4
Characteristics of Cordless Telephone, Second Generation (CT-2)

Specification	*CT-2*
Frequency	864.1–868.1 MHz
Access method	TDMA/TDD/FDMA
Modulation	GFSK
Filter (baseband)	0.5 Gaussian
Spectrum allocation	4 MHz
Voice coding	32 Kbps, ADPCM
Channel data rate	72 kHz
Coding	Cyclic, RS
Power	10 mW (max)
Number of carriers	40
Channels per carrier	1
Carrier separation	100 kHz
Handoff	No
Market	Public telepoint, originate only

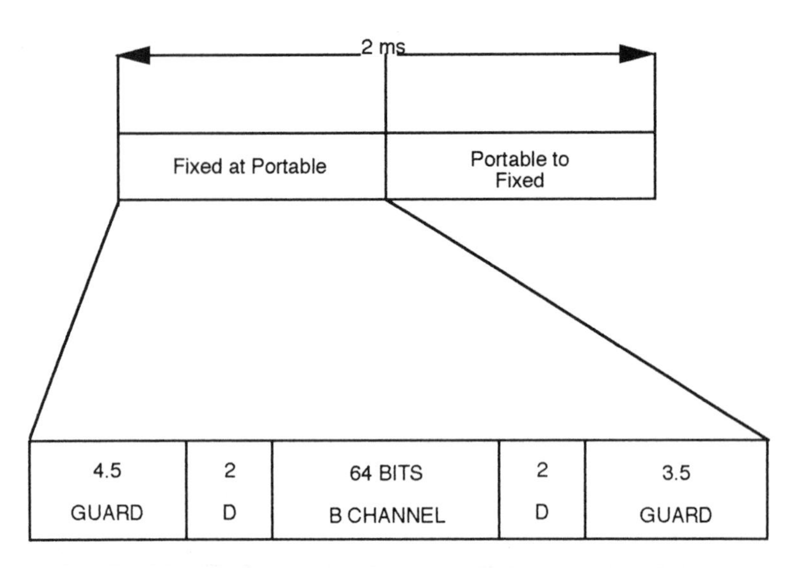

Figure 10.14 CT-2 time slot and frame.

for the other 1-ms, the base transmits to mobile. Each slot contains 64 bits for user information and 4 bits of control information, divided equally between two directions of transmissions. There are two guard bands, which adds an additional 8 bits. Thus, the frame of duration 2 ms contains a total of 144 bits providing an effective data rate of 72 Kbps. With the channel spacing of 100 kHz, the bandwidth efficiency is 0.72bits/sec/Hz. The modulation scheme used is MSK. The CT2 control channel is referred to as D channel. With 4 bits per frame, the control channel bit rate is 2 Kbps. Control messages consist of one to six 64-bit code words. Thus, a code word is distributed over 16 frames, and therefore has a delivery time of 32 ms. Each code word is protected by 16 bits of parity checks. Upon detecting transmission errors, the receiver requests retransmission of the code word.

Many cellular providers see CT-2 as a direct challenge to the cellular. However, there are some disadvantages of CT-2, including the following.

- Lack of complete coverage as provided by cellular. Cellular can be used all over the city, but telepoint is only located in certain areas and the subscribers have to know their locations to make calls.
- An incoming call is only permitted in very limited cases, although in the United Kingdom incoming call services aren't allowed for regulatory reasons. CT-2 technology, in reality permits incoming call. Of course, the caller has to know the location of the called party and the number of the base station. Therefore, CT-2 can be more effectively used with a paging service.
- CT-2 does not permit call handoff. For example, if you are in an area with

multiple telepoints your call will not be handed off once you leave the telepoint from where you originated your call. It would just disconnect.

10.3.4.1 Worldwide Status of CT-2

As stated in the introduction frequencies in the range of 800 MHz have been allocated in many countries in Europe including the U.K., France, and Germany. A Pan-European de facto standard for CT-2 has developed following the United Kingdom's allocation of spectrum for a new CT-2 service in the 864 to 868-MHz range. All European Community members have also signed a nonbinding memorandum of intent agreeing to implement CT-2 as a new service in this band. The CT-2 pilot system was launched in Strasbourg in early 1992 by France Telecom. This system uses the CAI standard and does not utilize telepoint concept. In this respect, the system differs somewhat with the basic CT-2 system developed in the U.K. More than 250 base stations have been installed to serve a population of 400,000 people, mostly pedestrian. France Telecom also is planning to launch a CT-2 system in Paris and plans to install more than 1,500 base stations. Both the German government and Deutsche Bundespost Telekom have also placed base stations on the German side of the Rhime River in the Strasbourg system. Germany is also implementing a pilot CT-2 system in Munich.

Among Asian countries, Hong Kong, Singapore, Malaysia, and Thailand have opted for CT-2 in this band. Four CT-2 operators have been licensed, and all 4 MHz will be available for use by all four operators. The Hong Kong system will use the common air interface (CAI) standard. The four licensees are Hong Kong Telecom CSL, Hutchinson Paging, Chevalier, and Personal Communications Ltd. Systems are already in operation with greater than 10,000 subscribers. Singapore launched the citywide CT-2 system in January 1992. The system will also use the CAI standard. Only one carrier will be operating, as the telecommunication system in Singapore is Government-owned. In Malaysia, a pilot CT-2 system is already in operation. Two operators have been approved thus far and more may be added at a future date. Thailand has allocated the 864 to 868-MHz band to CT-2 and has also adopted the CAI standard. A pilot system is in operation and a full CT-2 system is expected soon. There are other Asian countries, such as Indonesia, the Philippines, and Korea, which have available 800-MHz band that can be allocated for CT-2 services. In India, CT-2 may become a part of the proposal for GSM systems in four major metropolitan cities (Calcutta, Bombay, Delhi, and Madras).

Both Canada and Australia are also proposing to implement CT-2 systems. Canadian frequency allocation is from 944 MHz to 948 MHz, with an additional 4 MHz to be added in the range from 948 to 952 MHz at a future date if the demand exists. The Canadian system is CT-2 plus class IT technology, which is based on the CAI standard adopted in the United Kingdom with the addition of common channel signaling. The class II system developed by Motorola and Northern Telecom is said to have a higher traffic capacity, quicker call setup, and longer battery life. Licenses will be will be issued by the Department of Communications by October 1992. A minimum of two and a maximum of four licenses are proposed to be awarded. Australia's Department of Transportation and Telecommunications has allocated 12 MHz in the 800-MHz band for CT-2 service. CT-2 operators will share the band with operators of fixed microwave links.

In the United States, several companies filed for experimental licenses for CT-2 systems in late 1989. Bell Atlantic Mobile Systems has launched a technical trial of CT-2 technology in Philadelphia. The trial is not a commitment to a specific technology, but is a way to test the concept of a service offering. They want to integrate CT-2 within the spectrum of cellular. Advanced Cordless Technologies, which holds a license from the FCC to experiment with CT-2, is conducting trials in lower Manhattan. The company is also conducting trials in Monticello, NY.

10.3.5 Digital European Cordless Telecommunication (DECT) [9]

In 1990 and 1991, the European Community (EC) comprising Belgium, Denmark, France, Germany, Greece, Ireland, Italy, Luxembourg, Netherlands, Portugal, Spain, and the United Kingdom began the process of designating a frequency allocation for a coordinated introduction of digital European cordless telecommunication (DECT). The standard for DECT is being developed by the European Telecommunications Standard Institute (ETSI). DECT will include the services to residential cordless telephones, telepoints, and wireless branch exchange (PBXs), and in that respect will serve a larger subscriber base. DECT will be based on TDMA and time division duplexing (TDD), FDMA access protocol, an adaptive channel allocation concept, two-way talking capability, high capacity, and more sophisticated features than CT-2 system. As shown in Figure 10.15, each channel carrier frequency is divided into 24 slots, half of which are used for transmission from the base station and other half for transmission from the mobile.

ETSI and the European Conference of Postal and Telecommunications Administration (CEPT) have recommended a 20-MHz bandwidth in the frequency band from 1,080–1,900 MHz. EC directive 91/287/EEc, issued on June 3 1991, recommends that the member states designate the frequency band of 1,080–1,900 MHz by January 1, 1992. In accordance with the CEPT recommendation, DECT shall have priority over other services in the same band.

10.3.5.1 Architecture of DECT [9]

DECT applications include voice and data, service to simple residential CT, PABXs with wireless extensions, and wireless PABXs to Pan-European telepoints with

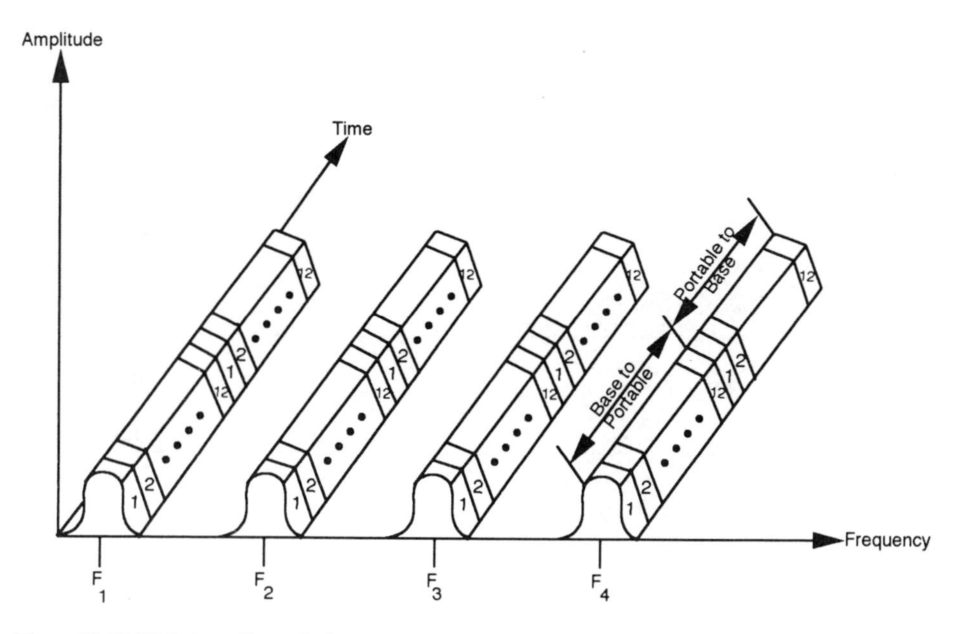

Figure 10.15 Digital cordless telephones.

both ingoing and outgoing calls. The conceptual architecture serving these three groups of users is shown in Figure 10.16. Wireless PABX serves users on the same and on different floors. Floors are divided into microcells having a preassigned number of channels. Floors having enough isolation can be assigned the same channels. Telepoint users can access the system within microcells. The system provides automatic handover between adjacent cells. Portable subscribers roaming in the service area are tracked by the system.

10.3.5.2 Technical Parameters of DECT

DECT employs time division duplex so that the information moves in both directions over the same carrier. In common with GSM, the system also uses GMSK. However, the premodulation normalized bandwidth of the filter is 0.5 ($BT = 0.5$). The number of channels per carrier is 12 with a carrier separation of 1.728 MHz. Speech is digitally encoded with 32-Kbps ADPCM modulation (ADPCM), with a channel data rate of 1152 Kbps over a channel bandwidth of 1.728 MHz that provides the bandwidth efficiency of 0.67 bits/sec/Hz, which is comparable to that of CT-2. Details of time slot and frame structure are shown in Figure 10.17. As shown in the figure, the total frame duration of 10 ms is divided equally between two 5-ms segments. Each 5-ms segment is divided between 12 time slots representing

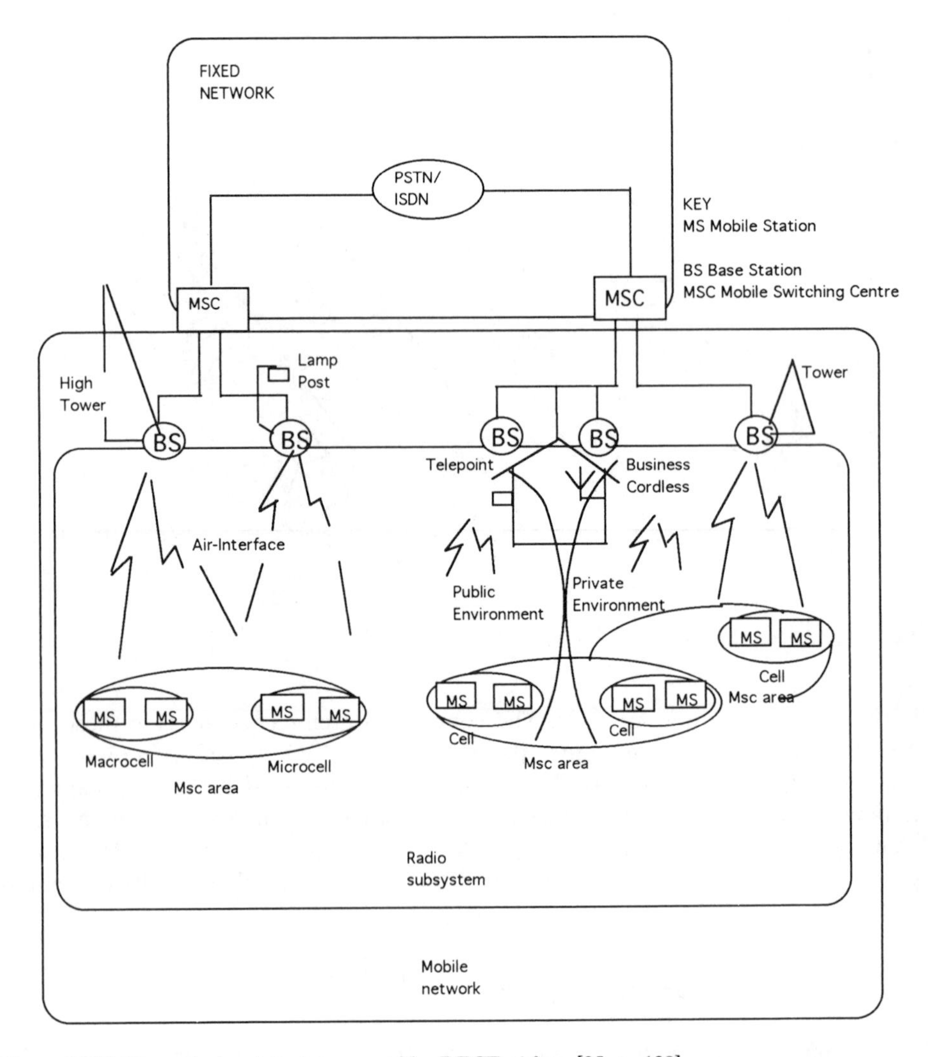

Figure 10.16 Conceptual architecture served by DECT. After: [25, p. 138]

12 channels each of duration 5/12 = 0.417 ms. Each channel data consists of 16 bits of preamble, 16 bits of synchronization, 64 control bits, 320 bits of information and 64 guard bits. With a total of 480 × 24 bits of user data over 10 ms, a total data rate of 1.152 Mbps is provided, which is higher than GSM data channel rate of 271 Kbps. Technical parameters are shown in Table 10.5. Each channel contains 64 bits of control data in which 16 bits are parity checks, leaving 48 bits of data with a header.

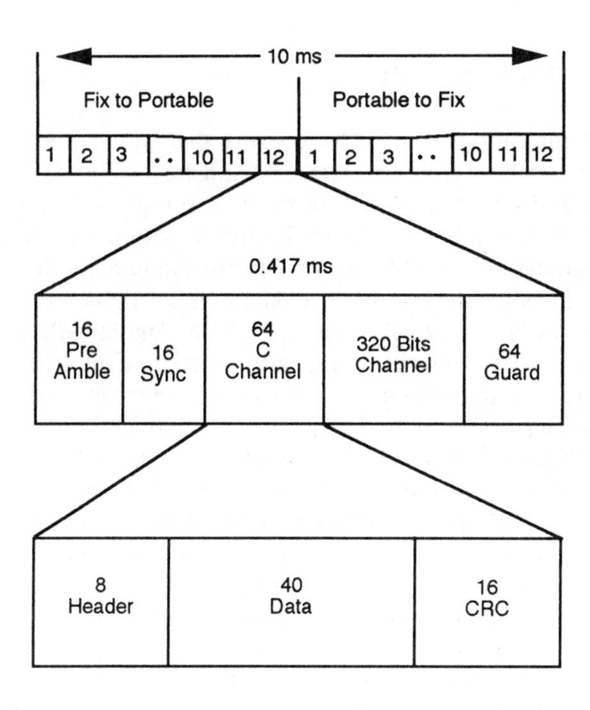

Figure 10.17 DECT slot and frame structure.

Table 10.5
Characteristics of Digital European Cordless Telecommunications

Specification	*DECT (CEPT)*
Frequency	1880–1990 MHz
Access method	TDMA/TDD/FDMA
Modulation	MSK
Filter (premodulation)	0.5 Gaussian
Spectrum allocation	20 MHz
Voice coding	32 Kpbs, ADPCM
Channel data rate	1152 Kbps
Coding	CRC − 16
Power	300 mW (max)
Carriers	132
Channels per carrier	12
Carrier separation	1.728 MHz
Handoff	Yes
Market	Business, telepoint, and residential

10.4 PCN SERVICES [25–32]

Personal communications is a service concept that will evolve from existing radio-based services, such as cellular, cordless, and paging, and will use existing and planned fixed and mobile infrastructures. This service will be provided by means of low-power, lightweight, wallet-sized wireless terminals connected to a radio-based microcell system utilizing a version of signaling system 7 protocol. Every conception of PCN means relatively inexpensive portable handsets. PCN will rely on many low-power, small-coverage cells. PCN service will also make use of a single portable handset and will eliminate the need for a multiple telephone devices. The key element of the system is its microcell structure service area, which will allow base stations to be situated wherever demand warrants. These base stations will be small; they can be placed, for example, on separate floors of office buildings, on walkways, and in residential neighborhoods. PCN transmitters will be smaller than cellular and will be situated closer together than transmitters of existing cellular telephone systems to reduce the size and cost of the phones. One revolutionary concept of PCN is that calls are made to people, not places. Customers will be assigned individual phone numbers, and the telephone network would route calls to them whether they were at home; in the office, or eating at a restaurant. PCN systems will not only carry voice but data also. To use PCN service, a caller may have to insert a "smart card" into the portable or mobile terminal handset. The card will identify the caller's location, the services and capabilities being purchased, and permit the services provided to perform billing. Once the smart card is inserted, a call will be transmitted by means of data packets to a microcell base station. This base station may serve a single subscriber or several subscribers in the form of a wireless PBX or centrex service.

Licenses have recently been issued in the U.K. and in the U.S. for wireless information networks. The cost of such service is expected to be somewhere between the cost of cellular and cordless telephone systems. In the U.K., three licenses have been issued, with the initial offering in 1992. All three networks will use the GSM radio transmission technique. In the U.S., a single service provider can operate in two cities on an experimental basis. PCN service in the U.S. will probably be based on CDMA techniques.

A typical layout of a PCN network is shown in Figure 10.18. The system consists of three main elements: pocket-sized handsets, microcells, and a nodal switching network. The main network is made up from a grid of microcells covering the country. However, there are additional cell types, such as cellular, included in the network to provide special services. A nodal switching network will provide routing, accessing other available network services, and interconnection to other networks.

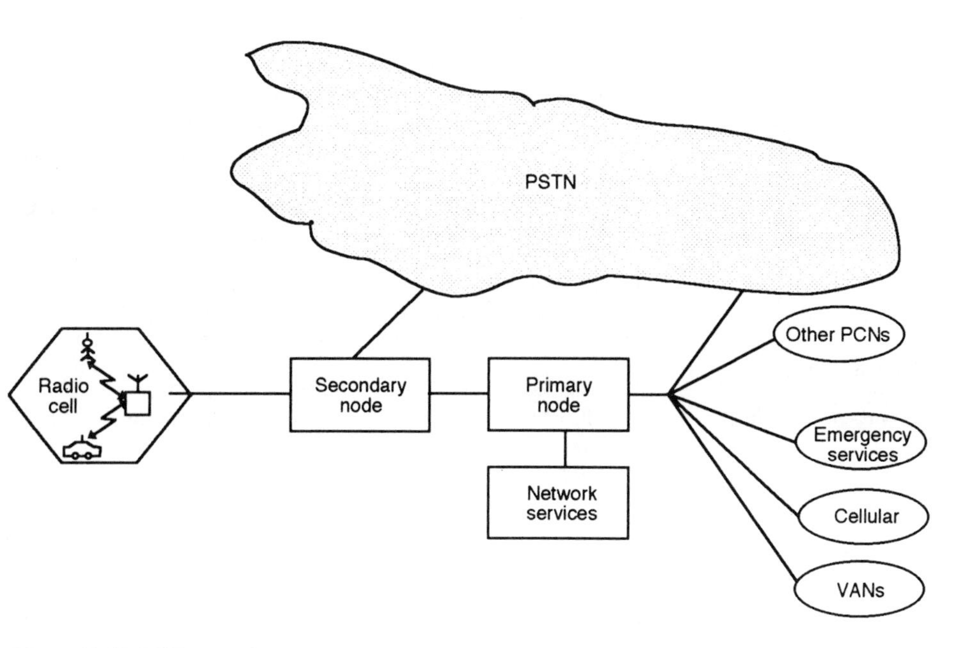

Figure 10.18 PCN network.

10.4.1 Comparing PCN to Cellular

The basic realization of PCN is on the same principles as cellular, and thus one expects that they should have some similarity. However, when one observes them closely, there are lots of differences. Table 10.6 provides comparisons of these two systems.

10.4.2 PCN in Europe

Even before GSM is up and running, the next-generation technology, the personal communications networks (PCNs), are being planned. Based on the same principle as GSM, PCNs are microcellular systems operating at 1,000 MHz. As a result the basic technology, the platform for PCN remains the same as GSM and thus is known as a digital cellular system at 1,000 MHz (DCS 1000). European market size is estimated to be worth \$1.1 billion. While 43% of this tally is expected to remain in the U.K., 11% will be generated in France, 8% in Germany, and the remaining 38% will be distributed throughout Europe. Instead of being oriented toward relatively high-powered vehicular terminals like the present cellular system,

Table 10.6
Comparison of PCN With Cellular

	PCN	*Cellular*
Use	Universal service; portable phone service	Vehicular mounted and portable phone service
User model	PCN service is universal; supposed to replace business and residential services	Cellular phone service used when on move; business and residential use also possible
Coverage area	Urban areas and transportation corridors	Continuous but not complete coverage throughout country; not cost effective in rural areas
Incoming calls	Yes	Yes
Handoff	Yes; high speed	Yes; slow speed
Network infrastructure	Switching and local transport separate, but interconnected to PSTN	Switching and local transport separate, but interconnected to PSTN

U.K. operators intend to use very low power pocket phones. The reduction in terminal size and power is possible because of the use of microradio cells that are between 0.4 km and 6 km in radius in urban areas. Since analog cellular uses cells that on the average are 15 km in size, the capacity of the DCS 1000 should be at least about 6.25 times the present capacity of the GSM system. Thus, PCN DCS 1000 should support some 6,000 users per km^2 in comparison to some 1,000 users per km^2 for the present GSM system. Coupled with advances, with cell management DCS 1000 will support tens of millions of users rather than the millions that are the limit of regular cellular systems. The technical characteristics of DCS 1000 are shown in Table 10.7.

Compared to GSM, the spectrum allocation is increased to 75 MHz, which provides 3,000 channels using the present coding rate of 13 Kbps. The capacity will double to 6,000 channels with half-rate speech coding, when the TDMA frame will be shared by 16 users.

A PCN pocket phone weighing slightly more than a microcassette tape recorder will use smart card technology and will be equipped with an internal antenna. At the end of 1989, the Department of Trade and Industry licensed three consortia—Mercury Personal Communications, Unitel, and Microtel—to provide two-way PCN services. Mercury includes Motorola and Telefonica; Unitel is made up STC, Thorn EMI, US West, and Deutsche Bundespost Telecom; and Microtel presently comprises of British Aerospace, Matra, Millicom, and Pacific Telesis.

10.4.3 PCN in the United States

As stated above, PCN service in the United States is most likely going to be based on spread-spectrum (SS) techniques. One of the main reasons for this is that there

Table 10.7
Characteristics of DCS 1000 (European PCN)

Parameters	*Characteristics DCS 1,000 (PCN)*
Coverage area	Europe
Frequency range	1.7–1.9 GHz
Technology	TDMA
Channels per carrier	8*, 16†
Modulation	0.3 GSM (baseband *BT* product)
Digitized speech rate	13 Kbps (RELP-LTP)
Mobile output power	250 mW to 2W
Total spectrum	75 MHz
Data rate	270.833 Kpbs
Baseband filter type	0.3 Gaussian
Channel spacing	200 KHz
Number of channels	3,000*, 6,000†
Expected year of operation	1992–1993

*Initially. †With half-rate coding.

are no unallocated frequencies in the U.S. that are practical for use in portable radio frequency (RF) devices, and spread-spectrum techniques offer the potential for sharing the radio spectrum with other systems. There are presently four companies in the U.S. that have been pioneering spread-spectrum techniques for wireless telephones: Qualcomm, Millicom, Omnipoint, and Cylink. There is a general believe that CDMA will be more efficient than frequency division multiple access or time division multiple access. Though there are many other advantages of SS, the following three advantages may play a predominant role:

- SS as a means of introducing new wireless telephone services by sharing radio spectrum with existing communications users, especially point-to-point microwave towers.
- SS as a means of introducing new wireless telephone products on an unlicensed basis under the existing FCC rules, which allow spread spectrum to transmit up to 1W of power for use with any application in certain RF bands.
- Any other approach which is not based on SS will take years to reallocate spectrum for PCN. In order to keep the competitive edge with the rest of the world, it is almost essential that the spectrum be allocated immediately, which also focuses towards SS.

Looking at cellular growth in the U.S., it is expected that there will be some 25–60 million people in urban areas who will subscribe for PCN service. It is expected that between 50–100 MHz of bandwidth will be required per PCN operator. Assuming two operators per area as in cellular, it is expected that the minimum bandwidth requirements will be of the order of 100 MHz. The first hurdle is to find 100 MHz of usable spectrum. For economic and practical reasons, the search has been concentrated within matured and inexpensive areas of technology, which is generally considered to be below 3 GHz. Within this range, the most attention has been focused on the 1,700–2,300 MHz band, where such services were granted spectrum in Europe. In the U.S., these frequencies are currently used by fixed point-to-point microwave towers. As a consequence, many experimental proposals to the FCC suggest sharing frequencies with microwave users. Within this range, the 1.85–1.99 GHz band has been the preferred choice of many as it is not as heavily populated in most cities and is used by certain private sector industries that PCN advocates believe may be more amenable to sharing than other microwave users. A further advantage to selecting this band is that it is the natural migration path for existing fixed users to the 1,710–1,850 MHz band, which is used predominantly by government-fixed users along with some mobile and space services. The migration can be tied to the growth of PCN and should allow sufficient time for planning by both nongovernment as well as government users of the two bands.

10.4.4 Expected Technical Characteristics of the CDMA System

Several important effects of choosing CDMA for PCNs are summarized here. These parameters are:

- Spectrum efficiency;
- Intelligent network;
- Wideband PCN environment;
- Soft handoff.

Spectrum efficiency increases both by reducing the cell size and by choosing spread-spectrum modulation. For achieving higher spectral efficiency, PCN systems will have to have a lower value of D/R ratio. This will permit frequent repeats of the channels, and thus an increase in capacity or the number of channels per km^2. Due to the reduced value of E_b/N_o and the speech activity factor, the capacity of the CDMA system can be increased substantially. The other property that increases the capacity of SS systems is the processing gain associated with the wideband to narrowband conversion process. Any interference present in the wideband channel is suppressed by processing the gain of the system.

Similar to digital cellular radio, PCN will need the support of an intelligent network so that the customer ID can be checked and a call can be delivered anywhere to the subscriber. The network will track the subscriber's location and update this information in its database as soon as the subscriber plugs his or her ID (smart card) into the PCN unit.

The wideband nature of the CDMA system makes it possible to resolve the multipath components, which can be for performance improvement in the form of diversity combining. The number of resolvable multipaths is given by

$$L \leq T_m/T_c + 1 \tag{10.6}$$

where T_m is the total multipath delay spread and is inversely proportional to the coherence bandwidth of the channel, and T_c is the chip duration. Thus, a maximum of L statistically independent paths can be resolved directly in the SS system. These L received signals can be combined in a diversity combiner to maximize the total wanted signal-to-noise energy at the demodulator input.

Soft handoff helps in reducing the unsuccessful handover to adjoining microcells. The technique allows the new connection to the microcell to be established before the old connection is released.

The expected characteristics of the PCN system are given in Table 10.8. Low-rate speech coding combined with linear DQPSK modulation will further aid in increasing the capacity of the system in terms of bits/sec/Hz. Due to low power mobile, linear amplification, and hence DQPSK, may be a good choice. Due to slow moving mobiles, the fast handoff requirement does not exist. At this moment it is not clear when the PCN service will be available in the United States as even the spectrum has not been allocated.

10.5 CONCLUSIONS

In order to meet the growing demand of mobile users in urban areas, the sizes of cells have to be reduced. Radio systems based on microcellular and low-power technology will satisfy the growing demand of residential as well as office users. The PCN system will assign a telephone number to an individual rather than to a place. The user will carry the telephone number wherever he or she goes. Smart-card technology will identify the user and not the instrument set. In the course of time, users will be able to make and receive calls anywhere in the world.

Table 10.8
Technical Parameters of a CDMA System

System Specification	*PCN (U.S.)*
Frequency	1,850–1,990 MHz
Access method	CDMA/FDMA
Modulation	DQPSK
Voice coding	Variable rate based on VCELP as in digital cellular
Power	Milliwatts range (10–20 mW)
Carrier separation	To be decided
Initial market	Residential, business, and mobile
Handoff	Yes (slow compared to cellular)
Expected service	Not known

PROBLEMS

10.1. a. Why do you think cellular cannot meet the future needs of personal communication networks?
b. What are the essential characteristics of PCN?
c. What are the differences between CT, PCN, and regular telephone?
d. Why is 1.7–2.2 GHz a good choice for PCN?

10.2. Answer the following questions with respect to microcellular systems.
a. Criterion for the selection of maximum and minimum radius of microcells.
b. Other than residential, office, and telepoint applications, name some other applications where microcellular technology can be used.
c. State reasons why the microcellular antenna height for urban application, as shown in Figure 10.2, cannot be increased to 100 ft.
d. Why is microscopic irregularity important?

10.3. Explain the following terms with regard to microcellular systems:
a. Channel segregation;
b. Multiple base station registrations;
c. Diversity transmission.
Can you come up with some other algorithm (not covered in Section 10.2.1.1) that will also achieve channel segregation?

10.4. Find the value of α, β, and K for (10.2) for Figure 10.9.

10.5. Justify that telepoint is a good substitute for pay phones in Europe. Do you think that it can also replace pay phones in U.S.?

10.6. State the advantages of wireless PABX over a wireline (ordinary) PABX.

10.7. Give reasons why echo delay and the absolute delay have to be controlled in CT systems. Why is time division duplex chosen for CT?

10.8. A metropolitan area of 10 km^2 has to be served by cordless telephones for residential use and wireless PBXs for office use. The residential area occupies 80% of the total area and generates only 5% of the total traffic, while the business generates 100 times the traffic of the residential users. Let the total number of households in the area be 100,000 and let each household subscriber generate 0.05 erlangs of traffic, out of which only 10% of the traffic is carried by CT. Estimate the total traffic in erlangs per km^2.

10.9. For CT-2, compute the data rate for B and D channels from fixed base to mobile and from mobile to fixed base. Find the spectral efficiency in terms of bits/sec/Hz. For a DECT system, compute the C channel data rate and the spectral efficiency of the system. Can I improve the spectral efficiency of these systems further? Suggest some techniques.

10.10. Provide reasons to justify why Europe is choosing TDMA for their PCN while the United States is going with CDMA.

REFERENCES

[1] Panzer, H. "Adaptive Resource Allocation in Metropolitan Area Cellular Mobile Radio Systems," *IEEE Vehicular Technology Conference*, 1990, pp. 638–645
[2] Bernhardt, R. C. "Performance Aspects of Two-Way Transmission In Portable Radio Systems," *IEEE Vehicular Technology Conference*, San Francisco, CA., May 1 1989.
[3] Kazeminejad, S., and R. Williams." Personal Communication—Where to Begin?," *Virginia Tech's First Symposium on Wireless Personal Communication*, June 3–5 1991.
[4] Okamura, K., et al. "In Building Portable Telephone System," *IEEE Vehicular Technology Conference*, 1990, pp. 299–304
[5] Akaiwa, Y. "A Conceptual Design of Microcellular Radio Communication System," *IEEE Vehicular Technology Conference*, 1990, pp. 156–160.
[6] Lotse, F., and A. Wejke. "Propagation Measurements for Microcells in Central Stockholm," *IEEE Vehicular Technology Conference*, 1990, pp. 539–541.
[7] Swain, R. S., and A. J. Motley. "Requirements of an Advanced Cordless Telecommunication Product," *IEEE Vehicular Technology Conference*, pp. 3.4.1–3.4.5
[8] White, P. H., et al. "A Full Digital 900 MHz Cordless Telephone," *IEEE Conference Publication,* No. 238, Sept. 1984.
[9] Cox, D. C. "Universal Digital Portable Radio Communication," *IRE Proc.*, Vol. 75, April 1987, pp. 436–477.
[10] Cox, D. C. "Universal Digital Portable Radio Communication," *IRE Proc.*, Vol. 75, April 1987, pp. 436–477.
[11] Hattori, T., A. Sasaki, and K. Monma. "A New Mobile Communication System Using Autonomous Radio Link Control With Decentralized Base Stations," *IEEE Vehicular Technology Conference*, 1987, pp. 579–586.
[12] Cheeseman, D. S., et al. "System Features—Next Generation Cellular Radio," *Proceedings 37th IEEE Vehicular Technology Conference*, June 1987.
[13] Whitteker, J. H. "Measurements of Path Loss at 910 MHz for Proposed Microcell Urban Mobile Systems," *IEEE Trans. on Vehicular Technology*, Vol. 37, No. 3, Aug. 1988.
[14] Kanai, T., et al. "A Handoff Control Process for Microcellular Systems," *Proceedings 38th IEEE Vehicular Technology Conference*, June 1988.
[15] Ito, S. "A Proposal of Portable Telephone System," *IEEE Vehicular Technology Conference*, June 1988, pp. 520–524.
[16] Ochsner, H. "DECT—Digital European Cordless Telecommunications," *39th IEEE Vehicular Technology Conference*, San Francisco, CA, May 1–3 1989.
[17] Ito, S. "System Design for Portable Telephones—Method to Enable Initiating and Receiving Calls From a Vehicle," *IEEE Vehicular Technology Conference,* May 1989, pp. 136–141.
[10] Callendar, M. "International Standards for Personal Communications," *IEEE Vehicular Technology Conference*, 1989, pp. 722–728.
[19] Goodman, D. J. "Second Generation Wireless Information Networks," *IEEE Trans. on Vehicular Technology*, May 1991, pp. 366–374.
[20] Sollenberger, N. R., et al. "Architecture and Implementation of an Efficient and Robust TDMA Frame Structure for Digital Portable Communications," *IEEE Vehicular Technology Conference*, San Francisco, Ca., April 28–May 3 1989, pp. 169–174.
[21] Bernhardt, R. C. "Time-Slot Management in Frequency Reuse Digital Portable Radio Systems," *IEEE Vehicular Technology Conference*, 1990, pp. 282–286.
[22] Kanai, T., et al. "Experimental Digital Cellular System for Microcellular Handoff," *IEEE Vehicular Technology Conference*, 1990, pp. 172–177.

[23] Horikawa, I., and M. Hirono. "A Multi-Carrier Switching TDMA-TDD Microcell Telecommunications System," *IEEE Vehicular Technology Conference*, 1990, pp. 167–171.
[24] Kohiyama, K., et al. "Advanced Personal Communication System," *IEEE Vehicular Technology Conference,* 1990, pp. 161–166.
[25] Grillo, D., and G. MacNamee. "European Perspectives on Third Generation Personal Communication Systems," *IEEE Vehicular Technology Conference,* 1990, pp. 135–140.
[26] Cox, D. C., et al. "Universal Digital Portable Communications: System Perspective," *IEEE Journal on Selected Areas Communication*, Vol. SAC-5, No. 5, pp. 764–773, May 1987.
[27] Cox, D. C., H. W. Arnold, and P. T. Porter. "Universal Digital Portable Communications: A System Perspective," *IEEE Journal on Selected Areas in Communication*: *Special Issue on Portable and Mobile Communications*, June 1987, pp. 764–773.
[28] Acampora, A. S., and J. H. Winters. "A Wireless Network for Wideband Indoor Communications," *IEEE Journal on Selected Areas in Communication: Special Issue on Portable and Mobile Communications*, June 1987.
[29] Bernhardt, R. C. "User Access in Portable Radio Systems in a Co-Channel Interference Environment," *IEEE Journal on Selected Areas in Communication*, Vol. SAC-7, Jan. 1989.
[30] Akerberg, "Properties of a TDMA Pico Cellular Office Communication System," *IEEE Globecom* '88, 1989, pp. 106–191.
[31] Harley, P. "Short Distance Attenuation Measurements at 900 MHz and 1.8 GHz using Low Antenna Heights for Microcells," *Journal on Selected areas in Communications*, Vol. 7, No. 1, Jan. 1989, pp. 5–11.
[32] Hattoti, T., et al. "Emerging Technology and Service Enhancement for Cordless Telephone Systems," *IEEE Communication Magazine*, Jan. 1988, pp. 53–58.

About the Author

Asha K. Mehrotra received his B.S. in electrical engineering from Bengal Engineering College, Sibpur, India, in 1961; his M.S. in electrical engineering from Nova Scotia Technical College, Halifax, Canada, in 1967; and his Ph.D. in electrical engineering from Polytechnic Institute of New York (formerly Brooklyn Polytechnic) in 1981. He is presently working as a member of the technical staff at The Analytic Sciences Corporation (TASC) in Reston, Virginia, where he deals with a large range of communication system engineering problems. In the past, he has worked on the MILSTAR satellite project at MITRE as a member of the technical staff and with TDRSS satellite at Space Communication Company as a manager of the Systems Analysis Group. His research experience includes HF communication, switching systems, and telephone transmission systems. His hardware experience spans the design of central office equipment, PBXs, and facsimile systems. Dr. Mehrotra has taught graduate courses in computer science and electrical engineering at Virginia Polytechnic Institute and State University and George Mason University. He has been an adjunct professorial lecturer at George Washington University for the past 13 years, teaching graduate courses in communication engineering. He is a member of IEEE Communication and Computer society.

Index

The Artech House Telecommunications Library

Vinton G. Cerf, Series Editor

Advanced Technology for Road Transport: IVHS and ATT, Ian Catling, editor

Advances in Computer Communications and Networking, Wesley W. Chu, editor

Advances in Computer Systems Security, Rein Turn, editor

Analysis and Synthesis of Logic Systems, Daniel Mange

Asynchronous Transfer Mode Networks: Performance Issues, Raif O. Onvural

A Bibliography of Telecommunications and Socio-Economic Development, Heather E. Hudson

Broadband: Business Services, Technologies, and Strategic Impact, David Wright

Broadband Network Analysis and Design, Daniel Minoli

Broadband Telecommunications Technology, Byeong Lee, Minho Kang, and Jonghee Lee

Cellular Radio: Analog and Digital Systems, Asha Mehrotra

Cellular Radio Systems, D. M. Balston and R. C. V. Macario, editors

Client/Server Computing: Architecture, Applications, and Distributed Systems Management, Bruce Elbert and Bobby Martyna

Codes for Error Control and Synchronization, Djimitri Wiggert

Communication Satellites in the Geostationary Orbit, Donald M. Jansky and Michel C. Jeruchim

Communications Directory, Manus Egan, editor

The Complete Guide to Buying a Telephone System, Paul Daubitz

Computer Telephone Integration, Rob Walters

The Corporate Cabling Guide, Mark W. McElroy

Corporate Networks: The Strategic Use of Telecommunications, Thomas Valovic

Current Advances in LANs, MANs, and ISDN, B. G. Kim, editor

Design and Prospects for the ISDN, G. Dicenet

Digital Cellular Radio, George Calhoun

Digital Hardware Testing: Transistor-Level Fault Modeling and Testing, Rochit Rajsuman, editor

Digital Signal Processing, Murat Kunt

Digital Switching Control Architectures, Giuseppe Fantauzzi

Digital Transmission Design and Jitter Analysis, Yoshitaka Takasaki

Distributed Multimedia Through Broadband Communications Services, Daniel Minoli and Robert Keinath

Distributed Processing Systems, Volume I, Wesley W. Chu, editor

Disaster Recovery Planning for Telecommunications, Leo A. Wrobel

Document Imaging Systems: Technology and Applications, Nathan J. Muller

EDI Security, Control, and Audit, Albert J. Marcella and Sally Chen

Enterprise Networking: Fractional T1 to SONET, Frame Relay to BISDN, Daniel Minoli

Expert Systems Applications in Integrated Network Management, E. C. Ericson, L. T. Ericson, and D. Minoli, editors

FAX: Digital Facsimile Technology and Applications, Second Edition, Dennis Bodson, Kenneth McConnell, and Richard Schaphorst

FDDI and FDDI-II: Architecture, Protocols, and Performance, Bernhard Albert and Anura P. Jayasumana

Fiber Network Service Survivability, Tsong-Ho Wu

Fiber Optics and CATV Business Strategy, Robert K. Yates *et al.*

A Guide to Fractional T1, J. E. Trulove

A Guide to the TCP/IP Protocol Suite, Floyd Wilder

Implementing EDI, Mike Hendry

Implementing X.400 and X.500: The PP and QUIPU Systems, Steve Kille

Inbound Call Centers: Design, Implementation, and Management, Robert A. Gable

Information Superhighways: The Economics of Advanced Public Communication Networks, Bruce Egan

Integrated Broadband Networks, Amit Bhargava

Integrated Services Digital Networks, Anthony M. Rutkowski

International Telecommunications Management, Bruce R. Elbert

International Telecommunication Standards Organizations, Andrew Macpherson

Internetworking LANs: Operation, Design, and Management, Robert Davidson and Nathan Muller

Introduction to Satellite Communication, Bruce R. Elbert

Introduction to T1/T3 Networking, Regis J. (Bud) Bates

Introduction to Telecommunication Electronics, A. Michael Noll

Introduction to Telephones and Telephone Systems, Second Edition, A. Michael Noll

An Introduction to U.S. Telecommunications Law, Charles H. Kennedy

Introduction to X.400, Cemil Betanov

The ITU in a Changing World, George A. Codding, Jr. and Anthony M. Rutkowski

Jitter in Digital Transmission Systems, Patrick R. Trischitta and Eve L. Varma

Land-Mobile Radio System Engineering, Garry C. Hess

LAN/WAN Optimization Techniques, Harrell Van Norman

LANs to WANs: Network Management in the 1990s, Nathan J. Muller and Robert P. Davidson

The Law and Regulation of International Space Communication, Harold M. White, Jr. and Rita Lauria White

Long Distance Services: A Buyer's Guide, Daniel D. Briere

Measurement of Optical Fibers and Devices, G. Cancellieri and U. Ravaioli

Meteor Burst Communication, Jacob Z. Schanker

Minimum Risk Strategy for Acquiring Communications Equipment and Services, Nathan J. Muller

Mobile Antenna Systems Handbook, K. Fujimoto, J. R. James

Mobile Communications in the U.S. and Europe: Regulation, Technology, and Markets, Michael Paetsch

Mobile Information Systems, John Walker

Narrowband Land-Mobile Radio Networks, Jean-Paul Linnartz

Networking Strategies for Information Technology, Bruce Elbert

Numerical Analysis of Linear Networks and Systems, Hermann Kremer *et al.*

Optimization of Digital Transmission Systems, K. Trondle and Gunter Soder

Packet Switching Evolution from Narrowband to Broadband ISDN, M. Smouts

Packet Video: Modeling and Signal Processing, Naohisa Ohta

The PP and QUIPU Implementation of X.400 and X.500, Stephen Kille

Principles of Secure Communication Systems, Second Edition, Don J. Torrieri

Principles of Signals and Systems: Deterministic Signals, B. Picinbono

Private Telecommunication Networks, Bruce Elbert

Radiodetermination Satellite Services and Standards, Martin Rothblatt

Residential Fiber Optic Networks: An Engineering and Economic Analysis, David Reed

Secure Data Networking, Michael Purser

Setting Global Telecommunication Standards: The Stakes, The Players, and The Process, Gerd Wallenstein

Smart Cards, José Luis Zoreda and José Manuel Otón

Secure Data Networking, Michael Purser

The Telecommunications Deregulation Sourcebook, Stuart N. Brotman, editor

Television Technology: Fundamentals and Future Prospects, A. Michael Noll

Telecommunications Technology Handbook, Daniel Minoli

Telephone Company and Cable Television Competition, Stuart N. Brotman

Teletraffic Technologies in ATM Networks, Hiroshi Saito

Terrestrial Digital Microwave Communciations, Ferdo Ivanek, editor

Transmission Networking: SONET and the SDH, Mike Sexton and Andy Reid

Transmission Performance of Evolving Telecommunications Networks, John Gruber and Godfrey Williams

Troposcatter Radio Links, G. Roda

UNIX Internetworking, Uday O. Pabrai

Virtual Networks: A Buyer's Guide, Daniel D. Briere

Voice Processing, Second Edition, Walt Tetschner

Voice Teletraffic System Engineering, James R. Boucher

Wireless Access and the Local Telephone Network, George Calhoun

Wireless LAN Systems, A. Santamaría and F. J. Lopez-Hernandez

Writing Disaster Recovery Plans for Telecommunications Networks and LANs, Leo A. Wrobel

X Window System User's Guide, Uday O. Pabrai

For further information on these and other Artech House titles, contact:

Artech House
685 Canton Street
Norwood, MA 01602
617-769-9750
Fax: 617-762-9230
Telex: 951-659
email: artech@world.std.com

Artech House
Portland House, Stag Place
London SW1E 5XA England
+44 (0) 71-973-8077
Fax: +44 (0) 71-630-0166
Telex: 951-659